# "Riddles" Update, February 2003

Since this book appeared, I've done additional field work to confirm or refute some conclusions. Three principal chang[illegible] see the Sierra Nevada as a Himalayan range that collapsed to around its present height about 80 million years ago; 2) Yosemite Valley's p[illegible] glaciers, as elsewhere, were only slightly larger than its Tahoe glaciers; there are no glacial deposits on the rims of Yosemite Valley including at Glacier Point and Turtleback Dome; and 3) there never was a Lake Yosemite. Changes made *after* the 2000 supplement are *italicized.*

*p. 20, 21, Feather Falls. It's actually 640 feet high, higher than Yosemite Valley's Bridalveil Fall.*

*p. 59-60, Figs. 9-12. These are hypothetical, based on a tale told to Brewer* before *he entered the Sierra. A miner's tale gave rise to the Sierra uplift myth!*

p. 125, Map 13: Glaciers originated on the slopes of Castle Peak as well as Basin Peak. The gentle slopes above Frog Lake Cliff lay under thin ice.

p. 126, end of section on North Fork Prosser Creek canyon: Although I found evidence around the Sierran crest above this canyon of a Tioga-age ice surface much higher than Birkeland's, in the Donner Pass area I derived a Tioga-age ice surface identical to his: 7900 feet.

p. 127, Map 14: The "seasonal lake just west of the lower part of Little Jamison Creek canyon's west-side moraine" (p. 128, col. 1) is a deep, permanent lake, and its basin was covered by a Tioga-age glacier.

p. 138-143, diachronous, (late-Tioga) lateral moraines. An excellent crosses Luther Pass, along Hwy. 89. If it had been Tioga-maximum, then it would abut against a slightly higher Tahoe-maximum lateral moraine—but no moraine exists. The Tioga-maximum moraine lies ¼ mile west.

p. 147-148, lateral moraine atop canyon rim, extending west from Camp Mather (from lower left corner of Map 21). Dodge and Calk's glacial mapping atop the rim is quite accurate; Alpha *et al.*'s interpretation is not. Much of their Tahoe is Tioga. The Tioga leaves the rim in the vicinity north of BM 4342; the Tahoe leaves it about 1 mile farther west, where Road 12 crosses a shallow rim saddle by BM 4236. Roadcuts by a gully 1.5 miles down the canyon's Road 17 expose Tahoe till and imply that the Tahoe glacier here was about 1200 feet thick. The Tioga glacier would have been slightly thinner. On this basis, I estimate that the Tioga and Tahoe glaciers advanced about 2-3 miles beyond the Tuolumne River's confluence with Cherry Creek, ending at an elevation of about 2000 feet, near the west edge of the Lake Eleanor 15′ quadrangle. Alpha *et al.* show the Tioga glacier ending about 2 miles before the confluence.

p. 147-150, Moraine Ridge, northeast of Laurel Lake. On USGS Map I-1874, this is the northern of two parallel ridges; on the Tower Peak 15′ quadrangle, it is the southern one. Map I-1874 shows both ridges capped with Tahoe till, and the upper slopes between them devoid of till. In the field one sees that both ridges are entirely capped with classic, fresh, boulder-rich Tioga till, and that sparse till occurs on the upper slopes between them. Furthermore, as is so common in the range, the till is late-Tioga. At its maximum, the Tioga glacier was thick enough to bury both ridges.

p. 181-182, Map 27, site 2. Matthes mapped 10 pre-Tahoe (his Glacier Point) erratics on slopes below Sentinel Dome. This area, including Glacier Point, never was glaciated, based on the following evidence. First, Matthes' erratics and dozens of other boulders all appear to have been locally derived. Second, Matthes' proposed source for his erratics is the Mt. Clark environs, transported to Glacier Point on the southern margin of the Merced glacier (his p. 63). However, any erratics deposited around Glacier Point would have been transported along the western margin of his north-flowing Illilouette glacier, but their source area has other rock types. His giant, if imaginary, Illilouette glacier would have flowed past Glacier Point and then west along the southern edge of the valley, preventing the Merced glacier from ever touching it. Third, most erratics are on slopes that are too steep and soil-rich for preservation. Pre-Tahoe erratics occur only on ridges, benches, and gentle, soil-poor slopes. Fourth, erratics should have been deposited around Glacier Point, but none exist. And fifth, Matthes' pre-Tahoe glacier here is about 2¼ times thicker than the Tahoe (his Wisconsin). My detailed mapping in the Stanislaus River drainage (Ch. 16) indicates that pre-Tahoe glaciers were only slightly thicker than Tahoe glaciers. Curiously, Matthes mapped erratics where there was a complex intertwining of two or more magmatic bodies, as here, and at Middle Cathedral Rock, Turtleback Dome, and west of El Capitan. One spot near Sentinel Dome, has four types of intrusive rock in a span of 10-yards. Earthquakes can easily move a boulder onto a different rock type, making it appear as an erratic.

p. 199-200, Map 30, Middle Cathedral Rock. Matthes' El Portal erratics are suspect for the same reasons just given. Additionally, like Matthes' pre-Tahoe Illilouette glacier, his pre-Tahoe Bridalveil Creek glacier was much smaller than he suggested, perhaps barely longer than his Wisconsin Bridalveil Creek glacier. There are no erratics or till where he mapped them near Dewey Point (south edge of Map 30).

p. 210-212, Turtleback Dome erratics and deposits. The boulder in Photo 124 may not be an erratic. The conspicuous dike spreads laterally and horizontally along the base, so one would not expect it to continue into the underlying bedrock. Also, the solitary Cathedral Peak granodiorite boulder in the deposit probably was placed there by someone since there is no way that it could have arrived there from its source area. Such diagnostic erratics should be common on the valley's north rim, but none have been found. Likewise, Mt. Clark granite, yellowish quartzite, and gray schist erratics should be in this deposit, but none have been found. The key to the deposits lies east of Turtleback Dome, about 200 yards west of "7" on Map 31. There, hundreds of boulders, forming a similar deposit, are in the process of weathering from a diorite sill parallel to the surface of the adjoining granitic bedrock. In reality, if these deposits were old, they should be boulder-poor and clay-rich, but clay is absent. Finally, there should be similar erratics and deposits on two prominent benches below BM 5729 on the north rim, but both benches are bare granite.

p. 213, Figure 30. Because there is no convincing proof that any pre-Tahoe glacier topped Turtleback Dome—or the north or south rims—I would adjust the size of my pre-Tahoe glacier downward to just slightly larger than my Tahoe glacier.

*p. 217. Like Yosemite Valley, Pickel Meadow and nearby Leavitt Meadow (both on Hwy. 108) have constricted stream flows and are flat and boulderless, due to floodplain sediments accumulating behind their constrictions. Hope Valley (Hwy. 89) lacks a constriction, so it slopes and it has boulders.*

*p. 217-226, and 356, Lake Yosemite. Lake sediments are dark brown organic clay. None exist (see p. 371). There never was a Lake Yosemite.*

p. 225, old glaciers eroded very weathered bedrock. This is an accommodating, poor rationalization on my part. By everyone's account all glaciers through Yosemite Valley lay below their ELAs, and therefore should have been depositing their transported sediments, not eroding the valley floor. Erosion should be primarily above a glacier's ELA, where its base flows across the bedrock floor. Nevertheless, contrary to long-established, well-known glacial physics, geoscientists continue to propose, either directly or indirectly, that erosion—mysteriously—is far greater below a glacier's ELA than above it.

p. 252-253, non-existent glacial deposits of Moore and Sisson: Another one of the many suspect USGS glacial deposits in the Sierra Nevada lies about 2 miles east of Hume Lake (Moore and Nokleberg's 1992 USGS Map GQ-1676: Tehipite Dome). Just southeast of the lake, take old (pre-1952) Road 13S05 3¼ miles to a roadcut through the first of three mapped lateral moraine crests. The cut exposes decomposing granitic bedrock; glacial deposits are absent.

p. 256-258, 271, V-shaped cross profiles. On the glacial map accompanying USGS Prof. Paper 329, Matthes indicated that the Wisconsin glacier (Tahoe/Tioga glaciers) flowing through a stretch of the Middle Fork San Joaquin River canyon between Devils Postpile National Monument and Fish Creek was between 2500 and 3000 feet thick, about twice that in Yosemite Valley. Nevertheless, there is no evidence of measurable glacial erosion. The canyon is distinctly V-shaped, and its floor is barely wide enough to hold the river—*in one spot only 30 feet wide.* Looking at a topo map, one would never conclude that this canyon had been glaciated. A V-shaped, glaciated canyon you can drive through, on Hwy. 89, is Carson Canyon, east of Luther Pass and Hope Valley.

p. 266, Lembert Dome. In his 1988 book, Sharp (Plate VI) mentions that the dome's lee face was plucked. However, it appears that at least along the lower part of this face, where evidence can be examined without climbing gear, the glaciers flowed alongside it, and so would not have plucked it. I revisited the face and found no evidence of plucking. This curving, lower lee face closely resembles curving Glacier Point Apron in eastern Yosemite Valley (Map 29, between Curry Village and Happy Isles). There, glaciers also flowed past the apron.

p. 287, Figure 41, and deposits 1200 feet thick. I revisited the deposits along both sides of the Dardanelles to more precisely determine the total thickness of volcanic deposits below the base of the Table Mountain Latite, Tst. These deposits span from 8200 to 6800 feet, a total thickness of 1400 feet, not 1200.

*p. 294, glacial erosion in the South Yuba River drainage. In upper Bear Valley—about ½ mile along Forest Route 18, which branches northeast from Hwy. 20—is the Sierra Discovery Trail, and it follows a short reach of the upper Bear River. Its channel has intact andesite, which originated near Donner Pass, where one rock was dated at 13 million years old. However, an intact sample I collected one mile west of the trail (beside Highway 20 just past and above its curve left) has a date of 18.5 million years old. In the trail's vicinity, Tioga and Tahoe glaciers were about 1100+ feet thick, yet they still did not remove all of this relatively nonresistant andesite. The bedrock-floored canyon actually was slightly deeper before volcanism than it is today, and back then the canyon already had its glacial looking cross-profile: steep-sided and broad, flat-floored.*

*p. 296-297. Windy Peak dacite. I collected two samples and both were andesite, dated at 4.1 million years. As one might have expected, all volcanic rocks are extrusive, none are intrusive. The USGS's designation of dacite as intrusive is solely to fit a shallow pre-glacial canyon theory-driven view.*

*p. 299+, Sierran Uplift. Recent published works suggest that the Colorado Plateau has been high for tens of millions of years, and that the Grand Canyon formed in response to extensional collapse of western lands. Before extension, both these lands and the Sierra Nevada would have been high.*

p. 316, col. 2, lowest volcanic deposits on the west side of Rancheria Mountain. Not only are river boulders absent, so too are the volcanic deposits. All I could find—here and elsewhere—were sparsely scattered volcanic clasts lying atop granitic bedrock. They likely got here either through glacial transport and/or downslope mass wasting. More important, Huber (1990) places the profile of his 10-million-year-old river at about 6000 feet elevation near Rancheria Falls. However, in a stretch of the Hetch Hetchy Reservoir trail about ¼ mile west of its junction with a spur trail to campsites, there are small remnants of *brecciated, hydrothermally altered granitic detritus transported to here by a south-flowing tuff.* These lie at about 4500 feet elevation, demonstrating that the ancient Tuolumne River canyon was at least 1500 feet deeper than Huber proposed. Because these lie on slopes, they do not represent the ancient canyon's floor, only its lower slopes. The floor would have been lower, and the ancient canyon would have been about as deep as today's despite millions of years of erosion.

## Changes and Additions

p. 8, 151: 10,800 radiocarbon years more nearly equals 12,500 calendar years, not 13,000 years.

p. 111, 313: A more-recent date for the Bishop tuff is 760,000 years ago.

p. 151: The photo caption should begin: "Doug Sornberger and Rudy Goldstein assembling ... "

*p. 259: Lake Chelan is one of several linear, deep lakes in n. Washington-s. British Columbia, and some (possibly all) lie in* grabens, *like east African lakes.*

p. 294, rhyolite. According to Hudson (GSA Bulletin, 1951, p. 934), there is 13 feet of basal mid-Eocene granitic gravel. Age is uncertain, but it is older than the rhyolite, indicating that the canyon is more than 33.2 million years old. *The gravel seems to be missing, perhaps due to the ski-area's construction.*

p. 304: Uplift method 5 was cited in Mary Hill's *Geology of the Sierra Nevada* (U. of Calif. Press, 1975). On p. 169 she says that in the last million years as much as 26,000 feet of uplift occurred, indicating that today's range could have been uplifted in ½ million years—easily the most rapid uplift ever proposed.

p. 304: About uplift method 6, David Jones concluded likewise in his "Uplift of the Sierra Nevada: fact or fancy" (Northern California Geological Society's *Guide to the Geology of the Western Sierra Nevada: Sacramento to the Crystal Basin*, Oct. 1997). He states (Intro., point 3): "The Sierra Nevada was not tilted as a rigid block; subsidence of the Great Valley exceeds uplift in the mountains. Consequently, dip of strata buried under or along the edge of the Great Valley cannot be projected eastward to establish topographic elevations of the mountains."

p. 342: In the same drainage, Segall *et al.* (*Geology*, 1990, p. 1248-1251) also suggested a similar age for joints.

p. 350, and parts of Ch. 24-26: I proposed that the last pulse of uplift in the southern Sierra was between 75 and 65 million years ago. After reviewing relevant literature, I now agree with David Jones (see p. 304 entry, above). He suggested that around 80 million years ago the brittle upper crust of the range was removed. Both of us see this accomplished through detachment faulting, and I suggest that the upper crust was faulted westward, then transported northward—a testable hypothesis. With the upper crust gone, the lower crust would have risen isostatically, although not to its pre-faulted height of perhaps 20,000 feet. Under this interpretation, my 80-65-million-year reconstruction of Yosemite Valley needs revision. Along with Jones, Martha House, Brian Wernicke, and Kenneth Farley also picture the range high during Late Cretaceous time. See their "Dating topography of the Sierra Nevada, California, using apatite (U-Th)/He ages" (*Nature*, 5 Nov 98).

p. 378, Love et al. Change MF-2031 to I-2031. p. 383, additions to Index. Alabama Hills 327 Bishop tuff 111, 313 Donner Pass rhyolite 294 Goldstein p. 151 Kingvale p. 294 Long Valley caldera 313 Middlebrook no p. 137 Sornberger 151

## Typos

Page 8, column 1, paragraph 3, line 1: 1993 > 1994 p. 8, col. 2, par. 2, l. 1: Violà! > Voilà! p. 10, col. 1, par. 1, l. 16: came > same p. 21, col. 2, par. 1, l. 8: layer > a layer p. 23, col. 1, par. 3, l. 4: It pristine > Its pristine p. 26, col. 1, last 2 l.: world > the world p. 45, Photo 18 caption, remove: an elevation of p. 64, col. 2, par. 4, l. 10: wait > to wait *p. 98, col. 1, par. 3, l. 14: 48/4 > 64/5.3* p. 103, col. 2, par. 3, l. 3: to the its > to its p. 108, col. 1, par. 3, l. 8: to road > the road p. 134, col. 1, par. 1, l. 3: all no > no p. 139, col. 1, par. 3, l. 22: sparse due, are > sparse, are p. 207, col. 2, par. 2, l. 1: Since throughout > Throughout p. 217, col. 1, par. 1, l. 6: (p. 423) > (his p. 423) p. 223, col. 2, par. 2, l. 7-8: accumulated > have accumulated p. 225, col. 2, par. 1, l. 1: pre-Tahoe > Tahoe p. 228, col. 2, par. 1, l. 3-4: inte-rpretation > inter-pretation p. 256, col. 1, par. 1, l. 18-19: traces > traces of p. 276, Photo 155 caption: 141 > 153 p. 303, col. 1, par. 3, l. 10-11: of > of his p. 304, col. 2, par. 2, l. 1-2: were > was p. 306, col. 1, par. 3, l. 3: it is > it p. 310, col. 1, par. 1, l. 3-4: opposite where > opposite p. 327, col. 1, par. 1, l. 12: Alan > Paul p. 348, Quote: mediochre > mediocre p. 349, col. 2, par. 2, l. 6: 6500 to feet > 6500 feet p. 351, col. 2, par. 1, l. 4: was > were p. 352, col. 1, par. 1, l. 14: valley' > valley's p. 354, col. 2, par. 1, l. 4-5: it > its p. 356, col. 1, par. 1, l. 14: closer > closer to p. 372, Table E, right col.: remove = sign p. 373, Axelrod, 1957: G.S.A. Bull. > Bull. of the G.S.A. p. 375: Elliott-Fiske > Elliott-Fisk

# The Geomorphic Evolution of the Yosemite Valley and Sierra Nevada Landscapes

## Solving the Riddles in the Rocks

Jeffrey P. Schaffer

Design by author
Figures, maps, and photos by author except as noted
Typesetting by Perrott Desktop Publishing, Sonoma, CA

Library of Congress Card Number 97-15708
International Standard Book Number 089997-219-5

Printed in the United States of America

Published by WILDERNESS PRESS
2440 Bancroft Way
Berkeley, CA 94704
Phone: (510) 843-8080
FAX: (510) 548-1355

Write for free catalog

**Library of Congress Cataloging-in-Publication Data**

Schaffer, Jeffrey P.
The geomorphic evolution of the Yosemite Valley and Sierra Nevada landscapes : solving the riddles in the rocks / Jeffrey P. Schaffer.
p. cm.
Includes bibliographical references and index.
ISBN 0-89997-219-5
1. Geomorphology—California—Yosemite Valley. 2. Geomorphology—Sierra Nevada (Calif. and Nev.) I. Title.
GB428.C3S33 1997
551.41'09794'47—dc21 97-15708
CIP

## Dedication

This book is dedicated to my twin brother, Gregory Schaffer, who long ago taught me the mountaineering skills necessary to explore the Sierra Nevada and to become a Yosemite Valley climber.

It is also equally dedicated to Thomas Winnett of Wilderness Press, who gave me book and map projects in the Sierra Nevada and other ranges, permitting me to see thousands of miles of trailside landscapes, glaciated and unglaciated, of varying rock types.

# Contents

**Acknowledgments**

I extend deep appreciation to those who have helped me in the field. In particular I acknowledge the members of my Laurel Lake trip: Greg Schaffer, Rudy Goldstein, Jeff Middlebrook, and Doug Sornberger. Without them I could not have gotten some 200 pounds of coring gear up to the lake to extract its sediments. According to the U.S. Geological Survey, that lake had to be hundreds of thousands of years old, using their method of relative dating. According to me, it could be no older than 16,000 years, using my preferred method. The radiocarbon date on the basal sediments, performed by Beta Analytic Inc., proved conclusively that for decades the USGS had been misdating glacial deposits. Thanks also go to Ken Ng, Eric Edlund, Peter Bernard, Scott Long, Eddie Matzger, and Rob Foster, all who accompanied me in the field. Working in the mountains can be both lonely and dangerous, so I greatly appreciate the camaraderie. During the summers of 1990 and '91, when I worked on the glacial history of the Stanislaus River drainage, Greg and Stephanie Stubbs allowed me unrestricted use of their mountain home as a base camp, which made the physical hardships of my daily work far more bearable. More importantly, I acknowledge the unflagging support of my wife, Bonnie Myhre. Without her support, my research and this book would not exist. At times when I got particularly demoralized—as when a prominent USGS geologist wrote to Cal Berkeley's Geography Department, strongly urging my dissertation committee to stop my research—I could be uplifted by reflecting on what is most important to me—our family, which includes my wife and our daughter, Mary Anne. I have also received constant encouragement from four friends: Elaine Hussey, Tom Winnett, Barbara Jackson, and Jeff Middlebrook. Finally, I would like to thank Noëlle Imperatore for her valuable advice on the production of this book and on past Wilderness Press books.

Nature conceals her secrets because she is sublime, not because she is a trickster.

—Albert Einstein

# Part I

# Background on the Sierra Nevada

# 1

# Riddles in the Rocks

Every year millions of people visit Yosemite Valley, and they encounter stunning scenery that can overwhelm their senses. This experience is as true today as it was when the valley was discovered. The first recorded discovery of Yosemite Valley by western civilization was made by Major James D. Savage's Mariposa Battalion, which, pursuing "hostile Indians" up the Merced River drainage, entered the valley on March 27, 1851. Early rumors about the valley had not prepared the men for the overpowering view now before them (Figure 1). Much later, in 1880,[1] that profound moment of discovery was recalled by Lafayette Bunnell, the battalion's doctor, who wrote:

> It has been said that "it is not easy to describe in words the precise impressions which great objects make upon us." I cannot describe how completely I realized this truth. None but those who have visited this most wonderful valley can even imagine the feelings with which I looked upon the view that was there presented. The grandeur of the scene was but softened by the haze that hung over the valley—light as gossamer—and by the clouds which partially dimmed the higher cliffs and mountains. This obscurity of vision but increased the awe with which I beheld it, and as I looked, a peculiar exalted sensation seemed to fill my whole being, and I found my eyes in tears with emotion.... I have here seen the power and glory of a Supreme being; the majesty of His handy-work is in that "Testimony of the Rocks." That mute appeal—pointing to El Capitan—illustrates it, with more convincing eloquence than can the most powerful arguments of surpliced priests.

How did this magnificent scenery originate? For some visitors the scenery was created by God. However, most visitors today accept a scientific explanation, one that has been around since the 1880s and '90s. It goes like this. In relatively recent geologic time, the Sierra Nevada block was tilted westward, greatly uplifting the east side of the block. Due to this uplift, Sierran rivers cut deep canyons, then later, glaciers advanced down them, widening and deepening them.

Early geologists could not agree on how much canyon cutting had been done by rivers and how much by glaciers. The answer to that problem was presented to the public in 1930 by François Emile Matthes (pronounced "Mat'-tees"). In 1913 he had been assigned the task of deciphering Yosemite Valley's geomorphic history—that is, how its landscape (and that of the Sierra Nevada) had evolved over time. His magnum opus, USGS (United States Geological Survey) Professional Paper 160, *Geologic History of Yosemite Valley,* has seemed to Sierran geologists to withstand the test of time. The only signifi-

---

[1]This book is based on the lengthy third draft of my Ph.D. dissertation (1995), which had 233 footnotes. Footnotes are distracting for the casual reader, so this is the only one you'll read. However, some may want to know the more important (of 400+) sources. Consequently I have identified authors and the dates of their publications, such as here (Bunnell/1880), and the full bibliographical entry appears under "References." In this citation, the actual entry is Bunnell, *1990,* a reprint of the *4th* edition of *1911.* I have cited this edition and reprint because it is readily available, whereas earlier editions and reprints are not.

**Figure 1. An early drawing of Yosemite Valley. Source: *Geological Survey of California,* J. D. Whitney, 1865.**

cant modification is that whereas Matthes proposed that the range had been raised by three short-lived, increasingly larger pulses of uplift, the most accepted modern view proposes that uplift began slowly and increased its rate over time. Both views posit similar amounts of uplift over similar amounts of time. The modern view is best presented by N. King Huber in his 1987 USGS Bulletin, *The Geologic Story of Yosemite National Park.* (See Map 1, which identifies the park's major features and important sites with crucial evidence.)

But if Matthes' and Huber's views are correct, why are there "mysteries," as the late Clyde Wahrhaftig (d. 1994) put it? For example, above the east end of Little Yosemite Valley, the descending crest of a Tahoe-age lateral moraine appears to dive beneath the crest of a Tioga-age one. (Glaciers, their deposits, and their ages will be discussed in later chapters. All you need to know here is that a glacier transports debris, and that some of this debris falls off the side of the glacier. If the slopes adjacent to the glacier aren't overly steep, the debris accumulates to form a lateral moraine. Quite obviously, the surface of the glacier that dumped this debris had to be at least as high as the moraine it left. Consequently, the crest of a lateral moraine is a good indication of the surface of its glacier.) Now, everywhere in the glaciated Sierra Nevada, both east and west of its crest, the earlier, Tahoe glaciers were a bit larger than the later, Tioga glaciers, and so they left slightly higher lateral moraines. Everywhere, that is, except east and above Little Yosemite Valley. There the crest of the Tahoe-age lateral moraine stands about 300 feet above the crest of the Tioga-age lateral moraine. Down-canyon the crest of a Tioga lateral moraine falls increasingly below that of a Tahoe. For some mysterious reason, the Tahoe-age lateral moraine suddenly does a nose-dive, its crest plunging rapidly below the Tioga's (Figure 2). As I've reconstructed it, using the 1987 glacier map (prepared by Clyde Wahrhaftig but cited as by Alpha *et al.*), the Tahoe glacier thinned more rapidly than the Tioga glacier, and barely reached Yosemite Valley's east end. The westernmost, lowest-elevation end moraine—deposits left at the front of a glacier—is, by everyone's judgment, Tioga, not Tahoe. Why was the Tahoe glacier shorter only here, in Yosemite Valley? During the time of the Tahoe glaciation, was the *local* climate over Little Yosemite Valley extremely warmer, causing excessive melting? Not possible. Yosemite Valley's Tahoe glacier also was mysteriously shorter than, and ended mysteriously higher than, the valley's largest pre-Tahoe

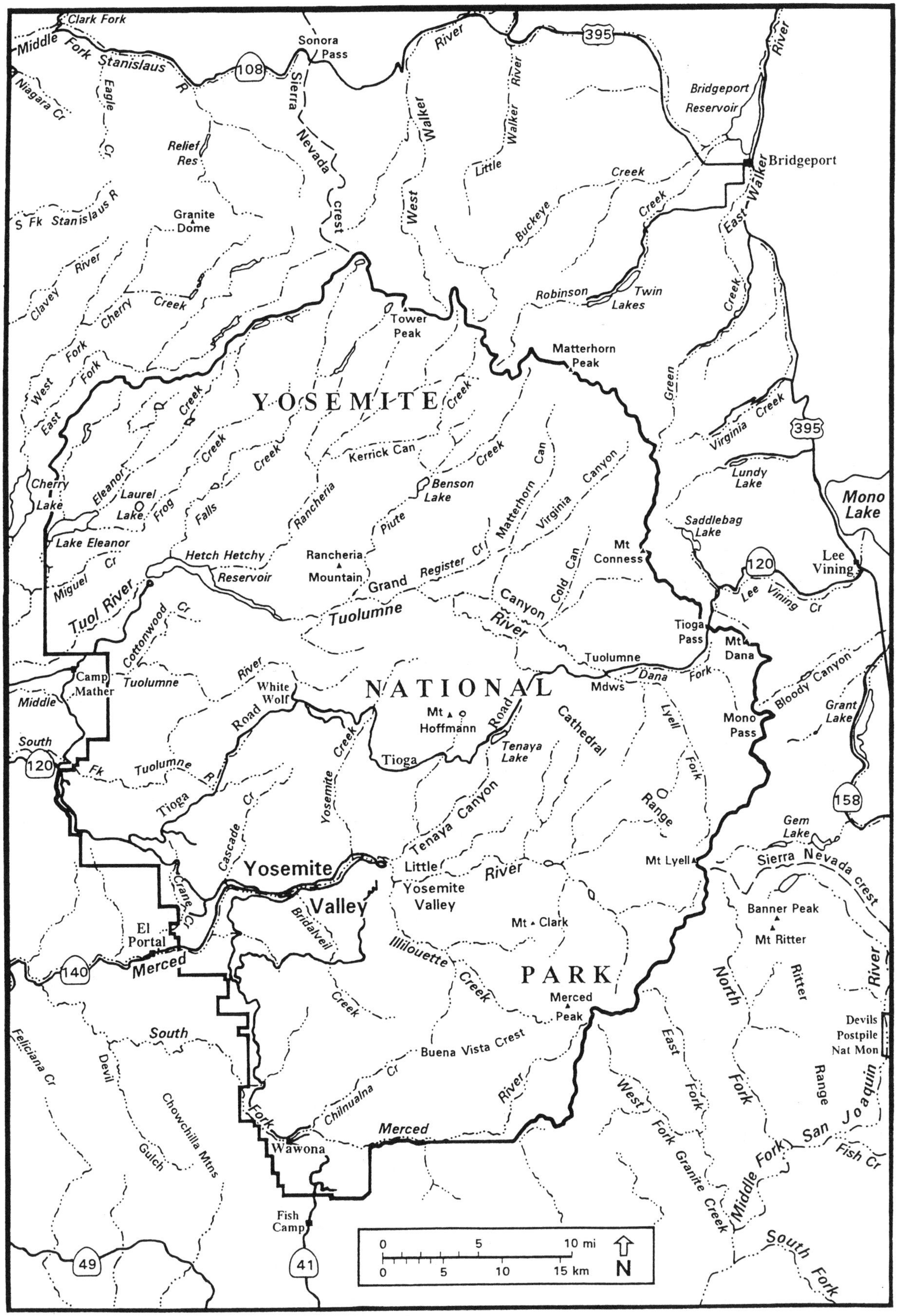


Map 1. Yosemite National Park and vicinity.

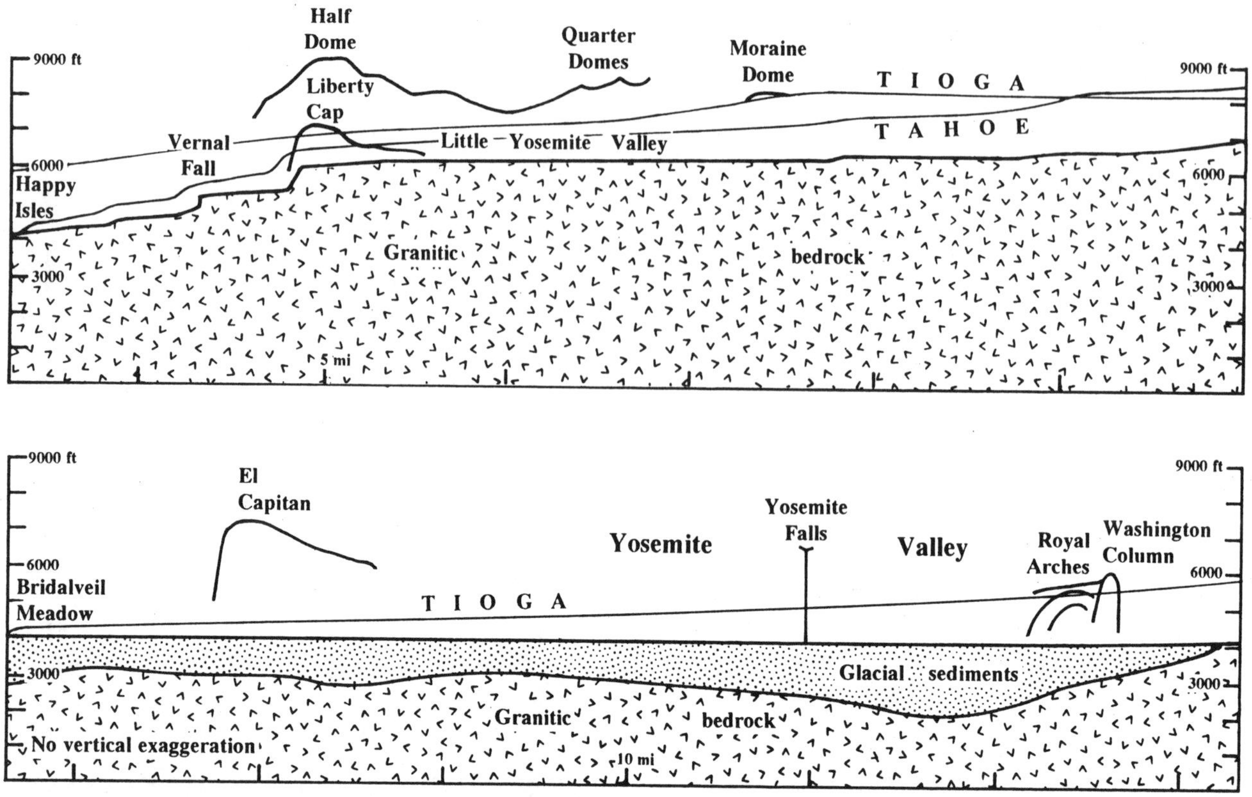


**Figure 2. Tioga and Tahoe glaciers through Little Yosemite and Yosemite valleys. The diving Tahoe moraine indicates that the Tahoe glacier thinned rapidly, barely reaching the floor of Yosemite Valley.**

glacier. In other river canyons, Tahoe glaciers were proportionally closer in length to their pre-Tahoe glaciers and the snouts of the Tahoe glaciers descended to elevations almost as low as those of the snouts of the pre-Tahoe glaciers.

Nor are the mysteries limited to the Merced River drainage. Before I had begun my serious research on Yosemite and the Sierra Nevada in 1990, no one had published a comparison of the uplift and the glacial histories of the major canyons. Had anyone done so, he or she would have encountered two major mysteries. First, by using universally accepted geologic assumptions and methods, one discovers that every river canyon of the Sierra Nevada has its own unique *uplift* history. Second, by using universally accepted geologic assumptions and methods, one discovers that every glaciated river canyon of the Sierra Nevada has its own unique *glacial* history. It is as if God had created each river canyon separately. For example, the Tuolumne River canyon is believed to have experienced major uplift between 10 and 2 million years ago and, despite possessing the Sierra's largest glaciers, is believed to have suffered minimal glacial erosion. In contrast, the Merced River canyon is believed to have experienced gradually increasing uplift beginning over 25 million years ago and, despite possessing glaciers half the length and thickness of the Tuolumne's, its Yosemite Valley suffered the range's greatest glacial erosion—even though it has the most-resistant granitic bedrock in the range. How can two adjacent river canyons have such discrepant uplift and glacial histories? If one canyon landscape is rising and the adjacent one is not, then there should be either a flexure or a fault running along the divide separating the two landscapes. None exist.

The conventional interpretations of the geomorphic history of Yosemite Valley and the Sierra Nevada raise more than mysteries; they pose riddles—*riddles in the rocks* (whence the book's subtitle). For example:

• How can three pairs of lateral moraines left by a 20,000-year-old glacier be dated at 20,000, 150,000, and 800,000 years old?

• When are voluminous glacial deposits—moraines—not the result of glacial erosion?

• Where do slabs of rock repeatedly fall from canyon cliffs over millions of years and yet the canyon never widens through cliff retreat?

• How does a steep-walled, flat-floored glacial valley develop in the absence of glaciers?

• How can a block of the earth's crust that was raised above sea level during the last 5 million years possess a mountain landscape that is over 60 million years old?

Under conventional assumptions, these riddles cannot be solved. For professional geologists, especially those familiar with the Sierra Nevada, these riddles may seem to be nonsense, but they are not. I have posed them for a specific reason: if you can provide logical answers to these riddles, you can then construct a geomorphic history for Yosemite Valley and the Sierra Nevada that not only is logical, but also fits all the field evidence. This evidence argues against conventional interpretations of both uplift and glaciation to an extreme degree. An example of each will suffice.

There are at least a dozen methods that purport to find that major Sierran uplift has occurred over the last 1 to 20+ million years. (But if it had occurred, geologists should have been able to pinpoint the commencement of uplift more precisely—for example, at 5 or at 10 million years ago.) Three of these methods are quite well known and are generally accepted. One method was proposed by Waldemar Lindgren in 1911 especially for the Yuba River drainage, a second by François Matthes in 1930 for the Merced River drainage, and a third by N. King Huber (who used Leslie Ransome's 1898 method) for the San Joaquin River drainage. If one ignores that in every article each geologist presented field evidence that contradicted both his uplift method and its results—an irresolvable problem in itself—then the three geologists seem to be in relative accord. From the Yuba drainage south past the Merced to the San Joaquin, the Sierran crest increases in elevation. Therefore, one would expect that of the three drainages the Yuba had the least amount of uplift and the San Joaquin the greatest. And that is what one sees. The maximum uplifts of these three drainages, using the corresponding three different methods, are roughly 4400 feet, 7000 feet, and 11,300 feet. But what happens if one uses each method in other drainages? The result is chaos. For example, Lindgren's method predicts only about 3100 feet of uplift at the crest for the Tuolumne River drainage, where 7,000-10,000 feet might be expected by Ransome's method. Far worse, for the adjacent Merced River drainage it predicts *subsidence*.

If uplift seems fraught with major problems, the ages, extents, and effects of past glaciers are equally fraught. Here is one example. On the 1989 geologic map of Yosemite National Park by N. King Huber, Paul C. Bateman, and Clyde Wahrhaftig, take a look at the Tioga and Tahoe lateral moraines on the north and south sides of Little Yosemite Valley. On the north side of the valley the uppermost Tioga moraine is at about 7100 feet elevation. The Tahoe moraine is believed to have dived well beneath it and so must lie lower, say at 7000 feet elevation (or considerably less). Because the floor of the valley is at about 6000 feet elevation, one can conclude that the Tahoe glacier in Little Yosemite Valley was about 1000 feet thick. Now look at the south side of the valley. The highest Tahoe lateral moraine is at about 8000 feet elevation. So one can conclude that the Tahoe glacier in Little Yosemite Valley was about 2000 feet thick. Riddle me this:

- How can a glacier be both 1000 feet and 2000 feet thick at the same time?

The data force one to conclude that the glacial ice on the north side of the valley melted far more rapidly than did that on its south side. Today, no glacier has such a laterally sloping surface, nor could such a glacier possibly exist on either theoretical or experimental grounds. Another riddle:

- If the ice surface on the north side of Little Yosemite Valley was 1000 feet lower than that on its south side, then how was it possible that a couple of miles up-canyon the ice surfaces on the north and south sides were both the same? (Check the moraines of Huber *et al.*'s geologic map north and south of Bunnell Point.)

Hopefully it is now quite obvious to you the reader that there are serious problems with conventional interpretations of both uplift and glaciation. As I said early in this chapter, these interpretations have existed more or less since the 1880s and '90s. Also as I said, the field evidence argues overwhelmingly against these interpretations. But if the field evidence is so strong, why wasn't it until my work in the 1990s that anyone realized there were serious problems? Is it because field work before then had not been done over a sufficiently large area or had not been done in sufficient detail? Definitely not. Principles of uplift and glaciation were known before the Whitney Survey made its first geologic reconnaissance of the Sierra Nevada in the 1860s. Fossils were collected in the 1860s, '70s, and '80s, and from these, paleontologists were able to deduce the range's past climates. And by the late 1890s the topography and geology of the northern half of the Sierra Nevada had been mapped in such detail that any qualified geologist, reviewing what had been published, could have reconstructed the uplift history of the range. Furthermore, if he had been assigned the task of reconstructing the extent of former glaciers and their impact on the landscape, he could have done so back then almost as easily as one can today. Instead, over a full century, where Sierran geologists had to make important conclusions with regard to uplift, to age of glacial deposits, to extent of glacial deposits of a given age, and to the degree of glacial erosion, far more often than not they made the wrong conclusions. Which leads us to perhaps the greatest mystery of all: How could a century of geoscientists (including the author during the 1960s, '70s, and '80s) *not* have figured it out? (The term geoscientists includes geologists, geophysicists, geomorphologists, geographers, paleontologists, etc.) Geologists in the early 1990s were still trying to solve the uplift and glaciation problems that their predecessors were working on in the 1890s. Over the century, no progress has been made toward a unified solution—answers that are valid across the entire range.

The answer to the greatest mystery appears to lie *in part* in the fact that processes which slowly transform mountain landscapes *appear* to leave very deceptive evidence. I stress "appear" because in fact the evidence is not at all deceptive. However, every geoscientist coming to the Sierra Nevada views it with a theoretical bias, and it is that bias that entraps you. How you determine the processes and rates that constitute the evolution of a mountain landscape such as the Sierra Nevada depends in large measure on your initial assumptions. And for at least a century the branch of science known as alpine geomorphology—the study of mountain landscapes—has operated under a set of false fundamental assumptions. Carrying these assumptions, we have misinterpreted the timing of uplift, the amount of uplift, the sizes of past glaciers, and the relative amounts of erosion due to streams and glaciers, not only for the Sierra Nevada, but probably for many of the world's ranges.

I had carried this intellectual baggage for a quarter century while hiking thousands of miles across the Sierra Nevada (and to a lesser extent, other ranges) before I finally dropped it. Why? When I finally got serious about examining field evidence of former glaciers in some detail, its implications about glacial erosion contradicted theory. Many others probably have been in a similar situation. As is reiterated in science (supposedly in jest, but unfortunately, sometimes it occurs in practice), "When evidence and theory conflict, discard the evidence." This time, in summer 1990, I chose to discard the theory. Over the next two summers of field work, I began formulating an alternate paradigm ("pair-a-dime": pattern or example) of uplift and glaciation based, not on theory, but on *field* evidence. (I stress "field" because it is the most reliable evidence available. It exists, and anyone can check it out and see if they agree with your interpretations. You can't do that with seismological evidence tens of miles down, with unrealistic one- or two-variable laboratory experiments, or with modeling on a computer.) By June 1993, my new paradigm was taking shape, and the field work I did in the summers of 1993, '94, '95, and '96 was to prove, improve, or disprove the paradigm. The third is most important. You must think of every conceivable way you could be wrong, and then find *unambiguous* evidence that either supports or refutes your assertions.

One incident in August 1993 is worth noting. I had convinced three friends and my twin brother to help carry about 250 pounds of gear up a grueling, hot ascent to Laurel Lake, in northwestern Yosemite National Park. The conventional paradigm dictated that the lake had existed for at least 200,000 years—more likely, for 800,000 years. My new paradigm dictated that the lake could be *no older than* 16,000 years. Hence, it was an ideal site to pit one paradigm against the other. From my self-designed portable drilling platform, we successfully extracted the lowest sediments in the lake—which, being lowest, were the oldest. A reputable, independent lab, Beta Analytic, then dated the carbon-14 in these sediments, which indicated they had been deposited about 13,000 years ago (11,000 years B.C.), that is, when the great Tioga glaciers finally disappeared. (The lab date is 10,800 years B.P.—before present, defined as 1950 A.D., the approximate year carbon-14 dating began. This dating process underestimates the true age of dates of this approximate age, hence the age discrepancy.)

Violà! A major part of the new paradigm is supported and the old counterpart is refuted. But that is not the point worth noting. What's worth noting is that on the ascent one of my friends, Rudy Goldstein, was making correct interpretations of the glacial evidence he saw. Did he have special talent? Well, in a way, he did. *He had no geological background.* Not knowing how glacial deposits should be dated, and not knowing what others previously had said about this area's glacial history, he was forced to consider only the evidence before him. Which he did. Geologists, on the other hand, have had the unfortunate (if self-imposed) burden of carrying erroneous uplift and glacial theories into the field, making objective observations well nigh impossible.

How have all geoscientists (including the author before 1992) concluded that major uplift of the Sierra Nevada has occurred mostly over the last few million years? The answer lies in the nature of past field work: it was largely piecemeal. As I said earlier, Lindgren, Matthes, and Huber each used a different method to determine timing and amount of uplift for one specific drainage, but they did not test it in other drainages. Each major west-side Sierran drainage appears to have had its own unique uplift history, dependent upon which method was used to decipher it. Until my research, no one had attempted to provide an uplift history for the entire range, that is, one that fits the evidence in *every* Sierran drainage.

The story for former glaciation is similar. One would expect that by now the entire Sierra Nevada had been examined in detail for evidence of former glaciation. This is not so, except for the east side of the Sierran crest, where glacial moraines are very conspicuous and are easily reached. In Yosemite National Park, with its 4+ million visitors per year, most of its lands have had at best only a glacial reconnaissance. The widespread distribution of glacial deposits on Huber *et al.*'s 1989 geologic map of Yosemite National Park makes one think that extensive glacial field work had been done. However, Matthes was the only person to do some detailed work, and then, only in and above Yosemite and Little Yosemite valleys. Elsewhere in the park, he did only a reconnaissance study. Others followed him, but being interested primarily in the bedrock, they minimized glacial studies. Finally, Clyde Wahrhaftig, a USGS geologist and Cal Berkeley professor, used aerial photographs to identify many of the park's glacial deposits, which allowed him to do a lot of field work without leaving his office. (He had a heart condition which prevented field work; nevertheless, other geologists could have—indeed, should have—done the work.) Aerial photographs are useful in helping you plan your field work, but they are no substitute for it. Despite having had extensive field experience with glacial deposits, I still find that my identifications of them on aerial photographs turn out in the field to be *wrong about three-fourths* of the time, and Wahrhaftig fared no better.

If Yosemite National Park has had patchy fieldwork done on former glaciations, other west-side areas have even less. Until my research, apparently no one had reviewed all that had been written and addressed the disparate results. Every glacial researcher had pieced together an internally consistent, believable account of former glaciations in his own area. Each area studied had its own glacial history. And one area had two glacial histories. In the 1950s Joseph Birman mapped in exquisite detail glacial moraines and deposits in part of the west side's South Fork San Joaquin River drainage eastward across a gap in the Sierran crest into the east side's Rock Creek drainage. Unlike others, in his 1964 publication he correlated his results with other parts of the Sierra Nevada. Unfortunately, he did not explain why his South Fork San Joaquin River results did not correlate with his Rock Creek results. Briefly, in the former area the Tioga glacier is much shorter than the Tahoe glacier; in the latter area, it is almost as long—why? No answer.

To answer why geologists have not succeeded, we need a historical perspective. How were geologists gradually led astray, especially in the Sierra Nevada? What areas did they actually visit? What evidence did they find? What evidence did they miss? What evidence did they ignore, suppress, or invent? What conclusions did they reach, and why did they reach them? Who influenced their thinking? Part II of this book answers some of these questions, although other answers appear under relevant topics, particularly glaciation and uplift.

Part III addresses glaciation. What are glaciers? How do they operate? How many episodes of glaciations have occurred? How many of those glaciations left evidence of their existence? What kind of evidence did each leave? How do you reconstruct the extents of former glaciers? And most importantly, how much erosion did they perform?

The answer to the last question is, contrary to widely accepted views over the last century, "Surprisingly little." But if glaciers didn't transform shallow canyons into deep ones, then canyons must have been deep before glaciation. If these canyons had been elevated many thousands of feet, then initially their floors must have

**Photo 1. Lake-coring crew resting in 90°F shade, half way up the strenuous 8½-mile trail that climbs to Laurel Lake. Three of five crew members are Rudy Goldstein (left), Jeff Middlebrook (center), and Doug Sornberger (stretching in background). The backpacks average 50 pounds in weight.**

been many thousands of feet lower, that is, below sea level. This problem is exacerbated when one realizes that over most of the time these canyons have existed, sea level was significantly higher than it is today, and so the canyon bottoms would have been under even more water. Because I had determined beyond reasonable doubt that glaciers perform very little erosion, I was obliged to address uplift. For the northern Sierra Nevada, up to about 2000 feet of uplift at the crest was permissible if the preglacial canyons had been deep; for the central Sierra Nevada, about 4000 feet. However, I was not prepared for the conclusion reached by a method I had devised for quantifying uplift based on field evidence: no measurable uplift in tens of millions of years. So I devised a second field method and then a third, all independent of each other, to quantify uplift. All indicated the came conclusion. Consequently, Part IV addresses uplift. When did it occur? Why did it occur? Why are so many widely accepted "proofs" of major uplift over the last few million years groundless? And most importantly, how do you measure uplift using only field evidence and no theoretical assumptions?

Finally, if you can demonstrate how much (or how little) glaciers eroded, and if you can demonstrate how the major canyons and their tributary canyons have been deepened over time by rivers and creeks, then you can reconstruct the landscape. For the Sierra Nevada, I have done this only in a general fashion. However, for Yosemite Valley, which to me is the Holy Grail of Sierran geomorphology, I have done so in considerable detail. When were the valley's first granitic rocks exposed? How deep was the granitic valley in the last days of the dinosaurs? (Conventional wisdom dictates it didn't even exist.) How did it appear immediately before glaciation? Much as it does today. Like today, the valley back then still would have been the rock climbers' Mecca of the world. And what of post-glacial Lake Yosemite—did it actually exist until a few thousand years ago? It was as real as the Loch Ness monster, the Bermuda Triangle, and UFO abductions. Most of these issues and more are addressed in Part V.

Upon completing this book, you, unlike Sierran geoscientists over the last century, should be able to answer the riddles in the rocks. This is true even if you don't have a geological background. This is true even if you didn't understand all that was presented in this chapter, for the story of how Yosemite Valley and the Sierra Nevada were formed is not that difficult. In the chapters that follow, I'll define terms and explain concepts when it becomes necessary to understand them. Some terms I mention early on without addressing them, for example, "plutonism" and "plutonic rocks" in the next chapter. These are given more for the benefit of geoscientists than for laypersons, as are specific names for volcanic and plutonic (granitic) rocks, such as trachybasalt and tonalite. Where a specific rock type is significant, I explain why, as for leucogranite in Chapter 18. For the benefit of laypersons, I have kept geological terms and phrases to a minimum. Like François Matthes, the Sierran landscape's foremost interpreter, I've a dislike for scientific jargon.

# 2

# Encounters with the Sierra Nevada

For Anglo-American immigrants traveling west toward California in the 1840s before the Gold Rush, the Sierra Nevada was a range that inspired both hope and anxiety. Hope, because their travails through the desert lands of Utah and Nevada were now over, and the promised land lay just ahead; anxiety, because the range stood in their way. It contained very few passes, as the earlier mountain men had learned, and the several that existed were difficult to surmount for those traveling in wagons. Indeed, the first two parties—Bidwell, 1841, and Chiles, 1843—had to abandon their wagons in order to surmount the Sierran crest at, respectively, Emigrant Pass in the central Sierra Nevada and Walker Pass in the southern Sierra Nevada. The Stevens party of 1844 was the first to take wagons across the crest, at what would become known as Donner Pass, northwest of Lake Tahoe (Map 2). Crossing in late November, they did not get far before being snowed in for the winter in this Sierra Nevada—the Snowy Range. All members, however, survived, and several 1845 parties crossing the now known route did so before heavy snows fell. The 1846 Donner Party arrived early enough to cross—barely—but rested a fateful five days, and by doing so were snowed in. They had plenty of stock with them which they could have butchered as they overwintered at Donner Lake, but they let the stock roam free, which left them without food, other than eating their own dead. Of 82 who had reached Donner Lake (earlier, the party had diminished through murder, other deaths, and desertion) only 47 survived to reach Sutter's Fort at Sacramento.

The fort, or Nueva Helvetica, as John A. Sutter originally named it upon receipt of his 48,839-acre Mexican land grant in 1839, was the goal of westbound immigrants who had survived the Sierran traverse. In 1847 Sutter needed lumber to finish his large flour mill, and he intended to sell lumber to others, particularly to those in the small village of Yerba Buena—San Francisco. For this he needed a sawmill, and serendipitously that spring James W. Marshall, a millright by trade, stopped by his fort to propose erecting one in the Sierran foothills. Marshall scouted the lands and decided that a good site would be the valley of Coloma. It had the South Fork American River for water power, nearby pines for lumber, and—most importantly to Sutter (due to previous failed attempts)—suitable terrain over which to construct a wagon road to the fort. The two men entered a partnership in August, and construction began in September. The mill was still under construction on an eventful morning of January 24, 1848, when Marshall looked down in the water and saw several small pieces of gold. Both Marshall and Sutter tried to keep the discovery a secret, but word soon got out, rumors spread, and on March 15 *The Californian,* a weekly San Francisco newspaper, announced the discovery. The sawmill commenced operation in mid-April, but after cutting a few thousand board feet of lumber, it was abandoned, as all hands were intent upon gold digging. The California Gold Rush had begun. By May, San Francisco had been reduced to five men to care for women and children, the rest of the men having gone to the foothills. Sutter's employees at the fort also abandoned him, causing him to lament:

> By this sudden discovery of the gold, all my great plans were destroyed. Had I succeeded with my mills and manufactories for a few years before the gold was discovered, I should have been the richest citizen on the Pacific shore; but it had to be different. Instead of being rich, I am ruined.

The Gold Rush at first was mostly by Californians, who by the end of 1848 were scouring the northern half of the

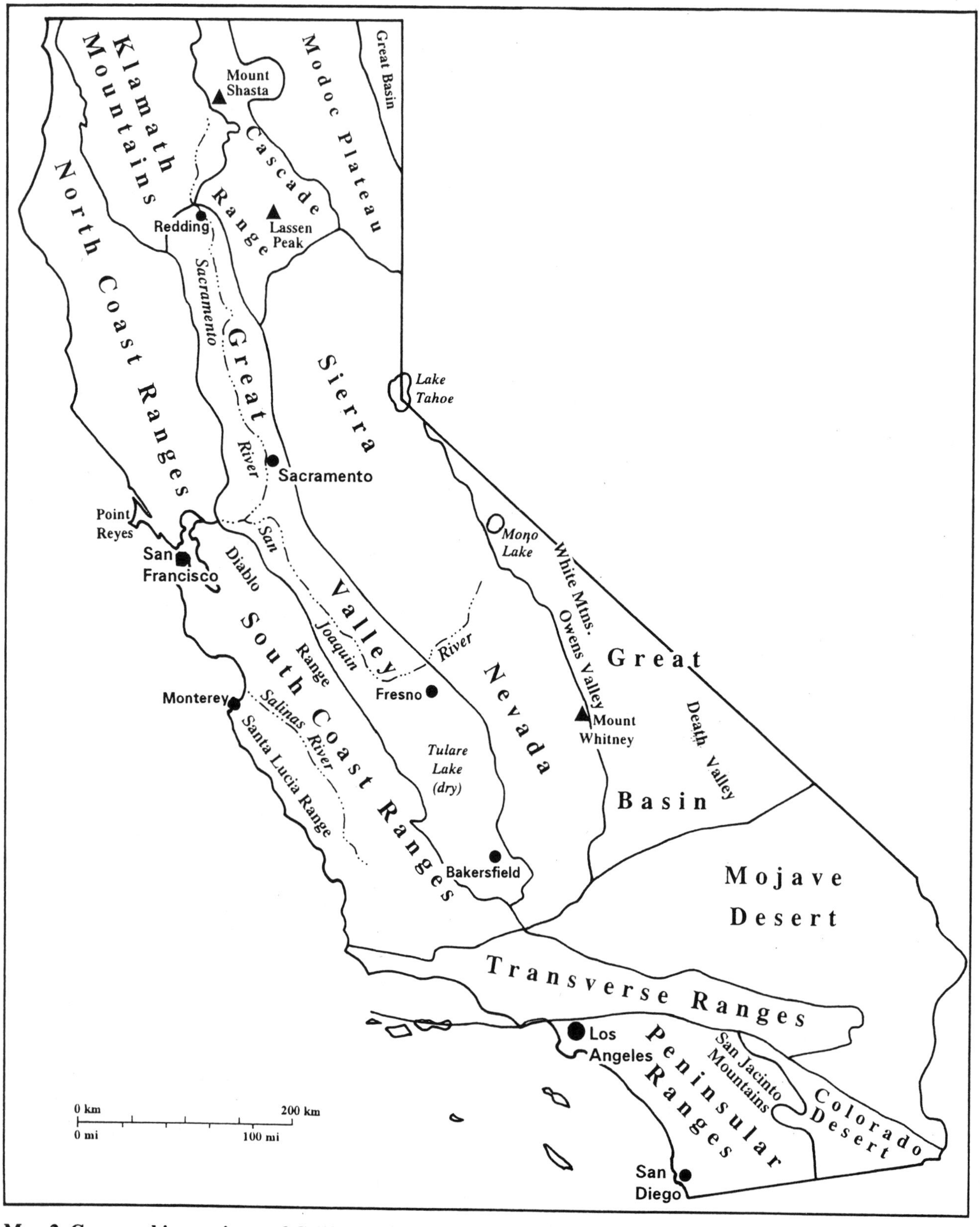


**Map 2. Geomorphic provinces of California. The boundaries differ slightly from those on older maps because in determining them the author used recent geological, geophysical, and plate tectonic data.**

Sierra Nevada from the Tuolumne River north to the Feather River. The main influx of immigrants occurred in 1849, the year synonymous with the Gold Rush, and by the end of it the state's population had swelled by at least 50,000. Like marauding army ants, the mostly inexperienced miners swarmed across the landscape, leaving no proverbial stone unturned. California gold production went from ½ ton in 1848, to 20½ tons in 1849, to 83 in 1850, to 153 in 1851, and to a peak of 164 in 1852—a pivotal year. By 1852 the placer gold, which lay within stream gravels, was becoming exhausted, and hydraulic mining began. The days of the solitary miner were numbered as well-financed big business stepped in. By 1855 hydraulic mining had become a major industry. This method required spraying vast amounts of high-pressure water against the gold-bearing gravels, which typically were located in waterless or nearly waterless areas. Consequently, reservoirs and canals had to be constructed. For the Malakoff mine, in Nevada County north of Nevada City, over 50 miles of canals had to be built to divert water from the Bowman Lake area down to reservoirs near the gravels. In 992-square-mile Nevada County, some 700 miles of canals were built for this purpose.

After 1854, annual gold production plummeted to about 36 tons in 1865, and then generally fluctuating between 25 and 60 tons annually until 1942, after which it dramatically dropped, averaging about 5 to 10 tons annually through 1960. An inexorable decline followed, and in 1968 the annual production dropped well below one ton. During the 1850s about 200 mining districts developed in the north half of the Sierra Nevada, these more or less confined to the range's main belt of metamorphic rock (Map 3). The few scattered districts beyond this belt *and* west of the Sierran crest are related to small remnants of metamorphic rock not shown on this map. Each district could have from dozens to thousands of miners in it. In less than a decade this area had gone from a largely unknown wilderness to a largely ravaged landscape. Mines, towns, shanties, and canals required tremendous amounts of timber, and miners tremendous amounts of food. Any forest creature from squirrel and quail up to mule deer and grizzly bear became fair game. And of course, the land became laced with roads and trails. Both were quite primitive.

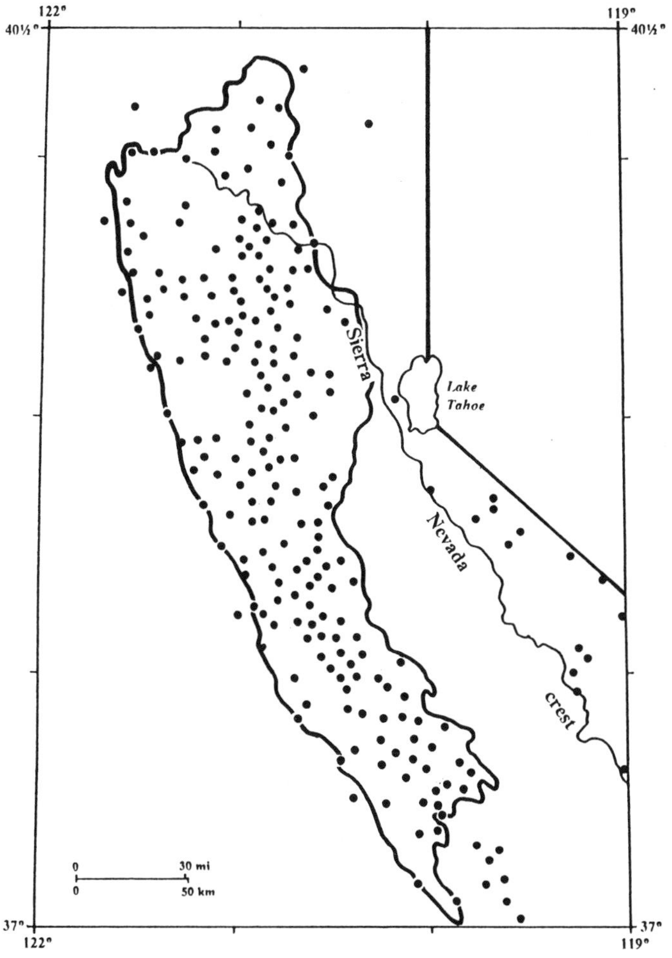


**Map 3. Gold districts (dots) and the metamorphic-rock belt (enclosed by the thick line) of the Sierra Nevada. Sources: Clark, 1970, and Jennings, 1977.**

For example, in 1852 Hangtown (Placerville), which had become the unofficial capital of northern California's gold-mining region, had an immigrant route of sorts built from it east to the Sierran crest at Johnson Pass, located ¾ mile north of today's Highway 50 crossing at Echo Summit, from where it dropped into Lake Valley (in the southern part of the Lake Tahoe Basin). Drop it did, so steeply that block and tackle had to be used to haul westbound wagons up it. An alternative grade had to be found. A route over Luther Pass, to the east, was surveyed in the winter of 1854 for the purpose of providing a wagon road to Sacramento and Hangtown that would be better than Johnson Pass and also shorter and easier than the primitive Carson Pass (today's Highway 88) route. That spring, Asa Hawley established a trading post in upper Lake Valley near a part of the Upper Truckee canyon's wall that quickly became known as Hawley's Hill. Construction soon began on a grade that would be gentle enough to safely accommodate wagons. Financed by private interest, this route—Hawley's Grade—was completed in 1857, making it the first conventional wagon road to cross the central Sierra. Combined with the recently constructed Luther Pass segment, this grade fast became *the* route for westbound would-be miners to take. In 1858 El Dorado and Sacramento counties improved western segments of this largely-one-lane toll road, making it far superior to the higher, longer-snowbound Carson Pass route to the south.

Timing couldn't have been better, for in 1859 silver was discovered in the Comstock Lode at Virginia Town, today's Virginia City, northeast of Lake Tahoe and southeast of Reno. Traffic was reversed on this road as a flood of miners from California's gold fields scrambled east over this toll road to try their luck at or near Virginia Town. Alas, even as Hawley's Grade was constructed to channel westbound miners and pioneers into California's Mother Lode country faster than was possible along the Carson Grade, so too were plans made to convey miners and others east to the Comstock by a faster route. By the summer of 1860, a wagon-and-stage toll road had been constructed down Meyer's Grade, then east to climb over Daggett Pass, situated above Tahoe's southeast shore. Hawley's Grade, briefly a shortcut that siphoned traffic from the Carson Pass route, now became the longer, unprofitable toll road. Investors, like most miners, lost money, as often was the case in these boom and bust times.

The year of 1860 is geologically significant, for on April 20 the state legislature passed an act to establish and fund a Geological Survey, whose purpose was

> to make an accurate and complete Geological Survey of the State, and to furnish maps and diagrams thereof, with a full and scientific description of the rocks, fossils, soils, and minerals, and of its botanical and zoological productions, together with specimens of the same.

The Act appointed Josiah D. Whitney the State Geologist and made him the leader of what became known as the Whitney Survey (Photo 2). Whitney was a graduate of Scientific School of Yale College (University), as was his right-hand man, William H. Brewer. Both had traveled extensively in Europe, including the Alps, and the geologic lessons they garnered over there they applied in their California explorations, which began in November 1860. Most of the work was performed from then through 1864, and in the following year Whitney produced *Geology of California,* volume 1.

Back in the 1860s geology was still a young science, with relatively few practicing geologists. Whitney was one of the best, and connections other geologists had with him with regard to the geologic history of the Sierra Nevada make an interesting story. Additionally, all were connected with either Yale or Harvard universities. German-born-and-educated Charles F. Hoffmann came to California in 1858 at the age of 20, already having served the previous year as a topographer for a wagon-road survey west to Honey Lake, in northeastern California. He joined the survey in July 1861, and next to Whitney stayed longer than anyone else, until its disbanding in 1874. In August 1863, Brewer ran into James T. Gardner (Gardiner in later life) and Clarence King, both 1862 graduates of the Scientific School. During the winter of 1862-63, King studied glaciology under Harvard's—indeed, North America's—glacial expert, Prof. Louis Agassiz. Hoffman taught Gardner and King the fine art of map making, and in 1865 they produced the first detailed topographic map (*sans* contours) of Yosemite Valley, which along with the Mariposa Grove of Big Trees had become a state park the previous year. King later performed valuable field work for the United States government, and then, in 1879, organized the United States Geological Survey and became its first director. While topographic maps with contours had existed previously, they were usually for small areas, such as a detailed map of San Francisco done by the U.S. Coast Survey in 1853. It was Hoffman who first produced a large-area topographic map with contours, his 1873 *Topographical Map of Central California together with a part of Nevada.*

Although the Whitney Survey was supposed to be a scientific endeavor, state legislators had hoped otherwise. In short, like today's insider traders, they had expected to profit from privileged information on locations of gold deposits. No such information was forthcoming, and likewise funds for the survey often were not forthcoming. At times Whitney, who had borrowed more money than any other survey member, augmented his income as a part-time professor at Harvard University, which in the 1860s was rivaling Yale for prominence in geology. Yale had Prof. James D. Dana, who in 1862 had completed his epic 800-page *Manual of Geology*, which in most respects is amazingly modern. Harvard had Prof. Louis Agassiz, a Swiss *zoologist* who in 1837, based on glacial reconnaissance of the previous year, expounded views on widespread former glaciation that laid the foundation for modern glaciology. In the summer of 1869 Whitney, aided with Brewer and Hoffman, taught a geology and topography field course to four students, two of whom would achieve great fame. The first was William Morris Davis, who became a professor at Harvard and devoted his time to the study of landscapes, becoming the father or American geomorphology. The second was Henry Gannett, who became the Chief Geographer of the USGS (and from 1910 to 1914 the president of the National Geographic Society). Working at the USGS as a topographer under Gannett at the turn of the century was another talented European émigré, François E. Matthes, who in 1904 learned of Davis, began writing to him, and during the school year of 1904-05 studied geomorphology under him. Davis had considerable field experience, but nevertheless derived his principles of geomorphology largely on logic. Matthes then used these logical principles to reconstruct the geomorphic history of Yosemite Valley (USGS Professional Paper 160), his work then being extrapolated for the entire Sierra Nevada and little modified since then—until now.

The final major player in the Yosemite-Sierra Nevada story was John Muir, who like Hoffmann was born in

**Photo 2. The California State Geological Survey, 1863. Josiah Whitney is second from left; William Brewer is third (in center). Yosemite Falls is in background. Source: Bancroft Library.**

1838, was an immigrant, though from Scotland, and eventually reached California and the Sierra Nevada in 1868. Unfortunately, he was outside the academic loop, having attended the backwoods "University" of Wisconsin (a few professors total) from 1860 to 1863, during the Civil War. Although he went for four years, he did not receive a degree, since rather than pursuing a major he took only courses that interested him. In doing so, he received instruction from several professors that profoundly influenced him. One in particular was a former student of

Agassiz, Professor Ezra Carr, who introduced him to glaciology. However, Carr apparently taught Muir that in earlier times an ice cap had originated at the North Pole and then grew equatorward—Agassiz' original view. In 1847 Agassiz had modified his view to one of earlier glaciers first covering northern Europe, followed by glaciers forming in the Alps. Not being tied to the Yale-Harvard connection, Muir missed this vital information. Muir, however, did have a Whitney connection: the two became bitter rivals over Yosemite Valley's origin, and Muir's Sierra Club in the early 1900s entreated Harvard-educated Matthes to vindicate Muir's interpretation, which essentially he did.

For geologists of the Whitney Survey, performing field work in some ways was not much more difficult than doing it today; then again, in some ways it was very taxing and frustrating. Obviously, transportation was slower back in the 1860s, and when the survey had a wagon along, it was restricted in where it could go. In the northern half of the Sierra Nevada, horse trails were abundant while wagon roads were sparse. In the generally gold-free southern half, both trails and roads were few. Much of the exploration of the High Sierra (the higher, glaciated lands of Yosemite, Kings Canyon, and Sequoia national parks and adjacent wildernesses) had to be done on horse back (or preferably, on sure-footed mule back), and in spots on foot. Today, the same is largely true. Food generally was not a problem. They could usually buy it in towns, in rural settlements, and at ranches and farms; in the mountains, they could shoot game birds and mammals and fish for trout. However, the geological team didn't have to procure and cook the food; one or two men were brought along for that purpose.

The Whitney Survey lacked at least three items available to twentieth-century geologists. In order of increasing importance, first was adequate sleeping gear—typically they had two blankets per man, which, like today's, were not waterproof. Brewer's journal indicates that at times he, like John Muir, reveled in sleeping out under the stars, but in rain, snow, or subfreezing temperatures, he suffered through the nights. The second item was money, previously mentioned. This can still be a problem for geologists today, especially after the 1995 downsizing of the USGS. The third was the most important: maps. In the Preface of his *Geology of California,* Whitney stated:

> In short, there was not a map of any portion of the State sufficiently correct in its details to allow of our laying down even the outlines of the geology upon it.

Brewer, in his journal, echoed similar frustration. Actually, members of the Whitney Survey had seen enough of the Sierra Nevada to correctly deduce its general geologic history, as summed up in King's introduction to the range in his 1872 *Mountaineering in the Sierra Nevada.* He identified five main periods: 1) ancient marine sedimentation; 2) metamorphism, plutonism, and uplift; 3) volcanism; 4) glaciation; and 5) modern climates and processes. Absolute dating did not exist back then, but his relative dating (based on fossils) was essentially correct.

Unfortunately, Whitney's *Geology of California* contained, unsurprisingly, a number of errors. The one most cited—which I personally view as unimportant—is his assertion that the gaping canyon of Yosemite Valley, unlike all other Sierran canyons, was fault-formed. When you are doing a pioneering reconnaissance of a vast state, you will see only a tiny fraction of specific evidence, from which you must make generalities. Only after you map the sedimentary, volcanic, granitic (plutonic), and metamorphic rocks in some detail can you reflect on the geologic history of the range and hope to have a high degree of confidence in your conclusions. Beginning in 1894 the USGS produced *Folios of the Geologic Atlas of the United States,* completing 212 by 1920 and then completing only 15 more up through 1945. Folios of the gold-producing lands of the Sierra Nevada—that is, most of the northern half of the range—were completed by 1900, allowing geologists to ask serious questions about the range's uplift and erosional and glacial histories. Each folio had a topographic map and at least two geologic maps (e.g., historic geology, economic geology), all at a scale of 1:125,000 (about ½ inch equals 1 mile). Furthermore, each was accompanied by an illustrated text. For most of the southern half of the range, only topographic maps were produced. The folios were replaced by *Geologic Quadrangles,* which are geologic maps, some with abbreviated text, produced at a scale of 1:62,500 (about 1 inch equals 1 mile). The first Sierran map at this scale was produced in 1956 by, surprisingly, the California Division of Mines, even though the geology was mapped by the USGS. It was for the Bishop Tungsten District, an area of strategic importance in the east-central Sierra Nevada. Elsewhere, there was no economic incentive, and geologic quadrangles were slow in coming, the first Sierran one being produced in 1965. Some 30 years later, the range still has not been completely mapped at this scale.

To return to Whitney's *Geology of California,* I find two errors that to me are extremely significant. First, today the central Sierra's Stanislaus River lies in a deep canyon well below remnants of a series of giant lava flows (the Table Mountain of Tuolumne County, Whitney, p. 243+; see my Photo 3). Whitney said that the flows buried a landscape of low hills, and then the Stanislaus River cut through parts of the flows to the underlying bedrock, and then cut through it, creating today's deep, bedrock canyon. Second, Whitney said that after the last glacier left Yosemite Valley, a giant, deep lake developed in a bedrock basin excavated by that glacier (p.

**Photo 3. Aerial view of Table Mountain of Tuolumne County, near Columbia.**

423). This lake basin eventually filled in with sediments, resulting in the valley's flat floor. Before my research, both of these highly logical interpretations were universally accepted.

The first is important because for major post-volcanic canyon cutting to have occurred, the Stanislaus River must have had its gradient greatly increased. That would have been possible only through major post-volcanic Sierran uplift or, as it is called today, *late Cenozoic Sierran uplift* (uplift from the Miocene epoch onward—the last 23 million years). Clarence King did not mention this uplift, either through omission or because he did not think it had occurred. But this is a very important issue: has major uplift (and associated canyon cutting) occurred during the late Cretaceous (before 65 million years ago), during the late Cenozoic (about the last 23 million years), or during both times? To accurately reconstruct the evolution of the Sierran landscape, you absolutely must know when uplifts occurred.

Whitney's second interpretation at first seems trivial. Who cares if a giant, deep lake developed after the last glacier left Yosemite Valley? The Sierra has a number of large post-glacial lakes, the largest being the Lake Tahoe Basin's Fallen Leaf Lake. But this is precisely my point—that lake still exists! At the rate it is receiving sediments, this lake will take about 200,000 years to be completely filled. Why then did Lake Yosemite, twice as large, fill in completely in a mere 10,000 years? During the last 10,000 years, why haven't thousands of Sierran lakelets and ponds become filled in if giant Lake Yosemite had? It takes only a pittance of sediments, relatively speaking, to fill them in. To answer this dilemma, in 1977 I proposed an explantion similar to one proposed by Stanford's Prof. Eliot Blackwelder in 1931: previous Lake Yosemites existed after some glaciers had retreated, but one did not exist after the last glacier had retreated. Therefore, the moraine that dammed the latest lake must have been left by the previous glacier (the Tahoe-age one). Then, during the most recent glaciation, the Tioga, glaciers from Tenaya Canyon and Little Yosemite Valley dropped prodigious amounts of sediments at the east end of Yosemite Valley, filling the lake. However, by the 1980s it had become obvious to geoscientists who had inspected the moraine that dammed the lake that it clearly was left by a Tioga glacier. I too, on viewing the moraine, had concluded this.

So I was faced with a paradox. On the one hand, the Tioga-age moraine dictated that latest Lake Yosemite must have come into existence as the Tioga glacier retreated from the valley. On the other hand, measured rates of the Merced River's sediment transport dictated that had the lake formed then, it still would exist almost unchanged today, since after some 14,000 years of post-Tioga sedimentation, the lake would have shallowed by only a few feet. I strongly believe that paradoxes do not occur in nature. If we perceive one, it is because we do not understand the problem. Either we are ignoring crucial evidence or we are unaware of it. One way out of this particular paradox was to assume that the moraines of the Merced River drainage had been either incorrectly dated and/or mismapped. But if so, then reconstructions of the extents and thicknesses of past glaciers could be in error. Any such error, in turn, means that our estimates of the ability of glaciers to erode resistant bedrock also could be in error. Until now, extents of former glaciers had been drawn to fit perceived "glaciated" topography; for example, you placed the lowest extent of glaciers where canyon cross profiles changed from a "glaciated" U-shape to an "unglaciated" V-shape. But if reconstructions showed that glaciers also occupied "unglaciated" topography, then the old paradigm that glaciers are tremendously effective erosional agents is in trouble, since this implies that glaciers did not substantially modify the topography. And that is exactly what I propose (and what abundant, unequivocal evidence dictates). And this has implications for every glaciated range in the world, since their deep canyons generally have been attributed in no small measure to the excavating power of glaciers.

# 3

# An Overview of the Sierra Nevada

For the Sierra Nevada we are faced with two basic problems: derive both an uplift history and a glacial history. And these two histories are what this book sets out to do. Once you can derive those histories, you then can realistically reconstruct the geomorphic evolution of the Sierra Nevada and, in particular, the geomorphic evolution of Yosemite Valley. For François Matthes, the latter quest was his Holy Grail, which all have assumed he had achieved. As this book documents, he fell far short of his goal, faring little better than his predecessors. He fared worse than James Mason Hutchings, a non-geologist who better than anyone understood the valley's formation.

To understand these two histories, you must know something about the nature of the Sierran landscape. Therefore, this chapter presents a brief overview. I list the river drainages of the range from north to south, starting with the Feather River and ending with the Kern (Map 4). For each drainage I mention specific issues that are relevant to the interpretation of the Sierra's uplift and glacial histories. I've excluded description of the east-side drainages because they are relatively unimportant with regard to these histories. One specific issue applies to the entire range. Its presumed late Cenozoic uplift has been characterized as a simple tilting of a very large block of the earth's crust. The westernmost part of this block lies buried beneath thousands of feet of sediments in the western part of the Great Valley (the Sacramento and San Joaquin valleys), while the easternmost part runs from the Sierran crest down its steep eastern slopes. The north half of the range, from about latitude 38° northward (and from about the Stanislaus River drainage northward) was buried under thick, volcanic sediments, this process beginning 30+ million years ago. When presumed late Cenozoic uplift occurred, new rivers should have developed across these sediments, and they should have flowed quite straight down slope along courses perpendicular to the crest of the range. They didn't. The North Fork Feather River has a large meander near its end, then flows parallel to the base of the range, not perpendicular to it. The North Yuba and North Fork American rivers likewise each have a reach that flows almost parallel to the base. The Tuolumne and Merced rivers both have large deviations in their lower parts, and the San Joaquin River in its lower canyon meanders excessively. Furthermore, the upper parts of some drainages are skewed to the south. These drainages are the Middle Fork Feather River, the South Fork American River, the Tuolumne River, and the San Joaquin River. Lastly, there is the matter of the Kern River, which drains not west, but south. Any proposed uplift history must address all these anomalous drainage patterns.

The **Feather River** drainage is unique, because its two major branches, the North and Middle forks, originate not at the Sierran crest, but rather well east of it. Each fork flows through a deep canyon that traverses the entire range. The North Fork's major east branch originates close to the California-Nevada border, while the North Fork proper originates on the slopes of Lassen Peak. For quite some time, that volcano and the much larger one north of it, Mt. Shasta—both part of the Cascade Range—were considered part of the Sierra Nevada. This is not surprising because from the Stanislaus River northward a thick cover of volcanic rocks, flows, and sediments blankets the underlying bedrock of granitic and metamorphic rocks, and there is no obvious boundary where one can see that the Sierra Nevada ends and the Cascade Range begins. However, more recent studies have shown that underlying granitic and metamorphic rocks do not continue beneath Lassen and Shasta. Rather, they continue in the Klamath Mountains, which were faulted westward tens of millions of years ago. Therefore, the northern boundary of the Sierra Nevada is the northernmost extent of granitic and metamorphic rocks. Lake Almanor and the North Fork Feather River drainage north from it lie within the exclusively volcanic Cascade Range.

A significant feature of the North Fork Feather River is the big bend it makes—appropriately, around Big Bend

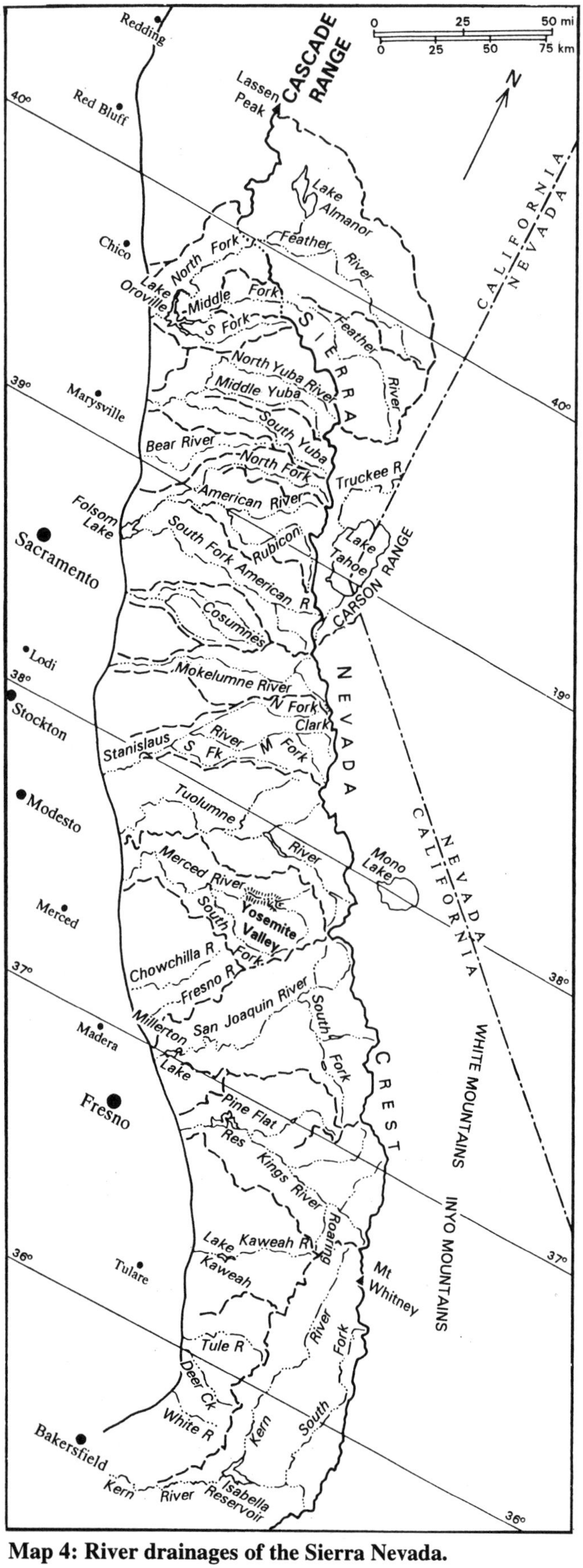


**Map 4: River drainages of the Sierra Nevada.**

Mountain—just before it reaches Lake Oroville in the Sierran foothills. These lands were buried under voluminous volcanic flows and deposits. As was just mentioned, in theory a new Feather River drainage should have developed in them, and the major river forks should have descended quite straight directly to the Sacramento Valley. Obviously, for the North Fork this did not occur. Any uplift history of the Sierra Nevada must address the North Fork's big bend, as well as the giant meanders in the lower part of the San Joaquin River drainage.

Whereas the North Fork in the vicinity of the Sierran crest is at an elevation of about 2300 feet elevation, the Middle Fork is at about 3600 feet. The former, with a lower gradient and flowing across a generally wider canyon bottom, was a logical emigrant route, one later adopted by the Western Pacific Railroad and State Highway 70. The scenic Middle Fork is decidedly wilder, as anyone attempting to traverse along it soon finds out. Within the Sierra Nevada there are only several roads that descend to the Middle Fork, and none that run alongside it. Therefore it was appropriate back in 1986 that most of the Middle Fork was declared a National Wild and Scenic River, starting from east of the actual range and running southwest across it down to Lake Oroville. Near that lower end is an important feature, a junction with a major tributary, Fall River, named for its 375-foot Feather Falls, located not far above the junction (Photo 4). There is a widespread misconception that major waterfalls occur only in glaciated areas, presumably because former glaciers cut deep canyons, leaving tributary canyons hanging hundreds, if not thousands, of feet above the floor of the trunk canyon. Glaciated Yosemite Valley has the country's highest, yet their origin has nothing to do with glaciation (although Bridalveil Fall was proclaimed by François Matthes, for much-needed theoretical considerations, as produced *solely* through glaciation). The world's highest waterfall is 3212-foot Angel Falls, located in the definitely unglaciated Venezuelan rain forest. As Sierran paleontologists have known since the 1870s, the vegetation of the ancient Sierra Nevada also was a rain forest. As we'll see much later, Sierran waterfalls owe their origins to tropical weathering, not to glaciation.

The upper part of the Fall River parallels a nearby upper part of the South Fork Feather River, and along the rims of their shallow river canyons are remnants of a once continuous and broad series of lava flows known as the Lovejoy basalts. These originated from fissures east of the Sierra Nevada and flowed southwest down these two drainages before curving west to cross what is today's Middle Fork and North Fork river canyons and then dying out along the east edge of the Sacramento Valley near Oroville. Like the flows comprising Table Mountain of Tuolumne County, the Lovejoy basalts have a similar history, as do the deep canyons lying beneath the flows.

Within the entire Feather River drainage, glaciation was minor. In the North Fork drainage a local ice cap developed on highlands west of the river, and small, local glaciers developed in high canyons east of it. No glacier, however, reached the river. In the Middle Fork drainage a local ice cap developed in the Eureka Peak-Lakes Basin-Gold Lake-Sierra Buttes area, and north-flowing glaciers barely reached the Middle Fork. A smattering of small glaciers also occurred in the drainage.

Like the Feather River, the **Yuba River** has three major branches. However, in deference to standardization, they were not named the North, Middle, and South forks, but rather the North, Middle, and South Yuba rivers. Historical quirks aside, this river drainage contains the greatest accumulation of prevolcanic river sediments, the so-called Tertiary auriferous (gold-bearing) gravels. On the rims above the North, Middle, and South Yuba river canyons there are thick volcanic deposits, and as elsewhere, one might logically conclude that each river canyon had been carved after their deposition. But in the South Yuba River drainage there is strong evidence against this interpretation. Locally the lowest auriferous gravels occur in tributary canyons not far above the floor of the South Yuba River canyon. Therefore these lowest gravels were deposited in a canyon about as deep as today's. The thick volcanic deposits, rather than being deposited in shallow river canyons carved in bedrock, instead were deposited atop the auriferous gravels. Waldemar Lindgren noted this back in 1911, but his exhortations (against rampant misinterpretations) went unheeded (both then and now).

In the North Yuba River drainage, glaciation was minor. The Sierra Buttes part of the Eureka Peak-Lakes Basin-Gold Lake-Sierra Buttes ice cap produced glaciers that barely reached the river. In contrast, the upper part of the Middle Yuba River drainage lay under an ice cap. However, glaciation was much greater in the South Yuba River drainage, which is the northernmost Sierran drainage to have experienced widespread glaciation. Most of the landscape from Donner Pass west down to Emigrant Gap lay under thick ice. This drainage contains a novel feature. From Donner Pass the South Yuba River descends west almost to Bear Valley, situated below Emigrant Gap. Logically, it should continue westward, as Interstate 80 descending from Donner Pass does. Rather, the southwest flowing **Bear River** originates in this vicinity. Just above Bear Valley the South Yuba River abruptly turns northwest and flows through a deep canyon cut across a ridge before turning westward. It has been suggested that a former Bear River originated at Donner Pass and descended to Bear Valley and thence down to the Sacramento Valley, as the modern Bear River does today. Then during a glaciation a lobe of a major glacier in this drainage spilled northwest across a gap in the ridge and into the South Yuba River drainage. This lobe is believed to have caused deep erosion, creating the deep canyon across the ridge. When the glacier retreated, the upper part of the Bear River took this course, and now this constitutes an upper part of the South Yuba River. This stream piracy by the South Yuba River is a logical, but not proven, assumption; a river can naturally cut across a ridge. For example, early on the river could have made a jog along layer of weak, overlying rock, and then over time the river could have continued to cut down through the resistant rock we see today. And piracy by glaciation is a novel explanation. Where major lobes are known to have overflowed through gaps in drainage divides, none has caused stream piracy. This is an important point to consider when one discusses the ability of glaciers to erode through bedrock.

**Photo 4. Aerial view of Feather Falls. It is located on the Fall River, about ½ mile above its junction with the Middle Fork Feather River.**

Another relevant point to consider is that although major glaciation occurred above the proposed stream-piracy site, major glacial erosion of those lands did not occur. A broad, low granitic bench that holds Kidd Lake and the Cascade Lakes (Photo 5), is mantled with a layer of rhyolite that is about 30 million years old. The lowest part of this rhyolite, just south and above Interstate 80's Kingvale exit, stands less than 200 feet above the floor of the South Yuba River valley, indicating less than 200 feet of erosion has occurred over the last 30 million years. Therefore, the net stream and glacial erosion of the valley floor, rather than being on the order of several hundred feet, more likely was on the order of several tens of feet. As I document later, dated volcanic flows in other drainages show that similarly minuscule amounts of erosion occurred over millions of years.

A final point to be made about the Yuba River drainage is about glacial studies. The glaciated lands of the South Yuba River drainage, through which Interstate 80 passes, comprise the most readily accessible area of major glaciation in the Sierra Nevada. Yet, a comprehensive glacial survey here has never been undertaken. Why not?

**Photo 5. Aerial view east up the shallow, glaciated South Yuba River valley. The Loch Leven Lakes are in the foreground, and Interstate 80 and the Southern Pacific Railway wind east toward the broad Donner Pass area. The conspicuous lake in the mid-ground is upper Cascade Lake, and left of it is Kidd Lake, close to the railroad tracks.**

Well, compared to the glaciated lands of Yosemite, Kings Canyon, and Sequoia national parks, the South Yuba's lands are lackluster: no alpine peaks, no steep-walled gorges, no spectacular waterfalls. Who wants to do mundane field work? The same applies to other lackluster drainages. One is the Stanislaus. In 1969 I had planned to do a Ph.D. dissertation on its glacial history. Years later, in 1990 and '91, I finally did the field work. No one else had touched it. Actually, I didn't want to do it, but because my proposed dissertation on reinterpreting Yosemite Valley was deemed unacceptable (*and,* anyway, impossible), I was asked to do the glacial history of the Stanislaus. This lackluster landscape turned out to be far more exciting than I would have imagined. Furthermore, it provided the evidence to solve the "impossible" Yosemite Valley problem. I suspect the South Yuba harbors similar exciting revelations.

True to form, the **American River** also has three main tributaries, the North Fork, Middle Fork, and South Fork. The North Fork drainage is one of the narrowest. It is also one of the deepest: Snow Mountain stands about 4000 feet above the floor of the river canyon's Royal Gorge. This rugged gorge rivals any in the central Sierra, and despite being only a few miles south of the Sierra's busiest highway, Interstate 80, it is seldom visited, for no trail descends into it.

In contrast, the larger drainage of the Middle Fork has a significantly shallower river canyon. A feature of interest is its major south tributary, the Rubicon River, whose upper half flows parallel to the adjacent Sierran crest. If you explore this upper drainage, the reason for its course becomes perfectly clear: the granitic rock has been fractured parallel to the crest, and the upper Rubicon River flows along this fracture zone. So, not only is uplift important in dictating a river's course, the structure is also important, as is found here. Finally, a river's course can be inherited, like the jog in the South Fork Yuba River may have been. For example, a river can initially develop a course directly downslope in response to uplift, which then is quickly modified by structural control of exposed bedrock. Presented with a choice of, say, between strong or weak beds of metamorphic rock (or between unfractured or fractured granitic rock), the river is more likely to erode down along the weaker rock. As the river incises through this rock, it becomes trapped in the canyon it has created, even if it reaches resistant bedrock. The North Fork Feather River's big bend is an example of inheritance. It cuts a gorge through various metamorphic rocks regardless of their structure or bedding.

The South Fork American River has two interesting aspects: its two principal branches. The main branch continues quite straight up-canyon to U.S. Highway 50's Echo Summit, where it *appears* to have been decapitated. (As with most Sierran pronouncements, this is an extremely logical assumption—but not proved.) In earlier times its headwaters may have been miles to the east, but with the downfaulting that formed the Lake Tahoe Basin, these highlands sank. But if indeed they sank, when? The unanimous answer has been: quite recently, within the last few million years, for a major, *presumably* youthful fault seems to run along the west edge of most—but, significantly, not all—of the basin (Photo 6). Verifiable rates of weathering of the granitic slopes rimming this basin dictate that it has existed little changed for tens of millions of years. Any geomorphic history of the range must include a history of the Lake Tahoe Basin that is supported by all of the field evidence.

**Photo 6. Steep slopes descend to the west shore of Lake Tahoe. A youthful fault may run along the bottom of the lake parallel to this shore. View north is from the 9735-foot summit of Mt. Tallac.**

The second principal branch is the Silver Fork American River, which up-canyon cuts southeast through a broad ridge of granitic bedrock. Its headwaters, which originate in the area of State Highway 88's Carson Pass, rightly should belong to the **Cosumnes River**, but seem to have been diverted into the Silver Fork. Is this stream piracy? More specifically, since glaciers have flowed down the Silver Fork, is this glacially induced stream piracy? As with the Rubicon River, the Silver Fork is merely flowing along a linear fracture system in the granitic bedrock. And as in the South Yuba River drainage, this one has old volcanic rocks on its slopes, which dictate that the Silver Fork canyon has changed little over the last 20+ million years. Lindgren, mentioned above, was aware of the depth and antiquity of the South Fork American River drainage, but his pronouncement did not accord with the paradigm of late Cenozoic Sierran uplift.

South of the Cosumnes River lies the **Mokelumne River**, which has but one major branch, the North Fork. There are many other branches, five of which are one third to one half its length. The two longest were designated the Middle Fork and the South Fork, while two others became Blue Creek and Forest Creek, and the last, which is the only one north of the North Fork, became, unfortunately, the Bear River—not to be confused with the real Bear River situated between the South Yuba and Cosumnes rivers.

Like the Middle Fork Feather River the North Fork Mokelumne River, set in a deep, rugged gorge, is wild and scenic but unfortunately is not currently so designated by law. It pristine character is marred by the Salt Springs Reservoir and by several miles of canyon-bottom roads leading east to its dam. Just upstream from the reservoir the canyon is at its deepest, and Mokelumne Peak, on its north side, rises some 4700 feet above the floor, much as Pyramid Peak, on the north side of the upper South Fork American River canyon, rises some 4100 feet above its floor. Both granitic summits stood above the voluminous volcanic sediments that buried the canyons, and both stood above the thickest glaciers. Both, I maintain, are extremely old mountains, having existed even in the last days of the dinosaurs some 65 million years ago. Should you climb either summit, you might imagine yourself standing up there with a friendly dinosaur, taking in the view. The principal difference would be the tropical vegetation.

For me, the most interesting aspects of the Mokelumne River drainage are two areas of lateral moraines. The first is located west of and below Mokelumne Peak, extending west to the Bear River (tributary). These moraines look like classic examples of Tahoe- and Tioga-age moraines, but on scrutiny, something is very awry. The second area is located south across the canyon, and it contains merely three small lateral moraines, each located only along the northeast side of each mouth of three severely hanging tributary canyons (Photo 7). Something is suspiciously amiss. Why aren't lateral moraines also along each mouth's southwest side? Any glacial history of the Sierra Nevada must address these perplexing lateral moraines.

**Photo 7. Brushy, descending lateral moraine at the mouth of Horse Valley. The mouth stands 1800 feet above the hidden floor of the North Fork Mokelumne River canyon. Lofty 9332-foot Mokelumne Peak, little more than a high point along the canyon's north wall, rises in the background.**

Perhaps my favorite river drainage is that of the **Stanislaus River**, up forks of which ascend State Highway 4 to Pacific Grade Summit and State Highway 108 to Sonora Pass. The Stanislaus has the compulsory three branches, the North, Middle, and South forks, the last much smaller than the first two. I like this area because I am better acquainted with the glacial evidence in this drainage than in any other, having spent a full 100 days in the field during 1990 and '91. This was on top of field work done for two books on the area, work done mostly in 1974 and 1986. Most importantly, the area contains the evidence needed to overturn both the accepted uplift and glacial histories of the Sierra Nevada. For the first, one need look no farther than the Dardanelles, a high volcanic ridge between the North and Middle forks. It is part of a 90-mile-long sequence of lava flows, of which Table Mountain of Tuolumne County is a lower part. The indisputable evidence of the thick, older deposits below the Dardanelles flows convincingly argues against late Cenozoic uplift. Ironically, in 1898 Leslie Ransome, mentioned in Chapter 1, used the Dardanelles flows to measure late Cenozoic uplift, despite having correctly mapped and described field evidence to the contrary. Down by Table Mountain, you might deduce late Cenozoic uplift, but only if you embrace some evidence while disregarding conflicting evidence. That was what the Whitney Survey did, and the uplift paradigm stuck. Perhaps with that fresh in his mind, Ransome saw what he believed, as did all others following him, including me. Perhaps like others, I believed in what had been written, and so did not bother to consult available geologic maps, whose volcanic units demanded a contrary interpretation. Furthermore, when you view the Dardanelles, the dramatic, nearly vertical lava cliffs make a strong impression on your mind, while the thick underlying sediments, veneered with talus and masked by an abundance of shrubs and trees, do not (Photo 8). In short: out of sight, out of mind.

The glacial evidence, like the drainage, at first looks lackluster. Superficially, nothing leads you to believe that the long-used method of relative dating will not work for the area's moraines. Consequently you can identify the Tioga- and Tahoe-age moraines locally about, say, Lake Moran, resting on a bench above the North Fork, and you can feel comfortable with your work. Only when you map the glacial evidence of the entire North Fork drainage (and parts of adjacent drainages) do you realize that what looked perfectly obvious must be in error. A cryptic clue occurs in the backwoods vicinity of Little Prather Meadow, three miles east of Lake Moran. There the distribution of erratics (boulders transported by glaciers and then left behind when they melted) defies explanation under conventional interpretations. To solve the riddle of their distribution, one must simply think about how a glacier flows, which has been known since the 1850s. With state-of-the-art 1850s glacial mechanics in mind, you are on your way to overturning the paradigm of Sierran glaciation, which includes both the relative dating of glacial deposits and the amount of erosion produced by glaciers.

To be able to distinguish between Tioga and Tahoe moraines, you have to view some of the boulders buried within them. Excavation is taxing and expensive, and moreover is illegal in most of the Sierra's glaciated lands, so generally you have to rely on natural excavation. Locally, the subsurface of a moraine may be exposed through minor gullying or other form of mass wasting, through erosion by a creek crossing it or flowing along its base, or through a root hole caused by a fallen, uprooted tree or by a severely burned one whose roots were consumed. Of course, the best opportunity is road cuts, and I feel the Sierra's best of these are through a pair of Tioga-Tahoe lateral moraines cut by the Dodge Ridge Road, in the Pinecrest Lake area.

For measuring the effectiveness of glaciers at eroding through granitic bedrock, patches of rhyolite, about 25 million years old, provide the answer. They show that the glaciated granitic landscape existing today is very similar to that of millions of years ago. Over 25 million years the net stream and glacial erosion through granite by the North Fork Stanislaus River was a *maximum* of about 30 feet, or about one foot per million years. Change has been so minor in the *granitic* lands that 25 million years ago you could have easily used today's topographic maps. The North Fork also has other evidence against the erosive power of glaciers. Like Turner and Ransome back in 1898, I placed the snout of the largest glacier down this canyon at 3600 feet. Interestingly, not far beyond this elevation, the river greatly increases its gradient for a rollicking descent to the Middle Fork. If glaciers—

**Photo 8. The Dardanelles, from Elephant Rock's summit. This ridge is a sequence of lava flows atop thick volcanic sediments.**

especially the ones "roaring" down major canyons—are so erosive, why does the floor of the glaciated part of the canyon stand so high above the floor of the unglaciated part?

South of the Stanislaus lies the **Tuolumne River**, whose branch names defy logic. The main branch is merely the Tuolumne River, and it has three relatively short forks, listed in decreasing length as the North, Middle, and South. Actually, the Middle is simply the Middle Tuolumne River, *sans* "Fork." Furthermore, this "river" is but a tributary of the shorter South Fork. Then there are longer forks. The Clavey *River* is longer than all three, and Cherry *Creek* is longer still. I suppose the North, Middle, and South forks were named first, because gold was discovered in each drainage (State Highway 120 skirts past most of the gold-bearing deposits), while none was found in the Clavey or Cherry drainage.

Four significant features of the Tuolumne River proper are four of its reaches: the Tuolumne Meadows, the cascading river dropping from them, the narrow Grand Canyon of the Tuolumne River, and Hetch Hetchy Valley (a now flooded "Yosemite" valley). In the Tuolumne River drainage, even more so than in the North Fork Stanislaus River drainage, the Tuolumne Meadows hang far too high. Lying in the prime erosional section of one of the Sierra's greatest glacial systems (the other being the San Joaquin's), the broad, flat-floored, spacious meadows have defied glacial downcutting. Much lower down, the river's Grand Canyon, repeatedly invaded by 4000-foot-thick glaciers, has resisted glacial widening. Knowledge of this anomalous glaciated topography has been known for about a century. A question arises: Why was the Grand Canyon barely modified by glaciers, while Yosemite Valley, in the next drainage south, was presumably greatly deepened and widened? The given answer has been in the nature of the Grand Canyon's bedrock—it is extremely resistant. But Yosemite Valley's bedrock is every bit as resistant. So this leads to a riddle: Why do glaciers barely erode resistant granitic bedrock in one canyon while they erode tremendous amounts of it in another?

Another feature of the Tuolumne River is Rancheria Mountain, which rises above the north wall of the river's Grand Canyon (Photo 9). It is the only significant mass of volcanic sediments within Yosemite National Park. A local outcrop of gravels at the base has been used as proof that about 10 million years ago the Tuolumne River left the gravels there, and since then the river has cut more than 3000 feet down to its present level. This raises another question: Why was the river so effective at downcutting, while monstrous glaciers were not? One can suggest that in the Grand Canyon (but not at Tuolumne Meadows) glaciers performed a lot of downcutting, but the authority on this matter, N. King Huber of the USGS, ruled out this interpretation in his 1990 paper. In it, he

**Photo 9. Aerial view of the Tuolumne River canyon's Hetch Hetchy Reservoir. Above it lies forested Rancheria Mountain. The Grand Canyon is hidden behind Smith Mountain's long west spur, on the right. Laurel Lake is out of sight, just left of the edge of the photo.**

concluded that the river's course just before glaciation was very close to its present course. This means that it was stepped: a low gradient across Tuolumne Meadows, a very high gradient down to the floor of the Grand Canyon, then a moderate gradient through it. However, the three primary methods of determining Sierran uplift, including Ransome's method expounded by Huber, *require* that before glaciation, each Sierran river necessarily had a nearly constant gradient along almost its entire length (except near the crest). You can't have it both ways. Either a preglacial river had either a smooth profile or a stepped one.

The final important features of this drainage are its moraines. About 3000 feet above Hetch Hetchy Reservoir's dam, from the end of Smith Mountain's long west spur, lateral moraines drop steeply to the hanging valley of Cottonwood Creek, run a level course beside the creek, then drop steeply to the Camp Mather bench, only to level off again. Presumably the glacier that left them had a matching ice surface. This stepped pattern for both moraines and ice surface would be logical only if the glacier were flowing down a matching, stepped canyon floor. But the Tuolumne River's gradient is quite regular and low and definitely is not stepped. Consequently one must explain why the enormous glacier, about 3600 feet thick at Smith Mountain's west spur thinned over a 2-mile distance by about 1000 feet, and then, 3 miles farther—after maintaining its thickness—thinned again by about 1000 feet, only to again maintain thickness for 2 miles down-canyon.

Finally, there are the moraines leading up the other side of the Tuolumne River's canyon to Laurel Lake. They all appear to be logically dated, by conventional relative dating, all the way up to the lake. Up there the moraines appear to be pre-Tahoe in age, that is, deposited at least 200,000 years ago, and more likely, about 800,000 years ago. The lake, therefore, should be at least 200,000 years old, but a date on some of its oldest sediments indicates the lake is vastly younger, having formed—like *all* other dated Sierran lakes—after the Tioga glacier retreated, which here would have been perhaps about 14,000 years ago. Something is definitely wrong with how we have been dating glacial deposits.

The **Merced River** drainage contains Yosemite Valley, the preeminent feature of the Sierra Nevada, the star player in the controversy of the range's uplift and glacial histories. More has been said about the origin of the valley than of any other Sierran landmark, and yet I cannot recall any landmark whose interpretation is so destitute of field evidence. For example, there is no field evidence for uplift here, although its uplift history has been constructed in detail—on entirely theoretical grounds. And field evidence for the lengths and thicknesses of the Tioga and Tahoe glaciers passing through the valley are limited to only two sites. François Matthes mapped in detail abundant glacial evidence, but much of what he perceived simply does not exist.

His work has been applied to glaciated ranges around world. How do we know glaciers severely erode their canyons? Evidence from Yosemite Valley proves it. As I have stated, I assert that glaciers are virtually powerless to erode resistant bedrock. Still, one can make a seemingly ironclad case for major glacial erosion based on evidence in the valley: its sediments, which unquestionably are up to 2000 feet thick. I say "seemingly" because the case is not as ironclad as everyone would believe. Tied to this case are two implicit assumptions. First, in other Sierran canyons glaciers of similar size were very erosive. And second, Yosemite Valley is a geologically young feature, largely formed in the last few million years, and under relatively modern climates. Both assumptions, as analyzed later, are demonstrably false by field evidence, most of it known since the 1890s.

In piecing together the geomorphic history of the valley, one has to explain several anomalous features. First, Tenaya Canyon is too deep for its seasonal creek (Photo 10). This was noted long ago by John Muir. In contrast, Little Yosemite Valley, located along the Merced River canyon just above Yosemite Valley, is too shallow. Also in that upper valley, the moraines around it defy explanation when interpreted by conventional relative-dating methods. Likewise, in the opposite direction, State Highway 140's Merced Gorge, below Yosemite Valley, is known to have been occupied by one or more large

**Photo 10. Tenaya Canyon, viewed from Glacier Point.**

glaciers, yet it completely lacks evidence of glacial erosion. Under the conventional interpretation, past glaciers appear to have been enormously effective in transforming Tenaya Canyon and Yosemite Valley, which have the most massive, resistant cliffs, and absolutely ineffective in transforming Little Yosemite Valley and the Merced Gorge, which have the least. This is not logical; the reverse should be true.

The Merced River drainage is one of the Sierra's smaller ones, and has only one major tributary, other than Tenaya Creek, and that is its South Fork. Along it, at a similar elevation of Yosemite Valley, lies what François Matthes termed the "Half-Yosemite at Wawona." It has steep slopes and a series of waterfalls on one side, reminiscent of Yosemite Valley, and on the other side of this broad-floored basin, it has subdued hills. Matthes explained its origin—correctly, I feel—solely in terms of the field evidence. That is, the variations of landforms are related to locally varying resistance of the granitic bedrock. In other words, massive bedrock results in cliffs, fractured bedrock results in subdued hills. The same holds true for Yosemite Valley, but that valley he interpreted far more with theory than with field evidence. Where field evidence conflicted with theory, he tossed the evidence.

Reading Matthes' 1930 Professional Paper 160 on Yosemite, anyone will be very impressed with the evidence presented. Matthes, however, knew that both the uplift evidence and the glacial evidence were weak. As I said, Yosemite Valley (and for that matter the entire Merced River drainage) contains no unambiguous evidence for uplift. Matthes needed a proxy, since much of the unglaciated Merced River had tributaries incised into metamorphic bedrock, and most of their profiles ran counter to his uplift hypothesis. So he turned to the next drainage south, the **San Joaquin River**, which has most of its trunk canyon and most of its tributaries incised in granitic bedrock. Here he found the tributary profiles he desired, and concluded, in a brief 1924 report, that "the analysis made in the Yosemite region accordingly stands fully confirmed." That statement was a bit premature, for the tributary canyon of Fish Creek, close to the Sierran crest, is more or less parallel to it. Therefore, the gradient of its creek shouldn't have been steepened with uplift. Consequently its junction with the Middle Fork San Joaquin should have been considerably hanging due to uplift, and even more so after glaciation. But it does not hang. Also, his methodology required that the Middle Fork originate well east of today's crest, since a large river was required to carve the deep canyons existing just west of the crest. Later, N. King Huber used Ransome's uplift methodology, and this too necessitated a large river. Interestingly, he located the river's former crest crossing a few miles north of Matthes' crest crossing. Which was the real site? Or, was there even a site at all? Perhaps the Middle Fork never originated east of the crest. For purposes of uplift and landscape evolution, we need to know.

The San Joaquin River drainage contains about five dozen remnants of lava flows, most about 3-4 million years old. They are scattered about the drainage, locally blanketing granitic bedrock, be it severely glaciated, weakly glaciated, or unglaciated. They show you what the landscape looked like at time of burial, and how much post-volcanic erosion of adjacent granitic bedrock has occurred. Altogether they indicate there are problems with Matthes' conclusion, as well as problems with conclusions in 1965 by Clyde Wahrhaftig and in 1981 by N. King Huber. The most important lava flow is the Kennedy Table flow, dated at about 10 million years, and having a history similar to the Feather's Lovejoy basalts and the Stanislaus' and Tuolumne's Table Mountain. This flow is believed to have advanced down the very broad floor of a part of the lower San Joaquin River canyon, but as I'll later show, that was not the case. The bulk of the lava flow lies on a broad granitic bench, Kennedy Table, but locally underlying the flow remnants at various sites are sediments. These readily erodible sediments are thin (up to 320 feet thick, generally much less), and both Wahrhaftig and Huber, while acknowledging them, disregarded them as insignificant. However, they could have been as much as 1500 feet thick. If they were, then, using Huber's method of uplift determination, no uplift has occurred. According to Huber, most of the uplift has occurred in the last 10 million years, and the San Joaquin River, having had its gradient increased, should have incised a relatively straight gorge. Instead, in the vicinity of Kennedy Table the river executes large meanders, much like the Feather River's Big Bend Mountain meander. Why? One cannot begin to construct an uplift history in this drainage without knowing the original thickness of the sediments and without addressing how the large meanders through resistant bedrock were created. Unfortunately, because of Huber's acclaimed Professional Paper, this drainage has uncritically become *the* "type locality" for quantifying major Sierran uplift, much as Yosemite Valley has become *the* "type locality" for quantifying major glacial erosion.

Rates of evolution of the Sierran landscape are linked to Wahrhaftig's paper on stepped topography. In this area from about Kennedy Table east to Shaver Lake and north past Huntington Lake (Photo 11) to Mammoth Pool, there are many irregular steps—cliffs from less than 500 feet to more than 2000 feet high—which cry for an explanation. These are best developed here, but they also occur to the north in the minor **Chowchilla River** and **Fresno River** drainages and to the south all the way along the range, in both major and minor drainages. These *stepped* western lands of the Sierra's southern half differ from *sloping* western lands of its northern half—an observation that goes back to early in this century. To explain this

**Photo 11. Aerial view northeast toward Huntington Lake. This reservoir lies in a gentle upland valley that is over 2000 feet above the floor of Big Creek canyon, in the foreground, and is bordered by cliffs on the west that rise about 4500 feet from the San Joaquin River.**

phenomenon, Mark Christensen in 1966 proposed that each half had its own uplift history: the northern lands were tilted westward, the southern lands were raised as a block. If there were two histories, then we should see a sheer zone or a fault zone roughly along the course of the San Joaquin River. But we don't. Both the stepped and sloped topography require explanations.

Wahrhaftig in part proposed one, but by ignoring the underlying gravels of unknown original thickness beneath the Kennedy Table flow, and by not reconciling the contradictory evidence in the minor river drainages, he concluded that the massive steps were formed largely during the last 10 million years. This flawed paper nevertheless was hailed as a landmark, and for it Wahrhaftig in 1967 received American geomorphology's highest honor, the Kirk Bryan Award. In determining how long and how fast the granitic Sierran landscape has been evolving, we need to devise a geomorphic history that is in accord with every bit of *field* evidence. I stress field evidence because evidence from other branches of geology can be quite weak and uncertain. For example, using the same plate-tectonic evidence, geoscientists have devised various, contradictory scenarios of Sierran uplift. Using seismology, geoscientists of the Southern Sierra Nevada Continental Dynamics Working Group recently were able to "look" tens of miles down into the Sierra's deep crust. Then they published two papers in 1996. One by Moritz Fliedner and Stanley Ruppert supported major late Cenozoic uplift, while one led by Brian Wernicke not only suggested no uplift, but hypothesized that major late Cenozoic subsidence occurred. Same research team; same data; diametrically opposed results. Personally, I prefer evidence not miles down, but on the surface, where everyone can see it, can take samples from it, and can accurately map it. You can't do that with seismology.

The Middle Fork San Joaquin River is one of the Sierra's wildest. Below Devils Postpile National Monument, which is located near the crest, the Middle Fork lacks a riverside trail all the way to Mammoth Pool Reservoir. Much like Feather River's Middle Fork, this one has only a few trails that descend its steep, narrow gorge to the river. In contrast, the South Fork San Joaquin River's valley is very broad, and so is more amenable to trails and roads. (It has State "Highway" 168, a narrow, barely 2-lane road, meandering through it.) The San Joaquin's drainage is asymmetric, its South Fork far greater in length and width than its North Fork. Both drainages, more or less parallel to the Sierran crest, have inner gorges, although the one on the South Fork is so broad that we must call it a valley rather than a gorge. Conspicuous lateral moraines stand on the rims of these gorges, just as they do in the Tuolumne River drainage in the Hetch Hetchy-Mather area. Why did past glaciers have a thickness just barely enough to drop their debris on these rims? Why shouldn't their ice surfaces have been above or below the rims? Is this sheer coincidence or intelligent design, or are we not understanding something?

Like the San Joaquin, the **Kings River** drainage is asymmetrical. And like the former, its North Fork is in the west, its Middle Fork is in the north, and its South Fork is in the east. The deepest canyons in the Sierra Nevada are within this drainage, the Middle Fork and South Fork canyons both having rim summits over 5000 feet above their canyon floors (Photo 12). The South Fork is the grandest, harboring Kings Canyon, which is another "Yosemite" valley. Unlike Yosemite, however, Kings Canyon, easily reached by Highway 180, is longer and narrower, has less steep cliffs, and is deficient in both waterfalls and visitors (a definite plus). It has the most challenging terrain in which to do glacial field work. Therefore, the canyon, despite being well known, lacks a detailed, published map of its glacial deposits.

**Photo 12. Aerial view east up the Kings River drainage.**

Except in the Roaring River drainage. This river flows north to Kings Canyon, cascading and roaring down to it, whence the name. This south-fork tributary drainage of the South Fork has several major, north trending canyons, each with dramatic lateral moraines. Nowhere else in the Sierra west of its crest is there such an assemblage, and in 1987 these Tioga- and Tahoe-age moraines were mapped by James Moore and Thomas Sisson. Trouble is, for a given canyon its pairs of Tioga and Tahoe moraines often do not match. A Tahoe moraine on one side may be hundreds of feet higher than its mate; or a Tahoe moraine may be paired with a Tioga moraine; or a moraine of one age may be present on one side, but absent on the other. With all its readily discernible moraines, the Roaring River drainage should be the Sierra's most easily mapped glaciated area, yet the mapped moraines defy logic. Obviously we have not been understanding the Sierra's moraines west of its crest. East of it, yes. And this raises another question: Why can we make sense of east-side moraines but not of west-side ones?

The **Kaweah River** drainage is abnormal because it is normal. Its topography is not stepped, but rather is unstepped, having "normal" slopes, as a Davisian landscape should. That is, the landscape exists as Prof. William Morris Davis had said it should. The unglaciated lands have moderate-sloped canyons that are V-shaped in cross profile, while the glaciated lands above them have steep-sided, flat floored canyons that are U-shaped in cross profile. Furthermore, in this drainage there is little doubt about the extent of former glaciers, since the lateral moraines are continuous and, unlike in most west-side drainages, those on one side of a canyon match those on the other side. However, there are a few abnormal features, a well-known one being Moro Rock (Photo 13). In truly Davisian landscapes, which best occur in geology and geography textbooks, there are no abnormalities. The trouble is, most of the granitic lands west of the Sierran crest are abnormal. But they are abnormal only with respect to the standard of what a normal landscape should be, as defined by Davis late in the nineteenth century and still largely accepted by American and European geomorphologists today. (Geomorphologists of tropical lands know otherwise.) By normal, I mean a landscape that looks like the Appalachian Mountains, in the eastern United States, where Davis drew many conclusions on mountain landscapes.

**Photo 13. Aerial view up the Middle Fork Kaweah River. Moro Rock is the conspicuous domelike rock.**

Actually, only the Middle Fork drainage is a Davisian landscape. To the west, the little known North Fork drainage, which is skirted by the Generals Highway, is stepped. Lying between the two is the Marble Fork, which has a hybrid landscape, and so are the East and South forks, lying south of the Middle Fork. The Middle Fork drainage raises two questions for us. First, why is the Middle Fork normal? Second, why are granitic east-side lands normal while most west-side ones are abnormal? Note that throughout the range, non-granitic lands usually produce normal landscapes. It is only granitic bedrock, which comprises the bulk of the Sierra Nevada, that gives rise to abnormal landscapes. And the South Fork, more of a creek than a river, flows through a very abnormal landscape. It originates in a small, gentle upland area about 15 square miles in area that was glaciated. It then descends 4000 feet in 5 miles, to the bottom of an enormous canyon some 4000 feet deep. Two more questions arise. First, except for two partly formed cirques in the headwaters, the glaciated upland area has a very unglaciated look to it. Furthermore, this upland stands more than 4000 feet above the unglaciated river canyon west of it. If glaciers are so erosive, why didn't they transform the landscape and why didn't they cut a deep canyon? Second, how did the minuscule South Fork cut the deep, enormous canyon?

A truly abnormal drainage, by Davisian standards, is that of the **Tule River**. Whereas the Kaweah River drainage, while not reaching the Sierran crest at least originates on the 12,000-foot Great Western Divide, the minor Tule River drainage falls short, originating on a not-so-great anonymous divide that stands mostly at about 8000-9000 feet elevation. The insignificant rivers of its North, Middle, and South forks have carved capacious canyons, 3000-4000 feet deep, rivaling those of any major Sierran river. Although west-side canyons generally have a V-shaped cross profile along most or all of their lower, unglaciated parts, there are exceptions. Perhaps the most striking exception is the never-glaciated North Fork Tule River drainage. The floor of its upper section is about as broad and flat as any west-side glaciated canyon. Below it the next section defies explanation by reference to conventional stream processes: a 3-mile-wide, gently sloping floor on knobby granitic bedrock that is locally buried under a veneer of sediments (Photo 14). This "floodplain" is worthy of the Colorado River, not of a wade-across creek. How did this topography originate? It

**Photo 14. Aerial view northeast up the lower North Fork Tule River drainage.**

cannot be explained in terms of modern climates and processes, as proposed by Wahrhaftig.

South of the Tule River drainage are even smaller ones, the drainages of **Deer Creek** and the **White River**. The creek is larger than the river, but neither is impressive. Both streams seasonally go dry before reaching the western edge of the range. Both have complex landscapes that have many landforms and, as above, these landscapes cannot be explained in terms of modern climates and processes.

South of them lies the last major Sierran drainage, the **Kern River**. Unlike all others, this river flows south, as does its major tributary, the South Fork. The south-flowing Kern River is quite straight, for it has eroded along an ancient fault, excavating over time a deep, narrow canyon. The South Fork is its antithesis. It meanders through several broad, shallow, flat-floored basins, these connected by reaches through increasingly larger gorges. Why the fork flows south at first appears to be a mystery. Major forks of a river usually converge. Instead, the fork flows south parallel to the main branch and then angles west to join it. The answer is that it is an inherited drainage. In the South Fork drainage there still are minor remnants of metamorphic rocks, these being former sediments that became tightly folded through compression. Before they were mostly eroded away, they probably resembled the sequence of parallel metamorphic rocks existing in the upper San Joaquin River drainage. There, the North and Middle forks flow south parallel to the sequence of metamorphic rocks and parallel to the crest.

Until now, the evolution of the South Fork Kern River landscape has defied explanation. Guided by François Matthes, Robert Webb gave it a shot back in 1946, but came up short—Davisian geomorphology failed. What needs to be resolved are how the flat-floored basins formed (Photo 15), why the lower, west-heading reach of the South Fork has created such a broad valley (one followed by Highway 178 to Walker Pass), and why topography with glacial morphology abounds, despite the absence of former glaciers. In the Kern River drainage, as elsewhere, the identification and relative dating of glacial deposits, and the reconstruction of the glaciers that left them, are in great error. Part of the error is due to the method of mapping the evidence. Apparently many glacial deposits have been "mapped" in the office. And part of the error is due to misinterpretation. Even if you have correctly mapped the glacial deposits, as Matthes generally had, the assumptions you make about that evidence will affect your glacial reconstruction.

As this chapter demonstrates, there are many unresolved issues in interpreting Sierran uplift and glacial histories. These concern the entire range, not just Yosemite Valley. All the issues I've raised can be resolved, but not under conventional alpine geomorphology. Especially in North America, mountain landscapes have been interpreted under the wrong set of assumptions. In the Sierra Nevada this has caused each major drainage to have its own unique uplift and glacial histories. If there were any merit to conventional alpine geomorphology, the entire range would have coherent uplift and glacial histories. During the 1990s, when I have been doing research for this book, geoscientists were still addressing uplift and glaciation problems that are essentially the same as those of the 1890s. Despite a century of research, the geologic riddles of the Sierra Nevada have not been solved. How you reach solutions depends on what questions you ask, and in the Sierra Nevada (and in other mountain ranges) geoscientists have asked the wrong ones. Why has it taken so long to decipher the Sierra Nevada? The answer lies with those who investigated it. The paramount issue was the origin of Yosemite Valley, and for that we should review what was said about it—and why they said it.

**Photo 15. Manter Creek basin, in Dome Land Wilderness. How could a step-across creek have carved out a 2-mile-broad basin?**

# Part II

# Previous Views of Yosemite Valley

# 4

# Whitney's Faulty Hypothesis

The first professionals to consider how Yosemite Valley was created were members of the Whitney Survey, who first explored Yosemite Valley from June 16 to 23, 1863. During that year, the party was composed of Josiah D. Whitney, William H. Brewer, Charles F. Hoffman and a hired hand. Their explorations included a climb east out of the valley up to the brink of Nevada Fall where, unlike below, glacial polish is abundant. Brewer makes no mention of Yosemite Valley's glacial evidence in his *Up and Down California in 1860-1864, The Journal of William H. Brewer.* Nor does any appear in his *Field Notes-Geology* or in his *Field Notes and Observations.* The first person credited with noticing the valley's previous glaciation may have been Clarence King, a later Whitney Survey member who visited the valley in October 1864 and who noted glacial evidence. (In 1855 Joshua E. Clayton of Mariposa, reporting to the California Academy of Natural Sciences, apparently had noticed glacial evidence in the upper Merced River drainage above the valley, but may not have noticed it within the valley.) Were glaciers important in shaping Yosemite Valley? Whitney acknowledged their former presence, though in his *Geology of California,* he thought another agent was important:

> Most of the great cañons and valleys of California have resulted from denudation, as has been already repeatedly stated in the course of this volume. The long-continued action of the tremendous torrents of water, rushing with impetuous velocity down the slopes of the Sierra, has excavated those prodigious gorges, by which the chain is furrowed to the depth of thousands of feet. But these eroded cañons, steep as they may be, have not vertical walls; neither have their sides the peculiar angular forms which the mass of El Capitan, for instance, has, where there are two perpendicular surfaces of smooth granite meeting at right-angles, and each over 3000 feet high [his Figure 63, my Figure 3].
>
> When Messrs. Gardner and King's map of the Yosemite is published, and farther investigations made to clear up some doubtful points, the theory of the formation of this valley may be discussed more intelligibly than it can be at present. It may, however, be stated, that it appears to us probable that this mighty chasm has been roughly hewn into its present form by the same kind of forces which have raised the crest of the Sierra and moulded the surface of the mountains into something like their present shape. The domes, and such masses as that of Mt. Broderick [today's Liberty Cap; his Plate III, my Figure 4], we conceive to have been formed by the process of upheaval itself, for we can discover nothing about them which looks like the result of ordinary denudation. The Half Dome seems, beyond a doubt, to have been split asunder in the middle, the lost half having gone down in what may truly be said to have been "the wreck of matter and the crush of worlds." It has been objected to this view, by some of the corps, that the bottom of the valley, in places where an engulfment must, according to this theory, have taken place, seems to be of solid granite, when there should be an unfathomable chasm, filled now, of course, with fragments, and not occupied by a solid bed of rock. To this it may be replied, in the first place, that the masses which have been engulfed may have been of such enormous size as to give the impression, where they are only imperfectly exposed, of perfect continuity and connection

**Figure 3. El Capitan, the Cathedral Rocks, and Bridalveil Fall.**

with the adjacent cliffs. But, again, this grand cataclysm may have taken place at a time when the granitic mass was still in a semi-plastic condition below, although, perhaps, quite consolidated at the surface and for some distance down. In this case it is not impossible, certainly, that the pressure from above may have united the yielding material together, so that all traces of the fracture would be lost, except in that portion of it which affected the upper crust. If the bottom of the Yosemite did "drop out," to use a homely but expressive phrase, it was not all done in one piece, or with one movement; there are evidences in the valley of fractures and cross-fractures at right-angles to these, and the different segments of the mass must have been of quite different sizes, and may have descended to unequal depths.

In the course of the explorations of Messrs. King and Gardner, they obtained ample evidence of the former existence of a glacier in the Yosemite Valley, and the cañons of all the streams entering it are also beautifully polished and grooved by glacial action. It does not appear, however, that the mass of ice ever filled the Yosemite to the upper edge of the cliffs; but Mr. King thinks it must have been at least a thousand feet thick. He also traced out four ridges in the valley which he considers to be, without a doubt, ancient moraines.

Whitney then describes the four moraines, including the westernmost one below Bridalveil Fall, and of it he says:

> It seems not unlikely that this moraine may have acted as a dam to retain the water within the valley, after the glacier had retreated to its upper end, and that it was while thus occupied by a lake that it was filled up with the comminuted materials arising from the grinding of the

> glaciers above, thus giving it its present nearly level surface.
>
> It is evident, from the fresh appearance of large masses of debris along the sides of the valley that these materials are now accumulating with considerable rapidity; and when we consider how small the whole quantity of talus is, as compared with the height and extent of the cliffs, we are forced to the conclusion that the time which has elapsed since the Yosemite was occupied by a glacier cannot have been very long.

For Whitney, the overriding process that shaped Yosemite Valley was down-faulting. Given the angularity of the valley's nearly vertical cliffs, especially El Capitan, which confronts you so forcefully upon entry to the valley, this was a logical conclusion. Whitney asserted (quite correctly, as my Part IV details) that a glacier merely occupied the valley; it did not widen it. When it retreated, it left a large lake that was subsequently replaced with sediments to create the nearly level valley floor that exists today. This "Lake Yosemite" would be recognized by virtually all investigators (other than Eliot Blackwelder in 1931 and me in 1977) as a *recent*, fleeting feature, one that formed after the latest glaciation and was filled with sediments before historic time. Also, like Whitney, all before me—other than James Mason Hutchings—would note the *apparent* paucity of the postglacial talus, which would be taken to imply: first, that glaciation was quite recent, and second, that mass wasting, especially through rockfall, was unimportant in the valley's creation. Actually the valley has far more talus than anyone expected.

Besides establishing the Lake Yosemite myth, Whitney established the Half Dome myth: Half Dome is the southeast half of a former dome, the northwest half having fallen into the valley. This myth fit very comfortably within his fault-formed hypothesis of Yosemite Valley, but as that hypothesis eventually fell into discredit, so too did the myth. Nevertheless, it is attractive, and seemingly logical. The myriad visitors to Glacier Point, which stands above the east end of the valley, look directly across it and see only half of a dome (Photo 16). For most unknowing visitors, it is perfectly obvious that the other half fell off. They have to be told otherwise.

Worth noting is that Whitney used "we," which implied complete agreement among the members of the survey. Actually, the hypothesis was his own, and disagreement can be seen in the clause, "It has been objected to this view, by some of the corps." The most likely objector was Clarence King, who did the most field work in and about

**Figure 4. The Merced River canyon between Yosemite and Little Yosemite valleys. Summits, from left to right, are: Half Dome, Mt. Broderick (Whitney's Cathedral Peak), and Liberty Cap (his Mt. Broderick). Vernal Fall lies below the gap between the last two.**

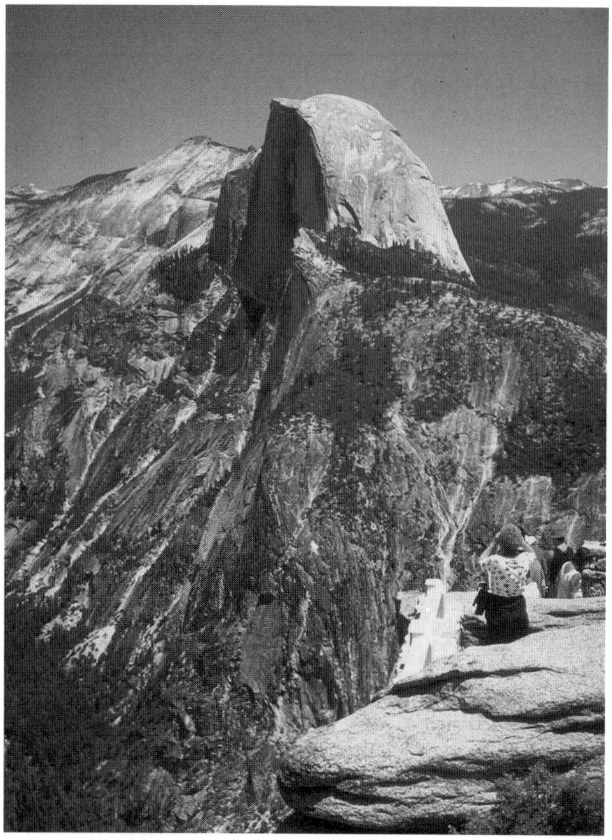

**Photo 16. Half Dome, across Yosemite Valley, viewed from Glacier Point.**

the valley, and so was the expert. Whitney also augmented his hypothesis with the unexplainable domes: "we can discover nothing about them which looks like the result of ordinary denudation," hence they must have been the result of something else, i.e., faulting. But William Brewer, who was Whitney's assistant, had a ready explanation. In 1863 he noted in his journal: "A great glacier once formed far back in the mountains and passed down the [Tuolumne River] valley, polishing and grooving the rocks for more than a thousand feet up on each side, rounding the granite hills into domes." Whitney, as the above quote shows, did not contest King's glacier through Yosemite Valley also being "at least a thousand feet thick." So both glaciers were perceived to have been of about the same thickness, but while the Tuolumne glacier had transformed blocky hills to domes, the Merced glacier had not. If it had, that would have implied significant glacial erosion, and a thick glacier through Yosemite Valley then just might have significantly eroded it.

Whitney's view of Yosemite Valley as originating through down-faulting soon would be challenged. Prof. William P. Blake, of the University of Arizona, visited the valley in 1866, and the following year he wrote (unfortunately in the French language in a technical French journal) that the valley was primarily sculpted by stream processes and secondarily by glacial processes. Blake, like other geologists of his time, had ascribed more erosion to the subglacial stream than to the actual glacier itself. Whitney attacked Blake's opposing view in his 1868 *Yosemite Book* and in his 1869 *Yosemite Guide-Book,* which would have later editions. But his attack soon would focus on John Muir. Muir, like Whitney, said that only a single process was responsible for the valley's formation, and for him it was glaciation.

As the controversy between the two raged, other men pondering the valley's origin more often than not sided with Muir, although none went to his extremes. The general view among "learned men" was that the Merced River and its glaciers both played erosional roles in carving the valley, some favoring glaciers as the more-important erosional agent; others, the river. After the survey disbanded, Whitney went back east, while Muir stayed in California. This gave Muir an edge, since he could lobby the local scientists, populace, and Yosemite Valley visitors. Especially with the formation of the Sierra Club in 1892, with Muir as its president, there developed an effective bully pulpit from which to lambaste Whitney.

Whitney had only one card stacked in its favor: he was a governmental authority; Muir was not. Good Lord, Muir wasn't even a geologist! (In his 1869 *Yosemite Guide-Book,* Whitney called Muir "a mere sheepherder, an ignoramus.") Governmental institutions tend to back the findings of other governmental institutions, and so it was with Yosemite. Yosemite Valley at first was part of a California State Park, so its origin was interpreted according to the findings of the California State Geological Survey—Whitney's. Interestingly, maps of the valley produced during this time were drawn, either intentionally or otherwise, with highly angular, unrealistic, "fault-formed" topography (see Maps 5-8). Such maps lent credence to Whitney's hypothesis that the valley was fault-formed. In 1890 the valley became part of Yosemite National Park, so its origin was then interpreted according to the findings of the U.S. Geological Survey, in this case, Israel Russell's findings. Using faulty logic, Russell reaffirmed Whitney's faulty hypothesis. Not until a new U.S. Geological Survey authority, François Matthes, appeared a generation later did the Park Service drop Whitney's hypothesis. As this book will demonstrate, Matthes' work was hopelessly flawed, and should have been abandoned in the 1960s, when its shortcomings first became manifest. But until an authority in the USGS makes a disclaimer, the Park Service will continue to present Matthes' views more or less intact.

Josiah D. Whitney has received a lot of bad press, both in the last century and even recently in this one. This was and still is undeserved. First, back in the 1860s, very little was known about the Sierra Nevada, so very little evidence was available to favor one hypothesis over an-

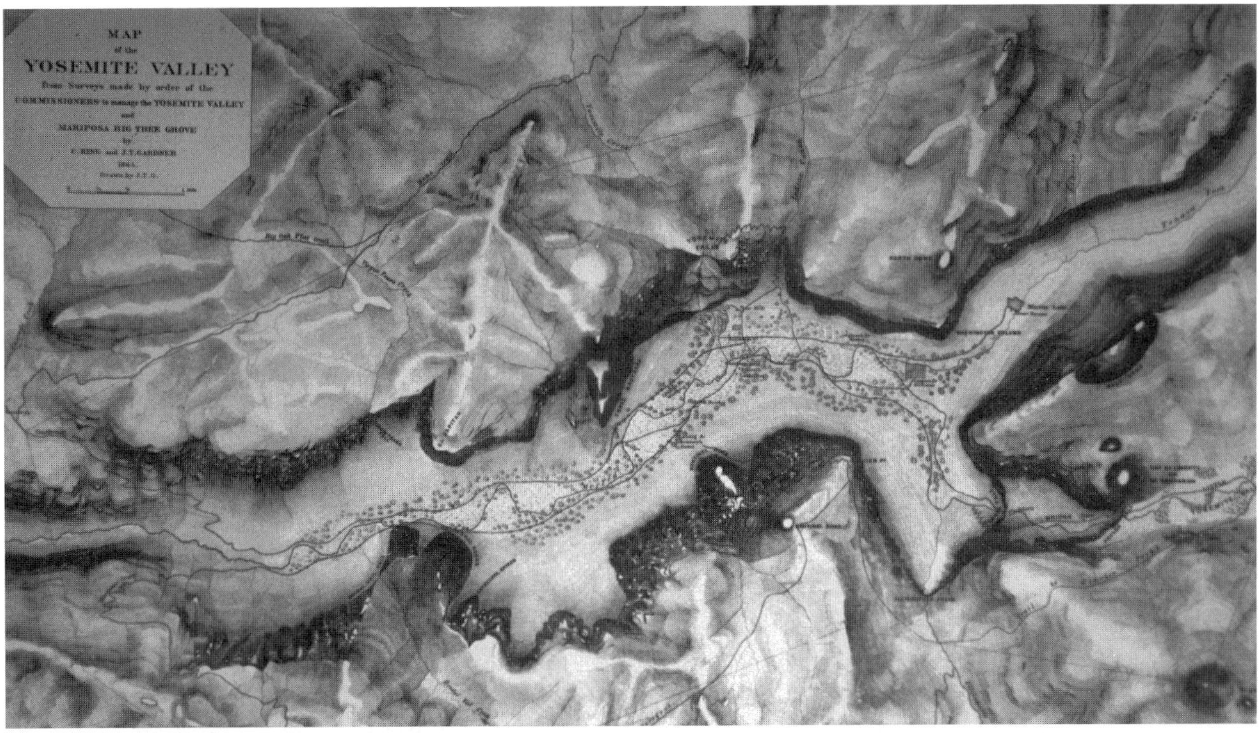


**Map 5. Yosemite Valley, King and Gardner's 1865 map.**

**Map 6. Yosemite Valley, A. L. Bancroft's 1872 map.**

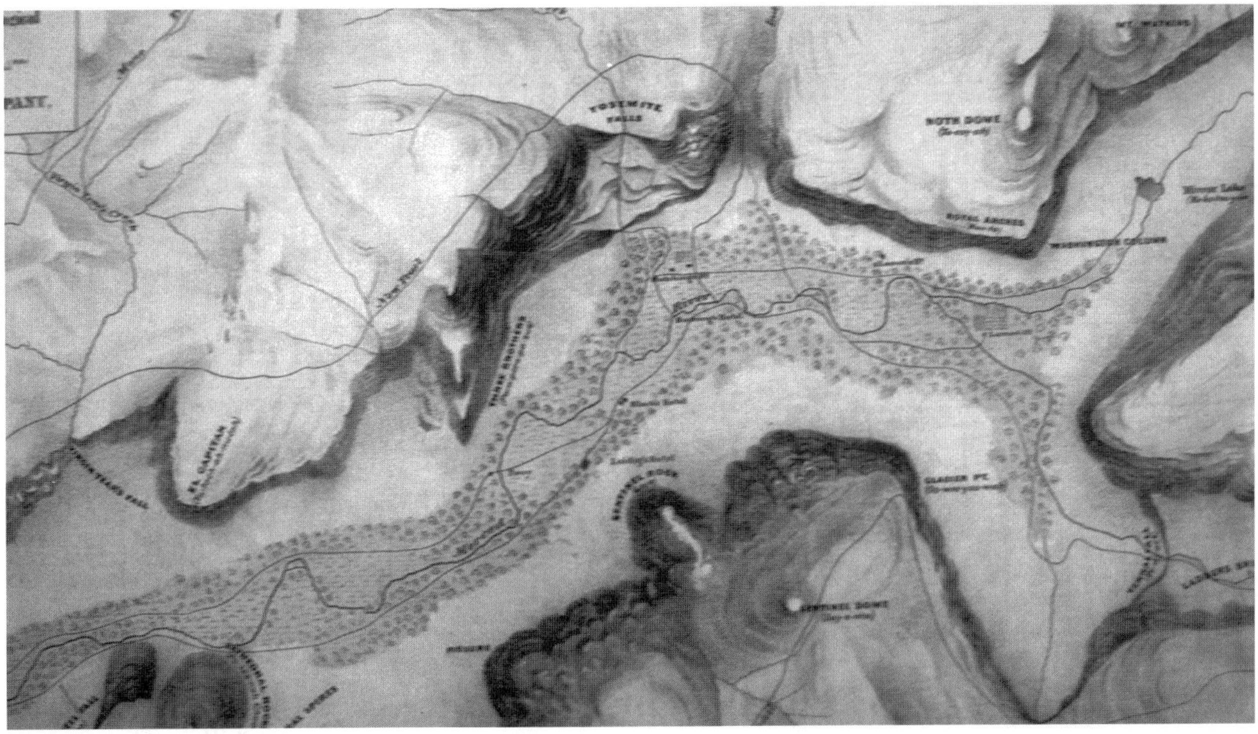


other. In 1872 the Owens Valley (Lone Pine) earthquake shook California, and the vertical displacement along the earthquake fault allowed the geological establishment to prove (incorrectly) that the Sierra Nevada had been raised. If a fault can exist there, why couldn't one exist within the range? As we now know, the Lake Tahoe Basin, within the range, is fault-formed. In the 1860s the geology of the Sierra Nevada was poorly known, and a number of undiscovered fault-formed valleys might have existed. (Kern Canyon, undiscovered back then, is a fault-

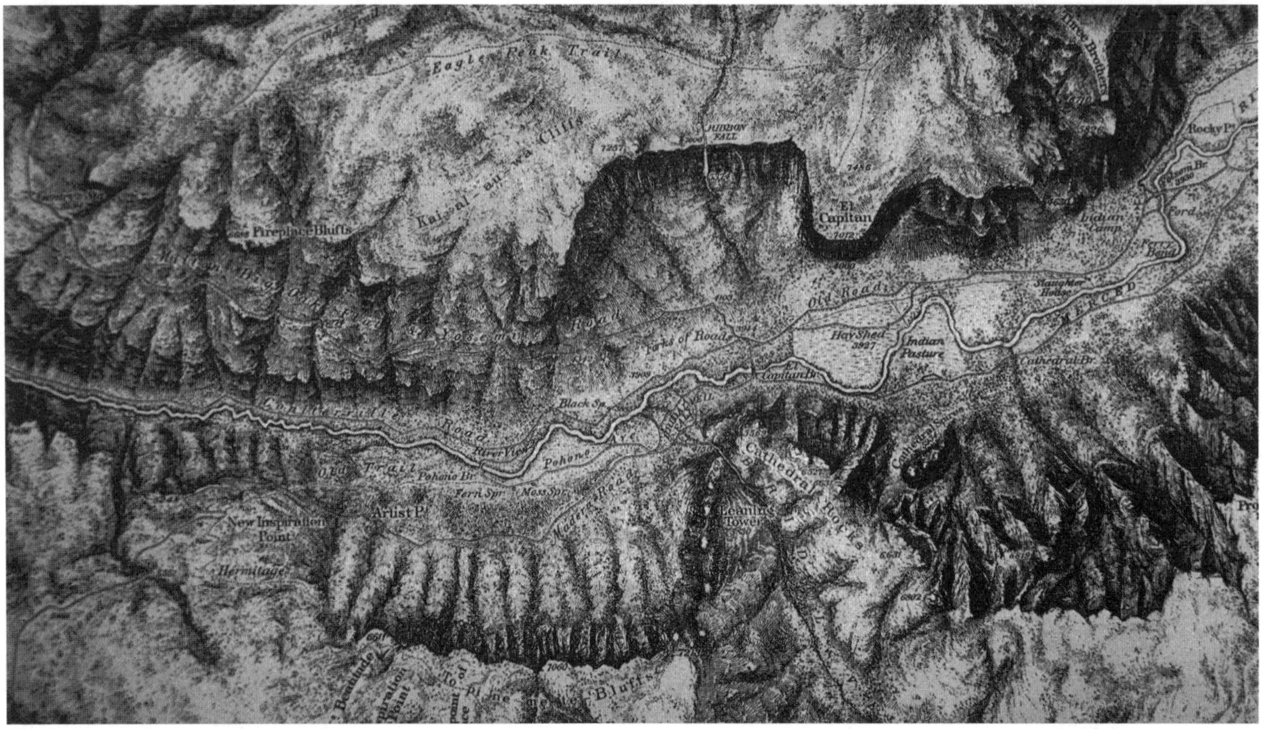


**Map 7. Western part of Yosemite Valley, Army Corps of Engineers' 1883 map.**

**Map 8. Yosemite Valley, modern USGS 1956 topographic map. Scale 1:62,500.**

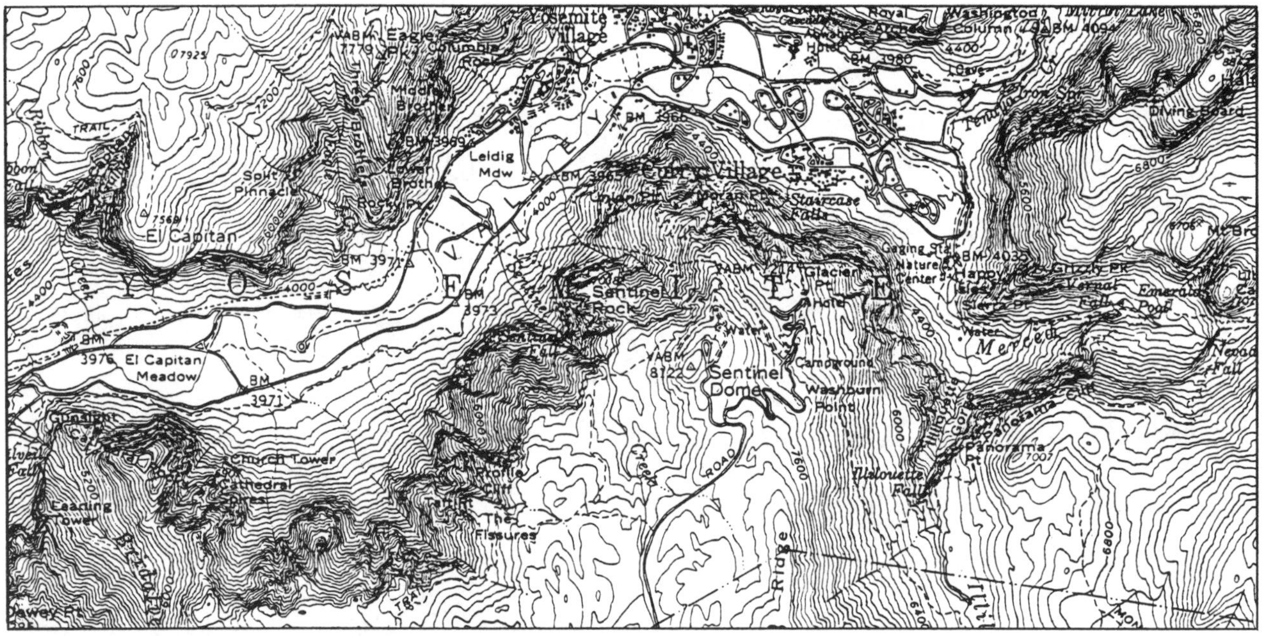


related valley two-thirds as deep as Yosemite Valley.) As this book will demonstrate, Yosemite Valley's enigmatic origin has very little to do with glaciation, but a lot to do with joints. Joints are fractures in rock along which very little movement has occurred. In the past we assumed that absolutely no movement occurred, but we now know that some does, even if it is a tiny fraction of an inch. In 1983 Paul Segall and David Pollard wrote two papers on minor offsets on some Sierran joints. Are these faults, that is, are they fractures in rock along which obvious movement has occurred? Faults may cause large or small offsets, and the offset grades down to the minuscule ones caused by joints. In some cases it is not at all clear whether a fracture is a fault or a joint. Consequently, Whitney's fault-formed hypothesis is not that far from a joint-formed one.

# 5

# Muir's Mighty Snow-Flowers

In 1868, Charles Lyell's *Principles of Geology* appeared in its tenth edition. Geology had come of age. Also in that year, on March 27, John Muir arrived by ship at San Francisco, just before his thirtieth birthday, coming of age as a young man in search of a calling (Photo 17). He arrived only four days after Governor Henry Haight signed an act creating the University of California (at Berkeley), which would recruit a former geology student of Harvard's Louis Agassiz, Joseph Le Conte. The next day, along with a man Muir befriended aboard ship, the two hiked eastward toward much heralded Yosemite Valley. They weren't in any hurry; the Gold Rush was over. In late May they finally arrived at the valley and spent eight days rambling about it. Muir made sketches and collected plant specimens. Needing money, he sought work, and to keep close to the valley (he abhorred cities), he secured employment under Pat Delaney, a sheep rancher living at the base of the Sierra Nevada, west of Yosemite. Both educated men, the two developed a fine relationship, and Delaney offered Muir a way to spend the upcoming 1869 summer in the Sierra Nevada. Delaney employed him as an overseer, not to tend his sheep, but to see that the shepherd did. Most of that summer Muir was free to wander Yosemite's highlands, and by late September he had seen more terrain—and in far more detail—than had anyone before him or long after him. He took notes, but they were not published until much later, in book form, in 1911, as *My first summer in the Sierra.* Muir studied everything—plants, animals, rocks—but giant glaciers came to be his prime focus. Everywhere in the Yosemite highlands he found abundant evidence of their former existence.

The following year Muir was back in Yosemite Valley, avidly preaching his glacial doctrines to any who would listen. This included the newly appointed University of California Professor Joseph Le Conte, who was leading a month-long excursion through the central Sierra Nevada. Muir accompanied the professor and his nine students from the valley to the Mono Basin, and explained his Yosemite Valley glacial views, of which Le Conte in his dairy, published in 1875, noted:

**Photo 17. John Muir in 1872, at age 34. Source: Bancroft Library.**

> He is really not only an intelligent man, as I saw at once, but a man of strong, earnest nature, and thoughtful, closely observing, and original mind. I have talked much with him today about the probable manner in which Yosemite was formed. He fully agrees with me that the peculiar cleavage of the rock is a most important point, which must not be left out of account. He further believes that the valley has been wholly formed by causes still in operation in the Sierra—that the Merced Glacier and the Merced River and its branches, when we take into consideration the peculiar cleavage, and also the rapidity with which the fallen and falling boulders from the cliffs are disintegrated into dust, have done the whole work. The perpendicularity is the result of cleavage; the want of talus is the result of the rapidity of disintegration and the recency of the disappearance of the glacier. I differ with him only in attributing far more to preglacial [stream] action.

In 1871 Muir was the first to discover a "living" glacier, one below the northwest face of Merced Peak, in the upper Illilouette Creek drainage. (This later diminished to a permanent snowfield and by October 1977, after a severely dry year, totally disappeared. Since then it has existed only as a seasonal snowfield.) Muir searched for other glaciers, and on August 21, 1872, measured the rate of movement of the center of the Maclure Glacier, in the upper Lyell Fork Tuolumne River drainage. Earlier that year he was in Yosemite Valley when the Owens Valley earthquake, centered around the town of Lone Pine, struck on March 26. The rockfalls generated by it in Yosemite Valley would lead him to conclude that the valley's talus slopes were mostly of earthquake origin. Then in late summer 1873 he explored the High Sierra lands south of Yosemite, including Kings Canyon, and he completed his southern explorations with an ascent of Mt. Whitney before descending into Owens Valley and heading north up it to Lake Tahoe.

John Muir went beyond describing the areal extent of glaciers; he gave them supernatural eroding powers. In writing for the general public, he wrote more in poetic language than in scientifically rigorous language. Perhaps his most famous passage on glacial erosion in the Sierra Nevada is the following, taken from his 1894 *Mountains of California*.

> It is hard without long and loving study to realize the magnitude of the work done on these mountains during the last glacial period by glaciers, which are only streams of closely compacted snow-crystals. Careful study of the phenomena presented goes to show that the preglacial condition of the range was comparatively simple: one vast wave of stone in which a thousand mountains, domes, cañons, ridge, etc., lay concealed. And in the development of these Nature chose for a tool not the earthquake or lightning to rend and split asunder [Whitney's views], not the stormy torrent or eroding rain [most of the remaining geologists], but the tender snow-flowers noiselessly falling through unnumbered centuries, the offspring of the sun and sea....
>
> Contemplating the works of these flowers of the sky, one may easily fancy them endowed with life: messengers sent down to work in the mountain mines on errands of divine love. Silently flying through the darkened air, swirling, glinting, to their appointed places, they seem to have taken counsel together, saying, "Come, we are feeble; let us help one another. We are many, and together we will be strong. Marching in close, deep ranks, let us roll away the stones from these mountain sepulchers, and set the landscapes free. Let us uncover these clustering domes. Here let us carve a lake basin; there, a Yosemite Valley; here, a channel for a river with fluted steps and brows for the plunge of songful cataracts. Yonder let us spread broad sheets of soil, that man and beast may be fed; and here pile trains of boulders for pines and giant Sequoias. Here make ground for a meadow; there, for a garden and grove, making it smooth and fine for small daisies and violets and beds of healthy bryanthus, spicing it well with crystals, garnet feldspar, and zircon." Thus and so on it has oftentimes seemed to me sang and planned and labored the hearty snow-flower crusaders; and nothing that I can write can possibly exaggerate the grandeur and beauty of their work.

Such a passage thrill the hearts of wilderness lovers, but hardly is food for thought for skeptical geologists. Nevertheless, directly or indirectly, he made an impact on them, for much of what is believed today about the erosive power of Sierran glaciers has its roots with Muir. He went beyond William Blake's position by asserting that *all* of the Yosemite Valley's excavation had been accomplished through glaciation. Early on he attacked Whitney's views, and the two became adversaries, in time becoming increasingly polarized over this issue. Whitney, taking a solitary stand, later denied that glaciers had ever entered Yosemite Valley. Muir, alone in his proclamation (with one exception, mentioned later), in 1894 stated that the California ice sheet discharged "fleets of icebergs into the sea" and in 1912, two years before his death, that "All California has been glaciated, the low plains and valleys

**Figure 5. Two of Muir's figures on glaciated, jointed rock. The upper figure shows vertical joints; the lower one, horizontal. Glacial sculpture has produced two different joint-controlled landforms.**

as well as the mountains." That Muir espoused glaciation is relatively well known among Yosemite researchers and interested lay persons, but what is little known are the details of his explanation, which are presented here. As a whole, his rather comprehensive explanation for the evolution of the Sierra Nevada and of Yosemite Valley was internally consistent—the conclusions fit the evidence *that he saw*. Furthermore, it did not seriously conflict with California's geology as it was understood in the 1870s. However, by the 1890s, when he was promoting his more-extreme views of statewide glaciation, he did so against widespread and abundant geological evidence to the contrary.

Having seen a great part of the *glaciated* Sierra Nevada, Muir wrote up his ideas on the creation of the Sierra Nevada and Yosemite Valley landscapes. From May 1874 through January 1875 the *Overland Monthly* published Muir's seven "Studies in the Sierra," which are summarized as follows.

### I. Mountain sculpture

Muir believed that the Sierra Nevada range was divinely created, and its evolution through glaciation was then directed by the "Master Builder." With regard to its creation, he states:

> we may now understand the Scripture: "He hath *builded* the mountains," as not merely a figurative but a literal expression.

With this divine creation Muir avoided the necessity of addressing the timing, amount, and pattern of Sierran uplift, a thorny issue that plagued other investigators. (Surprisingly, Muir, who had felt the 1872 Owens Valley earthquake while in Yosemite Valley, did not connect earthquakes, mentioned in his second study, with uplift.) Muir envisioned the original Sierra Nevada as having a smooth form, one devoid of any substantial landforms. But it also had myriad "planes of cleavage," at various angles and spacings, and extending downward to various depths. An ice sheet then mantled the entire range and over time removed virtually all of this jointed rock. Muir emphatically stressed the importance of the joint patterns (Figure 5):

> No matter how abundant the glacial force, *a vertical precipice cannot be produced unless its cleavage be vertical,* nor a dome without dome structure in the rock acted upon.

He also observed that where joints were lacking, glaciers were ineffective, but where joint planes extended deep (and as narrow, sinuous bands), canyons developed. According to Muir, late in the Ice Age the ice cover retreated, and then glaciers advanced only down these newly created canyons. It was this advance that Muir in 1876 suggested was responsible for the breakup of a continuous mid-elevation belt of giant sequoias into the groves that exist today between the major canyons. Muir, however, did not address where the giant sequoias grew while the ice sheet occupied all of the Sierra Nevada or, as he later claimed, all of California, in which case the sequoia populations must have migrated south to Mexico. This inference unfortunately went unnoticed by Prof. Daniel Axelrod of the University of California, who writing in 1959 and in later publications promulgated Muir's hypothesis. Actually, sequoias grew in large groves that mostly were on higher lands than are today's groves, but glaciers up there obliterated them and removed the thick soil they required, thereby preventing their reestablishment.

### II. Mountain sculpture, origin of Yosemite valleys

In this, Muir first attempts to discredit Whitney, who had proposed that *long-term* stream erosion was the major process creating the Sierran canyons, other than that part of the Merced River canyon known as Yosemite Valley. Muir says that he has not found any of Whitney's faults, despite five years of field observations, and that if all the canyons are old, as Whitney implies, then all should be choked with talus. Instead, in Yosemite Valley:

> large masses are loosened, from time to time, by the action of the atmosphere, and hurled to the bottom with such violence as to shake the whole valley; but the aggregate quantity which has been thus weathered off, so far from being sufficient to fill any great abyss [as suggested by Whitney], *forms but a small part of the debris slopes actually found on the surface,* all the larger angular taluses having been formed simultaneously by severe earthquake shocks that occurred three or four hundred years ago, as shown by their forms and the trees growing upon them.

In short, he argues that not enough talus has fallen to fill Whitney's hypothetical "graben." Actually, a tremendous amount of talus has fallen, but no one before me considered measuring its thickness and volume. Thus, Whitney is not disproved, Muir's exhortations to the contrary. As to long-term stream action, Muir correctly notes that in Yosemite Valley the streams have been depositing sediments, not transporting them away, and that above the valley the stream action on the glaciated bedrock has been minimal.

With earthquake faulting and stream erosion dismissed, Muir could conclude that glacial erosion was responsible for the Sierran topography. Yosemite-like valleys are special, occurring at mid-elevations. Furthermore:

> *All Yosemites occur at the junction of two or more glacial cañons.* Thus the greater and lesser Yosemites of the Merced, Hetch Hetchy, and those of the upper Tuolumne, those of Kings River, and the San Joaquin, all occur immediately below the confluences of their ancient glaciers.... *Furthermore, the trend of Yosemite valleys is always a direct resultant of the sizes, directions, and declivities of their confluent cañons,* modified by peculiarities of structure in their rocks.

The copious italics cannot save this statement, which is preached today even though it is not true, despite Muir's supporting illustrations of three Yosemites. These incorrectly show orientations of tributary canyons and the size and directions of their former glaciers (Figure 6). One must bear in mind, however, that no detailed, accurate maps of these "Yosemites" existed when Muir wrote this article. For Yosemite Valley there were King and Gardner's 1865 map (produced under Whitney's supervision) and A.L. Bancroft's 1872 map, both with excessively linear features that implied a fault origin (previous chapter, Maps 5 and 6). Muir could have identified many errors on maps such as these, and thus could have felt justified in sketching maps of Yosemite Valley and other valleys as he saw fit. Only Yosemite Valley had two large glaciers converging at its head; the other two did not. Furthermore, in the Sierra where two large glaciers have converged at mid-elevations, Yosemites failed to form. (Had Muir explored all of the Sierra Nevada's canyons instead of a glaciated few, perhaps he would have changed his views to reflect the evidence. Then again, he was known for never retracting his statements.) The remainder of Muir's article describes illustrations of glaciated features in the other Yosemites that resemble Yosemite Valley's Half Dome and El Capitan.

### III. Ancient glaciers and their pathways

Although Muir greatly overestimated the effectiveness of glaciers in the Sierra Nevada, he determined the length, width, and depth of the relatively recent Tahoe- and Tioga-age glaciers of the Merced River drainage about as accurately as could be expected, since topographic maps of the park did not exist in the 1870s. Half of his third study is devoted to their description. Of interest is that he noticed a general lack of end moraines in Tenaya Canyon, and so concluded:

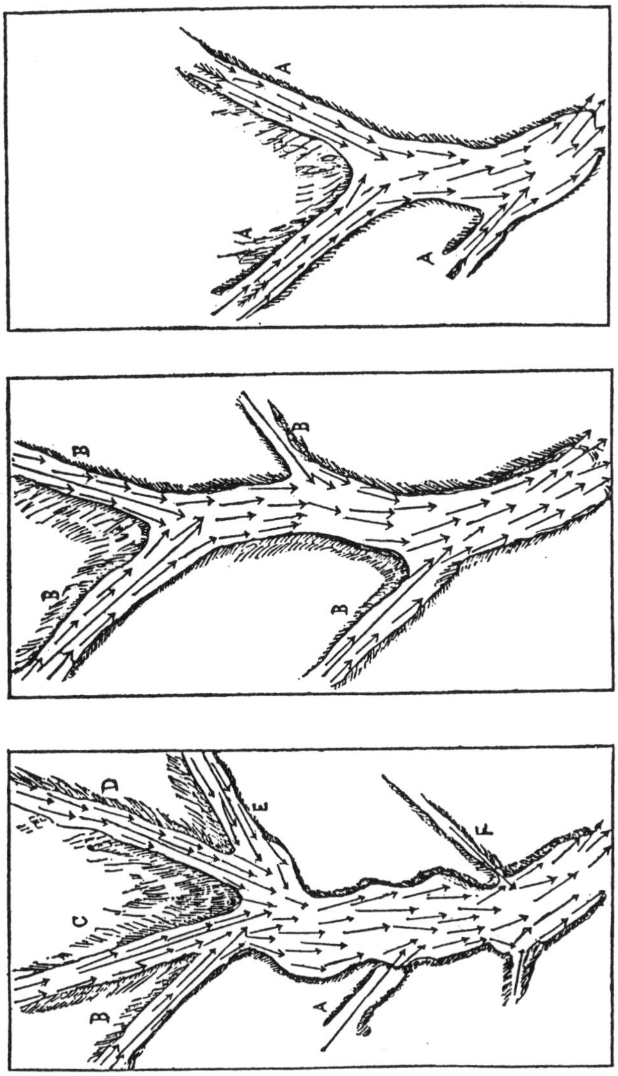

**Figure 6. Three of Muir's figures on converging glaciers in Yosemite-like valleys. Top to bottom, these are: the Tuolumne Yosemite (Hetch Hetchy Valley), the Kings River Yosemite (Kings Canyon), and the Merced Yosemite (Yosemite Valley).**

> the whole body of Tenaya became torpid, withering simultaneously from end to end, instead of dying gradually from the foot upward.

This was a perceptive observation, not that much different from a general statement made about a century later by Malcolm Clark in 1976, and accepted today—after much accumulated evidence—as true. Muir also noted that the Tuolumne glacier overflowed into the upper Merced River drainage, augmenting the Merced glacier, although he went too far in shouting:

> *Traces of a similar overflow from the northeast occur on the edges of the basins of all the Yosemite glaciers.*

This statement conforms with his view that an earlier glaciation swept southwest down across the entire range. Also in the upper Merced River drainage Muir saw "an imposing series of short residual glaciers," which he believed were remnants of the former glaciers. Such small Sierran glaciers have been shown later by investigators to be Neoglacial advances. In the Illilouette Creek drainage Muir recognized a difference in the amount of weathering between adjacent moraines, which indicated two stages. These moraines much later were mapped in the 1980s by the USGS as having been deposited during the Tahoe and Tioga glaciations.

In the remaining part of the article Muir argued, as he did in the first, that an ice sheet covered the range. He presented what he considered supporting evidence, including striations found at higher elevations that indicated glaciers flowed up and over the topography. (Glacial striations on bedrock are created by rocks and sediments dragged along the base and sides of the glacier. These dragged materials can also smooth the bedrock, creating glacial polish.) Locally glaciers flowed uphill, as one can verify by the striations in many localities, but he then incorrectly argued that such flow had occurred in the western part of Yosemite Valley, a glacier flowing, in part, up and over El Capitan, whose summit lacks both polish and striations (Figure 7). At least one early glacier flowed over the summits of the Cathedral Rocks, but left neither polish nor striations.

Muir also added supporting evidence to an issue raised in the second article, namely, that tributary glaciers lo-

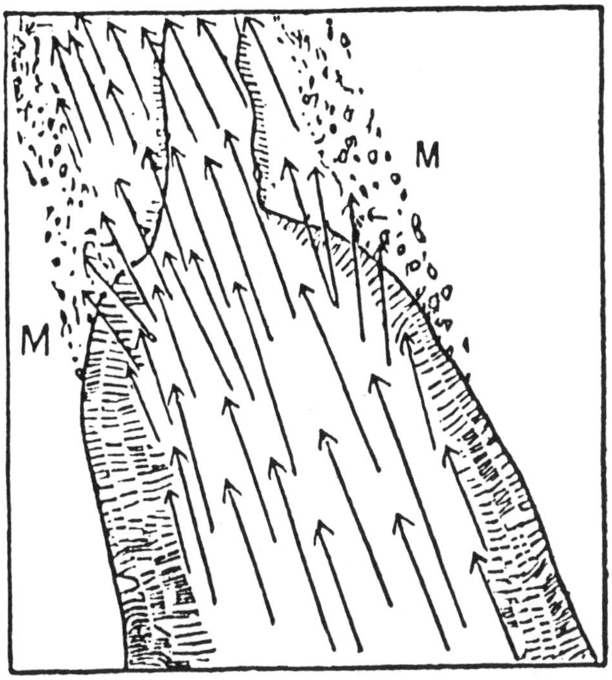


**Figure 7. Muir's proposed ice flow through lower Yosemite Valley. At the constriction El Capitan is on the right, the Cathedral Rocks on the left.**

cally directed the trend of a trunk canyon, such as the Tuolumne:

> throughout the whole period of its formation, the huge granite valley was lithe as a serpent, and winced tenderly to the touch of every tributary.

The Grand Canyon of the Tuolumne River indeed does meander, but it does so independently and in spite of its tributaries.

That stream erosion was not involved in the formation of Sierran canyons is demonstrated, claimed Muir, by the sheer enormity of Tenaya Canyon, which possesses a relatively minuscule, seasonal creek. Other investigators would disagree with Muir's overall picture of glacial erosion, but few have attempted to explain the sheer size of Tenaya Canyon. Muir recognized, perhaps more than anyone else, that any explanation of the Sierra's landscape, particularly that of Yosemite Valley, must address this enigmatic canyon.

### IV. Glacial denudation

Muir first reviewed glacial evidence: striations, polish, moraines, lake basins, etc. Then he gave a description of the west-slope's general geology, landscape, and glaciation from the crest to the foothills. In the foothills "only the laborious student can decipher even the most emphasized passages of the original manuscript" of glacial evidence, which unfortunately Muir did not discuss until later articles. Rather, he described the evidence for glacial erosion found at the indisputably glaciated middle and upper elevations.

Of importance in this study is its last section, the *quantity* of glacial denudation. He noted that metamorphic rocks ("slates") occur in the foothills, along the crest, and locally at intermediate elevations, indicating they once covered all the range (Figure 8). Then:

> from their known thickness in the places where they occur, we may approximate to the quantity removed where they are less abundant or wanting....
>
> If, therefore, we would restore this section of the range to its unglaciated condition, we would have, first, to fill up all the valleys and cañons. Secondly, all the granite domes and peaks would have to be buried until the surface reached the level of the line of contact of the slates. Thirdly, in the yet grander restoration of the missing portions of both granite and slates ... the maximum thickness of the restored rocks in the middle region would not be less than a mile and a half, and average a mile. But, because the summit peaks are only *sharp residual fragments,* and the foothills *rounded residual fragments,* ... we still have only partially reconstructed the range, for the summits may have towered many thousands of feet above their present heights.

The thickness of the foothills' metamorphic strata was not well known, and Muir had assumed it to be comparable to that in the crest area. Whatever the thickness of these fractured metamorphic strata, Muir claimed they had been entirely removed through glacial erosion, as had been the fractured granitic rock that lay beneath them. For Sierran geologists, this amount was too great to be credible, and they would spawn a field of hypotheses that would postulate differing, though lesser, amounts of glacial excavation.

### V. Post-glacial denudation

As in previous studies, Muir maintained that the old ice sheet covered the Sierra Nevada down to its base. The amount of post-glacial denudation is summarized at the end: no more than three inches in the higher, freshly pol-

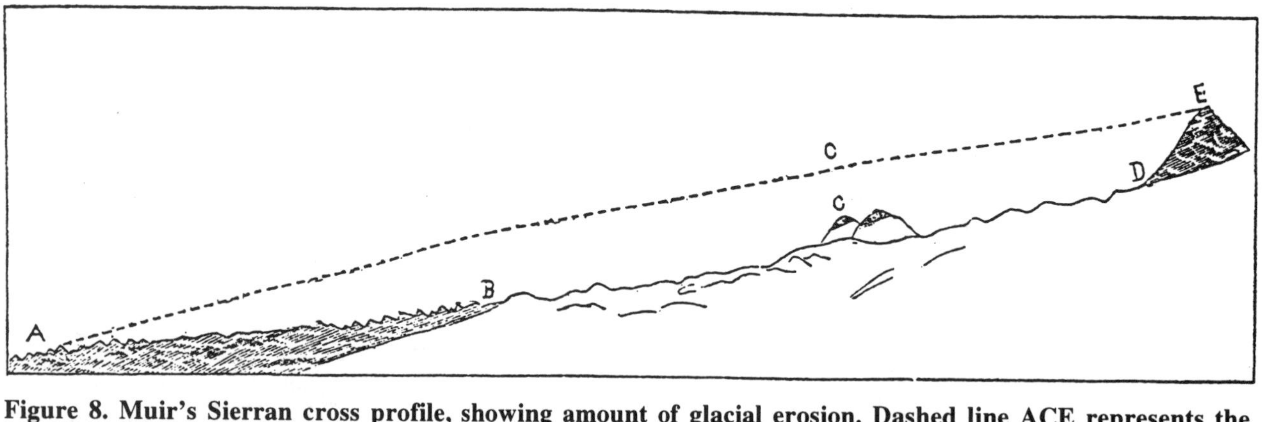


**Figure 8. Muir's Sierran cross profile, showing amount of glacial erosion. Dashed line ACE represents the minimum height of the preglacial landscape. Consequently the space between line ACE and line ABCDE (today's landscape) represents the minimum amount of glaciation. Muir suggested that glacial erosion could have been much more.**

ished lands, no more than a foot in the mid-elevations, and probably several feet in the lower elevations. Also in this article he explained why no evidence of the glaciation is found in the foothills—it has been weathered away over a long period of time:

> The ice sheet melted from the base of the range tens of thousands of years ere it melted from the upper regions. We find, accordingly, that the foothill region is heavily weathered and blurred....
>
> Atmospheric weathering has, after all, done more to blur and degrade the glacial features of the Sierra than all other agents combined, because of the universality of its scope.

Muir discussed at some length mass wasting through rock avalanches and snow avalanches, the latter, he concluded, not being significant. He observed that small rock avalanches occur in every month of the year, but of more importance are earthquake-generated rock avalanches, as mentioned in his second study. Also important are "land-slips," or earthflows. He noted that land-slips, in contrast to rock avalanches,

> slip in one mass, and, unless sheer cliffs lie in their paths, may come to rest right-side up and undivided. There is also a marked difference in their geographical distribution, land-slips being restricted to deeply eroded banks and hillsides of the lower half of the range, beginning just where rock avalanches cease.

It is this lower half where Muir's hypothetical older moraines were deposited. These were weathered over a long time, and eventually all moraines deposited on slopes sloughed away. But he was aware that some old erratics, boulders transported by glaciers, originally had been left by them on bedrock and now stood on pedestals. These pedestals,

> furnishing so admirable a means of gauging atmospheric erosion, occur throughout the middle granitic region in considerable numbers: some with their protecting boulders still poised in place, others naked, their boulders having rolled off on account of the stool having been eroded until too small for them to balance upon. It is because of this simple action that all very old, deeply weathered ridges and slopes are boulderless, Nature having thus leisurely rolled them off, giving each a whirling impulse as it fell from its pedestal once in hundreds or thousands of years.

Hence, down in the foothills, where erratics would have been exposed to weathering longer than higher up, *all* have disappeared over time (Photos 18 and 19). This is not unreasonable when one considers, for example, that at mid-elevations in the Stanislaus River drainage, where I have done extensive glacial work, the only evidence that remains from the early, pre-Tahoe glaciers are scattered

**Photo 18. Erratics of the North Fork Mokelumne River drainage. Three large granitic boulders are at an elevation of about 8890 feet elevation just below the Sierran crest at a site about 3.3 miles west-northwest of Ebbetts Pass (Ebbetts Pass 7.5' quadrangle). This site is relatively close to the head of the Mokelumne glacier. The three granitic boulders are obvious erratics. National in-line speed-skating champion Eddy Matzger provides scale.**

**Photo 19. *Possible* erratics of the North Fork Mokelumne River drainage. Four large volcanic boulders are at an elevation of about 7790 feet on slopes just below a ridge at a point about 0.4 mile southwest from Flagpole Point (Tamarack 7.5' quadrangle). This site is at about the maximum extent of glacial ice on this ridge, which divides the North Fork Mokelumne River drainage to the north from the North Fork Stanislaus River drainage to the south. The volcanic boulders are not certain erratics. Were they transported here by an ancient glacier, or did they weather from the underlying volcanic debris flow? Westward, the boulders become increasingly fewer and more enigmatic, and so Muir's assertion that erratics disappear altogether lower down, as age of glaciation increases, appears quite feasible.**

large erratics atop some gentle slopes and ridges. But what happened to the lower glacial deposits on *flat* surfaces? These are discussed in the next study.

### VI. Formation of soils

The title is misleading, for Muir mostly described the origin of sediments and of decomposed rock beneath soils, both of which he loosely equates to soil:

> If all the soils that now mantle the Sierra flanks were spread out in one sheet of uniform thickness, it would measure only a few feet in depth,... The largest beds rarely exceed a hundred feet in average thickness, and a very considerable proportion of the whole surface is naked. But we have seen that glaciers alone have ground the west flank of the range into soil to a depth of more than a mile.... It appears, therefore, that not the one-thousandth part of the whole quantity of soil eroded from the range since the beginning of the glacial epoch is now left upon its flanks.

The vast bulk of the "soil" ended up, as he stated near the end of his article, in the Sacramento, San Joaquin, Owens, Walker, and Carson valleys. In the 1870s no one knew either the thickness or the composition of these valleys' sediments, and so superficially his scheme could have seemed plausible. But there are many valleys and basins in California, such as the Los Angeles Basin and those in the Mojave Desert and the Great Basin, that closely resemble the ones Muir enumerated. For consistency all should have had a common origin, and perhaps because of this he later stated, as mentioned above, "All California has been glaciated, the low plains and valleys as well as the mountains."

Muir was not quite alone in believing that glaciers covered far more than parts of the Sierra Nevada, and that they even discharged fleets of icebergs into the Pacific Ocean. University of California's Professor Joseph Le Conte, even in the revised, relatively modern 1887 edition of his textbook, the *Elements of Geology,* stated:

> There seems no doubt that during the fullness of Glacial times the whole high Sierra region, as far as Southern California, was ice-sheeted. Whether this should be regarded as an extension of the northern ice-cap is perhaps doubtful. From the margins of this sheet stretched valley-extensions in the form of separate glaciers to the plains east

**Photo 20. Smooth, striated "glaciated" rock, Indian Rock Park, Berkeley.**

> and west. At the same time even the Coast Range was covered with perpetual snow, and glaciers ran down into the Bay of San Francisco.[2]

Le Conte's footnote 2 states: "Undoubted marks of ancient glaciers are found about Berkeley, 200 feet above the bay." The unstated evidence may be fault-caused slickensides at Berkeley's Indian Rock Park, which do resemble glacial striations and polish (Photo 20.) While Le Conte did not say that "all California has been glaciated," he did imply that most it was—far more than Whitney and other early researchers had suggested.

Muir in his third study had mentioned that the last glaciers disappeared suddenly and therefore left very scanty deposits, a point on which he elaborated here. In bolstering his arguments he first described the patterns of recessional moraines left by glaciers of the Sierra's east flank and stated their relevance to this matter. While those glaciers may have melted catastrophically, he said, for the vast ice sheet the melting must have been slow and gradual:

> The great forest soil-belt of the west flank has not been hitherto recognized as a moraine at all, because not only is it so immensely extended that general views of it cannot be easily obtained, but it has been weathered until the greater portion of its surface presents as smooth an appearance as a farmer's wheat-field.... Had the ice-sheet melted suddenly, leaving the flanks of the Sierra soilless, her far-famed forests would have had no existence. Numerous groves and thickets would undoubtedly have established themselves on lake and avalanche beds,... Yet the range, as a whole, would seem comparatively naked.... The mass of the Sierra forests indicates the extent and position of the moraine-beds far more accurately than it does lines of climate. No matter how advantageous the conditions of temperature and moisture, forests cannot exist without soil, and Sierra soils have been laid down upon the solid rock.

We now have a relatively complete picture of the California landscape according to Muir: deposits of the early glaciation filled the state's valleys, while those in its hills and mountains weathered to form the soils conducive to forests.

### VII. Mountain-building

This study, like the previous one, has a misleading title. There was no mountain building, for as Muir had stated in his first article, the Sierra Nevada range was created by God, then was eroded by glaciers to leave the mountain scenery. Muir reviewed the origin of cirques, lakes and other glacier-related features, as well as orientations of drainages, and he concluded—incorrectly, although very plausibly for his day—that only glaciation could have created the Sierran landscape. For those who doubted, he offered the following proof:

> Still, it may be conjectured that preglacial rivers furrowed the range long ere a glacier was born, and that when at length the ice-winter came on with its great skyfuls of snow, the young glaciers crept into these river channels, overflowing their banks, and deepening, widening, grooving, and polishing them without destroying their identity. For the destruction of this conjecture it is only necessary to observe that the trends of the present valleys are strictly glacial, and glacial trends are extremely different from water trends; preglacial rivers could not, therefore, have exercised any appreciable influence upon their formation.

The problem here is that the trends, or bearings, of the present valleys are *not* always glacial, despite Muir's exhortations and illustrations in this series of articles. Muir himself had noticed that glaciers in places spilled across gaps in mountain divides. By his reckoning, glaciers flowing across these divides should have greatly deepened and broadened them, but they show little glacial modification. Furthermore, striations abundantly show that glaciers did not precisely descend canyons, but in many places flowed around local, major obstructions, this indicating that the bedrock dictated the glacier's basal flow rather than the glacier's flow dictating the topography.

Muir's last paragraph summed up not only the last article, but also the whole series of "Studies in the Sierra:"

> Thus it appears that, no matter how the preglacial mass of the range came into existence, all the separate mountains distributed over its surface between latitude 36°30' and 39°, whether the lofty alps of the summit, or richly sculptured dome-clusters of the flank, or the burnished bosses and mountainets projecting from the sides of valleys—all owe their development to the ice-sheet of the great winter and the separate glaciers into which it afterward separated. In all this sublime fulfillment there was no upbuilding, but a universal razing and dismantling, and of this every mountain and valley is the record and monument.

There is one interesting peculiarity about this quote: why did Muir address only that part of the Sierra between latitude 36°30' and 39°? The major Kern River glacier extended south to about latitude 36°15', while at the other end, major Yuba River glaciers extended north to about latitude 39°30'. His given latitudes suspiciously bound the extent of his own Sierran explorations, that is, from Mt. Whitney north to southern Lake Tahoe. Lacking firsthand experience in the major North Fork and South Fork canyons of the Feather River, perhaps he chose not to address them, although written material was readily available. They posed one formidable problem: if glaciers sculpted every Sierran canyon, they must have sculpted the North Fork and South Fork canyons. But why do these two canyons traverse across the entire range, while the other major canyons south of them extend west only from the crest? These two trans-Sierran canyons defy a solely glacial origin.

# 6

# The Blind Men and the Elephant

For Gold Rush miners, the elephant was a symbol of California's entire gold-mining scene. Aspiring miners, leaving the port of San Francisco for the Sierra's gold fields, spoke of going to "see the elephant," to be a part of the Big Picture. During these hectic, frantic times, John Godfrey Saxe, residing on America's opposite shore, wrote a poem that would have admirably suited these amateur fortune hunters, *The Blind Men and the Elephant.* It begins:

It was six men of Indostan
To learning much inclined,
Who went to see the elephant
(Though all of them were blind),
That each by observation
Might satisfy his mind.

Each man feels only one part of the elephant's anatomy, and from it concludes the nature of the entire beast. The first, feeling its broad side, concludes it's like a wall; the second, its pointed tusk, a spear; the third, its squirming trunk, a snake; the fourth, its massive leg, a tree; the fifth, its flapping ear, a fan; and the sixth, its swinging tail, a rope. With these observations in mind, but not in sight:

And so these men of Indostan
Disputed loud and long,
Each in his own opinion
Exceeding stiff and strong,
Though each was partly in the right,
And all were in the wrong!

A Mr. J. D. Borthwick, of New York City, came to the gold fields in 1851 to make his fortune, which he did not, but he drew pictures and wrote accounts about "seeing the elephant." On his way from Sacramento east to Placerville, he made this observation:

> When about ten miles from the plains, I first saw the actual reality of gold-digging. Four or five men were working in a ravine by the roadside, digging holes like so many grave-diggers. I then considered myself fairly in "the mines", and experienced a disagreeable consciousness that we might be passing over huge masses of gold, only concealed from us by an inch or two of earth.

Perhaps the great majority of the miners felt likewise. The gold was waiting for *them* just out of sight. All they needed was a clue—any clue. A shiny rock, a piece of quartz. Better yet, a vein of quartz. Each clue suggested that the elephant was about to be revealed; it was *so* close you could almost smell it. A bonanza right there, beneath your feet, and all you had to do was dig a little deeper, a little farther, always a little more.

A dozen years later, William Brewer, of the Whitney Survey, headed east over the Sierran crest near Ebbetts Pass, and then descended to the town of Silver Mountain (Photo 21), lying near the foot of its namesake. Established by Scandinavian miners the previous year, it now hosted hundreds of men, and over three hundred claims were being worked. Brewer described this scene in some detail as "a good illustration of a *new* mining town." Of relevance to the elephant is his discourse with a miner:

> Perhaps half a dozen women and children complete that article of population, but there are hundreds of men, all active, busy, scampering like a nest of disturbed ants, as if they must get rich today for tomorrow they might die. One hears nothing but "feet," "lode," "indications," "rich rock," and similar mining terms. Nearly everyone is, in his belief, in the incipient states of immense wealth. [They "see" the elephant.]

**Photo 21. Silver Mountain City in 1867. Source: Bancroft Library.**

> One excited man says to me, "If we strike it as in Washoe, what a town you soon will see here!" "Yes—*if*," I reply. He looks at me in disgust. "Don't you think it will be?" he asks, as if it were already a sure thing. He is already the proprietor of many "town lots," now worth nothing except to speculate on, but when the town shall rival San Francisco or Virginia City, will *then* be unusually valuable. There are town lots and streets, although as yet no wagons. [No road to the settlement has yet been built.] I may say here that it is probably all a "bubble"—but little silver ore has been found at all—in nine-tenths of the "mines" not a particle has been seen—people are digging, *hoping* to "strike it." One or two mines *may* pay, the remaining three hundred will only *lose*.

But John Godfrey Saxe had not written about California miners attempting to see the elephant. The last stanza of his poem reveals to whom it was directed:

So, oft in theologic wars
The disputants, I ween,
Rail on in utter ignorance
Of what each other mean,
*And prate about an elephant*
*Not one of them has seen!*

Like theologians, who saw very little (if anything) of what they spoke of, so geologists in the late 1800s and early 1900s, educated in the best schools, each would make a brief visit to Yosemite Valley and, on the basis of a scanty investigation, each would deduce how the valley had been created. To paraphrase Saxe:

And there were sent from eastern schools
Great men so much refined,
Who went to see Yosemite
(Though all of them were blind),
That each by observation
Might satisfy his mind.

The first geologist of significance was Israel C. Russell, who, after several years of studying lands along the east side of the Sierra Nevada, in 1889 produced a classic monograph for the USGS, the *Quaternary History of Mono Valley, California*. This area today is called Mono Basin, and its centerpiece is Mono Lake. During glacial times the lake was much larger, as you can tell if you drive along Highway 395: you can see the former shorelines of glacial Lake Russell high above the shoreline of today's Mono Lake. Prof. Russell used lateral moraines to reconstruct the east-side Sierran glaciers that had transported and deposited debris to form them (Photo 22, Map 9). His reconstruction of these glaciers correctly showed that some of them had reached the glacial lake, and they calved icebergs into it.

Russell was impressed with the size of Mono Basin's prodigious lateral moraines, yet he concluded that overall glaciers had not performed that much erosion. He argued: take the volume of a canyon's moraines, place it back on the canyon's walls, and it still is a large canyon. Most of the canyon formation, he concluded, must have been due to very long-term stream erosion that had occurred prior to glaciation. Russell also had seen several glaciated west-side canyons, including Yosemite Valley, and this is what he had to say about it:

> Those who seek to account for the formation of the Yosemite and other similar valleys on the western slope of the Sierra by glacial erosion [i.e., Muir] should be required to point out the moraines deposited by the ice-streams that are supposed to have done the work. The glaciers of this region were so recent that all the coarser débris resulting from their action yet remains in the position in which it was left when the ice melted. If the magnificent valleys referred to are the result of glacial erosion, it is evident that moraines of great magnitude should be found about their lower extremities. Observation has shown that débris piles of the magnitude and character required by this hypothesis are notably absent.

With major glacial erosion dismissed, Prof. Russell supported Prof. Whitney, one of the few significant geologists to do so. Since Russell had visited both east- and west-side canyons, he must have known, but unfortunately seems to have overlooked, an important difference. On the east side in the Mono Basin vicinity, glaciers advanced beyond the mouths of their canyons; on the west side they did not. Within a canyon the bedrock slopes can be so steep that if a glacier leaves deposits on them, it will quickly slough off. This is the case in west-side canyons. The sloughed-off glacial deposits over time simply get transported downstream. Some east-side canyons also locally lacked lateral moraines due to the steepness of bedrock slopes. Such was the case along the

**Photo 22. Grant Lake, a reservoir along Rush Creek. The crests of lateral moraines, which flank both sides of the lake, indicate the approximate thickness of the Rush Creek glacier. View is from Reversed Peak, which was a massive obstacle that split the glacier into two lobes, the Rush Creek and the Reversed Creek (see lower part of Map 9).**

north bedrock slopes of the Lee Vining canyon and along most of the bedrock slopes on both sides of the Rush Creek canyon above Grant Lake. Russell correctly

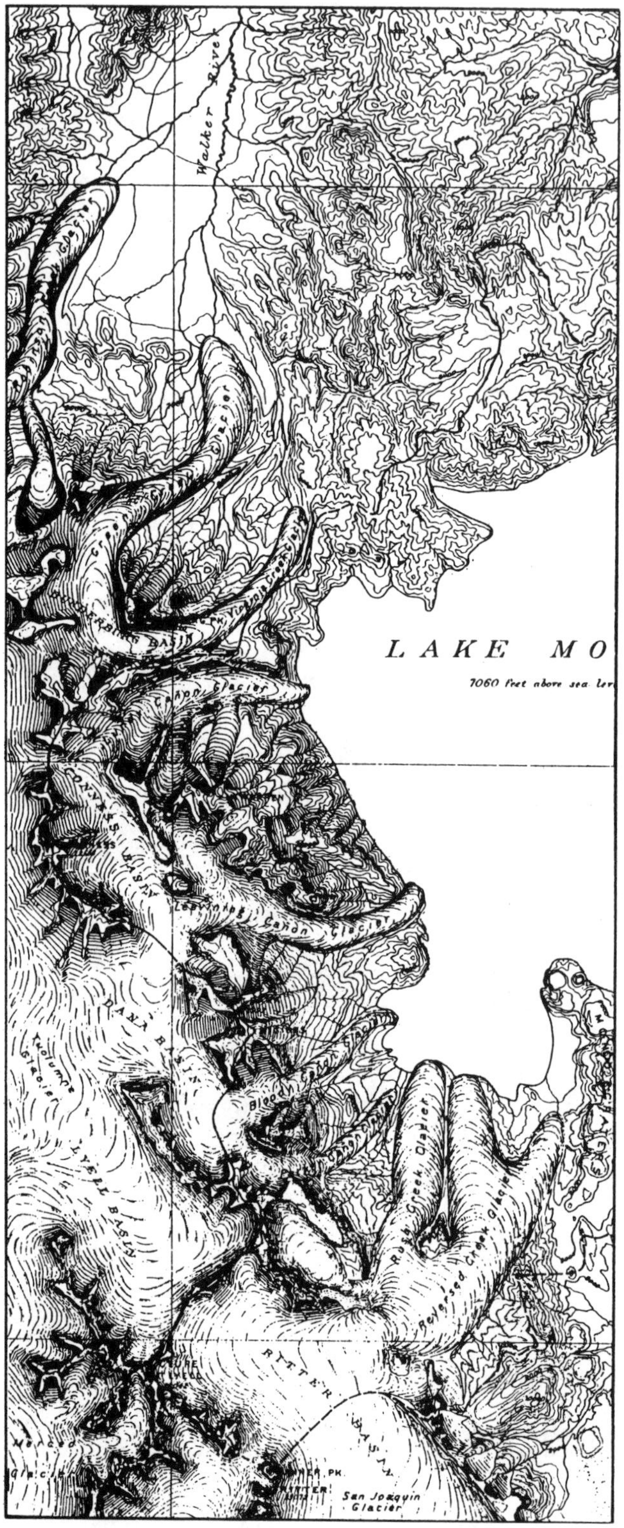


**Map 9. Part of Russell's Mono Basin map, showing former glaciers. Scale 1:250,000.**

mapped the moraines—and noted their absences—and he should have realized the obvious: moraines cannot form on overly steep slopes—ones steeper than about 35°. Later, much of what would be said about the extents of Tioga and Tahoe glaciers on the west side of the Sierra Nevada would be in error in large part due to the failure to recognize this simple fact. Absence of glacial moraines is not evidence for absence of glaciers.

In contrast to Russell, Henry W. Turner reached a very different conclusion. Other than Waldemar Lindgren, perhaps no one else was as familiar with the geology of the northern half of Sierra Nevada, for Turner had done field work for eight USGS Geologic Folios (30' longitude/latitude extent, 1:125,000 scale, published 1894-1898). These covered the eastern edge of the Great Valley and much of the lower and mid-elevation lands of the northern half of the Sierra Nevada from the Feather River drainage south to the Tuolumne River drainage. The Big Trees Geologic Folio, published in 1898, was the first serious geologic mapping of the western part of *glaciated* lands in the Stanislaus River drainage. In it Turner and Ransome suggested that Sierran uplift began about middle-Miocene time, that is, about 15 million years ago—a view generally in accord with modern uplift hypotheses. However, unlike the modern view, which generally postulates uplift now occurring at its greatest rate, they postulated uplift occurring only until the start of the Quaternary Period, or Ice Age.

With respect to glaciation, Turner and Ransome noted evidence of an early stage—in the form of weathered, isolated erratics on ridges beyond the extent of *till* (poorly sorted, amorphous deposits left by glaciers). These erratics implied that the early glaciers extended farther west than did later ones. They said the North Fork Stanislaus River glacier advanced to what is now Calaveras Big Trees State Park, and I concur. They believed the river canyons after that glaciation were shallow, existing essentially as they had before glaciation. Then, during an "interglacial epoch" between an early glacial period and a late one the Stanislaus River's North and Middle forks greatly deepened their canyons to about their present depths. The later glaciers then descended part way down these canyons, transforming them from V-shaped to U-shaped in cross profile. But if uplift had been occurring for millions of years before any glaciation, steepening the gradients of the Sierra's rivers, why didn't the rivers cut deep canyons then, instead of only between early and later glaciations? Also, if the early glaciers were larger than the later ones, they should have been more erosive. Why did only the smaller glaciers greatly widen the canyons?

With the extensive mapping experience that included some glaciated as well as much unglaciated lands, Turner observed that Yosemite Valley was similar in many respects to stream-cut canyons of the northern Sierra Nevada. He concluded that the valley, like other Sierran canyons, had been deeply cut by accelerated (if inexplicably delayed) stream erosion in response to the earlier pre-glacial uplift. Thus stream erosion and weathering processes, facilitated by the jointed structure of the granitic rock, were important. He concluded (as Matthes would) that the farthest glacier advanced below Yosemite Valley down to the El Portal area, but because that area lacks moraines, the glaciers must not have eroded much bedrock. In this respect he was echoing Russell by failing to recognize that very few deposits would have been preserved in the steep-walled canyon in the first place. Furthermore, over countless millennia, the deposits would have been eroded away. *Surprisingly,* Turner and Ransome had noted this in their folio, stating, "the glacial markings and scourings and the terminal moraines left by the retreating ice of the first epoch [of glaciation] would of course be obliterated."

Challenging Henry Turner, in 1901, was Henry Gannett, then the chief geographer of the USGS. His critique, a mere 1¼ pages long, has several misstatements, which indicates only a superficial study of Yosemite Valley, and furthermore the critique does not present any new field evidence or insight. One of his main points, the same one made by John Muir, was that the excavation of Tenaya Canyon could only be explained through glaciation. François Matthes, in his 1930 Professional Paper 160 (p. 5) sums up Gannett's paltry paper:

> Gannett, as a result of his [1898] studies on Lake Chelan, in the Cascade Range, had come to regard "hanging" side valleys as characteristic accessory features of deeply glaciated canyons, and contended that the height of such valleys affords a rough measure of the depth of glacial excavation in the main canyon. The Yosemite, he pointed out, has hanging side valleys of great height (the upland valleys from whose mouths the waterfalls pour into the chasm), and he therefore pronounced it to be "quite an ordinary and necessary product of glacial erosion."

I speculate that the only reason Matthes included this short, essentially worthless article was because Gannett had been his supervisor when he worked for the Topographic Branch of the USGS. Gannett's examination of Washington's Lake Chelan area must have been superficial, since the mouths of its tributary canyons do not hang consistently, but instead range from accordant (not hanging) to extremely hanging.

Another investigator was John C. Branner, who in 1903 effectively supported Henry Turner. However, his interest was only in why most of the valley's prominent falls leap off the tops of cliffs in seemingly illogical places. That is, Illilouette Creek leaps into the Illilouette Gorge not at its head, which is just beneath the railing of the Illilouette Fall viewpoint, but rather some 500 feet to the northeast. And Yosemite Creek leaps off a prominent cliff instead of more logically flowing down a major gully, to its west, which contains the switchbacking Yosemite Falls trail. Studying these localities, Branner concluded that the wearing done by the glacial ice was trivial as compared with the wearing done by the *glacial* streams, which flowed beneath the glaciers. This conclusion, however, says nothing about glacial erosion versus *nonglacial* stream erosion.

Also early in the twentieth century were two somewhat similar views that preglacial Yosemite Valley had been greatly deepened by glaciers through headward erosion. First, in 1910 E. C. Andrews proposed that Yosemite Valley is a gorge analogous to the Niagara Falls gorge. The latter was known to have formed by erosive action of the slowly receding Niagara Falls, and Andrews proposed that the valley was deepened by the erosive action of a plunging ice cascade—at least 2,000 feet in height—of the trunk glacier that flowed through the valley. As a result of the quarrying action of the ice cascade, the high cliff that supported it retreated up through the shallow, preglacial valley, leaving a deep, broad valley. The retreat up-canyon was incomplete, the quarrying getting only to the upper part of Tenaya Canyon and to the Vernal Fall-Nevada Fall vicinity in the Merced River canyon. Second, in the following year Douglas W. Johnson followed Gannett's idea out to its logical conclusion, deducing the depth of glacial excavation in Yosemite Valley from the heights and gradients of its hanging side valleys, and he estimated that glaciers had incised at least 2200 to 2500 feet. This is quite logical except for one problem. The only other major glaciated canyon that might have had a "retreating ice cliff" would be the Tuolumne River canyon, this located along the river's cascading reach west of Tuolumne Meadows, which has a similar size "retreating ice cliff." But none of the other major glaciated, west-draining canyons have a "retreating ice cliff." This implies that glaciers performed very little deepening in the Mokelumne, Stanislaus, San Joaquin, Kings, and Kaweah river canyons. Neither Andrews nor Johnson asked why glaciers were extremely effective in two canyons but impotent in others.

Finally, Lawson, in a chapter of the 1921 Ansel F. Hall's *Handbook of Yosemite National Park,* took a view contrary to what Matthes had been telling the public and the park staff for several years:

> We may thus picture to ourselves a pre-glacial Yosemite Valley, not as deep, nor as wide, nor as sheer-walled as the present Valley, but nevertheless a profound erosional gorge ending in spray filled *culs-de-sac* below both Little Yosemite and the high valley of the Illilouette, with great cascades in them not essentially dif-

> ferent from those we see to-day with so much pleasure and interest. Nevada, Vernal, and Illilouette are, therefore, from this point of view, falls which handed over their work of extending the cañon of the Merced into the High Sierra to the Merced Glacier for a geologically brief time, and have since resumed operations at nearly the old stand. The amount of recession effected by the glacier was probably not great, since the work must have been done chiefly, if not wholly, by the process of plucking, and the paucity of the moraines below Yosemite indicates but a small product.

Like Russell, Lawson concluded the lack of prodigious moraines indicated that the glaciers had excavated very little. Back in 1903 he had reached a similar conclusion for the upper Kern River basin in the southern Sierra Nevada. Taking the volume of the lower moraines and spreading it uniformly across the surface area of the Kern Canyon above them, he arrived of *an average glacial erosion of just 2 inches!* Like others before him (except John Muir), he failed to realize that most of the glacial products would have been transported down-canyon in the Kern River and/or weathered away.

From Josiah D. Whitney to geologists in the 1990s, all have assumed that Yosemite Valley and the other Sierran canyons have deepened significantly only in the last 10 or so million years, and have done so only under modern climates. But back in 1878 Leo Lesquereux reported on the fossil plants of the Sierra's gold-bearing gravels. The fossils were old, and they indicated a warm, wet climate. Because in the South Yuba River drainage the gravels exist at elevations almost down to the river, the canyon buried under those gravels originally must have been almost as deep as today's. Hence from the foregoing one should conclude that Sierran canyons are old features, and that they developed under warm, wet climates. Other than Waldemar Lindgren, who stressed this in 1911, no one addressing the Sierran landscape acknowledged Lesquereux's old fossil plants or those described later by others. It would appear that geologists quite consistently have been blind to the paleontological evidence. For over a century they have failed to "see the elephant."

# 7

# The Fateful Direction

There was only one man who was not blind to the evidence Yosemite Valley offered to the careful observer. He was not a geologist, nor was he Muir, the naturalist. Indeed, he was Muir's employer, James Mason Hutchings, and he was quite an entrepreneur. In 1855 he organized and led the first tourist party to Yosemite Valley, bringing along artist Thomas Ayres to record their discoveries. As soon as he had returned to Mariposa, Hutchings wrote up his adventures, which were published in the July 12 issue of the *Mariposa Gazette*. In 1855 California as a state was only five years old, but largely due to the clamor of the Gold Rush, it was an area of interest to many persons in the eastern states. Hutchings' article was copied in one form or another in a number of journals and newspapers, which, if nothing else, diverted readers' minds from the serious, burning issue of slavery.

Hutchings was not content to sit idle after his one article, and in July 1856 he began to publish a magazine, *Hutchings' Illustrated California Magazine,* which was devoted to the scenery of California. The first issue contained a lead article on "The Yo-Ham-i-te Valley," illustrated by none other than Thomas Ayers. Later, Hutchings elaborated on this article, producing "The Great Yo-Semite Valley" in a series of four installments, from October 1859 to March 1860. These installments then immediately appeared as part of his book, *Scenes of Wonder and Curiosity in California,* which stayed in print well into the 1870s. In April 1864 Hutchings acquired Upper House, one of the valley's first hotels, and two months later, the Congress and President Abraham Lincoln adopted legislation that made Yosemite Valley and the Mariposa Grove of giant sequoias a state park. Tourists came in increasing numbers, and Hutchings needed to expand his operations. To build cabins, he needed lumber, a sawmill, and a sawyer. Muir appeared in 1869 and performed sawing and carpentry for Hutchings when he was not exploring the countryside or leading tourists through it. In 1871 the two parted angrily. According to Yosemite historian Shirley Sargent, Muir had supplanted Hutchings, the "long time" resident and authority, as the *most knowledgeable* Yosemite Valley guide.

Before my research, James Mason Hutchings was the only person aware of the essential geomorphic processes that had created Yosemite Valley. As Muir's employer, he had become knowledgeable about glaciation, and both had witnessed rockfall in the valley generated by the 1872 Owens Valley earthquake. Both also had witnessed rockfalls due not to earthquakes but rather to freeze and thaw and to exfoliation through depressurization. But unlike Muir, Hutchings was one of the very few to have witnessed debris flows instigated by heavy precipitation, and wall-to-wall flooding of the valley's floor under 9 feet of water, all this occurring in an unprecedented storm of late 1867. This was mentioned in *Hutchings' Tourist's Guide to the Yo Semite Valley and the Big Tree Groves,* which first appeared in 1877. In it Hutchings speculated on the valley's origin, as follows.

> Down the South side of the Valley
>
> On our side down the valley, almost immediately opposite Pom-pom-pa-sus (the Three Brothers), on our left [south] there is upon the face of the mountain a white irregular spot, from which an avalanche fell, that, although of apparently insignificant dimensions, covered several acres with its *debris*. This will readily direct the eye to a point just below it, which bears the name of "Profile Mountain." [Profile Cliff.] Here can readily be recognized faces of all styles. But the most noticeable fact connected with this mountain is a remarkable fissure, or notch, or slot, or whatever it may be termed, that

is several hundred feet in depth, and only from three to five feet across it. [Actually there are five vertical, parallel fissures.] But for its rounding edges, one could stand on the one side and look down into its great depth, then step across it to the other. There seems to have been a narrow stratum of soft granite here, that was constantly removed by wind, or rain, or snow, or frost, until the hard walls only remained. There is not the smallest stream of water running through it. Several good sized boulders have dropped into it, and lodged about half way down [Photo 23]. This fissure was discovered and photographed by Mr. E. J. Muybridge, in 1868. Is not this a miniature illustration of the theory that the great valley itself may possibly have been similarly circumstanced in the friability of its rock, and formed in the same way—*by erosion.* And here may pertinently be introduced the enquiry,

## How the Yo Semite Valley was Formed?

Professor J.D. Whitney, is of the opinion that "the bottom of the valley sank down to an unknown depth." Prof. B. Silliman (of Yale College) thinks that "by some great volcanic convulsion, the mountains were reft assunder [sic], and an immense fissure formed." The writer differs from both of these, and believes that the probabilities are altogether in favor of the theory, that the Yo Semite valley was formed by *erosion.* There can be but little doubt that throughout the entire valley there existed soft strata of granite, that became easily acted upon by air, sunlight, moisture and frost: and so, that high water (much higher than has been seen or known in many centuries), created wild torrents, that swept through these soft strata with overwhelming force, and cut out the main plan of the valley. This again was widened and deepened by vast glaciers, thousands of feet in thickness; and which, by their irresistible attrition, soon removed the talus acted upon and brought down by the elements. We have been unable to find the first trace of any other creative power than *erosion.* Besides, this agency now, as in the past, is ceaselessly at work. This can be readily seen in the numerous and frequent avalanches that occur, especially in winter, when, during heavy snow storms, there will be one, upon an average, every ten minutes. Then, there are unmistakable evidences yet remaining of the passage down the Yo Semite of immence [sic] glaciers, that have grooved and rounded its mountain walls, and thus written the fact of their existence and mission beyond peradventure.

**Photo 23. A prominent fissure at Taft Point.**

Near here can be seen some of the effects of the great storm of Dec. 23, 1867, when the whole valley was a broad, foaming river; and rocks weighing many tons were hurled down steep mountain torrents with terrible power; the *talus,* when washed down, filled up the ravines, as you see, and buried the base of trees from two to twenty feet upward.... By evidences everywhere apparent, there has been no storm to equal it during the present century.

If Hutchings' had substituted "jointed granite" for "soft granite," then he would have come very close to my interpretation of how Yosemite Valley was formed. Perhaps he was careless in his terminology. It would have been impossible for him not to have heard Muir's views, which emphasized jointing. Muir referred to jointed and unjointed rock as weak and strong, respectively, and Hutchings might have used "soft" for "weak." Hutchings stood alone in recognizing the importance of mass wasting, stream erosion, and glacial erosion acting over time. But he was not a geologist, and unlike Muir he did not make friends with geologists in order to spread his interpretation. Because of his many publications, his interpretation may have become more widely known to the general public than that of Muir's seven "Studies in the Sierra," and certainly more known than the erroneous speculations by esteemed geologists in professional journals. Hutchings was more correct than was any contemporary or future geologist, including François Matthes and those who followed him, but he lacked the credentials to be taken seriously by professionals. Indeed, Matthes, in his review of former views on the formation of Yosemite Valley, did not mention Hutchings. Since Matthes was a strong admirer of Muir and was an active member in his Sierra Club, one might wonder if this disregard of Hutchings, whom Muir scorned, was not accidental. It appears that all other geologists and historians disregarded Hutchings' interpretation except for my first mention of him in my 1978 guidebook, *Yosemite National Park*. The U.S. Geological Survey ignored him entirely.

Instead of considering Hutchings' explanation, the branch of science known as alpine geomorphology—the study of mountain landscapes—embarked on a fateful direction. At stake was not only Yosemite Valley, not only the Sierra Nevada, not only the ranges of North America, but the ranges of the entire world. The geologists went astray in two respects: first, how to measure the ability of glaciers to erode; and second, how to identify the timing and magnitude of uplift of a mountain range.

The ability of glaciers to erode could have been solved early on, but fate decreed otherwise. Way back in 1830 Charles Lyell was on the right track in volume one of his classic *Principles of Geology,* when he remarked (p. 175-176):

> The glaciers also of alpine regions, formed of consolidated snow, bear down upon their surface a prodigious burden of rock and sand mixed with ice. These materials are generally arranged in long ridges, which sometimes in the Alps are thirty or forty feet high, running parallel to the borders of the glacier, like so many lines of intrenchment [sic]. These mounds of debris are sometimes three or more deep, and have generally been brought in by lateral glaciers: the whole accumulation is slowly conveyed to the lower valleys, where, on the melting of the glacier, it is swept away by rivers.

The long ridges are medial moraines, each forming when two glaciers merged together, as did their adjoining lateral moraines. A lateral moraine forms when debris that has been transported by a glacier falls off it to accumulate between the side of the glacier and the adjacent canyon wall. While the medial moraines eventually may be swept away by rivers, the lateral moraines, not addressed here by Lyell, stay more or less intact. But this is not the issue. The issue is the source of the debris—his rock and sand—that formed the moraines. Did the debris originate from rockfall atop the glacier, or did it originate from the glacier plucking pieces of bedrock from its canyon's floor and walls? In volume three, which appeared in 1833, Lyell chose the obvious and certainly correct answer (p. 149-150):

> Those naturalists who have seen the glaciers of Savoy, and who have beheld the prodigious magnitude of some fragments conveyed by them from the higher regions of Mont Blanc to the valleys below, to a distance of many leagues, will by prepared to appreciate the effects which a series of earthquakes might produce in this region, if the peaks or 'needles,' as they are called, of Mont Blanc were shaken as rudely as many parts of the Andes have been in our own times. The glaciers of Chamouni would immediately be covered under a prodigious load of rocky masses thrown down upon them.

The glaciers did not erode the debris; rather, the debris fell upon the glaciers (Photo 24). To paraphrase a 1990s bumper sticker, "Rockfall happens." Earthquakes are not required, nor are glaciers, for that matter. As Hutchings had observed in the above quote, rockfalls occur in Yosemite Valley. Visit the valley in late winter or early spring and spend a night there, listening to the rocks fall off the cliffs. As the rocks continue to fall off the cliffs, the valley widens. For example, say a large slab 10 feet thick falls off a cliff. Then at that spot the cliff has retreated by 10 feet and the valley has been widened by 10 feet. Perfectly obvious?

Not to more than a century of geoscientists, who came up with a "better" idea. Rockfall can occur year after year, millennium after millennium, eon after eon, and the canyon never widens. How so? Because *by definition* rockfall does not occur in glaciated mountains! How did they arrive at this absurd conclusion? I suspect it arose because of the way that many geologists go about doing their research. Like other scientists, they wanted to observe things—in this case processes that transform moun-

**Photo 24. Debris-covered snowfields at the base of North Peak. This location is just outside Yosemite National Park and is about 6 miles north-northwest of the park's Tioga Pass entrance station.**

tain landscapes. They could visit a mountain river and measure its velocity. Likewise, they could visit a mountain glacier and measure its velocity. But what about a rockfall? During the pleasant summer months (who wants to work in winter's storms?) they could sit for weeks, watching a cliff, waiting for a piece of it to break loose. Chances are that nothing might happen, and so, out of sight, out of mind. Worse, how do you quantify rockfalls? They are very messy; they seemingly occur at random times and in random sizes. Rivers and glaciers, on the other hand, readily respond to quantification and can be modeled with mathematical formulae. So apparently the mid- and late-nineteenth-century geologists tossed out that which they could not observe and quantify: mass wasting, in particular, rockfall. Look in any of their geologic textbooks. Mountain landscapes evolve solely by stream erosion and/or glacial erosion. There is no mention of mass wasting, which is the most obvious process to both alpinists and mountain residents.

Today's textbooks are little different. They will have a chapter or subchapter on mass wasting, but in their chapter on glaciation, the glacial debris—be it deposited as moraines or transported downstream—is *solely* the result of glacial erosion. Ironically, most chapters will have at least one photograph or drawing of an alpine glacier heavily laden with rockfall debris. The photograph I like best is one that has appeared several times. This is Austin Post's aerial photograph of a very massive rockfall—generated in the 1964 Good Friday earthquake—which blanketed the lower part of Alaska's Sherman glacier under about 30 million cubic yards of debris. Because the glacier has been advancing, this material has been transported and deposited. If you were a young, budding geologist fresh out of college, this rockfall would have occurred before you were born, and so you may not be aware of it. Should you visit the glacier and see its prodigious deposits, you would conclude, correctly so by standard convention, that the prodigious deposits were eroded by the glacier and they are evidence of major glacial erosion.

The second way in which geologists went astray was with regard to uplift of a mountain range. How do you measure the amount of uplift and how do you determine the span of time over which this uplift occurred? For the Sierra Nevada the answers lie in the late Cenozoic uplift paradigm which, generally put, states that major uplift has occurred over the last few million years. This myth probably had its origin in folk geology. Pioneers, confronted with the steep, fresh, eastern escarpment of the Sierra Nevada, could envision major, relatively recent uplift. Later, Gold Rush miners, working in the Stanislaus River and Tuolumne River drainages, became aware of gold-bearing gravels beneath a giant, meandering lava flow. It took little reason to deduce that this flow must have advanced down an ancient Sierran canyon, one with a river meandering through it. Prof. William Brewer, of the Whitney Survey, may have been the first to put this in writing, which he did in his November 2, 1862 journal entry:

> All through the Sierra there have been imense craters, or "vents," from which enormous quantities of lava have flowed, the streams streaking the slopes, and forming table mountains. This may be taken as a sample:

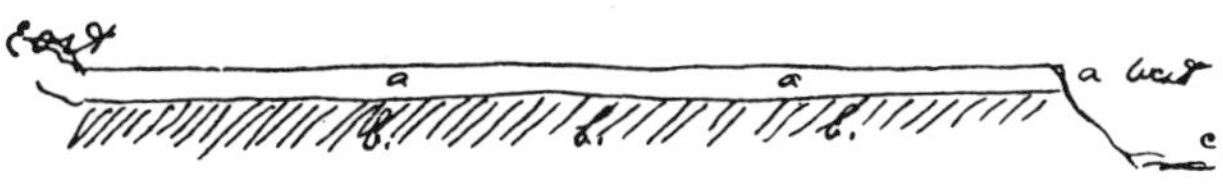

**[Figure 9. A table mountain, longitudinal profile.] *a-a-a,* table of lava stretching from the mountains back toward the valleys to the west; *b-b-b,* the gold-bearing slates, always standing more or less on edge and dipping east; *c,* valley of the interior of the state.**

> But these are *narrow* hills, and if cut across, their shape is thus:

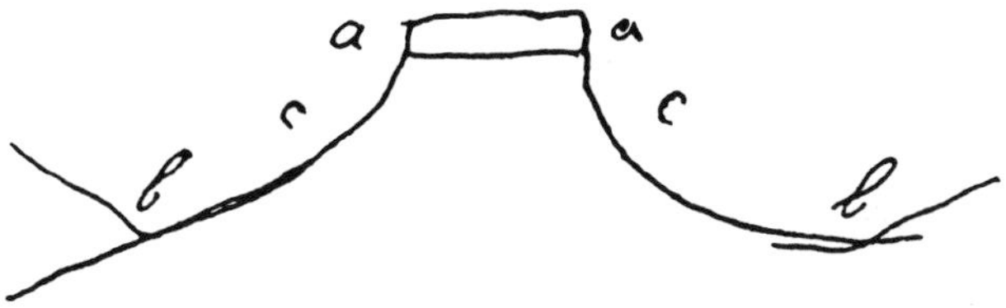


**[Figure 10. A table mountain, cross profile.] *a-a,* the flat, level top of the hill, naked rock, the edges abrupt precipices, from one hundred to six hundred feet thick; *b-b,* valleys of the present rivers and streams; *c-c,* steep sides, of gold-bearing slates, covered more or less with chaparral and bushes.**

> Now these are old lava streams, long and narrow. There is a table mountain in Tuolumne County that is *ninety miles long,* and never over a mile wide, generally less than half a mile! [Review photos 3 and 8.] How could it have been formed? you ask, and the answer is the extraordinary part of the story. This lava, now the top of a mountain, ran in a stream in the bottom of a valley! Those hills of softer slate have been worn down until they are the valleys, while the harder lava has withstood the elements and forms now the top of the mountain. Most extraordinary fact! Once the outline was the line *a-a-a-a* in this sketch; now it is the line *b-b-b-b:*

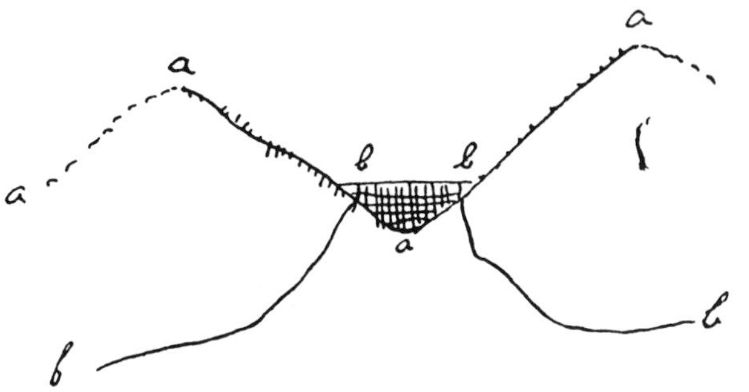


**[Figure 11. A table mountain originally forming on a canyon floor.]**

> Now, upon this fact is founded an important element in one kind of mining—"tunneling"—which proves that these streams of lava flowed down the valleys of old rivers. The mountain is tunneled, often at an enormous expense, to find the old river bed, and from that get the gold. The following sketch will show my meaning:

**[Figure 12. A table mountain, geologic cross section.] *a,* represents the lava top or "table"; *b-b,* the underlying gold-bearing slates, called the "bed rock," and also the "rim rock."**

Brewer goes on with his enthusiastic description, but we can stop here. What I want you to do is look at the lava flow in Figures 10 and 11. In Figure 10 its base is flat, and this is how the bases of the Sierra's table-mountain flows appear to be. But in Figure 11 its base is convex upward, which is how it should be if it had flowed along the floor of a narrow bedrock (e.g., slate) canyon. The first cross profile is real; the second, hypothetical, on which Brewer has constructed a geologic cross section (Figure 12). It is this geologic cross section of the Table Mountain of Tuolumne County that Whitney used in Figure 38 of his 1865 *Geology of California* (Figure 13):

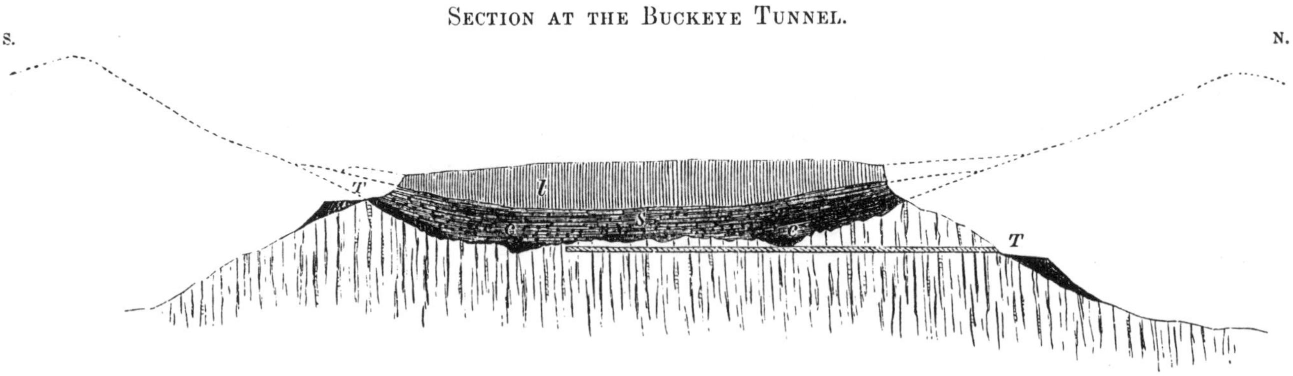


**Figure 13. Whitney's geologic cross section through Table Mountain. *l.* Lava forming the Table Mountain. *S.* Sandstone, underlying the lava. *T, T.* Tunnels, driven in from the north and south sides, to strike the two channels. *c, c.* The dotted lines indicate the probable former position of the surface, at the time the lava was erupted.**

Although Whitney's caption mentions two tunnels, only one is drawn, and it touches the base of the sandstone (the middle layer) only in one point, about midway along the tunnel. The cross profile of the sandstone's base is reconstructed on the basis of only one point! Thus, the cross profile of the base, which represents the floor of the bedrock canyon, is a hypothetical construct. Before going on to how this myth grew, I'd like to point out that Brewer's statement that the tunneling "proves that these streams of lava flowed down the valleys of old rivers" is untrue. What the field evidence shows is that streams of lava generally flowed across sediments and only locally flowed across bedrock benches. There is no evidence for a bedrock canyon, although the flow probably advanced down a canyon of unknown depth that had been carved out of sediments overlying the bedrock. Just how wide the canyon was is a matter of speculation, since the loose sediments that confined the flow have long since disappeared.

Whitney envisioned the Table Mountain lava flow descending along the floor of a shallow canyon carved in bedrock. Given that the Table Mountain is located in the western Sierra, which has lands of relatively low elevation and low relief (the difference in elevation between the high and low points), his reconstruction is logical, even if it is imaginary. In Figure 866 of Joseph Le Conte's 1887 *Elements of Geology* the geologic cross section has been embellished:

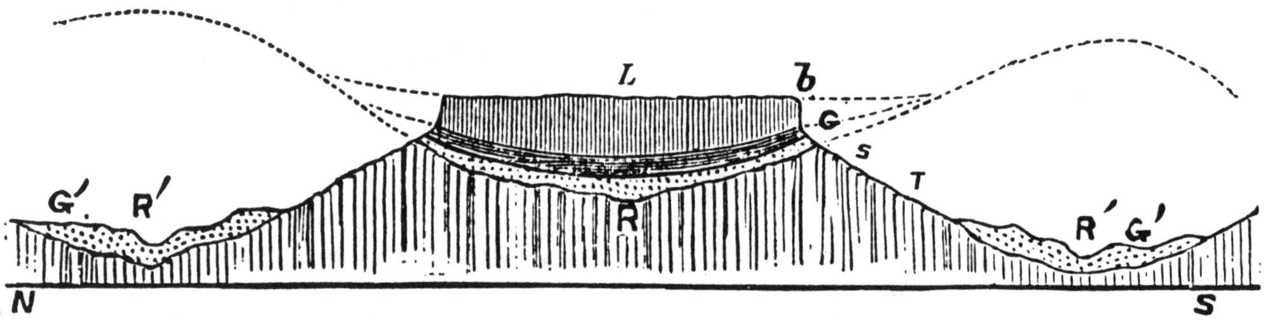


**Figure 14. Le Conte's geologic cross section through Table Mountain. Section across Table Mountain, Tuolumne County, California: L, lava; G, gravel; S, slate; R, old river-bed; R', present river-bed.**

Not only have the slopes of the previous canyon been slightly steepened, but the bases of the lava flow and the two underlying sedimentary layers have been redrawn to be more convex upward. If you've ever waded across *wide* streams or shallow rivers, then you know that from the banks the bottom quickly levels off and is mostly flat-floored. Table Mountain in Tuolumne County averages about 1500 feet wide, making it about five times wider than the Colorado River across the Colorado Plateau. Measurements there (Lees Ferry, Arizona) indicate that the floor beneath the Colorado is quite level, varying by only about 2 to 5 feet over a width of about 300 feet. The ancient river beneath Table Mountain should have behaved similarly. It is instructive to note that while the ancient rivers maintained *convex-upward* cross profiles—note the bases of the lower sedimentary layer, the upper sedimentary layer, and the lava flow—the floors of the two modern rivers are cut into *horizontal* stream terraces—deposits left not too long ago, geologically speaking. That is a correct interpretation of stream erosion across sediments; the convex-upward cross profile is not.

What has Table Mountain got to do with uplift? Well, in his 1898 USGS Bulletin 89, Leslie Ransome said that the meanders indicated that the river had a low gradient, much as the meandering lower Mississippi and meandering lower Colorado rivers do. Furthermore, in its longitudinal profile the ancient Stanislaus River had a fairly constant gradient. Therefore, Ransome could extrapolate this low gradient eastward toward the Sierran crest. He mapped the flows comprising Table Mountain from about Knights Ferry, near the base of the range, northeast about 70 miles to the Dardanelles. One could assume that the ancient river had a gradient of a few feet per mile, and so in the vicinity of the Dardanelles the canyon bottom may have been at only a few hundred feet elevation. Or, perhaps the gradient gradually increased eastward, so that the bottom was at 1000 feet elevation. Today the base of the lowest lava flow at the Dardanelles is at about 8000 feet elevation, which indicates perhaps some 7000 feet of uplift. The Dardanelles are not at the Sierran crest, which stands about 10 miles to the northeast. Up there the uplift would have been even greater, on the order of 8000+ feet.

Like Brewer and Whitney, Ransome asserted that the Table Mountain lava flow descended along the floor of a bedrock canyon. However, he correctly mapped the base of the lowest lava flow at the Dardanelles as standing about 1000 feet above a more-ancient remnant of rhyolite. How did the rhyolite get there? Obviously it must have been deposited on bedrock slopes that were existing back then, long before the lava flow. Furthermore, since we still don't know just what was the lowest level of the original rhyolite—it could have descended about 2000 feet almost to the floor of today's Middle Fork Stanislaus River canyon—we can't say how deep this canyon was when the rhyolite was deposited. Hence, rather than advancing along the floor of a bedrock canyon, the Table Mountain lava flows advanced across sediments from 1000 to 3000 feet thick. This uncertainty is troubling for geologists, who want concrete answers, which I provide later. As with the issue of glaciation, theory prevailed over field evidence that demanded an opposing conclusion. In the Sierra Nevada there is abundant evidence against major glacial erosion and against major late Cenozoic uplift, and a significant amount of it has been known since 1900. However, theory was in control. Over the decades the myth of late Cenozoic Sierran uplift (which implied major glacial erosion) grew as one geoscientist after another produced his (or in one case, her) "proof."

The exact contrary of what is generally believed is often the truth.

—Jean de la Bruyère

# 8

# Matthes, the Final Arbiter

The controversy over Yosemite Valley's geomorphic evolution lasted for more than a half century, not being satisfactorily "resolved" until the 1930 publication of François Emile Matthes' USGS Professional Paper 160, *Geologic History of the Yosemite Valley* (Photo 25). By 1913, when Matthes was assigned to resolve the "Yosemite problem," he seemed eminently qualified for the task. He had done topographic mapping of the Big Horn Mountains (1898-1899), of Glacier National Park (1900-1901, before it was a park), and part of Mt. Rainier (1910-1911), and he had written a glacial account for each area. Before the Mt. Rainier assignment, he made a brief visit to Yosemite Valley. His annual field diaries, which are now stored at the Bancroft Library (University of California, Berkeley), indicate that on June 9, 1903, he took a break from his Grand Canyon field surveying to visit Yosemite Valley. He arrived there by stage on June 14, and then on June 15, with his two survey partners, climbed to Glacier Point via Union Point, ascended Sentinel Dome, and then returned to the valley floor via Nevada and Vernal falls. On June 16 he climbed to the top of Upper Yosemite Fall and went over to nearby Yosemite Point. On June 17 he visited Mirror Lake and then left for the Grand Canyon the following day.

Upon completing his Grand Canyon field work in January 1904, he returned to the USGS in Washington, D.C., to begin drafting two topographic maps of the area. In February he received a form letter from Harvard's Prof. William Morris Davis regarding the inauguration of an association of geographers. The two men began corresponding, and then in September finally met at the Eighth International Geographic Congress, where Matthes gave talks on mapping the Grand Canyon and on glaciers of

**Photo 25. François Emile Matthes. Source: Bancroft Library.**

Montana's Lewis Range. Davis convinced him to come to Harvard for a master of science degree in geology, and his supervisors at the USGS granted him a leave of absence. During the school year of 1904-05, Matthes specialized in geomorphology, studying under Davis, who was the father of American geomorphology. Although Davis had considerable field experience, he nevertheless derived his principles of geomorphology far too much on logic rather than on evidence—and Matthes embraced the principles. Matthes did not get an M.S. degree, for on March 15, Hayes and Douglas of the USGS made him a definite offer to map Yosemite Valley, Matthes' cherished hope, and two days later he wrote a letter of acceptance. Over the next 2½ months, he frantically tried to finish drafting his Grand Canyon topographic maps, and he pored over papers on glaciation and on Yosemite Valley.

On June 5, 1905, Matthes left Washington, D.C., on a train bound for Sacramento, then took a mail stage to Wawona and Yosemite Valley. He surveyed Yosemite Valley from mid-June through late November, then returned home. In February he was elected to membership in John Muir's Sierra Club, then in March took a train to the University of California in Berkeley. While there, to assist geology professor Grove Karl Gilbert in his study of sediment transport by streams, Matthes experienced the San Francisco earthquake of April 18, 1906, and he quickly was assigned the task of mapping the newly discovered San Andreas Fault. From this experience he would develop close ties to the university, and ultimately would retire near it.

His return to Yosemite Valley had been postponed to late July, but nevertheless he was able to complete his surveying by late September. By the time he left the valley on October 24, he had become intimately acquainted with its topography and with much of its glacial evidence. Back in Washington, D.C., he completed his Yosemite Special map on March 27, 1907, and then only a few days later, on April 13, began working on his *Physical History of Yosemite Valley*. (Usually his diary explicitly or implicitly indicates if a project is assigned or not, but that is not the case here. I interpret his diary to indicate that he started working on this project unofficially in his spare time. However, the project became official by February 1909, if not sooner.) In late December 1907, he gave a talk on "The cliffs of the Yosemite Valley" at the Association of American Geographers' Annual Meeting in Chicago. His abstract stated in part:

> In as much as the rate of disintegration varies directly with the frequency of the fissuring, it follows that cliff recession must take place at greatly varying rates in such dissimilar materials.... Finally, it would appear that the entire Yos. Valley owes its exceptional character largely to the unusual structure of its rocks; stream and ice erosion have merely been the means to bring out its possibilities.

In a few years Matthes would place a much greater importance on stream and glacial erosion in transforming the landscape, guided in part through joint patterns. Mass wasting, such as rockfall, would be considered insignificant in the transformation, and cliff recession would be due not to rockfall but to glacial erosion.

Through 1908 and 1909 Matthes worked intermittently on a Yosemite paper, then gave it to his supervisor, R. B. Marshall, on December 21. In 1910 he spent two months revising it, and submitted it on April 21. But in 1911 he began a second draft and worked intermittently on it for 7 months, into early 1912. On March 26 he saw Waldemar Lindgren about his Yosemite paper, who arranged with Marshall that he should prepare a new, shorter paper for the Interior Department. This third draft (by far his shortest) took 2½ months to write, and was published in 1912 as *Sketch of Yosemite National Park and an account of the origin of the Yosemite and Hetch Hetchy valleys*.

Eighteen years would pass and a total of *13 drafts* would be written before Matthes finally completed his Professional Paper 160, which was published late in 1930. His field diaries indicate that over the years Matthes spent about 157 months working on these 13 *major* drafts, taking an average of a little over 12 months per draft. This time was spent writing; it does not include time spent on the original field work or on ensuing field work from Yosemite National Park south to Sequoia National Park. The public did not have wait until 1930 to hear his views, for he talked about them whenever he visited the valley, and then in 1922 an illustrated summary of his manuscript was printed on the back side of his topographic map of the valley.

What changed in those 18 years? First, in his 1912 "Sketch" he stated that the Sierra Nevada originally was the western part of a vast "dome" produced by the broad upwarping of the Great Basin of Nevada and Utah. These lands were faulted late in the Tertiary (that is, in the last few million years), leaving the Sierra Nevada to stand alone. This clearly follows Joseph Le Conte's 1886 interpretation, which almost a century later, in 1984, has been resurrected in a modified form by Peter Coney and Tekla Harms and then by others in the light of Basin and Range detachment faulting. (In short, rifting apart of the earth's horizontally extending crust caused a high, mountainous landscape to collapse, resulting in today's alternating basins and ranges.) Matthes, writing for the *Sierra Club Bulletin* in 1914, soon changed this view to an uplift of the range to its present height at the close of the Tertiary period—about a million years ago by his reckoning. But in 1930 he stated that the range rose as an isolated block in *three distinct, short, fairly well dated pulses,* the last one occurring at the close of the Tertiary period.

Second, Matthes in 1912 stated that streams initially had smooth profiles, but after glaciation these profiles became stepped, possessing alternating high- and low-gradient reaches. Later he stated that every stream quickly developed a smooth profile after each of the three pulses of uplift. Because Professional Paper 160 is many times longer than the "Sketch," it has much more detail, including analyses of minor points. For example, whereas like others he originally believed today's glaciers are remnants of a former glaciation, he later argued that they are remnants of the Little Ice Age, which ended in the 1800s. However, the crucial difference between the two studies is found on page 30 of his "Sketch":

> Many of the hanging valleys, it will be seen, debouch at heights ranging from 2,000 to 2,500 feet above the floor of the Yosemite Valley. It has been estimated from these figures, making due allowances for the lowering of the side valleys themselves and a number of other factors, that the Yosemite trough may have been excavated at least 2,000 feet. Whether all of this work is to be accredited to the ice, however, or in part also to stream erosion is a question that cannot yet be determined with certainty.

In Professional Paper 160, he claims to have determined it with certainty. His quantification and precision are apparent in the first three sentences of the professional paper's technical abstract:

> The problem of the origin of the Yosemite Valley inherently demands a solution in quantitative terms. Its essence is, To what extent is the valley a product of glacial action, to what extent a product of stream erosion? The principal result of the investigations upon which this report is based is the determination within narrow limits of the preglacial depth of the Yosemite Valley and of other facts concerning its preglacial development which permit fairly definite estimates of the proportionate shares of work performed by stream and by glacier.

Why was Matthes able to reach precise solutions where others could not? He "succeeded" because he used Prof. William Morris Davis' geomorphic principles. Davis' influence on Matthes cannot be overemphasized. On the one hand, Davis' methods of landform analysis allowed Matthes to reach precise solutions; on the other hand, this analysis ran counter to the abundant field evidence. The incompatibility of the two sets of evidence—theoretical and field—must have troubled him, and this may explain why his Yosemite paper was so long in appearing. Once he began working on his Yosemite paper in 1907, he spent the next 23 of his remaining 41 years—about half of his adult life—working intermittently on this problem. On the basis of his diary, his published writings, his correspondence with Davis, and praise by Kirk Bryan, Fritiof Fryxell, William Morris Davis, Robert Sharp, and Carl Sharsmith, and finally of my own experience (like him I was an identical twin with similar interests, similar life history, and similar problems), I suggest that only by reworking the manuscript year after year after painful year was he finally able to convince himself, through *subconscious* self-delusion, that he was correct. Others faced with his dilemma might have ignored the contradictory field evidence and would have completed the manuscript expeditiously. But Matthes as a perfectionist with the highest degree of integrity could not do this. Torn between loyalty to Davis (who got him into Harvard University and infused him with ideas) and the field evidence, Matthes had to endlessly rework his rationalizations to make evidence fit theory until he himself finally believed in them. One example of Matthes' rationalization will suffice. Gruss is decomposed granitic bedrock, but for Matthes it had two other interpretations. Where he thought ancient glaciers had flowed, such as on the slopes and benches below Mt. Starr King, the veneer of gruss became the remains of ancient moraines (Photo 26). Where he thought an ancient lake had existed in the Illilouette Creek drainage, the veneer of gruss on a granitic bench became lake sediments (Photo 27).

While Davis' geomorphic principles were accepted by many American geologists and geographers (at least during Matthes' time), they nevertheless were disputed by a few critics, notably by Prof. R. D. Salisbury of Chicago University. Ironically, when Davis suddenly resigned at Harvard University in 1912 (just before Matthes began to apply Davis' geomorphic principles), he was succeeded by William Atwood, a student of Salisbury's, rather than by his recommendation, Douglas W. Johnson, who was his most loyal student and a devout believer in Davisian ideas. Greater criticism came from German geologists, particularly after World War I from Albrecht Penck and his son, Walther, who was the severest critic. Had Matthes embraced Walther's geomorphic principles (which were equally inapplicable), he never would have been able to reach his precise solutions.

To return to our story of Matthes, on December 6, 1912, Leslie Ransome (the Sierran uplift progenitor) notified him of a request from the Secretary of the Interior for a more elaborate paper on Yosemite Valley. The Sierra Club had specifically requested the USGS to study the valley, and that by 1913 Matthes had already written four "Little studies in the Yosemite Valley" for the *Sierra Club Bulletin,* the series title resembling John Muir's "Studies in the Sierra." Just before Matthes' Yosemite

**Photo 26. Gruss-covered bedrock lip of a minor hanging valley. View is north toward summit 7007, which rises above Panorama Point.**

**Photo 27. Matthes' Lake Illilouette sediments: gruss.**

Valley assignment, Muir's book, *The Yosemite*, was published, in 1912. In it, Muir steadfastly maintained (as he had for decades) the long-abandoned view of the early nineteenth century that the world's ice caps originated at the poles and grew equatorward. Hence his statement (p. 131): "All California has been glaciated, the low plains and valleys as well as the mountains." As president of the Sierra Club from the time he founded it in 1892 until his death in 1914, Muir may have hoped that someone who was very credible in his field would vindicate his glacial views. Thus while Matthes probably would have been given the Yosemite Valley assignment on the merits of his previous work, it might never have come about were it not for John Muir, who had a personal stake in the matter. Matthes' favorable bias toward Muir is evident in Professional Paper 160, p. 4:

> It was John Muir, the *keen* student and ardent lover of nature, who *first saw clearly* that the glaciers themselves had done most of the excavating [Matthes had concluded that glaciers had done half].... Later, through his *charming* writings, he disseminated his ideas of glaciation far and wide, thereby uprooting in the minds of many people the *primitive* belief [of Whitney] in a catastrophic origin of the valley." (Italics mine.)

On July 1, 1913, Matthes was transferred from the Topographic Branch to the Geologic Branch, and he arrived in Yosemite Valley five days later. At this time there were 13 different hypotheses advanced to explain the valley's origin plus another in the *Handbook of the Yosemite National Park*. Matthes traced erratics to their sources, searched for and mapped moraines in the Merced River drainage and made a reconnaissance of parts of the Tuolumne River drainage, particularly around Hetch Hetchy and in and around Tuolumne Meadows. He concluded the season's field work in early December.

About this time, if not slightly earlier, he completed his "Studying the Yosemite problem," which appeared in the January 1914 issue of the *Sierra Club Bulletin*. It is most enlightening. He tells the public why yet another study of the valley's origin has been instigated: there now is an accurate, detailed topographic map of Yosemite Valley (his), on which can be mapped glacial evidence to show:

> how far the ice advanced and, if possible, how much excavational work it achieved.... The sculpture of the Yosemite region, however, is notably erratic and as an index of ice erosion quite deceptive.... the granites of the Yosemite region are peculiar in that they are not everywhere traversed by natural partings or "joints." ... While the glaciers might work to advantage in rocks divided into small joint blocks, they were relatively powerless when dealing with massive granite.... Former observers have not always made sufficient allowance for the exceptional nature of the granites of the Sierra. Perhaps they had in mind the work of glaciers in regions of more normally jointed rocks and thus were misled in assigning too great erosional achievements to the ice streams of the Yosemite region. The main lessons learned last summer in this regard are that except in certain restricted localities, such as the Yosemite Valley proper, the rock character precluded extensive remodeling by the ice, and that as a consequence the landscape, although bearing the unmistakable stamp of ice work, still retains very largely the features given it by normal weather-and-stream erosion prior to the advent of ice..... In the main it was found that exfoliation is rather more general in the Yosemite region than has commonly been supposed, and that many of the slabs and sheets clinging to cliff faces are really products of exfoliation, that is, of external influences, instead of being remnants of structures originally existing in the granite. It would appear therefore that in the Yosemite region massive rock prevails to a far greater degree than at first sight seems to be the case. And this fact must necessarily influence our views as to the amount of excavating to be accredited to the ice: in materials so unfavorable as massive granite it seems unlikely that glaciers would have been able to accomplish any great results.

This passage first implies that weathering, mass wasting, and stream erosion were important in the Sierra Nevada's geomorphic evolution, and that glacial erosion was not. Matthes then makes Yosemite Valley an exception, having been substantially remodeled by this ice. Then in the next paragraph he essentially contradicts this statement when he says that in the Yosemite region massive rock prevails. If so, then, as he said earlier in the passage, glaciers should not have extensively remodeled it. Yosemite Valley does contain some of the most massive granitic monoliths (especially El Capitan, Cathedral Rocks, Three Brothers, and Half Dome) in the Sierra Nevada. Therefore, to be consistent with his general statement, he should have concluded that the glaciers were relatively powerless when dealing with the massive granite found in Yosemite Valley.

Evidence to the contrary, Matthes sided with those who had said that glacial erosion in Yosemite Valley was effective. Matthes would say this was the case because there was easily quarried fractured, granitic bedrock. Although he doesn't state this explicitly in Professional Paper 160, his prose, his reconstructions, and his cross profiles indicate that a narrow Yosemite canyon was bounded by granite that was thoroughly laced with deep joints, and then this "loose" granite was easily removed by glaciers. Interestingly, the granitic bedrock did not achieve this highly fractured state until his canyon stage, which develops immediately before glaciation. Before then it was inexplicably much less fractured (compare his mountain-valley and canyon stages, illustrated in the next chapter).

In 1914 he resumed his field work in early June and completed it in late October, having filled in some details and expanded his areal coverage to include Bloody Canyon and Walker Lake, on the east side of the Sierra Nevada. In 1915 he did no field work, but spent most of the year writing his Yosemite paper. His diary indicates that on September 10 he studied notes on Prof. Davis' physiography lectures on Pennsylvania. Apparently he was studying "The rivers and valleys of Pennsylvania," written in 1889, or lecture notes on it. Davis wrote no other significant article on the area's geomorphology. This Matthes would apply as a model for the early, former Sierra Nevada landscape (Matthes, 1930, p. 27):

> Indeed, the similarity of its folded structure to that of the present Appalachian mountain system in the eastern United States enables the geologist to reconstruct its features with considerable confidence. Doubtless it was composed, like the Appalachian system, of long, roughly parallel ridges, separated by equally long, roughly parallel valley troughs.

On September 30 he was "drawing long sections of early-Quaternary Merced Valley," and on October 2 was "drawing cross sections to determine grade of early Quaternary Yos. Valley." This is his canyon stage; he has now begun to reconstruct the valley's geomorphic evolution strictly in accordance with Davis' views. Using Davis' methods, Matthes now had a way he believed would show how much excavation the glaciers had achieved. In late October he was "Making studies of various Sierra canyons to collect evidences of a second Sierra uplift." He reviewed Lawson's lengthy 1903 paper, "The geomorphogeny of the upper Kern Basin," which describes three distinct "geomorphic zones": high mountain, high valley, canyon. Matthes recognized three similar zones in Yosemite National Park and would name them broad valley, mountain valley, canyon. On November 1 he dictated a "long letter to Prof. Lindgren, giving instances of canyons-within-canyons in many different parts of the Sierra Nevada." In mid November he drew profiles of the Merced River as it might have existed in its different stages.

At this point it is only fair to state that throughout the central and southern Sierra Nevada nested canyons definitely exist. Logically one would conclude, as Matthes and others had, that the innermost ones are youngest, incised in older ones, and that each episode of incision was in response to a discrete uplift. Such an interpretation would be a vindication not only of Davis' revised cycle of erosion, but also of its *elementary* form. Both Albrecht and Walther Penck thought this cycle of erosion—rapid uplift followed by stream incision—was not realistic (varying rates of uplift and varying rates of incision were). In a very long letter to Albrecht Penck, dated April 3, 1921, Davis explained that his cycle is not as simple and unrealistic as his detractors make it out to be:

> In my Berlin paper, 1899, slow elevation is not explicitly considered; but possible complications are intimated in the statement regarding interruptions of a cycle: 'Only in this way can the theory of the cycle be made elastic enough to correspond to the variety of natural conditions', p. 288. Slow elevation not being considered there, I contributed a paper on 'Complications' at the Washington Congress, 1904, see p. 153. '*The elementary presentation of the ideal cycle usually postulates a rapid uplift of a land mass followed by a prolonged stand still.* The uplift may be of any kind and rate, but the simplest is one of uniform amount and rapid rate ... In my own treatment of the problem, *the postulate of rapid uplift is largely a matter of convenience, in order to gain ready entrance* ... It is therefore preferable to speak of rapid uplift in the first presentation of the problem, and afterwards to modify that elementary and temporary view by a nearer approach to the probable truth.' On the next page, 154, the case of an open valley eroded during a slow uplift is specifically considered. 'In such a case there would have been no early stage of dissection in which the streams were enclosed in narrow valleys with steep and rocky walls.' (Italics mine.)

Ironically, although Davis distanced himself from his 1899 paper on a cycle of erosion in its *elementary* form, Matthes embraced it, and crafted his Yosemite paper to conform to it.

By April 1916 Matthes' paper (now in its fourth draft) was nearly complete, and it was decided that Matthes should publish his paper in that year (three years after he officially had started it back in 1913). However, he returned to Yosemite National Park that summer and obtained new information. Next it was decided that Matthes' work, being major, should be a Professional Paper, and so in early January 1917 he began his fifth draft. Over the years he would make excursions to glaciated lands south of Yosemite, from the San Joaquin River drainage to the upper Kern River drainage. (It is unfortunate that like John Muir and others he did not investigate granitic landforms in the *unglaciated* parts of the northern and southern Sierra Nevada, which are difficult to reconcile with Davisian geomorphology.) These excursions plus readings and discussions would cause him to redraft his manuscript.

Reading through years of his diary, one finds countless instances that indicate he was obsessed with detail. For example, he would take days to write a one-page abstract, months to write relatively short articles, and for many years he worked intermittently at adding shading to his Yosemite Valley topographic map. This obsession with detail, causing him to redo a project again and again, slowed him tremendously, and must have been agonizing for his wife, whom he had married in 1911, back in the early days of the project. His Christmas 1929 entry is telling: "Presented Edith with statement of Survey editor showing that the Yosemite MS. has gone to the Gov't Printing Office, that being more cheering news than any conventional Christmas card I could find." Nevertheless, he continued working on it, taking half of February and most of March to write his Technical Abstract (pages 1-3 in the professional paper), and then taking five days in April to revise it. In January 1931, after its publication, he took four days to write a press release. His diary also indicates that beginning in March 1931 he started work on a popular account of Yosemite Valley, but he died—in 1948, some 17 years later—before completing it. Two other potential professional papers, on the central and southern Sierra Nevada, suffered the same fate, despite the field work done in the 1920s. His *Reconnaissance of the Geomorphology and Glacial Geology of the San Joaquin Basin, Sierra Nevada, California* was completed in April 1932 ready for publication. However, after a summer field season in Yosemite National Park, Matthes in September returned home to begin *rewriting* the San Joaquin manuscript, as he had done so many times with the Yosemite manuscript. His *Glacial reconnaissance of Sequoia National Park, California* also was delayed. Both professional papers were published posthumously.

In November 1925, Matthes had a conference with the Chief Geologist, W. C. Mendenhall, who declares his Yosemite paper "to be the best paper that has been produced by the Geol. Survey in some years." When his amended thirteenth draft finally was published as Professional Paper 160 in 1930, its intended audience (especially the Park Service) was more than ready for it. If there was any doubt that it had been an instant success (he hand-delivered a specially bound copy to President Hoover in January 1931), it would soon be declared so by Prof. Kirk Bryan, who in 1932 wrote:

> Occasionally in the history of science there appears a work so excellent, so comprehensive, that it becomes immediately a classic. Such a newborn classic is the long-awaited "professional paper" by Matthes on the Yosemite Valley. Much investigation of each minor detail of this unusually beautiful canyon is justified by a profound and widespread popular interest. It is fitting also that this information should be set forth in as simple and straightforward a style as possible. Matthes has demonstrated that clear and cogent reasoning can be expressed with a minimum of that scientific jargon used so unhappily by less careful writers.* Large issues of importance to every geomorphologist are discussed with great breadth of insight and a wealth of detail, largely of a quantitative nature. These issues are: (1) the geologic history of the Sierra Nevada, one of the great mountain ranges of the world; (2) the alternating stages of rapid uplift and intervening quiescence [Davis' *elementary* cycle of erosion]; (3) the threefold glaciation of the region; (4) the amount of glacial excavation; (5) the influence of jointing on erosion and the relative durability of unjointed monoliths.

My asterisk in the quote refers to a bit of irony. From Matthes' diary, June 24, 1935: "Spent most of forenoon reading Kirk Bryan's abominably written paper on pediments for the Internat. Geologic Congress."

Bryan then goes on to give a synopsis of the work, starting with the "Cretaceous peneplain." The presence of Davis' method of landscape analysis is unmistakable. It is unfortunate that Bryan wrote this glorious review, since first, he was highly partisan, and second, he was unqualified to evaluate the merits of the work. As to his partisanship, he and his wife Mary were friends of François and Edith Matthes, and Kirk worked with François as a colleague at the USGS from 1912 until 1926, when he was appointed assistant professor of physiography at Harvard University. This is the position that Davis had held earlier, and Bryan, in taking it, became a strong advocate of Davis' geomorphic principles, as exemplified in his 1935 memorial, "William Morris Davis—Leader in Geomorphology and Geography": "his greatest work .... was the formulation and elaboration of the doctrine of the Cycle of Erosion." (He then goes on to criticize Penck and "certain die-hards" who opposed the theory.) Second, as to Bryan's qualifications, he apparently never performed any serious Sierran field work, and the closest he got to the range was during two studies on ground water of the Sacramento Valley, which were published in 1915 and '23. His only glacial study was in 1928, the "Glacial climate in non-glaciated regions."

Furthermore, Matthes, who in his diary recorded every encounter with people, nowhere mentions that Bryan ever visited him in the field.

One might envision a Davis/Matthes/Bryan conspiracy here, since all were mutually reinforcing their joint beliefs, Bryan in particular using Matthes' work to show that Davis' elementary cycle-of-erosion theory was correct. However, a conspiracy seems unlikely. First, Matthes, a gentleman extraordinaire, had far too much integrity to have ever taken part in one. Second, Bryan apparently undertook the review on his own, as implied by Matthes' April 23, 1931 entry: "Rec'd note from Kirk Bryan saying that he is getting together material for a lengthy review of the Yosemite report for the Zeitschr. für Geomorphol." Still, even if others had reviewed Matthes' Professional Paper 160, they likely would have given it high praise. It is so extremely detailed and is so convincingly written that it was never *seriously* questioned. Amazingly, Matthes had drawn on his glacial map of Yosemite Valley an astounding amount of *nonexistent* glacial deposits. Since he was above deceit, he would not have intentionally added this false evidence. Therefore, as I implied earlier, he must have unconsciously deluded himself into believing his imaginary evidence actually existed. No one had cause to doubt his evidence, given his legendary reputation, and if anyone had noticed that some evidence on his map did not exist in the field, no one said so. This imaginary evidence went unnoticed until the 1990s, when I remapped glacial evidence in Yosemite. Matthes' most important conclusion, based purely on Davis' views, that Yosemite Valley has been greatly widened and deepened by glaciers, was spread worldwide, and it became one of the foremost "proofs" that glaciers perform major erosion in resistant bedrock.

# 9

# A Newborn Classic

Was Matthes' *Geologic History of the Yosemite Valley* a work so excellent, so comprehensive, that it warranted becoming immediately a classic? This professional paper has long been out of print, and where copies do exist, such as in some university libraries, they are falling apart, due to the relatively poor quality of the paper they were printed on. So for you to know what it said, I have reproduced its lengthy technical abstract. The work remains, after all, still the primary explanation for how the landforms of the Yosemite Valley and the Sierra Nevada evolved over time. I have also included his four drawings that illustrate the valley's geomorphic evolution. (These appeared not in the abstract, but rather in his text, as his Figures 13-16). Finally, in brackets I have critiqued what he has said. Because there are so many critiques, they are quite distracting, so it is better to read the Technical Abstract first to get a feel of what Matthes said, then go back and read it along with the critiques. You need not read the critiques; they are taken from evidence that is presented in later chapters. Most readers may not be familiar with geologic time, and because the Technical Abstract mentions geological periods and epochs, here is a short review of the subject.

## Geologic Time Scales

Conceptions of the earth's age have changed considerably since the first geological explorations in the Sierra Nevada in the 1860s. Before then, there was no acceptable age constraint, and estimates ranged from the Biblical few thousand years to William Conybeare's quadrillions of years. But beginning in 1862 and continuing through 1899—a time that was the heyday of pronouncements on the manner of origin of Yosemite Valley—acceptable (if erroneous) age constraints were set when William Thomson (Lord Kelvin) published a number of estimates of the earth's age. The first estimate was on the order of 100 million years, but the last ones were between 20 and 40 million years. These pronouncements were at odds with some ages based on how long it would take to produce all of the earth's sedimentary strata. The estimates ranged from 3 to 1584 million years, but most fell between 20 and 100 million years, that is, in accordance with Thomson's estimates. (Donald Prothero in 1990 suggested that most of the estimates were under 100 million years "probably because those who calculated them were all unconsciously influenced by the estimate of Lord Kelvin.") Although such a length of time seems sufficient to uplift and erode a mountain range, early investigators believed that the Sierra's uplift occurred late in the history of the earth, and therefore the carving of the Sierra's canyons must have been a relatively recent process. Whitney erroneously believed that the old gold-bearing gravels and the overlying volcanic deposits were quite recent—that is, Pliocene age—and then the river canyons may have been cut by extremely powerful streams produced by the melting of great glaciers. For Whitney, Kelvin's age of the earth allowed more than sufficient time for the Sierra's uplift and erosion.

A geologic time scale became possible only with the discovery of radioactivity and a way to measure it in igneous rocks. Radioactive decay was discovered in 1902, and by 1905 Bertram Boltwood, working in the United States with uranium-lead, obtained his first dates, which were quite accurate. By 1913, when François Matthes officially began his geomorphic field work in Yosemite Valley, sufficient dates had been performed to allow Arthur Holmes to devise the first true geologic time scale, which appeared in his book *The Age of the Earth.* Thus by the time Matthes completed his Professional Paper 160 in 1930, a relatively accurate scale existed for the Mesozoic and Cenozoic eras. Dates for the Paleozoic and Proterozoic eras were scarce, but then they were of no importance to the history of the Sierra Nevada. In Table 1, below, I present three geologic time scales for the Mesozoic and Cenozoic eras only: Matthes, Harland *et al.* (1990), and my own, which is an update of Harland *et al.* based on more-recent sources. Matthes' time scale is presented here for purposes of comparison because he refers to epochs that today have different upper and lower limits and different durations.

The date I use for the Plio(cene)-Pleistocene boundary is *not* the conventional date, which is based on the age of a sedimentary layer in southern Italy. Rather, I have used 2.48 million years, which is the date of the Gauss-Matuyama polarity reversal. (At that time, the polarity of the earth's magnetic field switched from normal, as it is today, to reversed, which is the opposite. During the Matuyama reverse epoch, the north magnetic pole existed as the south magnetic pole, and vice versa.) Support for this polarity-reversal date—and arguments against the conventional standard—were given by Roger Morrison in 1991. He gives four arguments in favor of it. First, it is essentially synchronous with the initiation of moderate-sized ice sheets in North America and Europe, after several million years without significant glaciation. Second, this drastic climatic change is well represented in sediments throughout the world. Third, this polarity reversal can be identified unambiguously in terrestrial and marine sequences throughout the world, much better than can the present boundary. And fourth, placing the Plio-Pleistocene boundary at 2.48 million years accommodates the two classic concepts of the Quaternary—the Ice Age and the Age of Man. Morrison also gives three arguments against the conventional boundary, to which could be added that its actual age is uncertain. He then cites others who have commented on its uselessness in the United States and elsewhere. Because my study is centered largely around the issue of glaciation, the 2.48 million-year Gauss-Matuyama polarity-reversal date is a far more logical candidate for the Plio-Pleistocene boundary. Furthermore, the Ice Age, composed of a series of ice ages, or glacial stages, has been long equated with the Quaternary period, its Recent epoch being the latest interglacial stage.

## Professional Paper 160—Technical Abstract

The problem of the origin of the Yosemite Valley inherently demands a solution in quantitative terms. Its essence is, To what extent is the valley a product of glacial action, to what extent a product of stream erosion? [*Chemical weathering, physical weathering, and mass wasting are not considered. These operated at varying degrees throughout the valley's history. Sierran field evidence indicates that they were more important than streams and glaciers combined, which overall merely transported away weathered and mass-wasted products.*] The principal result of the investigations upon which this report is based is the determination within narrow limits of the preglacial depth of the Yosemite Valley and of the other facts concerning its preglacial development which permit fairly definite estimates of the proportionate shares of work performed by stream and by glacier. The investigations comprise a detailed survey of the glacial and geomorphologic features of the Yosemite region and an equally intensive study of its rock formations, supplemented by reconnaissance work of both kinds in adjoining parts of the Sierra Nevada. The petrologic studies were made by Frank C. Calkins; the glacial and geomorphologic survey by François E. Matthes.

Detailed mapping of the morainal system of the ancient Yosemite Glacier has served not only to determine the farthest limits reached by that glacier but to throw new light on the significance of the hanging valleys of the Yosemite region. [*Hanging valleys would be significant for the reconstruction of former profiles of the Merced River during different stages only if their streams were accordant—not hanging—with the river during those stages. This often was not the case, as Clyde Wahrhaftig in 1965 recognized. Today's hanging valleys cannot be used to accurately reconstruct a river's pre-glacial profile, since all existed at the onset of glaciation.*] It is reasonably certain that the glacier never extended more than about a mile beyond the site of El Portal. The hanging side valleys of the Merced Canyon below El Portal, therefore, hang not because of any glacial deepening suffered by that canyon. [*By his own observations, the canyon should have been severely deepened. He concluded (p. 67) that the Merced River had incised through 50 feet of bedrock in postglacial time—the last 20,000 years, by his reckoning. So for the length of the Quaternary period, or Ice Age—1,000,000 years by his estimate—about 2500 feet of canyon incision should have occurred. Since the Merced River canyon between El Portal and Briceberg (from where Highway 140 climbs toward Midpines) averages about 3000 feet in depth, this canyon could have been incised almost entirely during the Quaternary period, if one assumes a constant rate of incision. Actually, Matthes is correct in stating that virtually no postglacial incision has occurred, but for the wrong reason. At the end of the last glaciation, the floor of the canyon was choked in outwash—sediments transported and left by the glacier's river. Over time most of these have been eroded away. Today all that remains are patches of sediments lying above the river's highest flood stage. The bases of these patches are up to 50 feet above the river's bedrock floor. So in postglacial time the Merced River has cut through 50 feet of glacial sediments, not 50 feet of bedrock.*] The explanation is offered that they hang because their streamlets have been unable to trench as rapidly as the Merced River since the rejuvenation of the Merced by the last uptilting of the Sierra Nevada. The streamlets were handicapped not only

**Table 1. Geologic time scales (except for Recent, time is in millions of years).**

| | Matthes Range / Duration | Harland *et al.* Range / Duration | This work Range / Duration |
|---|---|---|---|
| CENOZOIC era[1] | | | |
| Quaternary period | | | |
| Recent epoch | 20,000-0 / 20,000 yr | 10,000-0 / 10,000 yr | 10,000-0 / 10,000 yr |
| Pleistocene epoch | 1-0.02 / 1[2] | 1.64-0.01 / 1.63 | 2.48-0.01 / 2.47 |
| Neogene period | | | |
| Pliocene epoch | 8-1 / 7 | 5.2-1.64 / 3.56 | 5.2-2.48 / 2.72 |
| Miocene epoch | 20-8 / 12 | 23.3-5.2 / 18.1 | 23.3-5.2 / 18.1 |
| Paleogene period | | | |
| Oligocene epoch | 36-20 / 16 | 35.4-23.3 / 12.1 | 33.4-23.3 / 10.1[3] |
| Eocene epoch | 59-36 / 23 | 56.5-35.4 / 21.1 | 56.5-33.4 / 23.1 |
| Paleocene epoch | first part of Eocene[4] | 65.0-56.5 / 8.5 | 65.0-56.5 / 8.5 |
| MESOZOIC era | | | |
| Cretaceous period | 134-59 / 75 | 145.6-65.0 / 80.6 | 146-65 / 81[5] |
| Late Cretaceous epoch[6] | | | 97-65 |
| Early Cretaceous epoch | | | 146-97 |
| Jurassic period | 174-134 / 40 | 208.0-145.6 / 62.4 | 208-146 / 62 |
| Triassic period | 214-174 / 40 | 245.0-208.0 / 37.0 | 245-208 / 37 |

Notes:
[1]Harland *et al.*, along with the Geological Society of America, divide the Cenozoic era into the Neogene and Paleogene periods rather than into the older classification scheme of the Quaternary and Tertiary (Pliocene-Paleocene) periods, terms still common among stratigraphers and paleontologists.
[2]Rounded to the nearest million by Matthes.
[3]High-precision date obtained by McIntosh *et al.* (1992).
[4]The Paleocene was suggested by W. P. Schimper, a paleobotanist, in 1874, but was slow in gaining world-wide acceptance due to a paucity of recognized Paleocene fossils. As the number of these increased over time the Paleocene gained acceptance, but it was not formally recognized by USGS until 1939.
[5]Highly precise dates for the Mesozoic era are unnecessary for this study, although Swisher III *et al.* (1992) claimed a highly precise date of 65.0 Ma for the K-T (Cretaceous-Tertiary) boundary.
[6]The Cretaceous period is divided into Late and Early epochs because the Late epoch is important while the Early is not. It was during the Late Cretaceous that the central and southern Sierra Nevada experienced a major uplift and about then began to acquire a granitic character. The Early Cretaceous is not important with respect to the geomorphic evolution of Yosemite Valley and the Sierra Nevada. The terms "Early" and "Late" are taken from the DNAG 1983 Geologic Time Scale (Palmer, 1983).

by their comparatively small volume but also by the fact that their courses trend northwestward and southeastward, substantially at right angles to the direction of the tilting, and therefore have remained essentially unsteepened, whereas the Merced's course trends southwestward, directly down the slope of the Sierra block, and therefore has been appreciably steepened. [*While overall streams descending parallel to the Sierran crest have lower gradients than those descending perpendicular to it, there are too many exceptions to this generalization. The unglaciated South Fork Kern River descends southward parallel to the crest, yet its average gradient is as steep as westward-draining Sierran rivers. Furthermore, its gradient is highly variable, ranging from near zero to very high.*]

Projection of the longitudinal profiles of these hanging valleys forward to the axis of the Merced Canyon shows that they are closely accordant in height. [*Matthes was very selective in which tributary canyons he chose to produce the smooth profiles he needed for his three stages of the Merced River. He chose very few tributary canyons, which resulted in a great deal of uncertainty in the accuracy of his river profiles. Furthermore, he had to use a very inaccurate map for the most crucial part of his plotting. Using his method of analysis with modern, more accurate, maps, one arrives at totally implausible profiles.*] Their profiles indicate a series of points on a former profile of the Merced with respect to which the side streams had graded their courses prior to the last uplift. This old profile can be extended upward into the glaciated part of the Merced Canyon above El Portal and even into the profoundly glaciated Yosemite Valley, accordant points being furnished by a number of hanging side valleys (due allowance being made for glacial erosion suffered by those valleys). However, not all the hanging valleys of the Yosemite region are accordant with this set. Several of them, including the upland valley of Yosemite Creek, constitute a separate set indicating another old profile of the Merced at a level 600 to 1,000 feet higher than the first. Others, including the hanging gulch of lower Bridalveil Creek, point to an old profile of the Merced about 1,200 feet lower than the first. There are thus three distinct sets of hanging valleys produced in three cycles of stream erosion. [*Matthes determined three cycles of erosion that produced three surfaces: broad-valley, mountain-valley, and canyon. One cannot project former profiles headward, as Matthes did, because profiles in the Sierra's west-slope* granitic *canyons have been irregular—stepped—since their inception. Also, Wahrhaftig in 1965 demonstrated that Matthes' three surfaces are better explained by stepped topography, and all could have developed more or less at the same time, as I too have concluded. Unfortunately, due to misinterpreting the lower San Joaquin River's stratigraphy, Wahrhaftig believed that most of this topography developed in late Cenozoic time, essentially in the last 10 million years. Evidence indicates it was initiated during the Late Cretaceous epoch and was quite pronounced by early Cenozoic time.*] The valleys of the upper set, like those of the middle set, were left hanging as a result of rapid trenching by the Merced induced by an uplift of the range, there having been two such uplifts. Only the valleys of the lower set hang because of glacial deepening and widening of the Yosemite Valley, the cycle in which they were cut having been interrupted by the advent of the Pleistocene glaciers. [*The hanging valleys of the lower set* a priori *are assigned to the canyon stage. No tangible evidence is presented to demonstrate that they were accordant at the advent of glaciation. Matthes' method of quantifying uplift is purely hypothetical and unsupported.*] They consequently indicate the preglacial depth of the Yosemite Valley. That depth, measured from the brow of El Capitan, was about 2,400 feet; measured from the rim of Glacier Point it was about 2,000 feet. [*The valley was about this depth by either late in the Cretaceous or early in the Cenozoic. Furthermore, the Merced River through the valley had a nearly zero gradient, which resulted in greater relief in the east end, not in the west end, as Matthes concluded on the basis of a much steeper river gradient.*]

During that remote cycle of which the hanging valleys of the upper set and the undulating Yosemite upland are representative the Yosemite Valley itself was broad and shallow, past mature in form. [*In the first cycle of erosion Yosemite Valley itself was broad and shallow, past mature in form,* only *when analyzed according to Davis. I conclude that by early Cenozoic time it already had over half of its present relief and that its major features already were recognizable. As today, back then its slopes varied from moderate to vertical.*] That early stage in its development, accordingly, is called the broad-valley stage. [See Figure 15.] The deeper hanging valleys of the middle set were graded with respect to a deeper Yosemite of submature form which must have had the aspect of a mountain valley. That stage in its development is therefore called the mountain-valley stage. [See Figure 16.] The short, steep hanging valleys of the lower set and certain topographic features associated with them show that during the third cycle of erosion the Yosemite was a roughly V-shaped canyon with a narrow inner gorge. [*In Davisian analysis, stream-cut canyons by definition were V-shaped in cross profile. However, in reality many nonglaciated granitic canyons in the Sierra Nevada are highly variable in cross profile, some possessing stretches that are U-shaped in cross profile. Furthermore, cross profiles more or less have maintained their configuration over time. Hence, today's V-shaped canyons were V-shaped early on, and today's U-shaped canyons were U-shaped early on. There are V-shaped glaciated canyons and U-shaped unglaciated canyons.*] This stage of the Yosemite, which immediately preceded the glacial epoch, is therefore called the canyon stage. [See Figure 17.]

Correlation of the Yosemite upland with the upland of the Table Mountain district, between the canyons of the Tuolumne and Stanislaus Rivers, shows that it is in all probability a feature of late Miocene age; for fossil remains of plants and animals found in the lava-entombed stream channels on the upland near Table Mountain appear to be of late Miocene age, according to determinations made by Dr. Ralph W. Chaney and Dr. Chester Stock. [*The Yosemite upland cannot be correlated in age with the volcanic Table Mountain district. Most of the volcanic flows and sediments of that district are mid- and late-Miocene in age, but the earliest volcanic deposits there were laid on a mostly granitic Oligocene-age landscape. Farther north, in the South Yuba River drainage, the oldest deposits are Eocene or possibly Paleocene. They show that back then its granitic canyon already was as broad, and perhaps as deep, as it is today, and they imply that Yosemite Valley originated earlier, in the Late Cretaceous epoch. Additionally, known rates of unroofing of plutons—the removal of overlying bedrock*

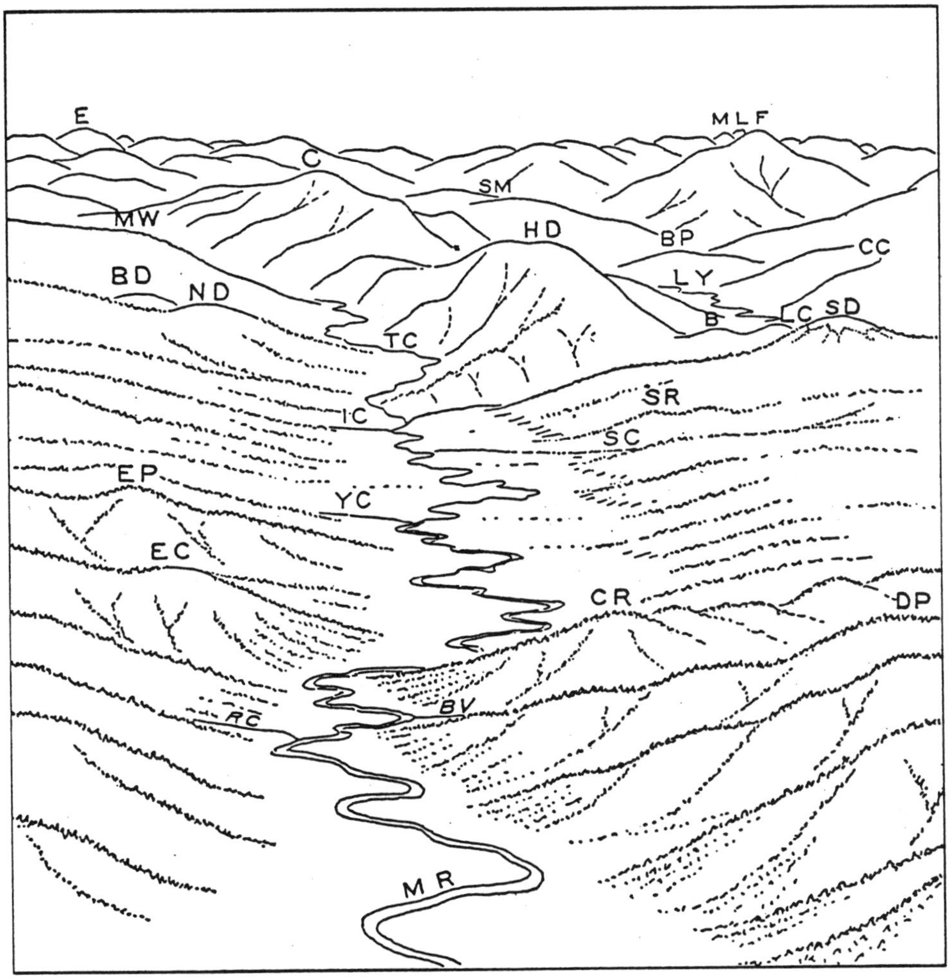


**Figure 15. Matthes' Yosemite Valley during the broad-valley stage. The labeled features on the figures are: B, Mt. Broderick; BV, Bridalveil Creek; C, Clouds Rest; CR, Cathedral Rocks; EC, El Capitan; EP, Eagle Peak; G, Glacier Point; HD, Half Dome; IC, Indian Creek; LC, Liberty Cap; LY, Little Yosemite Valley; MR, Merced River; ND, North Dome; R, Royal Arches; RC, Ribbon Creek; SC, Sentinel Creek; SR, Sentinel Rock; TC, Tenaya Creek; W, Washington Column; and YC, Yosemite Creek. There were three distinct episodes of westward tilting and uplift, each initiating a cycle of stream erosion. The first uplift, which occurred during the Paleocene epoch, was minor, and it raised the Sierran crest to about 4000 feet elevation. This uplift caused headward growth of the Merced River, which captured streams from a northeast-draining river that existed in a low, Appalachian-type range before any uplift. The Merced River drainage evolved over time to produce, during the Miocene epoch, the broad-valley stage. The river meandered sluggishly over a broad, nearly level valley floor flanked by rolling hills between 500 and 1000 feet in height. There were no cliffs or waterfalls, and the tributary streams all sloped gently down to the level of the Merced River.**

*above granitic masses—suggests that the valley's first exposures of granitic rock were back then, about 80 million years ago.*] Uplift of the Sierra Nevada at the end of the Miocene epoch initiated the next cycle, during which the mountain-valley stage was evolved. That cycle lasted presumably through most if not all of the Pliocene epoch. The canyon stage was produced in all probability wholly during the Quaternary period. [*This stage preceded glaciation. Yet on page 50 he states that glaciation presumably began* early *in the Quaternary period. This means that the incision to produce the canyon stage—by far the greatest incision of any of his three cycles—had to occur in an unrealistically short period of time. My equivalent of the canyon stage was reached either late in the Cretaceous or early in the Cenozoic.*]

The excellent preservation of the hanging valleys of the upper set, in spite of their great age, is explained by the exceedingly resistant nature of the massive granite that underlies them. The valleys of the middle set were carved in prevailingly jointed rocks that were less resistant to stream erosion, and the gulches of the lower set were carved in closely fractured rocks in which the streams eroded with relative ease. [*This is a theory-driven explanation, one implying that jointing in the area increases toward the central axis of Yosemite Valley in three quantized units (four, if one counts the valley as the most fractured, as Matthes indirectly implies in his reconstructions). This pattern of increased fracturing down a tributary canyon would cause a stream to erode increasingly more down-canyon. Hence, the three cycles of uplift are not needed to explain a stream's profile. David Burton Slemmons, in work for his 1953 dissertation, found more than three sets when he applied Matthes' methods to the Stanislaus River drainage, and Wahrhaftig (op. cit.) showed the sets are misaligned in the San Joaquin River drainage. In the Yosemite Valley area there are more hanging tributaries that lack the three sets than those that have them.*] The important part played by massive granite in the preservation of the upland valleys and the upland itself is most convincingly demonstrated near Wawona, in the valley of the South Fork of the Merced, which may be termed a half-yosemite. The north side of this valley, which is carved from prevailingly massive granite, has sheer cliffs, that reach up to a lofty upland, the analog of the Yosemite upland; and from the hanging valley of Chilnualna Creek, on this upland, leaps a waterfall similar to the falls that leap from the hanging side valleys of the Yosemite. The south side, which is carved from prevailingly jointed rocks, has sloping forms, and the side streams there descend to the level of the master stream without making any falls. [*The Wawona half-yosemite interpretation demonstrates that in a local area streams of the same age can vary from accordant to severely hanging, as is seen today in unglaciated, granitic Dome Land Wilderness (especially Rockhouse Basin). This realization is not applied to the interpretation of Yosemite Valley, since the Davisian scheme requires that the degree of hanging be correlated with a specific period of uplift. This is one of a number of instances of contradiction and lack of internal consistency found in Professional Paper 160.*]

In addition to the three sets of hanging valleys and the three old profiles of the Merced to which they point, there are at hand enough other topographic data to permit an approximate reconstruction of the configuration of the Yosemite Valley at each stage. Three bird's-eye views are shown, all drawn from the same point of view, in which the development of the significant features of the Yosemite Valley can be traced from stage to stage. A fourth bird's-eye view affords a direct comparison of the canyon stage, immediately preceding glaciation, with the U-trough stage at the end of the glacial epoch. [*Obviously the four bird's-eye views, reproduced here as Figures 15-18, are correct only if the analysis is correct, which it is not.*]

The gradients of the two higher profiles of the Merced, further, furnish data from which the amplitude of each of the two great uplifts of the Sierra Nevada can be calculated roughly. The uplift at the end of the Miocene epoch added about 3,000 feet to the height of the range; the uplift at the end of the Pliocene epoch added fully 6,000 feet more. Mount Lyell, which now stands about 13,090 feet above the sea, therefore stood at about 7,000 feet during the Pliocene epoch and about 4,000 feet during the Miocene epoch. [*The last major uplift was about 75-65 million years ago, during the Late Cretaceous epoch. There may have been minor uplift in the early Cenozoic, but there was no demonstrable uplift in the late Cenozoic, except perhaps for a minor amount in the northernmost part of the range. Uplift was greatest in the Mt. Whitney area (or perhaps south of it), and there it was about 8000 to 9500 feet. Amount of uplift decreased farther north, until in the Feather River drainage it was essentially zero. South of Mt. Whitney the Sierra Nevada subsided either late in the Cretaceous and/or early in the Cenozoic.*]

Of the earlier geomorphologic history of the Yosemite region a glimpse is afforded in the explanation of the origin of the southwesterly course of the Merced River and the arrangement of the lesser tributaries at right angles to it. The Merced established its course conformably to the southwesterly slant of the Sierra region, presumably early in the Tertiary period, when there still existed remnants of a system of northwestward-trending mountain ridges of Appalachian type, which had been formed at the end of the Jurassic period by the folding of sedimentary and volcanic strata of Paleozoic and Mesozoic age. [*The Merced River would have established its southwestern course after the Late Jurassic Nevadan orogeny—an episode of significant uplift—but the course would have been repeatedly changed due to massive volcanic eruptions that accompanied plutonism—the solidification of magma to form granitic rock beneath Sierran lands. Plutonism ceased about 80 million years ago, and about then the Sierra's rivers became entrenched in partly granitic-walled canyons. Overall, an Appalachian type of landscape did not exist. The orientation of Sierran streams correlates with joint patterns in the granitic rock, not with a former Appalachian-type fold belt.*] As it grew headward the Merced probably captured the drainage from the longitudinal valley troughs between these ridges. [*The Merced River did not grow headward early in the Tertiary period. Since about 80 million years ago (when*

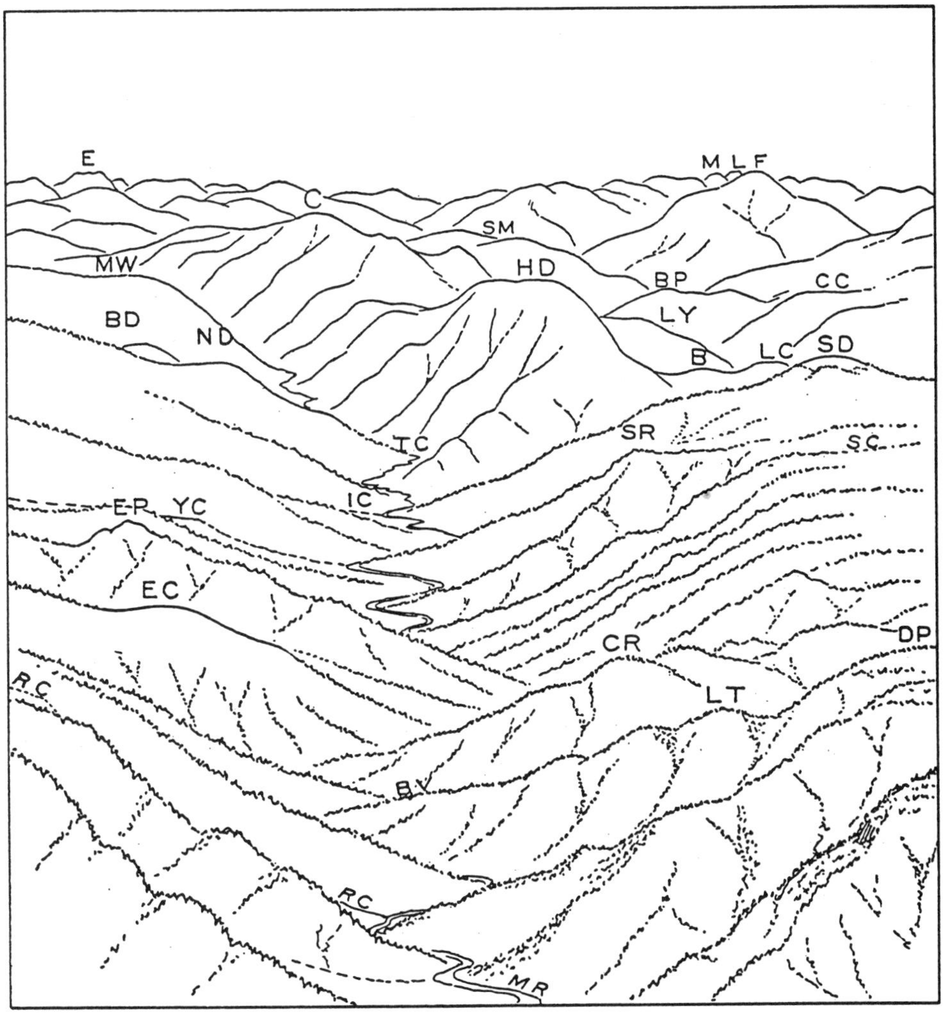


**Figure 16. Matthes' Yosemite Valley during the mountain-valley stage. At the end of the Miocene epoch a second uplift added 3000 feet to the crest of the range, and by late in the Pliocene epoch the mountain-valley stage had developed. The Merced River had deepened the valley by about 700 feet. It was flanked by uplands, and Ribbon Creek, Yosemite Creek, Sentinel Creek, Bridalveil Creek and (unseen) Meadow Brook cascaded steeply from the mouths of hanging valleys. Tenaya Creek, (unseen) Illilouette Creek, and Indian Creek, however, had incised sufficiently to be accordant with the Merced River.**

*volcanism ceased), if not sooner, the* long-existing *river system became fixed and then began incising, first through the overlying volcanic cover, then through the metamorphic rocks and down to the underlying granitic plutons.*] Below El Portal, on the lower slope of the Sierra Nevada, where the folded strata still remain in a broad belt, the lesser tributaries of the Merced are for the most part adjusted to the northwesterly strike of the beds. In the Yosemite region and the adjoining parts of the High Sierra, from which the folded strata are now stripped away, so that the granitic rocks are broadly exposed, the northwesterly and southeasterly trends of many of the streams are largely an inheritance by superposition from the drainage system of the now vanished older mountain system. [*In the western Sierra Nevada, the drainages clearly are adjusted to the folded strata, and this may have been the case—temporarily—as the Cretaceous Merced River in the Yosemite area incised into such strata. However, about Yosemite Valley and in most of the granitic range today the streams clearly are adjusted to the joint pattern of the granitic rock.*] It seems entirely probable that the northwesterly trend of the Clark Range, the Cathedral Range, and certain stretches of the main crest of the Sierra Nevada is likewise inherited from that ancient mountain system. [*The formerly metamorphic areas of northeastern Yosemite National Park clearly have streams adjusted to the joint pattern of today's existing granitic rock and oriented at roughly 90° from the trend of any folded strata. Furthermore, lands in Emigrant Wilderness and mid-elevation lands south of Highway 4 (Lake Alpine-Spicer Meadow Reservoir) show similar adjustment, despite having been exhumed from much of their volcanic cover only in the last few million years. Therefore it is uncertain that the northwest trend of the Clark Range, the Cathedral Range, and some stretches of the main crest of the Sierra Nevada is inherited from folded strata. Some topography and some streams may be, but the evidence is indecisive.*]

The mapping of the morainal system of the Yosemite Glacier has, further, led to the recognition of three stages of glaciation. [*Three stages of glaciation are recognized in the Yosemite Valley area, but the Wisconsin has been divided into the Tioga and the Tahoe (which* actually *is Illinoian, i.e., pre-Wisconsin), and the El Portal and Glacier Point have been combined into a catch-all term, the pre-Tahoe, which has generally been equated with the Sherwin. (I discuss all of these later, in Chapter 12.) In Matthes' following paragraph he recognizes but does not name (as Tioga and Tahoe) two glacial maxima.*] During the last or Wisconsin stage, which is recorded by well-preserved, sharp-crested moraines, the Yosemite Glacier advanced only as far as the Bridalveil Meadow. [*Matthes, as well as others before and after him,* arbitrarily *set the maximum extent of the Wisconsin glaciers at the Bridalveil Meadow moraine, none asking if it indeed is a terminal moraine (it's not) or if it is merely a recessional end moraine (it is). For example, none deny that in the next drainage to the north, the Tuolumne River, the Tioga and Tahoe glaciers extended miles below the westernmost end moraine in Hetch Hetchy Valley. Perhaps all of the major glaciations after the Sherwin glaciation advanced beyond Bridalveil Meadow. Both the Tioga and Tahoe advanced through some, if not all, of Merced Gorge, west of Yosemite Valley.*] During the preceding El Portal stage, which is recorded by relatively obscure, partly demolished moraines, the glacier reached as far as a point about a mile below El Portal, in the lower Merced Canyon. Remnants of the valley train of outwash material can be traced for a distance of 30 miles farther down the canyon. [*In the northern Sierra Nevada, rivers excavated virtually all of their hydraulic mining debris in about a century. It may be likely that Tioga and Tahoe outwash material could have survived tens of thousands of years of recurring catastrophic floods. The "outwash" deposits could be of much more recent origin, related to either catastrophic flooding and/or major river-damming (possibly earthquake-generated) rock falls. I have not seen nor read about outwash deposits in* other *glaciated canyons—why only here?*] Of the still more remote Glacier Point stage evidence is found only in erratic boulders that lie scattered at levels about 200 feet above the highest lateral moraines of the El Portal stage, notably near the west base of Sentinel Dome, 700 feet above Glacier Point, and on the broad divide east of Mount Starr King. [*Where I have spot-checked Matthes' El Portal (pre-Tahoe) lateral moraines, I found either large, scattered erratics or rocks developed from underlying bedrock.* All *his El Portal glacial deposits and El Portal lake sediments that I checked were merely gruss—decomposed granite no different from that found in unglaciated lands.* All *of his El Portal glacial deposits are suspect. The moraines on the broad divide east of Mount Starr King are Tioga and/or Tahoe. No undisputed pre-Tahoe moraines have been found west of the Sierran crest.*]

On both sides of the Little Yosemite the younger lateral moraines culminate in two parallel crests of about equal height, yet separated by a broad depression. [*These younger lateral moraines on the north and south slopes above the floor of Little Yosemite Valley do* not *make two parallel crests of about equal height. The highest southern moraine makes a relatively smooth drop in elevation westward. However, from Moraine Dome the highest northern moraine makes an erratic, fluctuating drop before making a long, nearly level traverse. This moraine is a* diachronous *(time-transgressive) late-Tioga lateral moraine. All of his Wisconsin (Tahoe and Tioga) lateral moraines in Little Yosemite Valley appear to be late-Tioga* recessional *moraines, and as such they do not represent the true thickness of the Tioga-maximum glacier. This is true of most lateral moraines west of the Sierran crest. Till (usually amorphous, glacial sediments) on a bedrock ridge that dams Laurel Lake, north of Hetch Hetchy, is classic pre-Tahoe by conventional interpretation. However, a radiometric date of 10,800 for the lake's basal sediments proves that the till must be Tioga. Mapping and age assignments of moraines in Yosemite National Park has yielded impossible ice surfaces (and basal shears values) for glaciers. The ages of almost all of the western Sierra Nevada's glacial deposits are suspect.*] The interval of time between the deposition of the outer and inner crests appears to have been much shorter than an interglacial stage; hence it is concluded that the Wisconsin stage was characterized by two glacial maxima. [*These maxima are the Tahoe and the Tioga. However, as was mentioned in the previous comment, the mapped moraines are recessional moraines. The Tioga-age glacier was about 2200 feet thick in central Little Yosemite Valley, not half that, as Matthes had proposed. Also, while pre-Tahoe Merced Canyon glaciers may have overflowed* up *into the lower Illilouette Creek canyon, the Tioga and Tahoe glaciers did not. The "till" in the lower part of the canyon is Tioga-age outwash left by the Illilouette Creek glacier's stream, and there are no moraines. One very moraine-like feature is a Tioga-age rockfall deposit.*] On the other hand, the interval of time

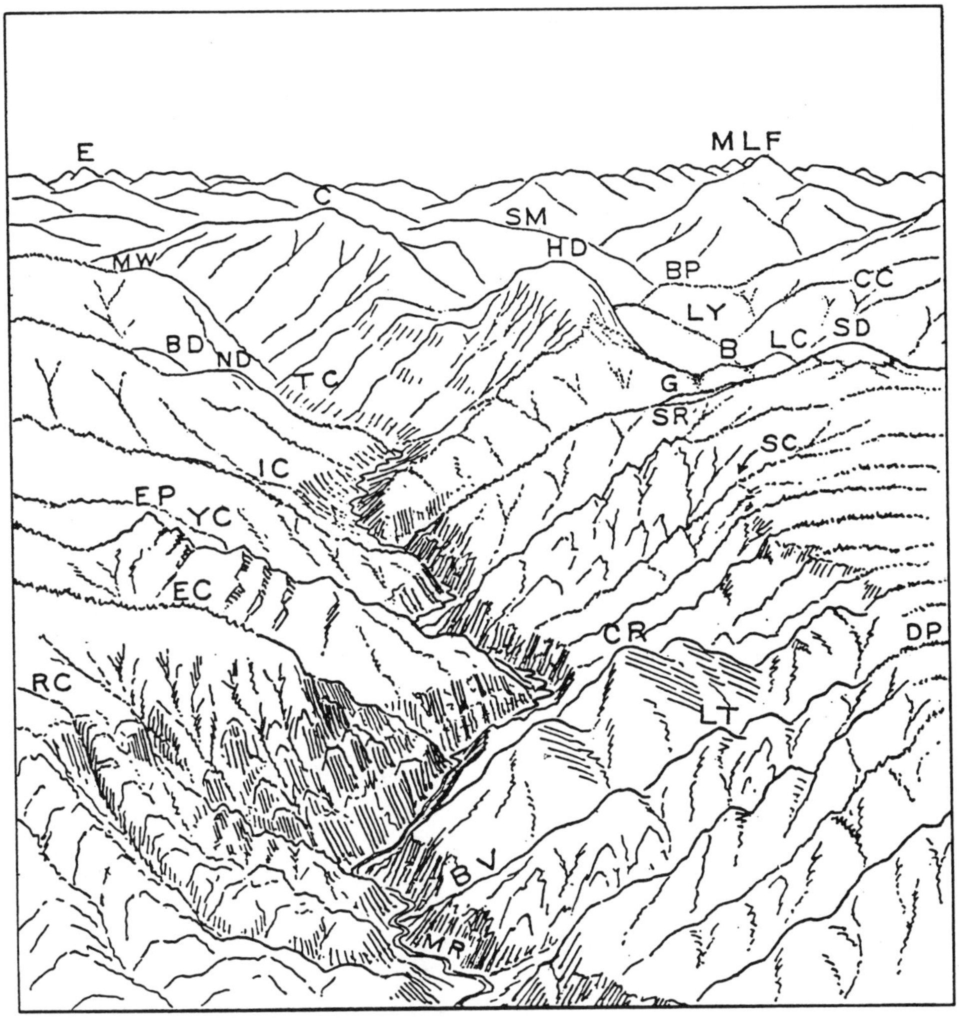


**Figure 17. Matthes' Yosemite Valley during the canyon stage. At the end of the Pliocene epoch (according to his technical abstract) or at the beginning of the Quaternary period (according to his text, p. 29) a third—and greatest—uplift added 6000 feet to the crest of the range to achieve its current height. The Merced River deepened the valley by about 1300 feet in the vicinity of El Capitan—less up-valley—to produce the canyon stage by initial glaciation about 100,000 to 300,000 years ago (different estimates on p. 72 and 106). The valley now had a V-shaped inner gorge and two generations of hanging valleys. Illilouette Creek and Indian Creek now also had cascades, but Tenaya Creek and lower Bridalveil Creek still incised sufficiently to be accordant with the Merced River.**

between the Wisconsin stage and the El Portal stage must have been very long, for whereas the massive granite on the sides of the Little Yosemite retains over large areas the polish that was imparted to it by the Wisconsin ice, it has lost all the polish that was imparted to it by the El Portal ice and in addition has disintegrated and been stripped by rain wash to a depth of several feet. The most reliable measures of the stripping are furnished by three dikes of slow-weathering aplite that project like little walls from the summit of Moraine Dome. Since the ice of the El Portal stage passed over the dome these dikes have come to stand out, by reason of the stripping of the surrounding granite, with heights of 7, 8, and 12 feet. [*A dike is a mass of solidified magma that was intruded along a fracture in bedrock. If it is composed of resistant minerals, it weathers more slowly than the adjacent bedrock, and so it protrudes above it. Matthes believed that Moraine Dome had not been overridden since the El Portal (Sherwin) glacier, so he calculated the amount of post-Sherwin weathering and erosion based on the heights of three dikes and on the height of a "pedestal" that held a large erratic boulder. However, the dome's summit was mantled by thin, stagnant Tioga ice, which lacked the power to erode the dikes. The "pedestal" is a recent feature. A large root of a nearby Jeffrey pine has pried away exfoliation shells next to those that comprise the "pedestal." Both dikes and "pedestals" exist on a ridge just east of the dome, which has deposits recognized by all as Tioga (late-Tioga, by my account). Their existence, not discussed by Matthes, demonstrates the hazard of using such features to determine post-glacial denudation.*] It is estimated, accordingly, that a period at least ten times and perhaps twenty times as long as the post-glacial interval has elapsed since the El Portal stage. That stage is therefore perhaps to be correlated with either the Illinoian or the Kansan stage of the continental glaciation. [*A good measure of post-glacial denudation occurs at Turtleback Dome, which has several large pre-Tahoe erratics. The average size of pedestals is about 4 inches, and if they were left by a Sherwin glacier, then the denudation rate would be about 5 inches per million years.*]

That the El Portal stage was in turn separated from the Glacier Point stage by a long interval of time is inferred from the fact that whereas the oldest lateral moraines of the El Portal stage still persist as continuous bodies, even on steep slopes, nothing is now left of the moraines of the Glacier Point stage save a few boulders of exceptionally durable quartzite and siliceous granite, even on nearly level surfaces where the conditions are particularly favorable for the preservation of moraines. [*Matthes' Glacier Point stage, which is represented only by large, resistant boulders, is no longer recognized. As was stated above, there are no pre-Tahoe moraines or till. The only deposit resembling a moraine is atop Turtleback Dome, which I suggest is the accumulation of locally mass-wasted, weathered bedrock that originally had several erratics on it. Matthes' assertion that El Portal moraines still exist on steep slopes is untrue. He mapped both Wisconsin and El Portal till along the lower part of the Four Mile Trail, but as any hiker can verify, only bedrock and talus exist. The curvature of the lower part of countless tree trunks attests to the instability of the talus slopes. Glacial deposits could not have survived very long on such slopes.*] The Glacier Point stage is therefore held to be comparable in age with the Kansan or possibly with the Nebraskan stage of the continental glaciation. [*Matthes equates his two old stages respectively to the Kansan and Nebraskan glaciations. These terms are no longer in use. As was mentioned above, his El Portal deposits are much-younger Tioga and Tahoe deposits.*]

The discussion of the glacial history of the Yosemite region is supplemented by a map of the valley on a scale of 1:24,000 on which all the moraines and the more significant erratic boulders are shown in detail, also by a map on a scale of 1:125,000 on which are delineated the entire Yosemite Glacier from its sources on Mount Lyell and in the Tuolumne Basin down to its terminus near El Portal and all the lesser ice bodies that lay within the area tributary to the Yosemite Valley. A brief description is given of each glacier as it appeared in the earlier and in the later ice stages. It is shown that Half Dome at no time was overtopped by the ice; also that the Yosemite Glacier at no time received a tributary ice stream from the Illilouette Valley, as has been commonly assumed. Instead, a lobe of the Merced Glacier pushed up into the Illilouette Valley. Imprisoned between that lobe and the Illilouette Glacier lay a temporary lake, whose extent is indicated by deposits of sand and gravel. [*Matthes' "lake sediments" are of two kinds. First, there is a discontinuous veneer of gruss on a nearly flat bench, the gruss being merely decomposed bedrock. And second, boulders and gravel along Illilouette Creek, these being outwash from an up-canyon Tioga-age glacier. Actual lake sediments (e.g., the silt and mud existing in today's post-glacial lakes) do not exist. Lake Illilouette is hypothetical. Because Matthes' interpretation of the evidence is incorrect, the actual extent of the largest early glacier in the Illilouette Creek drainage remains unknown.*]

The depth of glacial excavation in the Yosemite Valley is revealed in a longitudinal section affording a direct comparison of the preglacial and postglacial bottom profiles. It increases gradually from 500 feet at the lower end of the valley to a maximum of 1,500 feet opposite Glacier Point; thence it decreases abruptly to a minimum of 250 feet at the mouth of the Little Yosemite. [*Again, the Davisian method of reconstructing landscapes used by Matthes is based on the false assumptions that before glaciation the Merced River had a smooth profile and relatively constant gradient and that* selected *tributary streams were accordant with the river. Consequently, Matthes' depth of glacial excavation is imaginary, although the amount of glacial deepening in the eastern part of Yosemite Valley—at least 2,000 feet—was actually greater than Matthes had determined. Erosion-resistant sills (massive, joint-free bedrock floors) in the canyon above Merced Gorge and in Yosemite Valley plus buried bedrock floors of recesses along the valley proper both show that just before glaciation, the valley's bedrock floor was close to the elevation of the surface of the present-day sediments. The river's pre-glacial profile through the valley throughout most or all of the Cenozoic era before glaciation was nearly flat, perhaps interspersed with several reaches of rapids across resistant bedrock sills. Tens of millions of years of deep, subsurface weathering under tropical and semitropical climates transformed the valley-floor's highly jointed bedrock into decomposed rock, which was readily removed by large, early glaciers. Finally, Matthes' amount of glacial excavation at the mouth of the Little Yosemite may be more nearly correct, but still it is probably too high.*] The glacial widening is shown in a series of cross sections

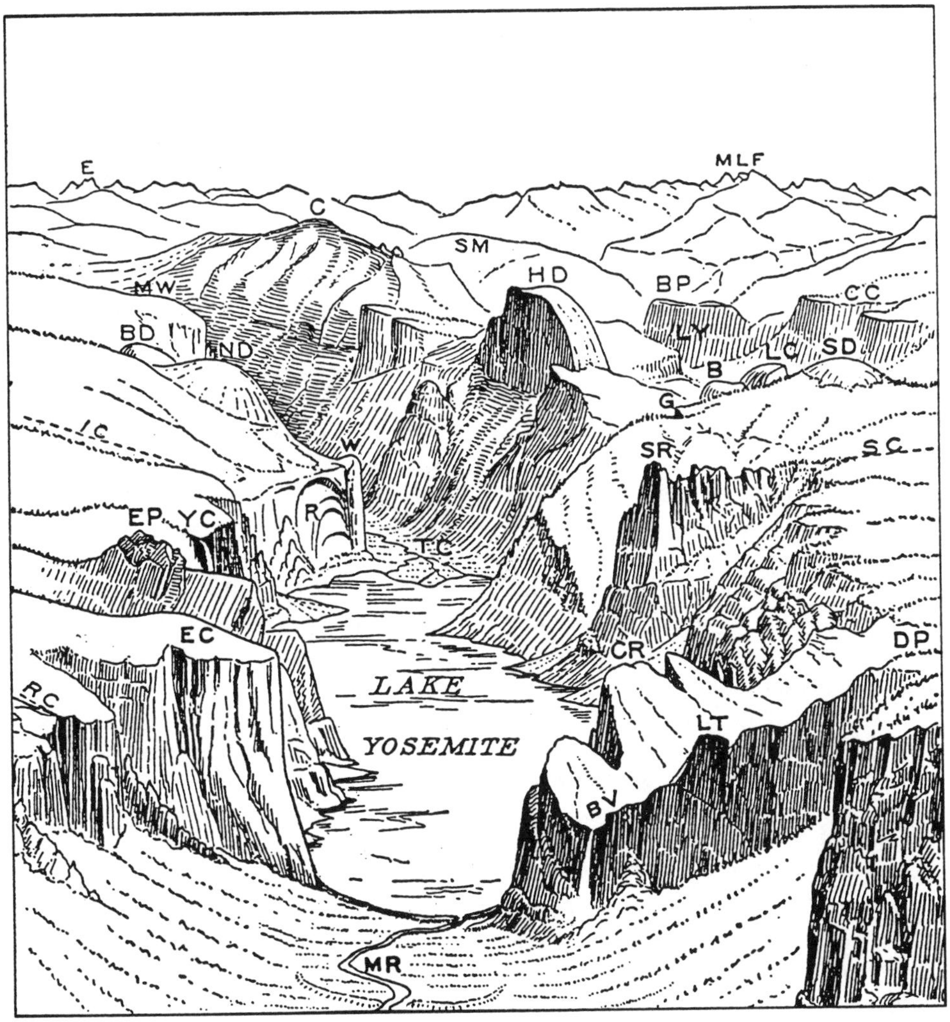


**Figure 18. Matthes' Yosemite Valley immediately after the last glaciation. The valley had been broadened and deepened to essentially its present proportions. Glaciers had removed 50% of the valley's eroded bedrock, the preglacial Merced River the other 50%. Glacial erosion had deepened the valley by 600 feet in the vicinity of El Capitan and by 1500 feet at the head of the valley. This deepening caused a third generation hanging valley—Bridalveil Creek—to form, and Bridalveil Fall developed. A lake 5.5 miles long occupied a basin gouged into the bedrock floor of the valley and dammed by a bedrock sill and an overlying moraine about 60 to 70 feet thick, situated between El Capitan and Lower Cathedral Rock. In the last 20,000 years, this Lake Yosemite, up to 300 feet deep, was completely filled with sediments, although upstream three much smaller lakes—Tenaya, Merced, and Washburn—were not.**

[*cross profiles*] in each of which the preglacial form is superimposed upon the postglacial form. [*Matthes was biased in choosing his cross profiles. He omitted ones that contradicted his views. And among those he chose there are serious problems. One is that in his El Capitan-Cathedral Rocks cross profile, he used Davisian analysis to reconstruct the pre-glacial south half (Cathedral Rocks) of the valley, but did not use it for the north half (El Capitan), where it absolutely does not work. For that half he added a series of hypothetical, parallel joints. Also, his explanation of his Figure 8 of the longitudinal profile of Indian Creek in itself refutes his entire uplift methodology.*] It exceeds the glacial deepening at all points and ranges from a maximum of 1,800 feet on each side in the upper half of the valley to a minimum of 500 feet on each side at the head of the lower Merced Gorge. [*Glacial deepening was more significant than glacial widening, which was insignificant. Like the determination of glacial deepening, the determination of major glacial widening was based upon Davisian analysis despite field evidence to the contrary, which shows that both glaciated* and *unglaciated lands had developed* U*-shaped cross profiles long before any glaciation had occurred in the range. Major glacial deepening occurred where subsurface weathering had been most effective. Glaciers essentially removed the weathered bedrock, which then was replaced with glacial and stream sediments, resulting in no significant change in the relief of the valley. El Capitan was as steep and high 2 million years ago, at the onset of major glaciation, as it is today.*] These marked variations are explained by the selective action of the glacier in rocks of widely varying structure. [*The marked variations in glacial excavation—both in deepening and in widening—certainly are due to the widely varying structure of the granitic rock. However, Matthes selectively used this observation where it fit the Davisian analysis and ignored it where it did not. Furthermore, also in order to fit the analysis, he was inconsistent with regard to the amount of glacial excavation of the valley's recesses, proposing minimal excavation in one, maximal in another, as mentioned below. Immediately before glaciation the valley looked very much as it does today. Minor widening was through retreat parallel to existing surfaces, not through steepening. The valley back then still would have been the rock climbers' Mecca of the world. My estimate of* average *widening during this time is about 460 feet, most of it due to mass wasting rather than to glacial erosion. However, retreat of thick glaciers reduced pressure against lower slopes, and this pressure release apparently induced mass wasting, so glaciers* indirectly *widened Yosemite Valley, albeit minimally.*] Where the rocks were thoroughly jointed the glacier excavated effectively by quarrying; where the rocks were too massive to be quarried the glacier only ground and polished. [*El Capitan is extremely massive and therefore the glaciers should have only minimally abraded it. However, Matthes shows the glaciers effectively excavating in this locality, this excavation made possible by a series of hypothetical joint planes in hypothetical bedrock that are believed to have existed just prior to glaciation. In like manner he proposed a hypothetical ridge from Half Dome west down the center of the eastern part of the valley in order to conform to the constraints imposed by Davisian analysis. He also proposed that Royal Arches originally was a half dome, but that half dome was removed by glaciation. Interestingly he did not draw it as such in his figure on the canyon stage. This book presents abundant evidence that glaciers are extremely ineffective at eroding resistant bedrock, despite a century of claims to the, contrary. The world's glaciologists and geomorphologists have analyzed the evolution of glaciated mountains solely in terms of glacial and stream processes. Mass wasting, weathering, and previous climates and their processes have not been considered.*] The capacious U form of the Yosemite Valley is therefore a product of wholesale quarrying in an area of well-jointed rocks. The gorges above and below the valley, on the other hand, have remained narrow because glacial abrasion has been able to effect but slight changes in their prevailingly massive rocks. [*The capacious* U *form is due to an area of well-jointed bedrock on its floor. However, as was mentioned above, this bedrock was deeply weathered long before the onset of glaciation. At the start of the Cenozoic era the valley had achieved one half to two thirds of its present relief, and all its major features were recognizable.*]

Highly significant in this connection is the fact that the Yosemite Valley lies in an area where a plexus of local intrusions composed mostly of granodiorite, diorite, and gabbro, all well-jointed rocks, breaks the continuity of the vast bodies of siliceous granite, generally massive in habit, that make up the central part of the batholith of the Sierra Nevada. [The range's plutons together comprise its batholith.] This plexus did not of itself give rise to the formation of the Yosemite, but it happened to lie in the path of the Merced River, which was superimposed upon it. [*The extent and characteristics of plutons across the Sierra Nevada were poorly known in Matthes' day. There is no more "a plexus of local intrusions" in the Yosemite area than elsewhere at mid-elevations of the western Sierra. The jointing, or lack thereof, is more important than the intrusive bodies. As Matthes knew, the very imposing, vertical-sided El Capitan and the unimposing slopes west of it are composed of the same plutonic rock, but the first is essentially unjointed, the second, highly jointed.*]

The stepwise mode of ascent of the floors of the Yosemite, the Little Yosemite, the upper Merced Canyon, and Tenaya Canyon is a characteristic result of glacial action. [*The stepwise mode is characteristic of tropical and unglaciated granitic landscapes. Glaciers merely accentuated the steps.*] However, the edges and risers of the steps are composed invariably of massive rock not susceptible of being quarried. [*Matthes states that the steps were* not *susceptible of being quarried (and that the steps were not backwasting up-canyon). Field evidence is in agreement, indicating little glacial excavation, but see the* next *comment. An implication here is that resistant, asymmetrical "domes" never had their down-canyon sides significantly quarried away—as implied by Matthes on p. 95-96 and Plate 44A. As Matthes argued, the resistant cliff over which Nevada Fall leaps is a non-migratory feature. This cliff makes up the base of the southwest face of Liberty Cap. Despite repeated glaciation for perhaps 2 million years, Liberty Cap has been only mildly quarried on its down-canyon side; it retains much of its* asymmetrical preglacial *morphology, like that which occurs in unglaciated granitic lands, such as the southern Sierra Nevada.*] They were therefore not migrant features that receded rapidly headward during the process of glaciation, in the manner implied by certain hypotheses that have been advanced in explanation of the development of glacial stairways. They were essentially fixed features definitely related to the structure of the rock. The canyon steps, accordingly, are conceived to have been produced by selective glacial quarrying. [*Matthes states that the steps* were *susceptible of being quarried, and extremely so—about 1000-1500 feet of deepening between Happy Isles and Vernal Fall. This contradiction of his above sentence is based on constraints of Davisian analysis as opposed to constraints based on field evidence.*] Each tread is essentially a basin quarried out in jointed rock; each edge is essentially a residual obstruction of unquarriable rock, smoothed on the upstream side by abrasion, steepened on the downstream side by the removal of jointed rock. Glacial excavation proceeds with greatest vigor at the head of each tread, because there the ice exerts the greatest force in consequence of its plunge from the step above and accumulates to greatest thickness. [*If glacial excavation had proceeded with greatest vigor at the head of each tread, then Vernal and Nevada falls should have been obliterated. They exist in spite of intense glaciation simply because glaciers could not effectively abrade or quarry the resistant bedrock. Unglaciated, granitic lands have creeks with stepped profiles. Feather Fall is a very prominent example. Glacial erosion is a superfluous hypothesis contradicted by field evidence.*]

This explains how the steeply rising preglacial floor of the Yosemite Valley was replaced by a nearly level, basined rock floor, and why the depth of excavation is three times as great at the head of the valley as at its lower end. [*As was mentioned in comments above, the preglacial floor of Yosemite Valley was not steeply rising headward but rather had gently graded reaches perhaps alternating with short reaches of rapids. Above the valley both Tenaya Creek and the Merced River had steeper reaches, composed of a series of falls, cascades, rapids, and pools.*] The ice descended into the head of the valley not merely by way of the giant stairway from whose steps the Vernal and Nevada Falls now leap, but during the culminating phases of glaciation it also plunged from the lofty platform at the southwest base of Half Dome in the form of a mighty glacial cataract. [*Matthes' longitudinal profile of the Merced River (his Plate 27) shows a smooth reach of the Merced River just prior to glaciation. A constant-gradient profile implies no irregularities in the bedrock, hence, a stepped profile could not be developed in this smooth reach of presumably uniform-spaced jointing. In the Sierra's unglaciated, granitic lands, irregularities in the bedrock result in irregularities in stream profiles. Matthes' glacial transformation from a smooth reach to a giant stairway is a theory-dictated hypothesis. My glacier reconstructions show that the Tioga and Tahoe glaciers overflowed the southwest base of Half Dome more than Matthes had imagined. Additionally, they also spilled through the gap between Half Dome and the Quarter Domes.*] The deep, walled-in heads of the Little Yosemite and Tenaya Canyon similarly were excavated mainly by great cataracts of ice. [*Just as there was a steep, irregular profile of the Merced River between Nevada Fall and Happy Isles that developed early in the Cenozoic era under tropical or nearly tropical conditions, so too there were such profiles at the heads of Little Yosemite Valley and Tenaya Canyon. In Little Yosemite Valley, glaciers* might *have excavated a deep basin, if deeply weathered bedrock had formerly existed, as was the case in Yosemite Valley. The same applies to lower Tenaya Canyon.*]

Structure control also has determined the level of each step. The high level of the Little Yosemite was determined by the height of the body of massive granite that forms the upper step of the giant stairway. The absence of a step at the mouth of Tenaya Canyon, on the other hand, is explained by the fact that glacial excavation there was facilitated by the presence of a belt of fractured rock. [*The same can be said of Yosemite Valley, which I believe was the case: both Tenaya Canyon and the valley had fractured bedrock floors. Preglacial Tenaya Creek would have cut down through its fractured floor, and preglacial Merced River would have cut down through its fractured floor. Major incision did not have to wait for glaciation; it would have begun tens of millions of years ago, late in the Cretaceous period.*]

The detailed sculpture of the walls of the Yosemite Valley is likewise a function of the structure of its rocks, the actions of the weathering processes having been sharply controlled by local variations in the jointing. Vertical master joints have determined the profile and orientation of most of the great cliff faces, including the sheer precipices over which the waterfalls leap. Northeasterly and northwesterly master joints account for much of the faceted sculpture. Easterly master joints have controlled the trend of the great precipice of the upper Yosemite Fall and of the famous cliff at Glacier Point. Oblique joint planes dominate the sculpture of the Three Brothers and of many lesser spurs. Prevailingly sparse jointing in the more siliceous rocks explains the predominance of massive rock forms. Narrow zones of intense fracturing, on the other hand, have given rise to deep recesses, even in places where no drainage descends or formerly descended from the uplands. [*Matthes assumes that today's deep recesses were occupied by jointed bedrock just prior to glaciation. There is no evidence for this, but Davisian analysis demands it. Field evidence indicates that glaciers merely occupied these recesses with stagnant or nearly stagnant ice, as Matthes so states for Indian Canyon. (He incorrectly states that minuscule Indian Canyon Creek excavated that gaping canyon, but* only *during the extremely short-lived canyon stage.) In contrast to the Indian Canyon recess, the Rockslides recess, the recess opposite it on the south side, and the recess east of the Cathedral Rocks all were greatly widened by glaciers, according to Matthes. Again, this view is based on constraints imposed by Davisian analysis, in contradiction to field evidence. When a glacier first advances through a canyon such as Yosemite, some of its ice spills laterally into the recesses. This ice becomes stagnant, and rather than eroding recesses it buries and protects them from other geomorphic processes.*] All the notches, gulches, and alcoves in the vicinity of the waterfalls at the mouths of hanging valleys and on the steps of the giant stairway are carved along fracture zones. Only a few have been produced in the manner explained by Branner, by torrents that flowed along the margins of the glaciers. [*Branner had proposed that streams flowing along the edges of glaciers incised short linear gulches. It is surprising that Matthes supported this view, since the gulches are clearly along fracture zones and they are located nowhere near the edges of former glaciers.*]

The domes of the Yosemite region have been evolved from giant monoliths by long-continued exfoliation due to expansion of the granite, presumably in consequence of relief from load by denudation. The irregularities of their curvature still betray to some extent the trend of the master fractures that originally bounded the monoliths. Half Dome is exceptional in that its sheer northwest side has been exposed only recently by glacial plucking and therefore still retains the plane form which it has inherited from a sheeted structure with northeasterly trend. [*By Matthes' reconstructions and by mine, the sheer northwest face* never *was glaciated by either Tahoe or Tioga glaciers, while the lower part of the rounded southeast face was. So why is the* less-glaciated *face sheer? If it still retains the preglacial plane form (vertical joints), which likely is the case, then the southeast face for consistency likely retains its preglacial rounded form. It apparently lacks a set of vertical joints. Glacial plucking has little to do with the shape of Half Dome. The sheer northwest face has experienced considerable retreat through exfoliation, not plucking.*] Exfoliation here and in certain other localities is producing essentially plane sheets. On cliffs ground concave by the glaciers, notably on the step above the Vernal Fall, it produces concave shells. [*Concave shells—arches—are not necessarily produced by glaciers excavating a concavity. Arches typically form on planar or convex surfaces, not concave ones. They form where glaciation has been recent, as at Royal Arches, or distant, as on North Dome and Half Dome, or where there was no glaciation, as on Sentinel Dome.*]

Examination of the débris piles at the bases of the cliffs dispels the belief of some of the earlier observers that 90 per cent of the material was precipitated by a single great postglacial earthquake. There is evidence that in addition to many small rock falls there have occurred several great rock avalanches, and that the intervals between those avalanches were of sufficient length to permit forests to grow up repeatedly on the talus slopes. Earthquake action appears to be indicated most definitely by far-flung hummocky masses of débris that contrast with the sloping taluses and that must have been precipitated from the cliff fronts in their entirety. [*Rockfalls occur in response to earthquakes, freeze-thaw cycles, rain, or unloading (sheeting) due to depressurization, the latter being the most likely cause. Earthquakes, freeze-thaw cycles, and rain would not generate rockfalls unless the rock was already quite loose. Far-flung hummocky masses of débris are not always an indication of earthquake-generated rockfalls. Great rain storms, such as the storm of December, 1867, can generate debris flows that advance out to the valley floor. Erosion over time then can reduce them to hummocky masses.*]

The greatest postglacial change in the appearance of the Yosemite region was brought about by the filling of the glacial-lake basins with stream-borne sediment. [*By and large today's flats have changed very little since glaciers left them. When the Tioga glaciers rapidly wasted away, they left very little debris behind. Mid-elevation to subalpine Sierran lakes have been filling in with sediment at a rate of about 1 foot per 1,000 years, which is far slower than that necessary to fill in all the areas that are flats today—unless each was initially a bog or marshy lake.*] The level sandy floors of the Yosemite and the Little Yosemite and the successive treads of Tenaya Canyon all replace glacial lakes. [*There is no evidence for a lake in Little Yosemite Valley. However, in Yosemite Valley, one or more lakes developed after even the early glaciations, the volume of water growing as the lake basins were deepened with each glaciation and ultimately coalesced. The maximum was achieved presumably just after the Sherwin glaciation, a large lake longer and much deeper than Matthes had imagined, and lower Tenaya Canyon became inundated by it. Smaller glaciers brought much glacial sediments, and the lake basin grew smaller, finally disappearing some time after—perhaps well after—the Tahoe glaciation. No lake formed after the Tioga glaciation. The flatness of the valley's floor in post-Tioga time was achieved mostly through floodplain deposits, not lake-bed deposits.*] The floor of the Yosemite Valley does not, however, indicate the exact level at which the water of ancient Lake Yosemite stood. It is a flood plain of the Merced River cut about 15 feet below the old lake level, which is indicated by terraces. [*The only terrace cited by Matthes (p. 104) is one that "overlooks Tenaya Creek near the head of the valley." This is along the creek just south of the Indian Cave area. The sediments exposed in the creek bank seem to be mostly coarse sand and certainly are not the fine organic-rich silts and clays typically found in Sierran lakes. Furthermore, the terrace, while nearly level in its western part, dips gently to the south and southwest in its eastern part. Finally, the terrace is anomalously absent along the south bank of the creek. It therefore appears to be the distal (lower) end of a former alluvial fan that now is mostly buried under the massive, centuries-old Mirror Lake rockfall.*] Mirror Lake is not a remnant of a glacial lake but was impounded by great rock avalanches that fell from the cliffs at the mouth of Tenaya Canyon, presumably as the result of an earthquake, some time after the glacial epoch. [*The Mirror Lake rockfall deposit, just mentioned above, is quite young. If it had been much older, then shallow Mirror Lake would have been completely filled in and covered with vegetation. It was not necessarily the result of an earthquake, although that is a likely conjecture.*]

An attempt is made in this volume to set forth these facts and interpretations in language intelligible to the general reader as well as to the scientist. The Yosemite Valley is treated not by itself but in its setting, as an erosional feature of the Sierra Nevada that came into being and was evolved by successive stages in consequence of certain epochal events in the orogenic history and in the glaciation of the range. [*The points that I contest in Matthes' Professional Paper 160 on Yosemite Valley apply equally as well to similar points raised in his Professional Paper 329 on the San Joaquin Basin (parts of Ansel Adams Wilderness, John Muir Wilderness, and Kings Canyon National Park) and in his Professional Paper 504-A on Sequoia National Park. Furthermore, since Robert W. Webb in 1946 used Matthes' (i.e., Davis') method of landscape analysis for the Kern River drainage, parts of that paper too are highly questionable. Using the Davisian method of landscape analysis, Webb was unable to determine the "origin of the extensive pattern of upland meadows so prominent in the southern Sierra" The pattern, however, is easy to explain if one understands the processes that operated in varying degrees throughout the geomorphic evolution of the landscape. In short, the geomorphic evolution of the entire western slope of the Sierra Nevada needs to be reevaluated in terms of physical and chemical weathering, mass wasting, stream erosion, and glacial erosion, not just in terms of the last two.*]

In the appendix the nature and significance of the remarkable complex of igneous intrusions into which the Yosemite Valley is hewn are outlined by Frank C. Calkins. A geologic map of the Yosemite region, the upper Merced Basin, and the upper Tuolumne Basin shows the complex in its relations to the vast intrusive masses that occupy the surrounding parts of the Sierra Nevada. The rocks described range from nearly white alaskite to nearly black hornblende gabbro, yet a strong family resemblance is visible in all. Two distinct series of intrusions are recognized—the biotite granite series of the Yosemite Valley and the Tuolumne intrusive series—and in addition there are several kinds of rock not definitely assignable to either of these series. [*The nature of the bedrock of the Yosemite region never has been controversial, and Calkins' detailed study, as shown on Professional Paper 160's Plate 51, has held up well. The classification of granitic rocks has changed since his time, so his gabbro is now classified with diorite, and his Bridalveil granite, Leaning Tower quartz monzonite, and Half Dome quartz monzonite are now classified as granodiorite. The contacts of the various units have not changed much around the valley, but in the high country to the east and northeast, where he did more of a reconnaissance than a detailed examination, a lot more complexity has been added by later geologists. Despite these additions, Calkins' broad classifications of the Yosemite Valley and the Tuolumne intrusive series are still accepted today.*]

## What Matthes could have known

Hindsight is wonderful. But the critiques I have added to Matthes' Technical Abstract are not the result of hindsight. Sure, geology has made great strides, especially from the 1960s onward, when plate tectonics became *the* foremost concept, much as evolution earlier had done for biology. But to solve the geomorphic riddles of Yosemite Valley and the Sierra Nevada, one needs neither high technology nor state-of-the-art geologic knowledge. The science of geology and the topographic and geologic mapping of the Sierra Nevada, as they existed in the 1910s and '20s, were perfectly adequate for solving the riddles. Starting in 1990, when I began serious Sierran glacial and geomorphic research, I came to the range with essentially the same set of misconceptions that had accompanied Matthes and others. One summer in the field forced me to choose between field evidence and prevailing theories. I chose the former. Matthes, like others before and after him, chose the latter. Matthes' diary indicates that he saw enough glacial and non-glacial landscapes to reach all the major conclusions I have advanced in this book.

Obtaining new facts may not be as crucial as discovering new ways of interpreting and linking them.

—Charles Rowland Twidale

# 10

# After Professional Paper 160

## Matthes after Professional Paper 160

While working on his thirteenth and final draft of Professional Paper 160, Matthes in May 1929 received a letter from Stanford's Prof. Eliot Blackwelder that might have been quite disturbing. In the 1920s Blackwelder had been examining evidence of past glaciations in the Sierra Nevada and Great Basin, and he had concluded that in them there were four, not three, major glaciations. From oldest to youngest, these were the McGee, Sherwin, Tahoe, and Tioga. Matthes could accept the first two, for they agreed with his Glacier Point and El Portal glaciations. He conferred with his supervisor, W. C. Alden (who had done significant field work elsewhere on former glaciations), and Alden let Matthes retain the names of Glacier Point and El Portal. The problem was with the other two, the Tahoe and Tioga, which Matthes had collectively lumped under the Wisconsin. If he wanted to identify both in the Yosemite area, he would have to spend at least another summer in the field and write at least one more draft. Either he chose not to do this, or his wife Edith and/or the USGS hierarchy told him enough is enough. (The USGS, remember, wanted to publish his work back in 1916.) He did not do additional research, but after seeing Professional Paper 160 off to the printer in spring 1930, he made arrangements to accompany Blackwelder in the field. He took a train west across the country to Reno, then took a "stage" (probably a bus) south to Minden, where on September 3 he met Blackwelder. In the prof's impressive Packard, the two of them made an eight-day trip that visited glacial features of the eastern Sierra from about the Bridgeport area south to the Rock Creek canyon, and glacial features in and around Tuolumne Meadows and in and around Yosemite Valley. Matthes came to accept the Tahoe and Tioga as two legitimate, separate glaciations.

For the XVI session of the 1933 International Geological Congress, held in northern California, Matthes wrote two articles for its Guidebook 16: Excursion C-1. The first addressed in general the geology of the Sierra Nevada, while the second concentrated on Yosemite Valley and provided a road itinerary through the park. In the first article Matthes added a second early Tertiary uplift, one that occurred later in the Eocene epoch, resulting in a total of four uplifts as opposed to the three presented in Professional Paper 160. It was after this second uplift that the Sierran rivers are believed to have entrenched themselves, creating canyons with floors that lay about 2000 feet below their drainage divides. Interestingly, while there were four supposed uplifts, there were still only three cycles of erosion, the earliest uplift having produced none.

An even more significant difference appeared in both of Matthes' articles. In Professional Paper 160 the Merced River had cut *down* rapidly during the canyon stage so as to maintain a fairly constant gradient up to Little Yosemite Valley. Matthes' newer interpretation was that the river, in response to uplift, was able to cut *headward only to the brink* of Little Yosemite Valley. That valley, he stated, belonged to the Pliocene cycle of erosion—the mountain-valley stage. Then the gentle uplands above it, such as the Starr King bench, must have belonged to the broad-valley stage. Furthermore, he said that the same pattern existed along the Tuolumne River, which cut headward only to near the west end of Tuolumne Meadows. This was a relict of the mountain-valley stage, and its north- and south-side benchlands had to belong to the broad-valley stage (Photo 28). In both areas, glaciers had barely eroded the landscape, in stark contrast to glacial erosion in Yosemite Valley. In both of his Guidebook 16 articles Matthes mentioned two later glaciations, believed to be early and late Wisconsin, which were Blackwelder's Tahoe and Tioga. However, the two men differed on the age of the Tahoe and the length of the Tioga in the Merced River drainage. These important differences are mentioned below under Blackwelder.

Most of Matthes' post-Professional Paper 160 works are not relevant to the discussion at hand. A relevant one was a chapter in *The Sierra Nevada: The Range of Light,* which was published in 1947, just one year before his death. Matthes stated (and illustrated with a geologic cross section) that around the Miocene-Pliocene boundary, a series of more or less parallel, north-trending faults, each downthrown on the west side, developed in the Sierra's western foothills. As a result of this faulting the western slope became broken by successive abrupt, descending steps. This faulting also caused the entire Great Valley of California to sink to a great depth below sea level. Thus the Great Valley, rather than Owens Valley, is the westernmost graben of the Great Basin. This is quite a change from the longstanding view, expressed in Professional Paper 160, and later in his 1937 paper on the Mt. Whitney area, that the Sierra Nevada was tilted to the west as a rigid block, its lower, western part becoming buried with sediments to become the Great Valley.

That view goes at least as far back as Whitney, who in 1865 noted that the valley's sediments near the lower part of the Kern River in places have been eroded away, showing "the original flooring of granite on which the Tertiary [sequence of sediments] was deposited." Muir, in his December 1874 study, while believing that the Great Valley's sediments were entirely glacial deposits, said they rested on the *flank* of the Sierra Nevada. A number of others assumed the sediments of the Great Valley buried the west part of the tilted Sierra Nevada. Israel Russell in 1889 referred to the Sierra Nevada as "a monoclinal ridge of the Great Basin type" and a "tilted block." Ransome then in 1896 presented a good discussion—and a remarkably modern view—of the relations between the Great Valley and the Sierra Nevada, considering "the whole western slope of the Sierra and the floor of the Great Valley as a single orographic block, partly tilted." Given that in the lower Kern River area petroleum had been discovered in 1899 and had been actively sought after since, perhaps Matthes had checked that area's extensive well logs (petroleum, natural gas, water) to verify the topography of the valley's bedrock floor. Along the eastern edge of the San Joaquin Valley there are indeed faults, the valley's floor downthrown as Matthes

**Photo 28. Aerial view west across Tuolumne Meadows. The Grand Canyon of the Tuolumne River is in the distance, while Lembert Dome is in foreground and Dog Lake is right of it. This lake rests on a broad-valley surface, while the flat-floored meadows are on the mountain-valley surface. According to Matthes (and to the author) this landscape was barely eroded by glaciers.**

had stated. The fault system dies out northward, before the Tule River, but Matthes implied that it continued along the whole western edge of the range.

Five lengthy works were published posthumously, all edited and/or compiled by Matthes' great admirer, Fritiof Fryxell. The first of two published in 1950 was *The Incomparable Valley; a geologic interpretation of the Yosemite,* the final product of the popular account "Dr." Matthes had begun back in March 1931. Matthes never obtained an M.S., much less a Ph.D. However, during commencement exercises in 1947, he received an *honorary* degree of L.L.D. from Robert Gordon Sproul, President of the University of California, which also happened to be the institution that published his two general books on Yosemite and Sequoia. With Fryxell elevating Matthes to "Dr.," he achieved a greater stature in the public's eye and his two books may have come to seen more authoritative. However, for some unknown reason he had been referred to as "Dr." much earlier. "The Story of the Yosemite Valley" (New York's American Museum of Natural History Guide Leaflet No. 60) was published in 1924 and authored by François E. Matthes, Ph.D.

The second book in 1950 was *Sequoia National Park; a geological album,* and it did for that park what the first did for Yosemite Valley (and more generally, for Yosemite National Park). A third, *François Matthes and the Marks of Time; Yosemite and the High Sierra,* was published considerably later, in 1962. This book included a memorial to Matthes and had a collection of 15 of Matthes' articles. All three books served to make Matthes the geomorphic and glacial authority on the Sierra Nevada. The two remaining works were both USGS professional papers: *Reconnaissance of the Geomorphology and Glacial Geology of the San Joaquin Basin, Sierra Nevada, California*, published in 1960, and *Glacial Reconnaissance of Sequoia National Park, California*, published in 1965. (Matthes had initiated field work for the San Joaquin drainage in 1921 and for Sequoia-Kings Canyon National Parks in 1925.) Each publication included a topographic map that showed the distribution of early and later glaciers. The first also included Miocene (broad-valley stage) and Pliocene (mountain-valley stage) profiles for the Middle and South forks of the San Joaquin River (the North Fork is minor). Like the Merced and Tuolumne rivers in his Guidebook 16 Excursion C-1 article, the San Joaquin's forks during the early Pleistocene's canyon stage had not cut headward all the way to the crest. The second publication, being a glacial reconnaissance, made no attempt at addressing the area's geomorphology, which, as Fryxell explained in the "Foreword," had been covered primarily in his 1950 popular account on Sequoia and in two earlier articles: "The geologic history of Mt. Whitney," and "Avalanche sculpture in the Sierra Nevada," respectively appearing in 1937 and '38. Both articles appear in Matthes' posthumous 1962 book. The first article was an attempt to explain the upper Kern River drainage in cycles of uplift and erosion, much as he had done for Yosemite Valley. The February 24, 1937, entry of Matthes' field diary stated that the Whitney article "grows to unforeseen proportions and develops many difficult problems." Such vexing problems appear to have plagued him in all his Sierran research, and I suggest they were due to his continued usage of Davis' unworkable cycles of uplift and erosion. The second article addressed avalanche chutes in the Mt. Whitney area and the use of their bases to reconstruct ice surfaces of the Sierra's Wisconsin glaciers. He suggested that past glaciers eroded the sides of canyons, thereby eroding the lower parts of avalanche chutes. Consequently, the bases of chutes today should mark the ice surfaces of former glaciers. As we'll see later, this assumption led Fryxell to reconstruct some mighty peculiar ice surfaces for glaciers in Sequoia National Park.

## Other Geoscientists

Other geoscientists are covered only in brief here, in part because their work was not specifically directed at Yosemite Valley, and also because they are discussed fully in later chapters. Many more have addressed in varying degrees the Sierra's bedrock geology, geomorphology, uplift history, and/or glaciation, and they are mentioned where appropriate.

First, Blackwelder in the 1920s found evidence of four glaciations on the eastern side of the Sierra Nevada, and he attempted to correlate them with glaciations on the western side as well as with those in Montana, Wyoming, Colorado, Utah, and Nevada. In an attempt to correlate glaciations on the two sides of the Sierra's crest, Matthes and Blackwelder made a field trip, mentioned above, in order to examine glacial evidence on the eastern side of the Sierra, up on the Dana Plateau above Tioga Pass, and in and about Yosemite Valley. With respect to Yosemite Valley they disagreed on several significant points. Blackwelder in his 1931 paper concluded that a Tioga-age glacier advanced only to just above the lower part of Tenaya Canyon, and another Tioga-age glacier advanced through the upper Merced River canyon to not much below Nevada Fall. The rapidly descending, fresh lateral-moraine crests on the north slopes of the eastern part of Little Yosemite Valley, which are absent from Yosemite Valley, lend credence to this interpretation. Both glaciers failed to reach Yosemite Valley proper, he said, and therefore Lake Yosemite must have formed behind a Tahoe moraine, not a Tioga one. In the South Yuba River drainage, originating at Donner Pass, Blackwelder said the Tioga-age glacier also was much smaller than the Tahoe-age one. This west-side interpretation conflicts with Blackwelder's east-side interpretation, where the Tioga-age glaciers were almost as long as the Tahoe-age ones. Blackwelder unfortunately did not advance a cause for this discrepancy. Later, Joseph Birman, mentioned below, employed Blackwelder's east-west glacial discrepancy in the San Joaquin River drainage.

From April 26 to May 6, 1935 and September 15 to 23, 1937, Beno Gutenberg and John P. Buwalda, both of Cal Tech, conducted geophysical investigations in Yosemite Valley to determine the depth of its sediments and the topography of its bedrock floor. They mentioned their results in a one-page abstract in 1938, which basically went unnoticed. With their seismic-reflection data they reconstructed a bedrock floor far different than the one proposed by Matthes, arriving at a maximum depth of about 2000 feet rather than of about 300 feet. Robert P. Sharp, a glacier expert at Cal Tech, later joined with them for a detailed paper, published in 1956, eight years after Matthes' death. This late publication probably was for the better since, as discussed later, their conclusions essentially refuted the Davisian analysis that Matthes had used. Therefore, rather than dash Matthes' world-famous reputation, Buwalda, who delayed the report for years, perhaps had decided not to upset the aging gentleman-scholar. Matthes, incidentally, had visited Gutenberg and Buwalda at Cal Tech on November 13, 1935, and then visited them in Yosemite Valley on September 17 and 18, 1937, to observe their field work. Sharp told me in later years that not only did Matthes not like their results, he never accepted them. Knowing their implications, Matthes could not have accepted them without repudiating his "newborn classic."

David Burton Slemmons, completing a Ph.D. dissertation from Cal Berkeley in 1953, studied the geology of the Sonora Pass region. This region lies in the first major east and west drainages north of Yosemite National Park. He used Matthes' method of analysis to determine the number of uplift-erosion cycles in this part of the Sierra Nevada, and concluded there were three additional cycles. Rather than questioning Matthes' underlying assumptions for the cycles, Slemmons attempted to make his additional cycles fit into Matthes' classification scheme. He also addressed Tahoe and Tioga glaciers. The former were believably long, while the latter were unbelievably short—typically about a mile or so long. Nevertheless, he said they "conform to the description given by Blackwelder," which was not true.

At the same time that Slemmons was in the field, Joseph Birman was in the field, but he completed his Ph.D. dissertation from UCLA later, in 1957. Out of it came a major publication, in 1964, which included a detailed map of glacial moraines stretching across the crest of the Sierra Nevada. He correlated those of the east slope's Rock Creek with those of the west slope's Mono Creek, South Fork San Joaquin River, and Huntington Lake. The San Joaquin is in the first major drainage south of Yosemite National Park. He studied some of Matthes' "type localities" of the glacial stages found about Yosemite Valley, but unfortunately did not visit Matthes' preferred "type locality" El Portal moraines on the Starr King bench south of and above Little Yosemite Valley. Those moraines, which actually are much younger than Matthes had imagined, differ greatly from those of Yosemite Valley. Birman's glacial reconstructions for the Tahoe and Tioga glaciations somewhat paralleled Blackwelder's west-slope reconstructions: Tioga trunk glaciers were much shorter than Tahoe trunk glaciers, but still many miles long. Inexplicably, west-slope Tioga tributary glaciers were, like east-slope ones, only slightly shorter than Tahoe tributary glaciers.

Clyde Wahrhaftig, of Cal Berkeley and the USGS, in 1965 identified stepped topography (cliffs and benches) found at low-to-mid-elevations from near the southern boundary of Yosemite National Park southeast through the San Joaquin River drainage to the edge of the Kings River. This topography, which has developed only on granitic rock, is best developed in this area, but also occurs elsewhere in the Sierra Nevada. He concluded that the topography resulted primarily from the much more rapid weathering of buried granitic rocks than of exposed ones. He further suggested that this topography developed in response to initial westward tilting of the range, which he believed was in Oligocene or Miocene time, about 20-30 million years ago. (I place episodes of tilting at about 80-160+ million years ago, during and after the so-called Nevadan orogeny.) The stepped topography, he said, did not exist before tilting. Also, he stated that this development occurred under essentially modern summer-dry climates, not tropical, glacial, or periglacial ones. He rejected other hypotheses of stepped-topography morphogenesis, including Benjamin Hake's 1928 fault-scarp hypothesis, and he questioned the validity of Matthes' ancient erosion surfaces, since dated lava flows within the San Joaquin River drainage dictate otherwise.

Because both Matthes' and Wahrhaftig's visions of Sierra Nevada landscape evolution depend on westward tilt, knowledge of the range's uplift history is important. In 1966 a young, tall, and dashing Mark Christensen, recently made a professor at Cal Berkeley, reviewed the published evidence and concluded that the works of Lindgren (1911) and Matthes (1930) appeared to be correct, even though he noted that Wahrhaftig's criticism of Matthes' erosion surfaces had merit. He also concluded, as Matthes had, that Sierran uplift essentially was complete by late Pliocene time, before the onset of glaciation. In contrast, N. King Huber in 1981 presented data from the San Joaquin River drainage that indicated an increasing rate of uplift from mid-Cenozoic time up to the present. Furthermore, his total amount of uplift significantly exceeded others' estimates, such as that in a 1979 paper by Slemmons *et al.* for the Stanislaus River drainage. They calculated an uplift about half that of

Huber's determination, even though the present Sierran crest in their area is not significantly lower than the crest in Huber's area. Huber addressed only structural uplift and did not factor in denudation (essentially, erosion), and this accounts for some—though not all—of the discrepancy.

Others have addressed Sierran uplift, and they are mentioned in Part IV. Timing and magnitude are vitally important, as is the rate of a stream's incision in response to uplift. For example, could a seasonal stream such as Tenaya Creek be sufficiently accelerated in uplift to have eroded most of Tenaya Canyon? This canyon gave some investigators trouble, including me back in 1986, when I proposed that a giant Tenaya *River* had carved Tenaya Canyon, the river later having been beheaded through river piracy late in the Cenozoic by the Tuolumne River. I had concluded that since most of Tenaya Canyon was not dramatically altered by glaciers, a large canyon had existed in pre-glacial time, which required a large river to carve it. This actually was a desperate attempt to salvage Matthes' work, not to destroy it. But since I, like all others, was interpreting the Yosemite Valley landscape solely in terms of the actions by streams and glaciers, this proposal too had its problems. After my article was criticized by Huber in 1990, I found Matthes' interpretation to be irreconcilable with the valley's topography, and was forced to abandon it.

Whereas I made an unsuccessful attempt to explain how a river could have carved Tenaya Canyon (I explain its real origin later), neither Wahrhaftig nor Huber made an attempt to explain the origin of the broad benchlands bordering the San Joaquin River canyon in the vicinity of Kennedy Table, Squaw Leap, and Table Mountain, all above Millerton Lake (Photo 29). According to both geologists, these two lava-flow remnants, plus other remnants, plus the thin, discontinuous sediments that in places underlie them, together define an ancient valley floor of the San Joaquin River some 10 or 11 million years ago. These remnants occur on low granitic ridges rising from broad granitic benches. Today the remnants stretch about 8 miles in a direction perpendicular to the flow of the river, indicating an ancient valley floor of at least that width. But that is only half the story. If the remnants are extrapolated northwest to Ward Mountain and southeast to Black Mountain, then the ancient valley floor would have been 15 miles wide. Can you imagine a river that wide? That is the approximate width of the lower Amazon River, a river that today discharges more water in one day than the San Joaquin River does in an entire year.

One lame explanation, about the only one available to conventional American geoscientists, is that before 10 or 11 million years ago, the gradient of the river was nearly flat, so it began to meander, much as does the lower Mississippi River. But there is a difference: the Mississippi meanders across loose sediments, the San Joaquin

**Photo 29. Table Mountain, east of Millerton Lake. The dark, upper layer is 10-million-year-old trachyandesite lava, the gray rocks in the lower half are 114-million-year-old granitic bedrock, and the grassy layer between the two is 11-million-year-old alluvial pumice. Photo is taken from Auberry Road at its junction with Frontier.**

would have had to meander across resistant bedrock. How does a river, perhaps 100 yards wide, erode its adjacent bedrock sides until the river's floodplain is 15 miles wide? It simply cannot be done. On this planet, no river is doing this today.

This proposed beveling through meandering is similar to the proposed formation of pediments in the American Southwest deserts. A pediment there is a *bedrock* surface that slopes gently from the base of a desert range. The traditional explanation, still found in many geology and geography texts, is that the pediment's ephemeral streams meander back and forth, planing the surface smooth. This explanation simply does not work. Edward Tarbuck and Frederick Lutgens, in the fourth (1993) edition of their introductory geology textbook, *The Earth,* avoided explaining how pedimentation actually occurs, ending their paragraph on pediments with the sentence: "Just how the water carves the pediment, however, is unclear and still a matter for debate." I wrote them about pediments, and in their fifth (1996) edition their response was to remove the paragraph! However, pediments still are identified in their illustrations on the evolution of a desert landscape. I had told them how these sloping pediments form: they are the result of exhumed detachment faults—extensional faults that originate more or less horizontally 6+ miles beneath the earth's surface, but over time are bowed up to it, exposed as a sloping plane. In their fifth edition, the authors redrew their diagrams, which reflect detachment faulting. However, their redrawn pediments absolutely defy explanation. If as a college instructor I knew nothing about detachment faults, I would be at a complete loss to explain the diagrams to my students. (But I do know quite a bit about detachment faults and *still* cannot explain their diagrams.)

The matter of pediments is the subject of another book, not this one. Desert hard-rock geomorphology is just as problem-ridden as mountain hard-rock geomorphology, both having roots in theoretical constructs by none other than William Morris Davis. What you need to know is that the American Southwest has two types of pediments, the first being those resulting from surfaced detachment faults. The second type occurs in areas of warm, wet climates, where over millions of years the percolating ground water causes very deep subsurface weathering, decomposing the bedrock—as much as 2000 feet below the surface of the land! (Remember this when we look at the 2000 feet of sediments in Yosemite Valley.) Warm, wet climates existed in the American Southwest until about 33 million years ago, and in such climates granitic benchlands develop, such as in southern California's Joshua Tree National Park and in the Sierra Nevada. The San Joaquin River's benchlands from Ward Mountain southeast to Black Mountain are such a pediment. Many exist in the range, most of them small, but some quite large, especially in the Kern River drainage. It is most unfortunate that Wahrhaftig side-stepped pedimentation in his acclaimed "classic" on stepped topography of the Sierra Nevada.

At stake is more than the evolution of Yosemite Valley, since most of the central and southern Sierran landscape west of the crest has been interpreted according to Matthes' cycles of uplift and erosion and his glacial mapping. (The only major glaciated areas not covered by him are the Middle and South forks of the Kings River, which nevertheless he had visited.) This west-slope area extends from the Stanislaus River drainage south to the Kern River drainage, the latter's geomorphology having been elucidated in 1946 by Webb, mentioned earlier, who used Matthes' erosion surfaces and glacial data. Using this approach Webb was unable to explain either the topography of the South Fork Kern River drainage or the southern Sierra's many broad, flat-floored upland meadows (i.e., pediments that originated tens of millions of years ago).

Even though Matthes' underlying premise—the cycles of uplift and erosion—was criticized by Wahrhaftig, his interpretations have persisted, especially as presented by Huber. In his 1987 USGS Bulletin, *The Geologic Story of Yosemite National Park,* Huber acknowledges that the three distinct pulses of uplift proposed by Matthes were more nearly continuous than he had envisioned. The three cycles thus become illusory, but Huber nevertheless recapitulates Matthes, and even includes Matthes' six illustrations of the geomorphic evolution of Yosemite Valley, which unconditionally required the discredited pulses of uplift.

From the foregoing discourse one can see that there are a number of problems with regard to the uplift, geomorphic history, and glacial history of the Sierra's western slopes as interpreted by Matthes and later investigators, but that the old paradigm dies hard—for lack of a new one. This book presents a new one, which shows that all major west-side canyons in the Sierra Nevada had similar uplift, geomorphic, and glacial histories.

My book differs from Professional Paper 160 in that rather than trying to interpret the landscape according to a prevailing model of landscape evolution (e.g., Davis), the interpretation is based on two major, unquestionable assumptions. The first is that every geomorphic process is more effective on fractured rock than on unfractured rock. A corollary would be that rocks of a similar type and of a similar joint pattern should be equally transformed by any process that operates on them with equal strength. This seems obvious, but Matthes used it only where convenient and ignored it where it was a problem. One example is how glaciation transformed the recess immediately east of the Cathedral Rocks and the one immediately west of them. Matthes' reconstructions (his Figures 15 and 16) indicate that although the eastern recess was exposed to thicker ice, it was barely excavated, while the western one, which did not even exist before glaciation, was excavated enormously. My second assumption, which is unnecessary for Yosemite Valley by itself but is necessary for the Sierra Nevada as a whole, is that at any given time all of the range's west-side drainages were experiencing similar processes and similar climates (other than major volcanism, which occurred only in the northern half of the range). Thus drainages of similar area, similar morphology, and with similar crest elevations should have had similar-size glaciers during any glaciation. However, as the glacial histories of the Stanislaus, Tuolumne, Merced, and San Joaquin drainages have been mapped, each drainage has a unique pre-Tahoe/Tahoe/Tioga glacial history that is irreconcilable with the others. In short, the west-slope canyons should have had similar glacial histories, just as the east-slope canyons have had similar glacial histories.

# Part III

# Sierran Glaciation

# 11

# Introduction to Glaciology

It is mid-April, Spring Break, and you're a graduate student driving along Highway 395 through Owens Valley with your friend to climb Mt. Whitney. He's a rather inexperienced mountaineer, but you're a rank novice, although one with stamina and promising talent. You're ascending the Class 2 Mountaineer's Route, which is easy enough to do without a rope, but you've brought one along anyway. You're glad you did, for near the top you have to cross a snowy, wet, sloping, drop-off ledge, which later will be dry and safe for the hordes of summer mountaineers. The day is quickly fading, so you and your friend take each other's photos on the windswept, 14,494-foot summit (Photo 30), then hurry south on a crest-hugging trail to a notch, Trail Crest, where you confront an enormous, steep snowfield. Your friend tests the snow, declares it safe, and leaps off onto it and starts a rollicking descent. After a few seconds of hesitation, you too leap onto the snowfield and try to keep your head above your feet. In under five minutes you slide down more than 1000 feet, to where the gradient rapidly eases off. Except for a few bruises and minor cuts from collisions with minor rockfall boulders that pepper the snowfield, you

**Photo 30. Author atop Mt. Whitney, April 1971.**

make it down in fine shape. The next day is R & R, and you plan something more challenging: the Class 3 East Face of Middle Palisade Peak. The guidebook says this route is the easiest way to the 14,040-foot summit and, unlike on slightly higher North Palisade Peak, you needn't worry about dodging rockfall left and right. You spend several hours floundering sometimes waist-deep in snow before finally reaching a steepening of gradient just below the base of the peak. This part of your snowfield is the Middle Palisade Glacier, and it looks no worse than the snowfield you slid down two days ago. After about 30 minutes, and after about a dozen breaks for shortness of breath, you near the glacier's midpoint. You start once again, then immediately cry out as you plunge through soft snow atop a hidden crevasse. Fortunately, it is a minor one, and you've fallen only 40 feet before wedging to a halt. Also fortunately, your friend—rather than you—has carried the lightweight rope, and after much effort is able to haul you out. You have just learned your first lesson about glaciers. Cal Tech geologist and glaciologist Robert Sharp calls a glacier "living ice," and it can devour you with a gaping mouth and then years later expel you at its lower end.

A mountaineer doesn't have to know everything about how a glacier functions in order to safely ascend it. Likewise, one does not have to know everything about how a glacier functions in order to understand past Sierran glaciers. The principal issue we must understand is flow of glacier ice. If you want to know much more about glaciers, I recommend two books. A non-technical, richly illustrated, enjoyably readable book is Robert Sharp's *Living Ice*. A technical, dry, math- and graph-laden one is David Drewry's *Glacial Geologic Processes*. Both authors, like apparently all others who have written about glaciation since the 1860s, assert that glaciers perform major erosion in resistant bedrock. Sharp presents the field evidence; Drewry adds the complex mathematics. But as I show in Chapter 21, The Case for Major Glacial Erosion, their evidence, which at first seems irrefutable, actually is very weak. Glaciology, which is the study of the mechanics, or physics, of glaciers, was first discussed in rudimentary form by Charles Lyell in 1830. The understanding of glacier flow came later, getting its start with Louis Agassiz, who had observed glaciers intermittently from 1836 through the mid-'40s. Then John Tyndall, a physicist and alpinist, performed more detailed observations in the mid-'50s. Whereas Lyell, Agassiz, and Tyndall recognized that rocks fall on glaciers, and then are transported and eventually deposited by them, others from about the 1860s onward gradually came to see glacial deposits as all the result of glacial erosion of bedrock, and none the result of rockfall transport. If glacial geologists—those who study glacial erosion, glacial deposition, and glacial landforms—would envision how a glacier flows, they would be forced to rethink their branch of science.

Glacial geology? Geology, simply put, is the study of rocks. There are three major rock types: sedimentary, igneous, and metamorphic. Is a glacier composed of any of these? In a sense it is composed of all three. To see this we must examine how a glacier originates. A glacier develops in areas where the annual amount of snowfall exceeds the annual amount of wastage due to melting and sublimation (a direct change from the solid form, snow and ice, to the gas form, water vapor). Gradually the snow accumulates, forming additional layers year after year. Sedimentary rocks, such as sandstone, are composed of deposited layers known as strata (singular: stratum). Likewise, the developing snowfield is composed of deposited layers, one for each year. Over the years the snowfield becomes thicker, and the snow lower down is compressed. Snowflakes are transformed into granular grains of ice, which the French call névé and the Germans firn. As the snowfield continues to grow thicker, the firn is transformed into ice. Igneous rocks, such as granite, are composed of crystals that formed from molten rock, the melt known as magma. In an analogous way, the grains of firn melt and refreeze to become ice crystals. Finally, the snowfield becomes so thick that the air spaces between ice crystals disappear. The ice crystals grow in size, up to a few inches in diameter. The density has increased from about 0.1 with fresh snow on the surface to about 0.9 for the crystals (the same density as ice cubes; the density of fresh water is 1.0). More importantly, the crystals can recrystalize, deforming in the process, which results in slow flow. Metamorphic rocks, such as gneiss, recrystalize and deform, so a glacier is essentially a large mass of "rapidly" deforming metamorphic rock.

To recapitulate, if we were to tunnel down through a glacier, we would first pass through "sedimentary rock" (snow and firn), "igneous rock" (ice crystals), and finally "metamorphic rock" (deforming ice crystals). The "sedimentary and igneous rocks" comprise a glacier's brittle upper layer. This layer is passive; it merely rides atop the underlying flowing, plastic layer. Changes in that layer's velocity can cause cracks—crevasses—to develop in the overlying brittle ice. Also within a glacier, the velocity varies, and this causes stresses that can result in crevasses. Crevasses are rarely over 100 feet deep, because at about that depth the brittle ice can give way to plastic ice. If a glacier advances across almost level terrain, the flow may be very slow, and the brittle ice may extend as far down as 200 feet. In this book I use an average value for the thickness of the brittle ice of the Sierra's former glaciers: 150 feet.

This thickness is greater than that of today's typical Sierran glaciers, which are minuscule by comparison. These small glaciers lie in deep half-bowls—cirques—at the heads of canyons, and would not flow if they laid on gentle slopes of a canyon floor. They flow simply because they are on steep slopes and so experience strong gravitational forces. For example, the average gradient of the Dana Glacier, which is situated below the north face of Yosemite's Mt. Dana, lies on slopes averaging about 40°. This angle is even too steep to hold talus, which accumulates on slopes up to about 35°. Indeed, without rock-climbing shoes, you probably could not stand on such a steep slope. These cirque glaciers, flowing in response to strong gravitational forces, are unsuitable analogs for the Sierra's former large, low-gradient glaciers.

A large glacier, which can be thought of as deforming rock, can also be thought of as a fluid, although a very viscous one. The higher the viscosity of a fluid, the slower it flows. Water has low viscosity; it flows freely. Maple syrup has a higher viscosity, and toothpaste and peanut butter even higher. Deforming glacial ice has a very high viscosity, although not as high as that of the earth's lower continental crust, which can flow an inch or so per year. Glaciers in mountains typically flow from about 10 to 1000 feet per year, which is from about a fraction of an inch to several feet per day. In Alaska, some large glaciers periodically surge, flowing over 100 feet in a day. This happens when they accumulate considerable amounts of water along their base, which allows a second method of glacier movement, basal sliding.

One's general perception of a glacier is that it is a mass of solid ice. Large glaciers, such as today's Alaskan glaciers, are very good analogs for the Sierra's large former glaciers. Large Alaskan glaciers can have super-glacial streams flowing along their surfaces, englacial streams flowing within them, and subglacial streams flowing beneath them, and the large Sierran glaciers probably had similar streams. These streams are possible because such a glacier has a temperature close to freezing rather than well below it. Heat, rising from deep within the earth, reaches the bedrock surface and warms the base of a glacier, causing it to melt. The amount of water that forms beneath a glacier is minimal, averaging about ¼ inch in thickness, but nonetheless it is very important, for basal slip is much more difficult without its presence. Additionally, when flowing glacier ice meets protruding bedrock, be it a few feet high or a few hundred feet high, the ice is compressed against it, and the resulting increase in pressure causes the ice to melt. The meltwater flows around the obstacle and then, in a zone of lower pressure on the lee (down-canyon) side, it refreezes, this melt-and-freeze process being known as regelation. If the floor of a glacier-occupied canyon has a hummocky bottom, then between the base of the glacier and the bedrock there can develop pools of subsurface water, which facilitate basal slip. In glacial areas where the climate is very cold, such as in Antarctica, northern Greenland, and very high mountains, the bases of glaciers are too cold to melt, and the bases are frozen to the bedrock. Generally, the thicker the glacier, the faster it flows, but in Antarctica, where ice can exceed three miles in thickness, surface velocities can be on the order of a mere 10 feet per year.

The variable rates of flow within a glacier are much like those within a stream. In both, surface flow is greater than basal flow, and flow in the center is greater than flow along the sides. This is illustrated in Figure 19. Notice in the three left drawings the correlation between the rate of flow of the ice and the thickness of the ice: the thicker the ice, the greater the rate of flow. In the right drawing, a section through a glacier, notice that over half of the movement is by basal flow, and the remainder by internal flow within the plastic layer. In a surging glacier, the

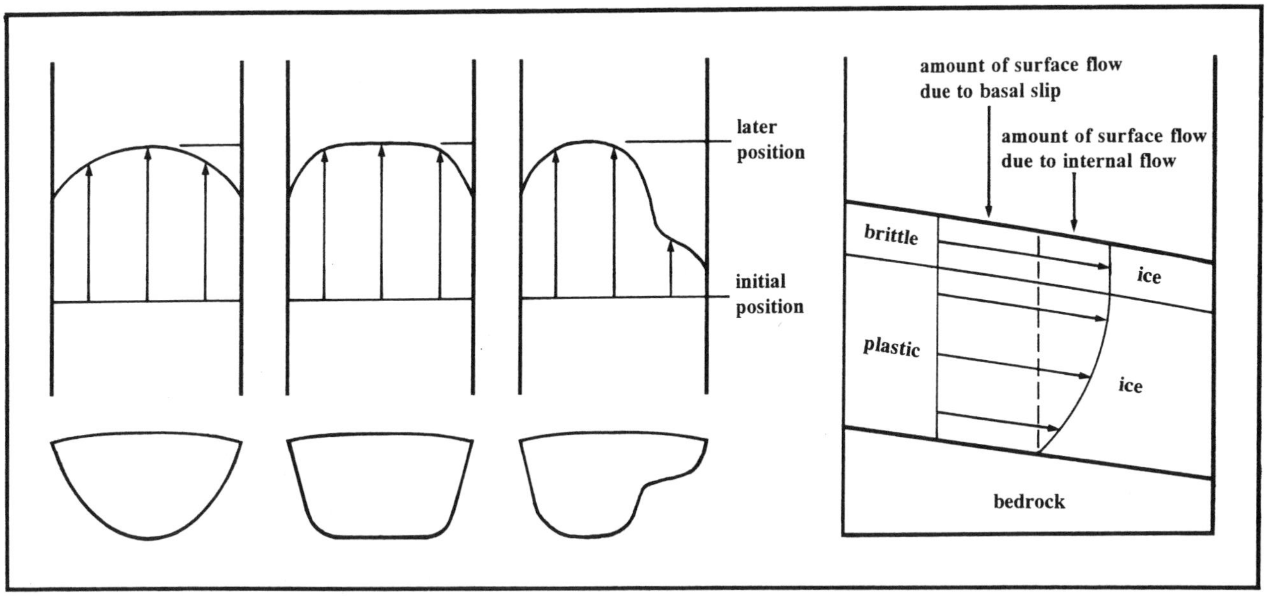


**Figure 19. Variable rates of flow for glaciers. Three left drawings: three glaciers with differing patterns of variable surface flow. Beneath each is a cross profile of the canyon it flows through. Right drawing: an example of variable flow within a typical glacier.**

basal flow would be vastly greater, and would be almost as great as the surface flow. In contrast, in an Antarctic glacier, basal flow would be zero, and plastic flow, which is slow, would account for all of the surface flow.

The thicker the plastic ice, the faster it can deform, so plastic ice at or near the base should deform and flow faster than ice higher up, where pressures are less. Nevertheless, the plastic ice just beneath the brittle ice moves farther than the plastic ice at the base. How come? To understand why, let's arbitrarily break the plastic ice into three layers: lower, middle, and upper. The lowest plastic ice, under the greatest pressure, flows fastest; the middle, under less, flows less; and the upper, under the least, flows the least. For example, in one day the lower layer may flow 6 inches, the middle layer 3 inches, and the upper layer 1 inch. However, the middle layer, which of itself flows 3 inches, is carried 6 inches by the lower layer, for a total movement of 9 inches. The upper layer, which of itself flows 1 inch, is carried 6 inches by the lower layer and 3 inches by the middle layer, for a total movement of 10 inches.

There are two general laws about the velocity of the surface of a glacier. The first is that the surface velocity is proportional to the fourth power of the ice thickness, and the second is that it is proportional to the third power of the surface gradient. Let's examine the first. Take two glaciers, the only difference between them being that the second is twice as thick as the first. If the velocity of the ice at the surface midway across the first glacier is 4 inches per day, what will be the velocity of the ice at the surface midway across the second glacier? All other variables being equal, the velocity of the ice of the second glacier, which is twice as thick, will be 2 to the fourth power, that is, 2 x 2 x 2 x 2, or 16 times faster, which equals 48 inches, or 4 feet, per day. Let's take two other glaciers, the first 200 feet thick, the second 2000 feet thick. A 200-foot thick glacier has only a thin layer of plastic ice, and so it will flow very slowly, say 1/100 inch per day. What about the second glacier? All other variables being equal, the velocity of the ice of the second glacier, which is 10 times as thick, will be 10 to the fourth power, that is, 10 x 10 x 10 x 10, or 10,000 times faster! Then, 1/100 inch per day times 10,000 equals 100 inches per day, or 8 1/3 feet. That is quite some difference. Remember this example, because some former Sierran glaciers locally were so thick that they overflowed their canyons, spilling shallow ice onto adjacent, nearly flat lands. This is especially true of Little Yosemite Valley.

The second general law is that the surface velocity is proportional to the third power of the surface gradient. We'll examine two cases similar to the ones just above. Say over a horizontal distance of 100 feet the ice surface of the first glacier drops a vertical distance of 5 feet, while the ice surface of a second glacier drops 10 feet. All other variables being equal, the velocity of the ice of the second glacier, which has twice the gradient, will be 2 to the third power, that is, 2 x 2 x 2, or 8 times faster. What if the ice surface of the second glacier dropped 20 feet? All other variables being equal, the velocity of the ice of the second glacier, whose ice surface was four times steeper than the first, will be 4 to the third power, that is, 4 x 4 x 4, or 64 times faster. What this means is that when a glacier spills over a cliff, it greatly accelerates its velocity. This was the case for former large glaciers advancing through Little Yosemite Valley and then on the descent to Yosemite Valley. They first spilled over the Nevada Fall cliff then spilled over the Vernal Fall cliff. While a cliff may be dead vertical—and the base of the glacier descends along it—the glacier's ice surface may be much less steep. With regard to issues discussed in this book, this second law is of minor importance.

We can now complete our brief introduction to glaciology with an explanation of Figure 20. A glacier can be divided longitudinally into two parts: that above the equilibrium line, known as the zone of accumulation, and that below it, known as the zone of wastage (or technically, of ablation), as shown in Figure 20, top. The equilibrium line, which is more of a zone than a discrete line, is where on the average the amount of annual snowfall equals the amount of annual snow loss (through melting and sublimation). Increasingly farther above the equilibrium line, the snowfall increasingly exceeds the snow loss. Increasingly farther below the equilibrium line, the reverse is true: the snow loss increasingly exceeds the snowfall. The zone of maximum snowfall is not necessarily above the equilibrium line. In the Sierra Nevada, its past mammoth glaciers experienced maximum precipitation, mostly as snow, below their equilibrium lines. And in today's few small cirque glaciers, the range's elevation zone of maximum precipitation lies well below any of these glaciers.

Another relevant point about glaciers is the direction of ice flow within them. Near the bergschrund, ice forming at or near the surface descends relatively steeply toward the canyon's floor as it flows down-canyon. Closer to the equilibrium line, the ice forming at or near the surface descends less steeply and to a lesser depth. At the equilibrium line, the ice at all depths flows parallel to the surface. Not far beyond the equilibrium line, subsurface ice flows slightly upward toward the surface. And finally, near the snout the subsurface ice flows steeply upward toward the surface. You need not worry about the mechanics of this well-verified pattern of flow. However, what is extremely important is what happens to debris that falls onto the glacier's surface. Well above the equilibrium line, where rockfall is most abundant, it can be transported down to the base of the glacier, and this debris, scraping along the canyon floor, *theoretically* should be quite effective at abrading and eroding it. I emphasize theoretically, because as you will see in

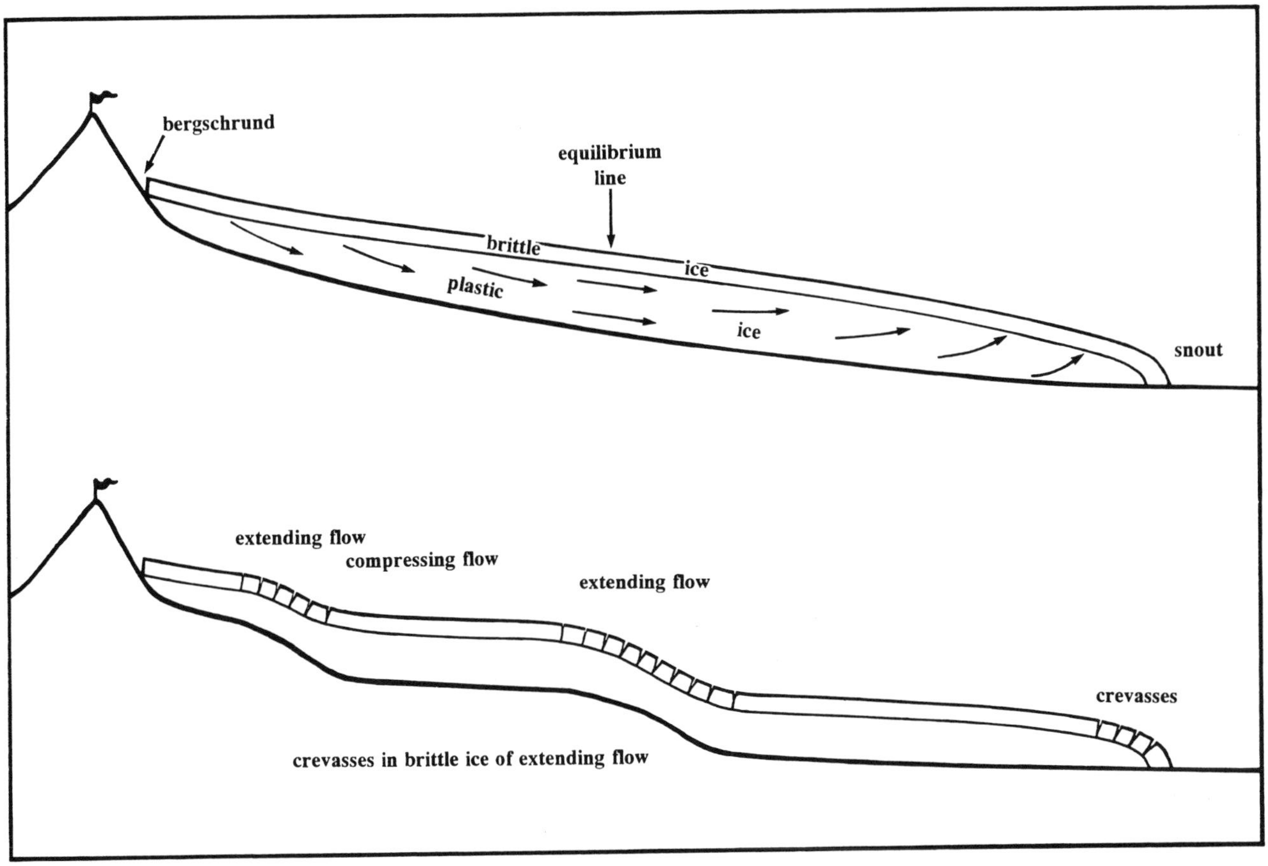


**Figure 20. Longitudinal sections of two alpine glaciers. Top: glacier with smooth canyon floor of relatively constant gradient. Bottom: glacier with irregular canyon floor of variable gradients. Where the gradient of the canyon's floor increases, the glacier accelerates and thins, and numerous crevasses develop in the brittle ice. Where the gradient decreases, the glacier decelerates and thickens.**

Chapter 21, theory and lab experiments dictate profound erosion, field evidence does not.

Below the equilibrium line, debris that falls onto the surface cannot get to the bottom. However, debris can accumulate there, this transported from sources well above the equilibrium line. As the ice closer to the snout increasingly melts, debris increasingly accumulates. Consequently, after the glacier finally melts away, it leaves this debris—a ground moraine—which near the equilibrium line is thin and discontinuous, but becomes thicker toward the snout. There are two useful observations about this ground moraine. First, increasingly below the equilibrium line the glacier is increasingly flowing across its transported-and-deposited debris. Thus the base of the ice flows not across bedrock, but across sediments. Therefore, glacial erosion of the underlying bedrock is impossible, since the glacier does not touch the bedrock. Nevertheless, especially in the Sierra Nevada, geoscientists have proposed maximum amounts of glacial erosion of bedrock well below the equilibrium line. Yosemite Valley is the prime example.

The second observation is that the ground moraine's upper end *should* serve to mark the approximate location of the former glacier's equilibrium line. Indeed, this is one of the ways geoscientists have determined its location. But there is a problem. If the glacier was at its maximum size and then suddenly disappeared, one could use the highest ground-moraine deposits as a proxy for the equilibrium line. But in the Sierra Nevada, as elsewhere in the glaciated western United States, alpine glaciers were at a maximum about 20,000 to 18,000 years ago, but then diminished in size over several thousand years before rapidly melting away by about 14,000 to 13,000 years ago. Therefore, what a ground moraine's upper end represents is not the equilibrium line of the glacier at its maximum extent, but rather that of the glacier just before it rapidly melted away. Amazingly, no one doing glacial research in the Sierra Nevada seems to have realized this.

Lateral moraines have a similar problem. As a rule, the upper end of a lateral moraine is believed to originate close to the equilibrium line, so the elevation of that upper end can be used as a proxy for the equilibrium line. In the Yosemite area, this may be around 8400 feet elevation for the Tioga-age glaciers that existed some 20,000 years ago. However, as those waning glaciers retreated ever higher toward their sources up at around 10,000 to 12,000 feet, their equilibrium-line altitudes (ELAs, in technical jargon) also increased in elevation, and with the ELAs, the upper ends of lateral moraines. I estimate that when the waning Tioga-age glaciers were creating their major lateral moraines, their ELAs were at about 9000 feet elevation, not 8400 feet. Again, no one doing glacial research in the Sierra Nevada seems to have realized this. All that I have read treat the upper ends of lateral moraines as having been deposited when the glaciers were at their maximum extent. This assumption has created problems. For example, the upper end of a lateral moraine of the Lee Vining Creek canyon's Tioga-age glacier lies atop the southern edge of the Dana Plateau, opposite Mt. Dana, and the elevation of that upper end is at about 12,000 feet. This spot is slightly less than ½ mile from the head of that glacier, one which extended about 12 miles down to glacial Lake Russell (of which today's Mono Lake is a mere remnant). So by using the upper end of the highest lateral moraines, we arrive at an ELA located only 4% along the length of the glacier. Most glaciers have ELAs located on the order of 40% along their lengths, that is, about ten times farther.

The typical alpine glacier originates at the head of a canyon, which usually is half-bowl in shape and is called a cirque. Note in the figure that the head of the glacier is not in contact with the wall of the cirque. There is a gap, the uppermost crevasse, called a bergschrund. Many geoscientists believe that the most effective way a glacier erodes is through plucking away loosened bedrock. Too many texts have taken this view one step farther, saying that the heads of canyons are enlarged to cirques through glaciers repeatedly plucking away at them. However, it is impossible for a glacier to pluck away at bedrock if it is not touching it! What actually goes on may be seen by a quote from Willard D. Johnson, who in 1904 descended the bergschrund at the head of the Lyell glacier, Yosemite National Park's largest glacier, lying below the park's highest peak, 13,114-foot Mt. Lyell. While he was doing this risky work, a fellow USGS employee, François Matthes, was mapping the topography of Yosemite Valley. Johnson wrote of his descent:

> ... in the last twenty or thirty feet, rock replaced ice in the up-canyon [cirque] wall. The schrund opened to the cliff foot. I cannot say that the floor there was of sound rock, or that it was level; but there was a floor to stand upon, and not a steeply inclined talus. It was somewhat cumbered with blocks, both of ice and of rock; and I was at the disadvantage, for close observation, of having to clamber over these, with a candle, in a dripping rain [from melting snow and ice above], but there seemed to be definitely presented a line of glacier base, removed from five to ten feet from the foot of what was here a literally vertical cliff.
>
> The glacier side of the crevasse presented the more clearly defined wall. The rock face, though hard and undecayed, was much riven, its fracture planes outlining sharply angular masses in all stages of displacement and dislodgment. Several blocks were tipped forward and rested against the opposite wall of ice; others, quite removed across the gap, were incorporated in the glacier mass at its base.

Basically, a lot of water in cracks of the cirque's rocky wall repeatedly freezes and thaws, prying off a lot of rocks. Many of these eventually can become attached to the glacier. This can occur, for example, by first having winter's snow partly fill the bergschrund, then later having spring meltwater descending into it, which commingles with the snow and changes it to ice that becomes attached to the glacier.

From the bergschrund the glacier quite steadily thickens to the equilibrium line, the rate of ice flow increasing by the fourth power of the thickening. From the equilibrium line onward, the rate of ice flow decreases according to the same law. Also, where the canyon's floor steepens, the ice surface steepens to some extent, and flow rate there increases and the glacier thins (Figure 20, bottom). Crevasses form here, but only in the brittle ice, which is riding passively atop the plastic ice.

An important observation is that glaciers thicken rapidly near their head and thin rapidly near their snout, and that in between they gradually thicken up to the ELA then gradually thin below it. If the floor of the canyon is quite smooth, the surface of the glacier also is quite smooth (Figure 20, top). But if the floor drops, the ice surface also drops (Figure 20, bottom). What does not occur is a major drop of the ice surface where the canyon floor is level (that is, dramatic thinning), or conversely, no drop of the ice surface where the canyon floor drops dramatically (dramatic thickening). Also what does not occur is a glacier having the ice surface along one of its sides lying hundreds of feet lower than the ice surface of its opposite side. Finally what does not occur is the ice-surface of a glacier fluctuating wildly, including an increase in thickness below the ELA. Glaciologists are well aware of these matters, but in the Sierra Nevada, unlike in other ranges, they nevertheless have reconstructed past glaciers that possess these impossible attributes. Why they did and how they got away with it for so long is a riddle—one answered later in this book.

Glaciers transport debris and then deposit it as sediments. These have two fates: either to be carried away from the glacier by its streams—these sediments called outwash—or to be deposited beside, in front of, or beneath the glacier—these sediments respectively called

**Photo 31. The back side of Liberty Cap, which is a large roche moutonnée.**

lateral moraines, end (or frontal) moraines, and ground moraines. (All of the sediments deposited directly by a glacier are called till. This excludes outwash.) The end moraine at the snout of the glacier is called its terminal moraine. As the glacier recedes, it may periodically stagnate, and so leave several recessional end moraines. The terminal moraine is larger than the recessional end moraines because the snout of the glacier has stayed there much longer, and so has deposited a lot more sediment. As the glacier recedes it can also leave a series of recessional lateral moraines, whose crests lie beneath the crest of the larger lateral moraine that forms on each side of the glacier during the glacial maximum. Lateral moraines and end moraines are not separate entities. Where a canyon's sides are not overly steep, a lateral moraine on each side will begin near the equilibrium line and then descend quite uniformly to the snout of the glacier, where it wraps around the snout as an end moraine. This is an important point, for in the western Sierra Nevada the moraines descend *erratically,* requiring an explanation.

Finally, ground moraines form beneath the glacier in the zone usually from about the equilibrium line down to the snout. (The upper edge of a ground moraine *approximates* the position of a former glacier's equilibrium line.) Where the canyon floor is quite smooth, such as in the lower Lee Vining Creek canyon, which descends east to Mono Lake, these deposits are likely to completely mantle it. Where the floor is undulating, such as on the broad, corrugated bench between the North Fork Stanislaus River and its major tributary, Highland Creek, the deposits can be locally deep in the small hollows and absent on adjacent small ridges. Protruding obstacles can serve as local dams behind which sediment accumulates. This sediment, called *lodgement* till, can develop above a glacier's equilibrium line, which appears to be the case for the back side of Yosemite's Liberty Cap, where it is plastered on the lower northeast slopes between about 6300 and 6500 feet elevation. This resistant bedrock feature is a large roche moutonnée (Photo 31), the name for any asymmetrical, glaciated bedrock feature, be it a few feet high or, like Liberty Cap and Tuolumne Meadow's Lembert Dome, on the order of a thousand feet high. These features actually are asymmetrical ridges that from their upper end rise, usually gently, before steeply descending. In theory, both the stoss (up-canyon) and lee (down-canyon) slopes were gentle before glaciation, but then the glaciers plucked away the gentle lee slopes to create the cliffs. This implies substantial glacial erosion. But what geoscientists have consistently ignored is that roches moutonnées exist in *un*glaciated lands. Asymmetrical ridges are a common landform in granitic landscapes, be they glaciated or unglaciated. As we'll see again and again, where theory and field evidence conflict, theory has been embraced and field evidence has been discarded.

# 12

# Ice Ages and Glaciations

Our current Ice Age began perhaps about 2½ million years ago—or about 15 million years ago, or about 33 million years ago. The great majority of geoscientists would say the first date is the correct one, but it really depends upon your point of view. About 33 million years ago, there was a major buildup of ice in Antarctica, concomitant with the formation of cold, deep water around that continent, and overall, the world cooled. Donald Prothero has called this Paradise Lost, the greenhouse world replaced by the icehouse world. Cold water contains a lot more dissolved oxygen than warm water—the equatorial waters, except near coastlines, are biological deserts. Consequently, in the southern ocean, which began serious chilling, there arose vast populations of small marine invertebrates (which proved to be a boon for evolving whales, other marine mammals, and penguins). Major glaciation about 33 million years ago may have been largely restricted to that continent. After some warming, the world cooled again, beginning about 15 million years ago. California's southbound coastal current turned cool, and there arose vast populations of small marine invertebrates... Uncountable numbers of them died and were buried in coastal sediments, which compacted over time, to become the oil-rich Monterey formation. Major glaciation may have been a bit more widespread than at 33 million years ago, now including large glaciers in the world's high ranges. The Sierra Nevada may have had small, cirque glaciers, or longer, upper-canyon glaciers, or in brief cold snaps maybe even long, major, canyon glaciers. Finally, a recognizable, world-wide Ice Age began about 2½ million years ago, which in the northern hemisphere produced very large ice sheets. Virtually all of Canada and virtually all of Europe from the Alps northward both lay in frozen tombs. Our Ice Age, sometimes referred to as the Pleistocene Ice Age (see Chapter 9's Table 1, Geological Time Scales), so far has had about four dozen glaciations, each one an episode lasting about 10,000-40,000 years.

Our current Ice Age is not unique. Every so often the world has one. Two well-defined ones were centered around 250 and 650 million years ago. There may have been two others centered around 450 million years ago and 2100 million years ago. Before that, the geologic record is too incomplete; erosion has removed most of the world's 2-5 billion-year-old rocks. These Ice Ages appear to have lasted perhaps 20 to 50+ million years, which says something about our current Ice Age. If it began 33 million years ago (glacial sediments back then would have been deposited off Antarctica), then perhaps we are nearing the end of our Ice Age. But if it began 2½ million years ago, then we are in for a very long chill.

One might wonder, what causes an Ice Age, and what causes the glaciations within an ice age? The answers appear to be, respectively, plate tectonics and the Milankovitch astronomical cycles. The what?

## The Pleistocene Ice Age

If the earth were an "ideal" planet, it would *circle* the sun with its axis (a line through its north and south poles) perpendicular to the its orbital path, the plane of the ecliptic. In this situation, every spot on the earth's surface would have the same daily length of night and day, 12 hours each. There would be no seasons. But as every school child learns, the earth's axis is tilted at an angle about 23½° from the perpendicular, and so the earth has seasons. Here in the United States in summer, our days are longer and our nights are shorter, while in winter, our days are shorter and our nights are longer. The farther you head toward either pole, the more extreme the difference. In northern Alaska, just north of the Arctic circle (66½° north latitude), the sun will shine at the start of summer 24 hours a day. Six months later, night will last 24 hours a day.

In the above paragraph, I emphasized "circle." But the earth does not make a perfect circle as it revolves around the sun. Instead it makes an elliptical orbit, and over time this orbit varies from nearly circular to less nearly circular, then back again. This is the first of three Milankovitch astronomical cycles. Milutin Milankovitch was a Yugoslavian astronomer who from 1921 to 1941 worked on an astronomical theory of glaciations, which was based on long-term variations of three properties of the earth: its eccentricity, tilt, and precession. In astronomy, eccentricity is the measure of the deviation of a planet's orbit from a perfect circle. Currently, the earth's orbit is close to a circle. In early January the earth is about 91½ million miles from the sun, then in early July it is about 94½ million. Between these two times the distance is intermediate, and the average is about 93 million. Over tens of thousands of years the orbit will become more eccentric, having extremes of about 89 million and 97 million miles from the sun. Then the orbit gradually will become more circular. The whole cycle, from nearly circular to quite elliptical and back again, takes about 100,000 years to complete. With low eccentricity, the earth's climate will be less seasonal; with high eccentricity, it will be more seasonal.

The second relative property affecting climate change is the tilt of the earth's axis. Now it is at about 23½°, but this varies from about 22½° to 24½° over a cycle of about 41,000 years. We are about halfway through this cycle. The less the tilt, the less the seasonality; the greater the tilt, the greater the seasonality.

The third relative property is the precession of the earth's axis. The axis itself slowly rotates, much like the axis of a child's spinning, wobbling top. This cycle takes about 26,000 years to complete. Currently in the dead of winter the northern hemisphere is tilted away from the sun when it is closest to it. Being a bit closer to the sun than on average, we get more radiation from it. About half way through the cycle, some 13,000 years hence, in the dead of winter the northern hemisphere will be tilted toward the sun when it is farthest from it. Our winters then will be colder (and our summers hotter).

When the three cycles interact in such a way that northern hemisphere snowfall exceeds snowmelt (most of the earth's land mass lies in this hemisphere), an episode of glaciation begins. Conversely, when they interact in such a way that snowmelt exceeds snowfall, the episode comes to a close. Currently we are in an interglacial episode, the Holocene epoch, which is defined as having begun 10,000 years ago. Mountain glaciers and continental ice sheets, however, don't end at the same time; the Sierra's giant glaciers disappeared about 13,000 years ago, while the last ice over Canada disappeared about 6000 years ago.

The Pleistocene Ice Age began about 2½ million years ago, but the Milankovitch astronomical cycles have been with us in one form or another apparently since the formation of the earth-moon system some 5 billion years ago. So what caused our Ice Age to begin? The answer to that lies in the sizes, shapes, and relative locations of the world's continents, which in turn dictate the ocean currents and global weather. Continents ride atop a few large, moving plates ("continental drift"), which interact with one another. Their motions are called plate tectonics. Each plate can increase or diminish in size. Each can move away from another one, can dive beneath another one, or can slide past another one. Before 33 million years ago the continents on these plates were in different positions than they are today. North America and South America were widely separated by water, for Central America as we know it did not exist. Also, Africa was farther from Eurasia. In both cases, warm water circulated westward between the two sets of continents.

India, riding on a plate, began drifting north from Antarctica early on, and about 43 million years ago collided with southern Eurasia. Continental crust is lighter than oceanic crust, and while the denser oceanic crust in front of northern India's coastline could dive beneath the lighter crust of the continent of Eurasia, the subcontinent of India could not. The collision of the two masses of continental crust resulted in the creation of the Himalayan range. The world's climate, however, did not appreciably change. Australia, on a slower moving eastern part of the Australian-Indian plate, took longer to distance itself from Antarctica. By about 33 million years ago, however, it had moved far enough away from Antarctica for today's cold, east-flowing current around that continent to originate, and on that continent, ice began to form. By about 15 million years ago, South America had drifted far enough north from Antarctica to allow this current to strengthen to about today's size. The continent then became buried under ice and basically has stayed that way since, although the volume of ice has oscillated over these last 15 million years.

The modern California climate began about 15 million years ago. Seasons became more pronounced; in particular, summers became drier. If the Sierra Nevada has been a high range for tens of millions of years, as I set out to demonstrate in Part IV, then it might have experienced glaciation from about 15 million years onward, when the combination of the three Milankovitch cycles was conducive for excess snow accumulation. Evidence for or against glaciation might be found in the thick volcanic sediments near the Sierran crest from Sonora Pass north to Carson Pass. Unfortunately, no one has studied them, so we are completely ignorant on this matter. My personal conjecture is that small glaciers might have existed in the higher lands (above 10,000 feet) of the range. Major glaciers had to wait until 2½ million years ago, when a major change to the earth's surface must have taken place. What was it?

There is a general view that an Ice Age cannot occur unless a lot of land is located at or close to one or both of the earth's poles. This is partly true. Antarctica has been sitting over the south pole for about 100 million years, but there was no ice age until 2½ million years ago. So some have suggested that either North America or Eurasia, or both, have drifted far enough north to instigate an Ice Age. But both were farther north in the last days of the dinosaurs, some 65 million years ago, and have since drifted away from the north pole. I, like some others, suggest that what instigated the Ice Age was the final growth of Central America, linking North America and South America together, which put an end to the warm, west-drifting equatorial current. Our current Ice Age and the others before them had two properties in common: a lot of land at or near a pole, and land blocking equatorial and tropical currents.

There is a recurring Ice Age mechanism that seems to refuse to die. This is that major uplift of the Himalayan range has altered the world's climate. However, by 2½ million years ago, the range had been rising for some 40 million years, and probably had been high for most of this time. The range today is not rising, but rather is falling apart, splitting laterally both west and east along major faults. Southeast Asia has been "spit" out from the range. Continental crust can thicken, as through continent-continent collision, only to a certain extent. It floats on underlying, denser, plastic continental crust (and underlying, plastic mantle), much as a glacier's brittle ice lies atop its plastic ice. When the continental crust gets too thick, the plastic rock flows, carrying with it the overlying crust. Today the Rocky Mountains and the Great Basin are also falling apart. Their crust had been overthickened through mountain-building compression in late Cretaceous and early Cenozoic time, and now the "crevassing" brittle crust is being thinned through extensional faulting that is rifting it apart.

## The Sierra Nevada's Glacial Record

Although certain fossils in the ocean's sediments indicate there have been about four dozen glaciations over the last 2½ million years, in the Sierra Nevada there is evidence of only several of them. Glacial deposits *west* of the Sierran crest have serious dating problems. The deposits, mostly moraines, typically have abundant granitic boulders in them. With several methods of radiometric dating (age determinations based on a constant rate of radioactive decay of isotopes of atoms), we can *accurately* determine when a given granitic rock solidified. (We can also do this for volcanic rocks.) Unfortunately, we can't do it for the glacial deposits; we have to rely on relative dating. This method is based on how quite a number of characteristics of a moraine—such as the abundance of its surface boulders and the shape of the moraine—change with increasing age. Most of these characteristics, it turns out, are useless, which is why a number of glaciations "identified" through relative dating have been very misdated. When push comes to shove, there is west-side evidence for only three legitimate glaciations, or glacial stages, as some call them: the Tioga, the Tahoe, and the pre-Tahoe. The pre-Tahoe deposits essentially are large, infrequent erratics (glacier-transported boulders). These erratics may be from one glaciation, the Sherwin, which seems to have produced the range's largest glaciers, or they may be from several old glaciations. The named Pleistocene glaciations I discard are: Hilgard and Tenaya (both are late-Tioga); McGee (weathered bedrock); and Deadman Pass (non-glacial deposits). The ages of these glaciations, real or imagined, are given in Table 2. This table is a much abridged, slightly modified form of a chart constructed by David Fullerton, of the USGS, in 1986. These glaciations are for the *eastern* Sierra Nevada, where more glaciations have been identified. These serve as a basis of discussion for the construction of my western Sierra Nevada glacial chronology presented in Table 3. In the text of his article Fullerton gives a detailed account of

**Table 2. Glaciations in the eastern Sierra Nevada.**

| | |
|---|---|
| Holocene (the current *interglacial*; the last 10,000 years; 8000 B.C. to present) | |
| **Little Ice Age** | |
| **Hilgard** | |
| Late Pleistocene: Wisconsin (122,000-10,000 years ago) | |
| **Tioga** | Matthes' younger Wisconsin |
| **Tenaya** | |
| **Tahoe** | Matthes' older Wisconsin |
| Sangamon *interglacial* (132,000-122,000 years ago) | |
| Late Middle Pleistocene: Illinoian (302,000-132,000 years ago) | |
| **pre-Tahoe** and **Mono Basin** | |
| Middle Middle Pleistocene: Pre-Illinoian (610-302,000 years ago) | |
| **Casa Diablo** | |
| Early Middle Pleistocene: Pre-Illinoian (788,000-610,000 years ago) | |
| **Reds Meadow** and **Rock Creek** | |
| Early Pleistocene: Pre-Illinoian (1,650,000-788,000 years ago) | |
| **Sherwin** | Matthes' El Portal |
| Pliocene: Pre-Illinoian (slightly more than 2,710,000 years ago) | |
| **McGee** | Matthes' Glacier Point |
| Pliocene: Pre-Illinoian (3,090,000-2,810,000 years ago) | |
| **Deadman Pass** | |

each, including numerous citations, which are useful for those who want to further investigate this subject. As was just stated above, in the western Sierra Nevada, we can identify only three major categories—Tioga, Tahoe, and pre-Tahoe. In Table 3, however, I have added qualified subdivisions.

## Little Ice Age

We know the Little Ice Age existed, since over the last few centuries our ancestors lived through it, and a few people indirectly wrote about it—for example, by mentioning the miserable weather or the failed crops. Modern biogeographers have also recognized it, by analyzing tree rings and plant pollen. Some trees can live to be thousands of years old, and the variable spacing of their concentric, annual growth rings reflects the climates they grew in. In milder climates, growth is faster, so the rings are broader than those in harsher climates. Pollen from trees, shrubs, and herbs collects in bodies of water such as marshes and lakes. As climate changes, so does the vegetation. (For example, in California during the last major glaciation, the Tioga, forests and woodlands expanded their ranges, while chaparral and desert plants retreated.) The pollen collecting in bodies of water also changes, and by identifying the pollen of dozens of plant species found at any depth of sediment, we can reconstruct the local climate that existed for those species.

The Little Ice Age did not suddenly begin or end, and around the world it was more intense and longer lasting in some localities than in others. Therefore, its duration is hard to define. Here are three estimates, taken from three biogeographers: 1350-1870 A.D., 1400-1800 A.D., and 1550-1800 A.D. Take your pick. I (very subjectively) prefer the first, because then John Muir got to see the tail end of the Little Ice Age. In the Sierra Nevada the glaciers were small, typically a mile or less in length, and most originated in cirques whose floors were above 10,000 feet elevation (Photo 32). These cirque glaciers obviously had no effect on most of the range's lands. However, each would have increased the intensity of freeze and thaw in its own cirque. This brings up an important point. Over the last 2½ million years the head of each glaciated canyon has been glaciated considerably more often than has the bulk of the canyon. Consequently it is the part most likely to have been transformed by glaciers.

**Photo 32. Gray moraines, in mid-ground, left by an Ice Age glacier. Dana Lake, at 11,100 feet, is in foreground. In the upper right is the lower part of the north face of Mt. Dana. At 13,053 feet, this is Yosemite National Park's second highest peak.**

## Hilgard

Joseph Birman, in his detailed 1964 report and map of glacial deposits across the crest of the Sierra Nevada, formally proposed the Hilgard glaciation as one that had occurred early in the Holocene epoch, about 10,000 years ago or shortly thereafter. His map was for part of the glaciated South Fork San Joaquin River drainage on the west side, and for all of the glaciated Rock Creek drainage on the east side. In 1989 Roy Bailey produced an equally detailed map of an east-side area that included Mono Basin and Long Valley. This also included the lower end of Rock Creek canyon, so Bailey identified Hilgard deposits on his map. Following Birman, he delineated fairly long Hilgard glaciers in the Mammoth Mountain area and just east of it in Sherwin Creek canyon, both in the east-side central Sierra Nevada north of Rock Creek canyon. Inexplicably, Hilgard glaciers are absent from most other canyons which, for consistency, should have had them. Something is amiss.

One of Bailey's Hilgard-age lakes is Lake Barrett, the smallest of the (not so) Mammoth Lakes. For the basal sediments from this lake, Scott Anderson obtained a radiocarbon date of about 11,730 years B.P. (before present—before 1950 A.D.). As was mentioned in Chapter 1, this dating process underestimates the true age of the date, which would have been about 13,000 years ago (11,000 years B.C.), when the *Tioga* glaciers finally disappeared. Therefore, this is a Tioga lake; therefore, the Hilgard glaciation is not a separate, early Holocene glaciation, but rather is the late (i.e., recessional) Tioga glaciation. In 1995, the father and son team of Malcolm and Douglas Clark concluded the same, and urged that the Hilgard glaciation be abandoned. I had earlier reached this conclusion, based on dates I've obtained in 1990 on basal sediments extracted from "Hilgard" lakes in the western Sierra (Photo 33). As with others who have dated such western lakes, all mine have pre-Holocene dates.

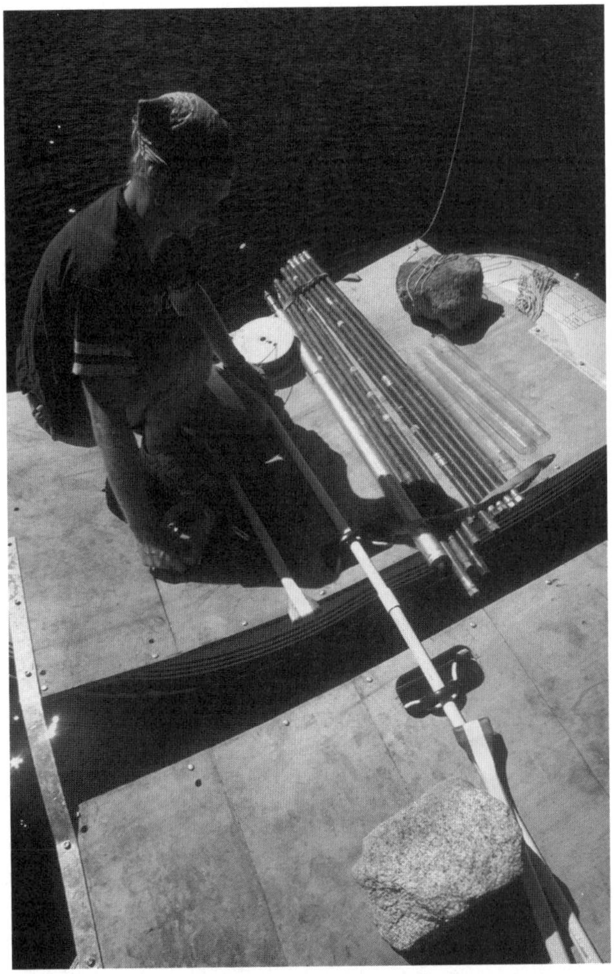

**Photo 33. Eric Edlund on my portable lake-coring platform, Summit Lake. This is located at 7068 feet elevation in the North Fork Stanislaus River drainage. In 13½ feet of water we cored through 14¼ feet of sediments. The lowest sediments had a minimum-age radiocarbon date of about 11,700 years B.P.**

## Tioga

The Tioga glaciation is the only stage in the Sierra Nevada whose age is *not* controversial. It is late Wisconsin, and its duration is approximately synonymous with oxygen-isotope stage 2, which ended about 13,000 years ago. We are currently in oxygen-isotope stage 1, an interglacial period. The Wisconsin glaciation, which actually is composed of several periods of glaciation, occurred in stages 2-5d. The Sangamon interglaciation occurred in stage 5e. To better comprehend oxygen-isotope stages, you need to know about isotopes. Atoms are the fundamental particles of which matter is composed. They in turn are composed of electrons circling a nucleus of protons and (except for the common form of hydrogen) neutrons. Usually the numbers of protons and of neutrons are the same, but for some elements the number of neutrons varies. The resulting different forms of an element are called isotopes. Oxygen has three isotopes. All have 8 protons, but on the average 99.8% have 8 neutrons, 0.2% have 10 neutrons, and a smidgen have 9 neutrons. The protons and neutrons comprise an element's atomic mass, so oxygen atoms with 8 neutrons have an atomic mass of 16 while oxygen atoms with 10 neutrons have an atomic mass of 18. Water is composed of one oxygen atom and two hydrogen atoms. Especially over the oceans, where most of the earth's surface water lies, water is evaporated, but proportionally more of oxygen 16 than 18 is evaporated, since it is lighter. The oxygen-18 water molecules that are evaporated, being denser, tend to precipitate from clouds proportionately faster than do the oxygen-16 ones. Consequently, after some rain or snow has fallen, the clouds farther inland have become relatively enriched in oxygen-16 water molecules. During glacial times, a lot of these accumulate in continental ice sheets, leaving the oceans relatively enriched with the denser oxygen-18 water molecules. Scientists date layer after layer of the ocean's sediments, and measure the ratio of oxygen 18 to oxygen 16, producing an oxygen-isotope ratio. By determining this ratio (in a mass spectrometer), scientists can determine if a sediment was deposited in a glacial or an interglacial time. Additionally, some small marine organisms fossilized in the sediments are temperature-sensitive, and because some exist near the ocean's surface while some exist along its bottom, we can determine an ocean's temperatures both near its surface and near its bottom. The temperatures of these two extremes (as well as those of water between them) change over time, and a general pattern of temperatures prevailing over time makes up an oxygen-isotope stage.

Back to the Tioga glaciation. This may have begun about 35,000 years ago, then reached its greatest intensity—the glacial maximum—about 20,000-18,000 years ago. The glaciers then began to melt, significantly so from about 16,000 to 14,000 or 13,500 years ago. I have put the Tioga glaciation/Holocene interglaciation boundary at 13,000 years ago to conform with the end of oxygen-isotope stage 2, even though in the Sierra, major glaciation may have ended at 14,000 years ago. Possibly the highest lakes formed about 13,000 years ago—or even later, as some dates suggest. But without high-precision dating, the actual time that glaciers left the higher-elevation lake basins is unknown. These remnant glaciers, however, would have melted away long before the start of the Little Ice Age, for from about 9-8,000 years ago until about 5-4,000 years ago, the climate was warmer than it is today. This warmer period has been called a number of names, including the hypsithermal, the altithermal, the xerothermal, the thermal optimum, and the mid-Holocene warm period.

## Tenaya

Joseph Birman, who proposed the Hilgard glaciation, also proposed the Tenaya glaciation. He said it had occurred between the Tahoe and Tioga glaciations. Its type locality—that is, the location where a geologist believes certain field evidence is best observed—is Manzanita Ridge. This is the broad ridge, unnamed on topographic maps, that is immediately west of Lake Edison, in the South Fork San Joaquin River drainage. Because to road from Huntington Lake to Lake Edison is quite narrow and winding, Birman assigned a much more accessible reference locality to this glaciation: the floor of Yosemite Valley from Bridalveil Meadow east to the mouth of Tenaya Canyon. However, because no suitable unused geographic names were available in the valley, Birman chose the term "Tenaya" standing for Tenaya Canyon, "within which moraines of this glacial advance are widely distributed." Ironically, Tenaya Canyon proper is devoid of moraines! None appear on maps of the area by François Matthes (1930), Ronald Kistler (1973), Paul Bateman *et al.* (1983), or N. King Huber *et al.* (1989), nor did I find any when I traversed through the canyon in 1977, or in 1993, while doing field work up higher, in the vicinity of Tenaya Lake.

On the floor of Yosemite Valley, where its few moraines are readily observed, there is now a general consensus that all but the westernmost of the valley's moraines are late Tioga. As I'll discuss later, I am also assigning the westernmost (Bridalveil Meadow) moraine to the late Tioga rather than to the maximum Tioga. Clyde Wahrhaftig, in a 1984 report, found no evidence of a Tenaya glaciation in either Yosemite or Sequoia national parks, and Huber *et al.* also do not recognize a Tenaya glaciation on their geologic map of Yosemite National Park and vicinity.

A number of geologists have identified "Tenaya" moraines on the east slopes of the Sierra Nevada. In 1972 Robert Sharp presented evidence that this stage was a valid and distinct glaciation, albeit short-lived, between the Tahoe and the Tioga. A decade later, Alan Gillespie assigned it a Middle Wisconsin age of 35-45,000 years. The next year, however, with Marcus Bursik he suggested that the Tenaya may have occurred as little as 5000 years before the maximum Tioga, which would make it early Tioga. However, in 1994 Margaret Berry, looking at soils of different moraines of some of these east-slope drainages, could not identify a distinct Tenaya glaciation. She was able to identify only three glaciations: Tioga, Tahoe, and pre-Tahoe—the ones used in this book. Still, it is possible that in the Sierra Nevada, glaciers may have occurred during the Middle Wisconsin (oxygen-isotope stage 3, about 65-35,000 years ago), the Early Wisconsin (stage 4, about 79-65,000 years ago), and/or the Eowisconsin (stages 5d-5a, about 122-79,000 years ago). This suggestion has some support from research by Alan Mix, who in his oxygen-isotope analyses of the Wisconsin suggested that minor glaciations were centered at about 75-68, 98-92, and 118-112,000 years ago.

Evidence for a minor glaciation during the Middle Wisconsin occurs outside the Sierra Nevada, but not in it. Perhaps on both sides of the Sierran crest there were one or more "Tenaya" glaciations, but because they were smaller than the Tioga glaciers, their evidence would have been destroyed. Evidence for increasing size of glaciers during the Wisconsin occurs in the deep-ocean oxygen-isotope record. Because the Tenaya glaciation does not occur in its type locality or in its reference localities, usage of "Tenaya" should be abandoned, and a new name (or names) should be proposed for the relatively small glaciation (or glaciations) that may have occurred between the Tioga and Tahoe glaciations.

## Tahoe

As you drive north on Highway 395 through Owens Valley, Mono Basin, and Bridgeport Basin, you see pairs of large lateral moraines advancing from some of the Sierra's east-side canyons. These were left by Tahoe glaciers. Following a tradition begun by Blackwelder and Matthes and continued by Birman, geologists have generally assigned the range's Tahoe glaciation to the Early Wisconsin rather than to the preceding glaciation, the Illinoian. (More precisely, Blackwelder and Birman assigned the Tahoe glaciation to the Iowan stage, which later became designated as Early Wisconsin.) In all of the Sierra Nevada, there is only one site where the maximum possible age of a "Tahoe" moraine is known through radiometric dating of a basalt flow; a minimum-age limit is absent, since no flow overlies the moraine. The crucial evidence is in Sawmill Canyon, on the east slopes northwest of the Owens Valley town of Independence. Nested moraines occur in this canyon, and the key moraine in question is a lateral moraine identified on the Mt. Pinchot 15' topographic map as "The Hogsback." Alan Gillespie dated an *interglacial* basalt flow, one that formed between two glaciations, as having erupted 118,000 years ago. This flow is partly overlain by a younger Hogsback moraine and it partly overlies an older moraine. He dated another basalt flow beneath the older moraine at 460,000 years old. He presumed these two moraines to be, respectively, Tahoe and pre-Tahoe, the Tahoe therefore being post-Sangamon, that is, either the Eowisconsin or the Early Wisconsin. Previously, James Moore in 1963 and Raymond Burke and Peter Birkeland in 1979 had called these two moraines, respectively, Tioga and Tahoe. Who was right?

Although the evidence seems to favor Gillespie's interpretation of an Eowisconsin or Early Wisconsin Tahoe glacier, there is a problem. Gillespie stated that "the crest of the Hogsback is not uniform, but contains exceptionally fresh and bouldery regions alternating with patches of grusy [weathered-to-gravel] boulders," which implies that it is both Tioga (fresh) and Tahoe (grusy). Furthermore, he stated that the Hogsback appears to be a composite moraine with a massive Tahoe core underlying thin layers of younger till. Finally, he states that the age of the interglacial basalt does *not* contradict Burke and Birkeland's assignment of the Hogsback to the Tioga glaciation, but that it also *permits* a Tahoe designation. Since the evidence at the Hogsback is ambiguous, the case for the Tahoe glaciation having existed early in the Wisconsin is weak. There are at least six lines of evidence to indicate a Late Illinoian age (oxygen-isotope stage 6) for the Tahoe glaciation. The first three are specific, the last three are general.

First, Brian Atwater *et al.* in a 1986 paper studied southern San Joaquin Valley's Tulare Lake, which is dammed behind an alluvial fan (a fan-shaped wedge of sediments). They concluded that it was produced by the penultimate major glaciation, that is, the Tahoe. This created a lake that originated no later than 100,000 years ago, indicating that the Tahoe glaciation was probably pre-Wisconsin.

Second, Stan Soles *et al.* in 1993 dated volcanic shards (fragments of volcanic glass) in the soil of a Tahoe moraine at Bloody Canyon, in east-central Sierra Nevada, just east of Yosemite National Park. On the basis of their dates they concluded that the soil formed on the Tahoe moraine is at least 75-95,000 years in age, and thus the underlying till must be at least as old as oxygen-isotope stage 6.

Third, sediment characteristics for Owens Lake indicate past climates and glaciation-interglaciation cycles in the stretch of the eastern Sierra Nevada above Owens Valley. Between 120,000 and 70,000 years ago (essentially during oxygen-isotope stage 5) the lake had saline diatoms, and J. Platt Bradbury in 1994 took this to reflect warm, dry (nonglacial) climates. But between 170,000 and 120,000 years ago (essentially during oxygen-isotope stage 6) the lake had freshwater diatoms, which is indicative of a wetter and/or cooler (glacial, i.e., Tahoe) climate.

Fourth, the Sierra's Tioga and Tahoe lateral moraines are similar in both relative length and relative form to sets of other similar moraines in other ranges of the conterminous United States. For example, Eliot Blackwelder correlated the Tioga and Tahoe moraines to, respectively, the Pinedale and Bull Lake moraines of the Rocky Mountains. However, the Bull Lake glaciation has been dated radiometrically as pre-Sangamon (i.e., Illinoian). In contrast, very few moraines and till in the ranges outside of California have been assigned to either Eowisconsin or Early Wisconsin. The few that have been generally have uncertain ages. The sole exception appears to be in Yellowstone National Park, where radiometric dates argue in favor of several possible glaciations (two Eowisconsin and possibly one Early Wisconsin, and possibly one Middle Wisconsin) between the Pinedale and the Bull Lake glaciations. Those might correlate with the eastern Sierra's "Tenaya" glaciations. If one assumes that the two major later glaciations in the Sierra Nevada correlate with the two major later ones elsewhere in the Cordilleran ranges, then the Tahoe glaciation must correlate with the Bull Lake and other identified glaciations of Late Illinoian age.

Fifth, oxygen-isotope records from the northern Atlantic Ocean and the equatorial Pacific Ocean, together with electrical conductivity and deuterium (heavy hydrogen) variations from Antarctica's Vostok ice cores all show that major ocean cooling comparable in size to that of the Late Wisconsin (Tioga) occurred during oxygen-isotope stage 6, and not during stage 4. In general the oxygen-isotope record suggests glaciation during stage 4 was somewhat less voluminous than that in either stage 2 or stage 6. If so, then stage 4 moraines would have been overridden by stage 2 (Tioga) glaciers.

And sixth, the degree of weathering of *subsurface* boulders is extremely different in Tioga and Tahoe moraines. Clyde Wahrhaftig, in his 1984 report, stated:

> The Tahoe till is much more weathered, both underground and at the surface, than the Tioga-Tenaya till. [In this report, Wahrhaftig considered the Tioga and Tenaya tills as products of the same glaciation.] In a few localities elsewhere (most notably in a road cut in the east moraine of Cascade Lake in the Tahoe basin of the northern Sierra) it is possible to see that this weathering took place before the deposition of the Tioga-Tenaya till, for there the fresh younger till (mapped as Tioga) rests on the weathered Tahoe. The impression that these exposures give is that the interval of time between Tahoe and Tenaya-Tioga times is much greater—possibly an order of magnitude [10x] greater—than the interval since the Tioga was deposited. If the end of the Tioga is 10,000 years ago, then the Tahoe should be 100,000 years old or older, and probably correlates with the Illinoian glaciation of the mid-continent, and should coincide with the penultimate glacial period identified in deep-sea cores, which reached its maximum approximately 150,000 years ago. This age is difficult to bring into agreement with the latest determination of the age of basalt from beneath till identified as Tahoe at Sawmill Canyon near Big Pine of about 90,000 years ago (Gillespie, 1982, p. 189-380 [actually, p. 573-617]), or with the age of the Tahoe as determined from weathering rinds on andesite boulders in the northern Sierra Nevada (Colman and Pierce, 1981).

Wahrhaftig wrote this with respect to his glacial mapping in the Wolverton and Crescent Meadow areas (the very popular western part of Sequoia National Park (west central part of Triple Divide Peak 15' quadrangle). These two areas were at the lower end of the Tioga and Tahoe glaciers. Gillespie's work has already been addressed. Steven Colman and Kenneth Pierce made age determinations on volcanic stones at a number of sites in the western United States, the *sole* northern Sierra Nevada site being near Truckee, below and east of Donner Pass. The only other principal sampling area in California was in Lassen Volcanic National Park, and the age of the Tahoe moraines there conflicts with the age of those at Truckee. Their Truckee results do not present a problem to Wahrhaftig's assessment of the Tahoe age.

Nevertheless, Colman and Pierce stand by their Lassen results, based on mapping by Kane in 1975, who concluded that the Tahoe deposits there are early Wisconsin. However, Michael Clynne, who during the 1980s did extensive geologic mapping of the park and adjoining lands for the U.S. Geological Survey, in a conversation told me that much of Kane's glacial mapping and interpretations were incorrect, and that the glacial mapping needs to be redone. Therefore, the Lassen deposits are suspect; they may not be Tahoe and likewise may not present a problem to Wahrhaftig's assessment of the Tahoe age.

As I mentioned earlier, the Tioga began its retreat from western Sierran canyons about 16,000 years ago, and the Wolverton and Crescent Meadow areas would have been devoid of glaciers by 14,000 years ago, if not by 15,000 years ago. Higher up, near the Sierran crest, the glaciers disappeared completely by 13,000 years ago. Therefore, an order of magnitude (10x) greater makes the retreat of the Tahoe at about 140,000-150,000 years ago, and its glaciers would have disappeared completely by 130,000 years ago. Marine isotope records indicate the Illinoian ended about 130,000 years ago, and so Wahrhaftig's scenario may be better than he had stated. Also worth noting is that the Sierra Nevada has experienced negligible uplift between the last two glaciations, while California's central Coast Ranges have had measurable uplift, as indicated by their uplifted marine terraces, perhaps enough to cast a greater rain shadow on the Sierra Nevada, so that the Tioga glaciers were slightly shorter than the Tahoe glaciers. However, the difference is not that great, and therefore over the last 150,000+ years the two glacial stages had similar climates and the two interglacial ones also had similar climates. Thus Wahrhaftig's extrapolation of the weathering rates back in time as being relatively constant appears legitimate. In the light of the foregoing, I have assigned the Tahoe glaciation to oxygen-isotope stage 6, especially since I made similar observations in "Tahoe" road cuts in the western Sierra Nevada, and independently have reached the same conclusions as did Wahrhaftig about the nature and age of Tahoe moraines and till.

## Pre-Tahoe

The initiation of moderate-sized ice sheets in North America and Europe occurred about 2½ million years ago, and so notable Sierran glaciation should have begun at the same time. The pre-Tahoe, then, is an excessively long time, occupying about 95% of the Pleistocene glacial record if one makes the conventional assumption that the Tahoe is Early Wisconsin, or about 88% if one assumes it includes all of the Illinoian. Therefore, I feel uncomfortable lumping all deposits earlier than the Tahoe into one catch-all category, the pre-Tahoe. That is equivalent to lumping all Pre-Mesozoic rocks (about 95% of the earth's age) or all Pre-Cambrian rocks (about 88%

of it) into one category, which today is unthinkable. On the other hand, as yet there is no way to determine the age of the western Sierra's pre-Tahoe deposits.

These "deposits" are radically different from those of the Tahoe and the Tioga in that they lack both moraines and till. Overwhelmingly, the pre-Tahoe deposits consist of scattered large erratics (Photo 34). However, in the Merced River drainage—but *not* necessarily elsewhere—there is a minor deposit on the summit of Turtleback Dome and one west of and below it on the summit of Elephant Rock. Both landmarks are easily reached from the Wawona Road, which cuts between them above the west end of Yosemite Valley. In his Professional Paper 160 Matthes mapped abundant El Portal (pre-Tahoe) moraines and till, but as is explained later, these mapped features are either nonexistent or are of Tahoe or Tioga age.

The summits of Turtleback Dome and Elephant Rock are respectively some 650 to 1000 feet below Matthes' ice surface of the El Portal glacier through Yosemite Valley (his Plate 29). Consequently, their erratics could be interpreted as having been deposited by a glacier that was smaller and later than the largest known pre-Tahoe glacier (the Sherwin, discussed below), but was larger and earlier than the Tahoe glacier. No such intermediate glacial stage has been recognized on the west slopes, although on the east slopes, two *may* exist. The younger one, the Casa Diablo, is based on a glacial-looking deposit whose age is bracketed by underlying and overlying basalt flows at between 453,000 and 288,000 years in age. But Bierman

**Photo 34. A large erratic on Dodge Ridge, east of Highway 108. This 10-foot-high pre-Tahoe erratic lies immediately east of and below a small ridge summit, 7175 feet in elevation. A large slab has broken off, exposing the erratic's very weathered interior.**

*et al.*, in a 1991 field-trip guide, question the age and nature of this deposit. The unnamed older stage, at Reds Meadow near Devils Postpile National Monument, is bracketed between dates of 738,000 and 647,000 years.

It is possible that the ice surface of the El Portal/Sherwin glacier through Yosemite Valley may have been lower than Matthes had envisioned, and if so then the summits of Turtleback Dome and Elephant Rock may have been at or close to that ice surface. Despite a concerted effort, I was not able to find any till or moraines on the extensive slopes southeast above Turtleback Dome, where Matthes had mapped abundant evidence. Nor should any be expected, since these slopes are moderately steep, and therefore weathering, mass wasting, and erosion, acting over about 800,000 years, should have removed every trace of evidence.

## Sherwin

On the east side of the Sierra Nevada, pre-Tahoe moraines and till do occur back to the Sherwin glaciation, which has been equated with Matthes' El Portal glaciation. Because they occur there, they have been assumed to occur also on the west side. For example, Birman, like Matthes, mapped Sherwin (El Portal) moraines and till, but a road cut through his Sherwin moraine on the saddle south of and above Huntington Lake shows fresh and grussified (decomposed) subsurface boulders, which makes the moraine unambiguously Tahoe in age according to both Wahrhaftig's and my criteria. Apparently past workers, ignoring or unaware of Blackwelder's 1931 caveat on differential weathering, have not appreciated that the Sherwin erratics on western Sierran lands lie in an altitudinal zone that currently receives several times as much precipitation as Sherwin moraines and till on eastern lands. This pattern likely has persisted to some degree throughout post-Sherwin time. Because the wetter western lands had more precipitation, they also must have had deeper, more mature soils as well as more biomass per unit area. All these factors would have caused the Sherwin moraines and till on western lands to weather much faster than did those on eastern lands, and this would explain their absence today.

The Sherwin's age locally in the eastern Sierra is fairly well known at its type locality, where it underlies the Bishop tuff. This is a widespread volcanic unit that formed from super-hot, incandescent ash, which first was expelled high into the atmosphere, then it settled and coalesced. This unit has reversed polarity, meaning that when it solidified the earth's magnetic field was the reverse of today's—the magnetic poles occasionally flip. That reversal occurred about 790-780,000 years ago. Both Robert Sharp in 1968 and Peter Birkeland *et al.* in 1980 estimated the Sherwin till to be at least 50,000 years older than the Bishop tuff, given its deep weathering. Thus the Sherman till would be over 800,000 years old. David

Fullerton in 1986 *tentatively* correlated this till with the reversed-polarity continental till in Iowa and Nebraska that apparently was deposited 890-800,000 years ago.

## McGee and Deadman Pass

The McGee glaciation was proposed by Eliot Blackwelder, and by doing so he arrived at four glaciations—the Tioga, Tahoe, Sherwin, and McGee—the first two relatively young and the second two relatively old. This fit the pattern perceived by geologists of his day that there were at least two relatively old glaciations. One has to wonder whether, if only one old glaciation had been perceived, Blackwelder would have proposed two.

The McGee has been an unchallenged stage that is considered much older than the Sherwin. It has been identified only in two locations. In Sequoia National Park of the western Sierra Nevada, Matthes in his Professional Paper 504-A called his oldest deposits the Glacier Point stage, which he correlated with Blackwelder's McGee stage. Today the Glacier Point stage, whose type locality is at Glacier Point on the rim of Yosemite Valley, is not a valid stage. Rather, it has been combined with Matthes' El Portal stage and is considered merely pre-Tahoe. Finally, some of these deposits are moraines, but since no unquestionable west-side moraines exist for the younger Sherwin glaciation, none should exist for the older McGee. Therefore, the true age of these Sequoia National Park deposits are unknown.

This leaves only the type locality, which consists of boulders that are located almost entirely on McGee Mountain's gentle slopes from about 9400 to 10,700 feet in elevation. This mountain is one whose base you skirt as you drive northwest on Highway 395 from Lake Crowley toward Convict Lake (Map 10). Some of the boulders were believed to locally lie *atop* remnants of a 2.71 million-year-old basalt flow. However, since there was no upper flow, the McGee deposits have no upper age limit, and so they could be younger. Much younger.

Nevertheless, Eliot Blackwelder and others after him—such as Dean Rinehart and Donald Ross, who mapped the geology of the area—have concluded that the McGee glaciation is very old. The reasoning is this. Large granitic boulders, presumably erratics, must have originated from granitic bedrock to the south, across and beyond the McGee Creek canyon, at the beginning of the Ice Age, when that canyon was quite shallow. Today its floor lies about 2500 feet below the McGee's erratics. To create such a deep canyon after the McGee glacier left its boulders must have taken a long time; hence the boulders must be quite old. Furthermore, tremendous deepening by following glaciers must have occurred to arrive at today's great depth of the McGee Creek canyon. In this book's Chapter 22 I present considerable evidence that major glaciers erode very little in resistant granitic bedrock canyons, downcutting perhaps a few tens of feet over

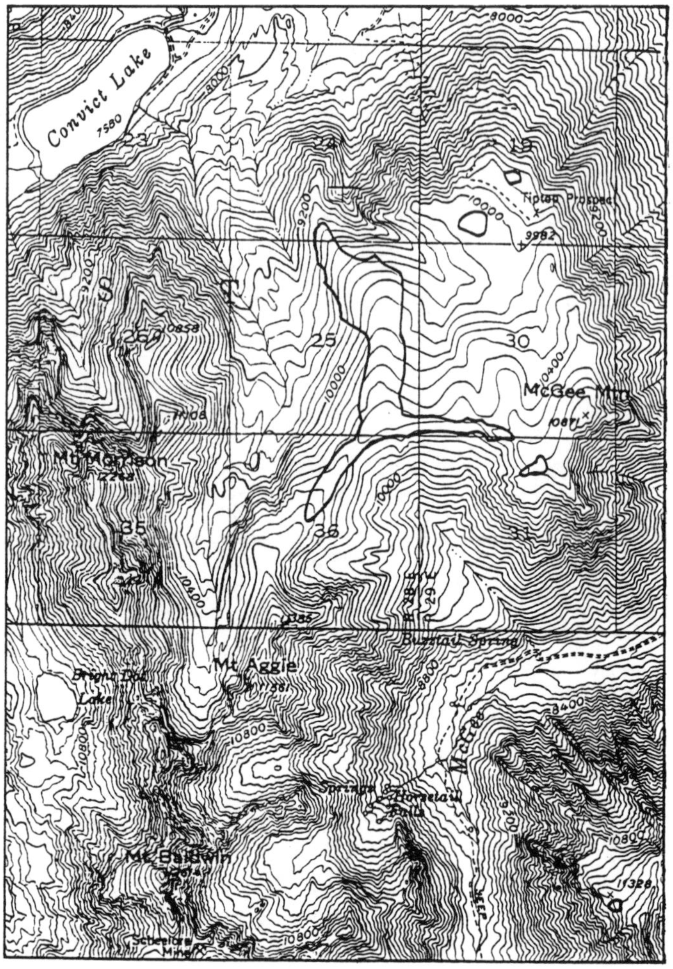


**Map 10. Distribution of McGee boulder deposits at the type locality. Scale 1:62,500.**

some 2+ million years of glaciation. Does the deep McGee Creek canyon dictate otherwise?

The relative inaccessibility of McGee Mountain probably has kept skeptics away. However, it must be visited. If you do, you will discover that the McGee boulder deposits have been totally misinterpreted. First, the dated lava flow, which is in the southwest corner of section 19, has granitic boulders both above and below it (Photo 35). Consequently, if they are erratics, some are older than 2.71 million years, some are younger, and hence represent deposits from two glaciations. But the granitic boulders were not left by glaciers. Rather, they have developed in place, weathering from underlying granitic bedrock. Literally thousands of granitic boulders (I would estimate 10,000+) lie on gentle slopes of McGee Mountain, as well as on fairly steep slopes in the northeast part of Section 25. There is no way that boulders on those slopes could have remained in place for millions of years. Indeed, despite extensive searching for pre-Tahoe erratics on fairly steep Sierran slopes, I have found not one. They lie only on horizontal surfaces or gentle slopes.

**Photo 35. Scott Long sitting on a formerly buried granitic boulder. Such boulders were in existence when the flow buried them 2.71 million years ago. The flow here is merely a lag deposit—a veneer of remnant rocks. However, in the background, it is still largely intact.**

For me the most interesting site is the east-west ridge of myriad erratics along the southern edge of Section 30 (Photo 36). A curious situation exists here. Roughly half of the McGee Creek canyon's drainage—the source area for these supposed erratics—lies on metamorphic bedrock, so roughly half of the "erratics" should be metamorphic. None are, and this holds true for all of the boulder-deposit sites. You can find a very minuscule amount of metamorphic pebbles and cobbles, but no boulders. These small rocks, which make up perhaps one ten-thousandth (or less) of the total mass of the deposits can be explained as having been derived locally from adjacent metamorphic bedrock. It is curious that the McGee glacier entrained only large granitic boulders and no metamorphic ones, and furthermore that it dropped virtually most of them along an east-west ridge and a north-south one. Outside the mapped deposits shown on the map, there are no scattered erratics. But as I detail later, when a glacier spills over a ridge, it leaves extremely few boulders—only the very few that lie atop the last ice to melt away. Most would have been deposited at the greatest extent reached by the glacier, which would have been the northern part of the McGee Mountain plateau. But there the boulders are least. I have not seen a single moraine in the Sierra Nevada that has such an abundance of large granitic boulders as found at this site.

On the other hand, in unglaciated southern California's granitic lands I have seen extremely bouldery sites similar to those of McGee Mountain. You don't have to make a grueling hike to visit such sites. For example, in the Peninsular Ranges along Interstate 5 from Escondido north to Temecula—a 28-mile stretch—or in lands to the east, such as at Mt. Woodson (Photo 37), there are many bouldery hills. Large outcrops and large roadcuts show that the superabundant boulders are weathering from underlying bedrock, and that they were not deposited by an ancient glacier. (John Muir emphatically maintained that all of California was glaciated—perhaps this was some of his glacial evidence for southern California?) On some of this stretch's smaller, bouldery hills, there are no obvious outcrops and no cuts, and on those, like atop McGee, one could conclude—incorrectly at both places—that the boulders are glacial deposits.

Additional evidence against a glacial origin for these boulders is at a small, resistant, metamorphic knoll that lies just south of and below the east-west ridge, in the northeast corner of Section 36 (Photo 38). Given that 1000+ large granitic boulders were deposited on the ridge, at least a dozen or so should have been deposited on the metamorphic knoll and on the slopes below it. None exist. Furthermore, the ancient glacier, spilling northward across the ridge, should have plucked metamorphic blocks from the knoll and deposited them on the ridge. None exist. Furthermore, here, as at each granitic-boulder area, the soil is granitic, not metamorphic. If these granitic boulders originally were glacial deposits left on

**Photo 36. Resistant boulders atop McGee Mountain's east-west ridge. Notice that here they occur in two restricted locations and are completely absent elsewhere in the field of view.**

**Photo 37. Very bouldery Mt. Woodson, located between Poway and Ramona.**

metamorphic bedrock, the developing soil should be metamorphic, not granitic. In all the glaciated lands of the Sierra that have erratics, the soils are derived from the bedrock, not from the erratics, which only stubbornly release coarse grains of sand.

The greatest number of boulders lie on and below the north-south ridge (Photo 39). The slopes there are steep enough that the boulders, if they had been ancient erratics, should have been buried by downward creeping soil, and then should have slowly disintegrated into granitic sand. Additionally, this ridge, which would be interpreted as a lateral moraine, should have been derived from a glacier

**Photo 38. Metamorphic knoll lacking granitic boulders. Such boulders are extremely abundant on the east-west ridge, from which this photo was taken.**

**Photo 39. Thousands of boulders lie on McGee Mountain's north-south ridge.**

flowing north down the canyon that originates near Mt. Aggie. However, while the large boulders of the north-south ridge are exclusively granitic, the bedrock of the canyon is almost exclusively metamorphic (Photo 40). In short, Blackwelder and Matthes, by seeing all deposits as glacially derived, saw what they wanted to see. Later geologists uncritically accepted their rationalized conclusions. There is no McGee till.

But because Blackwelder suggested two other sites for McGee till, they must be addressed. One is Ebbetts Pass (which Highway 4 crosses), and the other is north of Minaret Summit (which the Devils Postpile road crosses). In my glacial mapping of the Stanislaus River drainage, I covered the Ebbetts Pass area in detail and found only fresh Tioga erratics. No Tahoe, no Sherwin, no McGee. Earlier glaciations undoubtedly occurred but left no surviving evidence. North of the Minaret Summit area is a deposit known as the Deadman Pass till. Recently Roy Bailey, N. King Huber, and Robert Curry visited the site, and in a 1990 paper all concluded that the deposit had a non-glacial origin and so is not a till. Consequently, neither the McGee nor the Deadman Pass are verifiable glaciations.

Many pre-Sherwin glaciations likely occurred, given the sedimentary records of the Northern Hemisphere. During them and also during the Sherwin, precipitation reaching the range likely was substantially greater, since central California's inner and outer Coast Ranges were low. Yosemite Valley and comparable mid-elevation stretches of western canyons may have experienced the effects of larger glaciers by about 2 million years ago, this date based on the onset of significant glaciation recorded in the sediments of Searles Lake, east of the southern Sierra Nevada, and also recorded in the sediments of the San Joaquin Valley, west of the southern Sierra Nevada. During post-Sherwin time the Coast Ranges were raised enough to cast a significant rain shadow, as evidenced by post-Sherwin major uplift in the Diablo Range and by post-Sherwin marine terraces fringing the Santa Cruz Mountains. The increasing rain shadow may explain why all post-Sherwin glaciers, such as those of the Tahoe and Tioga stages, seem to have been smaller than the Sherwin glaciers.

With all the foregoing analysis of glacial stages in mind, I have constructed a new glacial chronology, Table 3, that may be suitable for the western Sierra Nevada. Since the Sherwin appears to be the only uncontested, recognized stage, I have used it as a convenient, if arbitrary, marker to divide the pre-Tahoe into three substages: late pre-Tahoe, Sherwin (middle pre-Tahoe), and early pre-Tahoe.

**Photo 40. Granitic boulders on north-south ridge contrast with metamorphic "source area" for these "erratics."**

**Table 3. Major Quaternary glaciations of the western Sierra Nevada.**

| Glacial stage | Interglacial stage | Oxygen-isotope stage | Age (x1000 years) |
|---|---|---|---|
| | **Holocene** | 1 | 13-0 |
| **Tioga** | | 2 | 35-13 |
| (includes late Tioga deglaciation, 16-13 ka) | | | |
| | **Middle Wisconsin** | 3 | 65-35 |
| **Early Wisconsin** | | 4 | 79-65 |
| **Eowisconsin** | | 5d-5a | 122-79 |
| | **Sangamon** | 5e | 132-122 |
| **Tahoe** (Late Illinoian) | | 6 | 198-132 |
| | **Middle Illinoian** | 7 | 252-198 |
| **Early Tahoe** (Early Illinoian) | | 8 | 302-252 |
| Older glaciations—no interglacial stages or oxygen-isotope stages given | | | |
| **late pre-Tahoe** | | | 780-302 |
| **Sherwin** (middle pre-Tahoe) | | | 900-780 |
| **early pre-Tahoe** | | | 2500-900 |

# 13

# Plausible Moraines and Plausible Glaciers

When you view large lateral moraines projecting from the front of any mountain range, it is not hard for you to envisage what the glacier that left them must have looked like. The Rocky Mountains have some fine examples, especially in Wyoming's Grand Teton National Park (Photo 41) and Idaho's Sawtooth Range (Photo 42). Also in the Rocky Mountains and elsewhere, such as in the Cascade Range of Oregon and Washington, there are large lateral moraines that fail to project out from the front, but nevertheless seem to give a good indication of the sizes of the glaciers that left them. Indeed, moraines, both lateral and end, have long been used to reconstruct past glaciers. And the Sierra Nevada shouldn't be any different. Along the east side from the Mt. Whitney area north to the West Walker River drainage, near the junction of Highways 108 and 395, there are conspicuous lateral and terminal moraines that seem to be good indicators for the size of former glaciers. All of the lateral moraines are logical. By this I mean that any pair of lateral moraines begins at a credible location for the past glacier's equilibrium line, and down-canyon from it the moraines diminish in height quite steadily all the way to the terminal moraine. Furthermore, any point on the crest of a lateral moraine matches in elevation a point on the crest of the lateral moraine on the opposite side of the canyon. If you were to reconstruct a past glacier's longitudinal profile by using either crest of a canyon's two lateral moraines, both would produce very similar results. Not only are the moraines plausible, so too is the glacier reconstructed from them.

**Photo 41. Taggart Creek canyon's terminal moraine, Grand Teton National Park.**

**Photo 42. Pettit Lake canyon's lateral and terminal moraines, Sawtooth Range.**

Superficially, west-side lateral moraines resemble the classic east-side ones, but in actuality they are significantly different. Not all east-side ones are classic. Some located in the northern part of the range have interesting, irregular profiles. We'll first look at two classic pairs of lateral moraines, then two anomalous pairs. Each example contains a topographic map of the moraines plus longitudinal profiles of one or more moraine crests and a longitudinal profile of the canyon floor. If you are uncomfortable using topographic maps, you can skip them. They are provided for professionals who may want to know exactly which crests are used to produce the longitudinal profiles. The important issue for most readers is the shape of each profile. Does it represent a plausible ice surface of a glacier, such as the one shown in Chapter 11's Figure 20, or does it represent an implausible ice surface?

Only the highest lateral moraine of the Tahoe or Tioga stage is identified, for logically it best represents each glacial stage at its maximum. The floor profile is necessary, since a sudden drop along it would cause a sudden drop along the adjacent moraine crests. This may seem self-evident, but on the west-side the moraine-crests' profiles sometimes fluctuate independently of the floor's profile. This is aberrant behavior, for we know that the ice surface of a glacier, like its adjacent lateral moraines, makes a sudden, major drop only if the canyon floor makes a sudden, major drop.

## Lee Vining Creek canyon

The first example is one of the best known: the conspicuous lateral moraines of lower Lee Vining Creek canyon (Photo 43). These are readily observed from Highway 120, which ascends the canyon from Highway 395 west up to Tioga Pass, the east entrance to Yosemite National Park. Map 11 shows the canyon's maximum Tioga and Tahoe moraines, while Figure 21 shows profiles of these two moraines, which descend along the south (better preserved) side of the lower canyon. Bedrock outcrops on both sides of the canyon indicate that the glacier was confined by the canyon's walls. In the lower part of this canyon, the sides were gentle enough (less than about 33° steep) that materials could be deposited and be preserved as part of a moraine. At the bend up-canyon, the sides were too steep, so a continuous morainal cover is absent. The profiles of the two moraines in Figure 21 are similar. Both moraines drop somewhat regularly to near the end, where they plunge and then slacken off as they transform into an end moraine. The profiles show that relatively quickly up-canyon the crest of the Tioga lateral moraine essentially merges with that of the Tahoe. This is a common characteristic.

## Bloody Canyon

The second example, which is only about 3 miles south of the first, is of the conspicuous lateral moraines of lower Bloody Canyon (Photo 44), which are among the most studied. There have been at least eight studies, and not all agree, in part because some show a Tenaya moraine, while others map it as Tahoe or Tioga. The following map and profiles are based on Ronald Kistler's 1966 map, which has Tahoe and Tioga moraines that closely match those of Roy Bailey, of Alan Gillespie, and of Sharp and Birman, except that Kistler shows their hypothetical "Tenaya" moraines as recessional Tahoe moraines. (Even in an area where mapping of moraines is about as simple as one could hope for, geologists still cannot agree.) Map 12 shows Bloody Canyon's maximum Tioga and Tahoe moraines, while Figure 22 shows profiles of these two moraines, which descend along the southeast (better preserved) side of the lower canyon. In contrast with the first example, glaciers descending this canyon apparently were not confined by its walls. The lower canyon indeed is defined by its lateral moraines, not by bedrock.

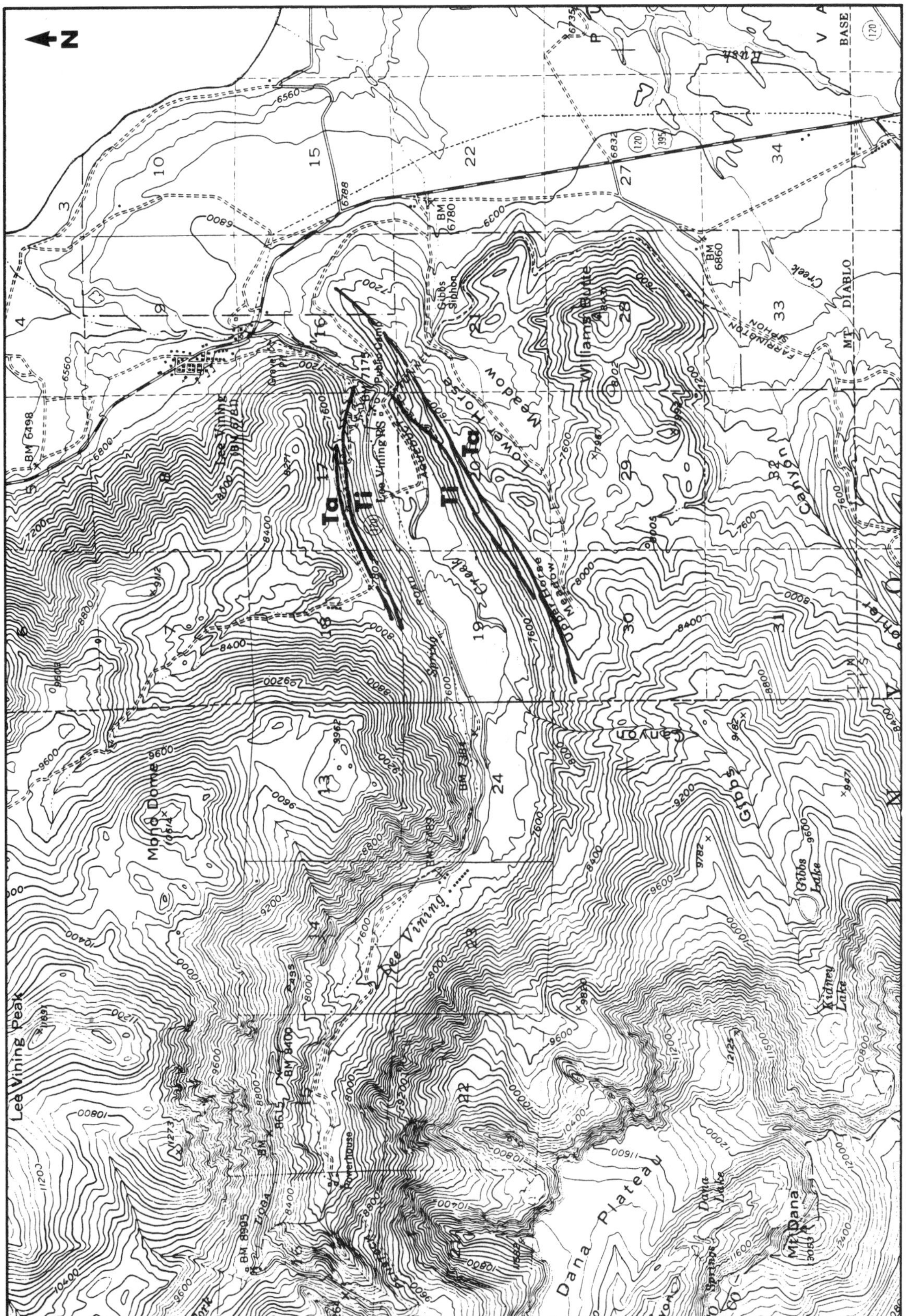


**Map 11. Tioga and Tahoe lateral-moraine crests of Lee Vining Creek canyon. Scale 1:62,500.**

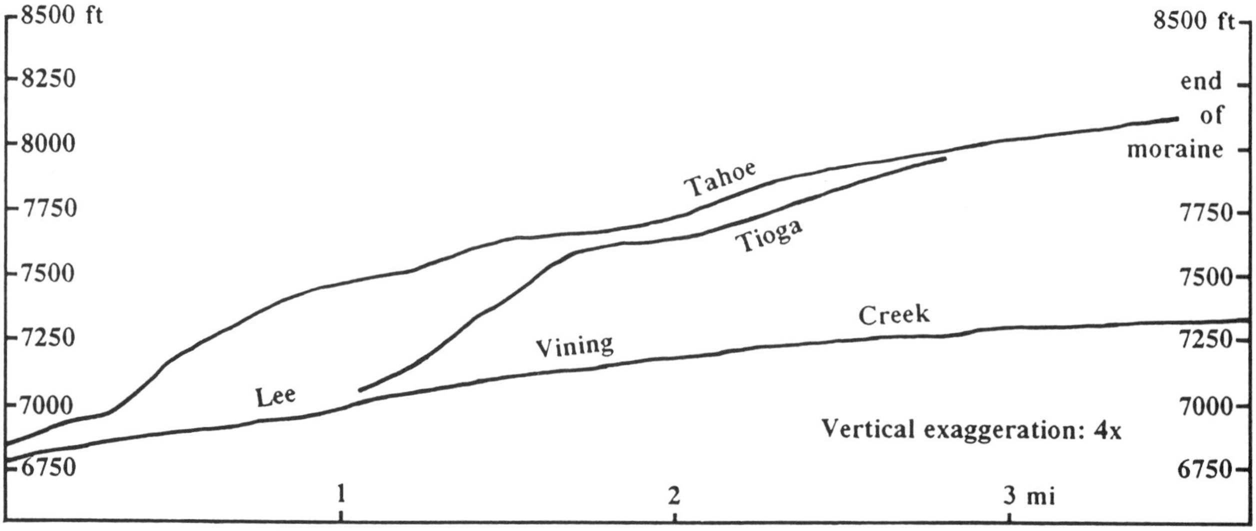


**Figure 21. Profiles of Tioga and Tahoe lateral-moraine crests of Lee Vining Creek canyon.**

Bedrock outcrops are absent, although possibly some bedrock lies just beneath a thin veneer of the northwest moraine. That glaciers were not confined is shown by past fluctuations. A conspicuous pair of lateral moraines, defining Sawmill Canyon and buried in part by the southern Bloody Canyon lateral moraine, indicates that an earlier glacier took a very different path. Andrew Bach in 1992 noted that along the east slopes of the Sierra Nevada from Bishop north to Bridgeport there are 17 major drainages in which past glaciers took different courses. As in the first example, the profiles of Bloody Canyon's Tioga and Tahoe moraines are similar: they slacken off as

**Photo 43. Aerial view of Lee Vining Creek canyon.**

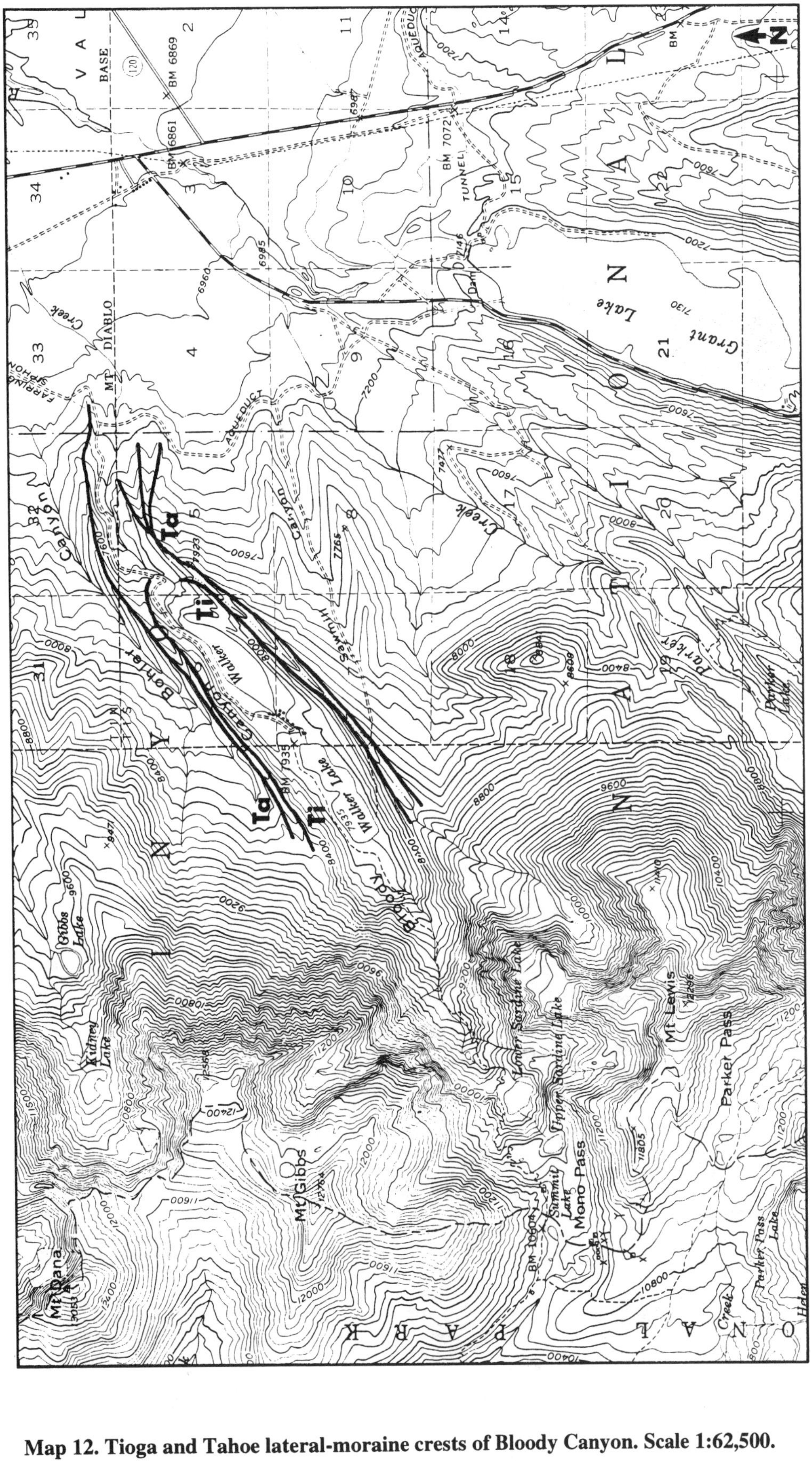


Map 12. Tioga and Tahoe lateral-moraine crests of Bloody Canyon. Scale 1:62,500.

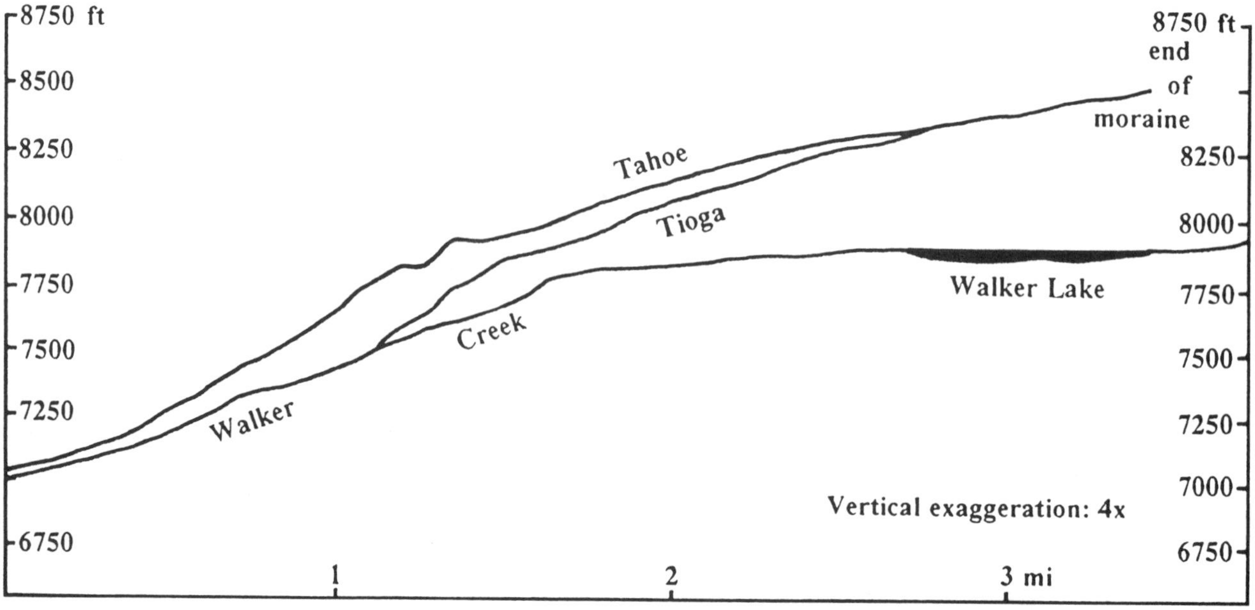


**Figure 22. Profiles of Tioga and Tioga lateral-moraine crests of Bloody Canyon.**

they transform into an end moraine, and up-canyon the crest of the Tioga lateral moraine essentially merges with that of the Tahoe. However, on the Tahoe crest there are two minor high points that project higher than the depression immediately up-canyon from each, which is characteristic of a fair number of crests of east-side lateral moraines.

On the east slopes of the Sierra Nevada the glaciated canyons have pairs of lateral moraines similar to those in these two examples regardless of whether the glacier was confined, partly confined, or unconfined. Also, a point atop one lateral moraine is at about the same elevation as a point atop the lateral moraine across the canyon. Therefore, one can feel confident reconstructing the ice surface of any canyon's Tioga or Tahoe glacier based on the crest of the lateral moraines. The unvarying pattern is that both the Tahoe and Tioga glaciers were about the same thickness at the ELA, and both thinned down-canyon, each Tioga glacier thinning faster and ending not far behind the terminus of the preceding Tahoe glacier.

**Photo 44. Tioga and Tahoe lateral moraines of lower Bloody Canyon. Walker Lake lies below them.**

Mapping by researchers from Eliot Blackwelder onward indicates that the length of the Tioga glaciers relative to the length of the Tahoe glaciers is about 80-95%. These patterns exist along the east slopes north to about the vicinity of Donner Pass.

## Donner Pass area

On east slopes in the Donner Pass (Interstate 80) area, patterns, as mapped in detail by Peter Birkeland in his 1964 paper, begin to break down. The northernmost canyon he mapped was the North Fork Prosser Creek canyon, where Tioga and earlier glaciers originated on a broad Sierran-crest saddle that now separates Paradise Lake from Warren Lake (Map 13). Whereas at first the Tioga ice reconstruction looks plausible, when it is plotted in longitudinal profile (Figure 23)—which Birkeland did not do—it is seen to have problems. For about 3.0 miles in the middle part of its 6.8-mile length, its thickness is nearly constant, then over a distance of about 1.0 mile, between about miles 2 and 1, it rapidly diminishes in thickness before continuing as a thin glacier about 0.8 mile farther. Unlike in the two previous profiles of moraines (Figures 21 and 22), which curved as end moraines over to their creeks, this last part of the profile does not represent a lateral-moraine-to-end-moraine transition; all but the very last bit represents the crest of a lateral moraine, which in turn represents the longitudinal ice surface. Glaciers at their maximum do not have such an ice surface near their terminus. An ice surface can be quite steep if the broad-floored canyon down which the glacier flows also is quite steep. But what is readily apparent in Figure 23 is that the canyon floor is nearly flat. Drastic thinning should not have occurred. Did it?

There are limits to how steep the ice surface can be, and one can test the validity of a reconstructed ice surface by performing a basal shear calculation. (Basal shear stress, calculated in metric units, equals ice density times gravity times ice thickness times the sine of the angle of the ice surface.) Kenneth Pierce in his 1979 Professional Paper 729-F performed calculations for past glaciers in Yellowstone National Park, and anyone interested in the rationale and mathematics of basal shear calculations should read his pages 69-73. The gradients along the ice surfaces of modern glaciers have basal shear values between 0.5 and 1.5 bars. (A bar is a unit of pressure equal to one million dynes per square centimeter; one bar is approximately equal to the earth's atmospheric pressure at sea level. One dyne is the force required to accelerate a mass of one gram at one centimeter per second per second.) Because Pierce's values lie within 0.5 and 1.5 bars, his glacial reconstruction appears good. But this is not necessarily so, since for any Sierran canyon I can reconstruct a number of totally imaginary glacial reconstructions, all with values between 0.5 and 1.5 bars. However, Pierce's accumulation-area ratio—the area of the ice surface above the equilibrium line divided by the total area of the ice surface—is abnormally high. He rationalizes its value and attempts to prove its validity with a mass balance (in a system of any kind, outputs equal imputs). All you need to know about this method is that it is extensively used, and like statistical methods, it allows you to prove anything, real or imaginary. By selecting an abnormally low amount of precipitation—one below his range of acceptable values that he himself mentions—Pierce was able to come up with a mass balance that "validated" his abnormally high accumulation-area ratio. The bottom line is that mass balances and basal shear calculations cannot prove you are right, they can only prove you wrong. If your basal shear values are below 0.5 bars or above 1.5 bars, your glacial reconstruction is questionable.

With the preceding in mind, we can make some basal shear calculations for the North Fork Prosser Creek canyon's glacier. In Figure 23 between miles 4.7 and 1.7, the thickness of the ice is relatively constant, averaging about 900 feet. The angle of the ice surface in the down-canyon direction is about 1°, and the resulting basal shear stress is about 0.4 bars. This is a bit on the low side, but not fatally so. However, over the next mile, from miles 1.7 to 0.7, the glacier thins dramatically. The angle of the ice surface is about 7.5°, and the basal shear stresses at the start, midway through, and at the end of this stretch are respectively 3.1, 2.1, and 0.8 bars. Only the last value is within the range; the two others are impossibly high. As you will see later, reconstructions of past glaciers on the western slopes of the Sierra Nevada have stretches with impossibly high basal shear values.

Another problem with Birkeland's areal reconstruction is with the absolute and relative lengths of his glaciers. The Tioga glacier of the South Fork Prosser Creek canyon is anomalously short, about 4.0 miles long, despite being fed in large part by the Donner Pass ice cap. East-flowing ice from this ice cap mostly flowed down the Donner Lake canyon, but despite this massive source, the length of its Tioga glacier was essentially that of the North Fork Prosser Creek canyon's Tioga glacier, which lacked an ice cap. One would expect the South Fork glacier, which had this source, would have been longer than the North Fork glacier. Furthermore, the lengths of the Tioga glaciers, expressed as a percent of the lengths of their Tahoe counterparts, also show an anomalously low value for the South Fork Prosser Creek canyon's Tioga glacier: Donner Lake canyon—84%; North Fork Prosser Creek canyon—83%; South Fork Prosser Creek canyon—58%.

A final problem with Birkeland's areal reconstruction is that in it the ice at the Sierran crest is too thin. For example, at the deep saddle (Photo 45) at the head of North Fork Prosser Creek canyon the Tioga ice is on the order of 100 feet thick. Being this thin and having essentially a zero-gradient ice surface, the ice would be brittle and non-

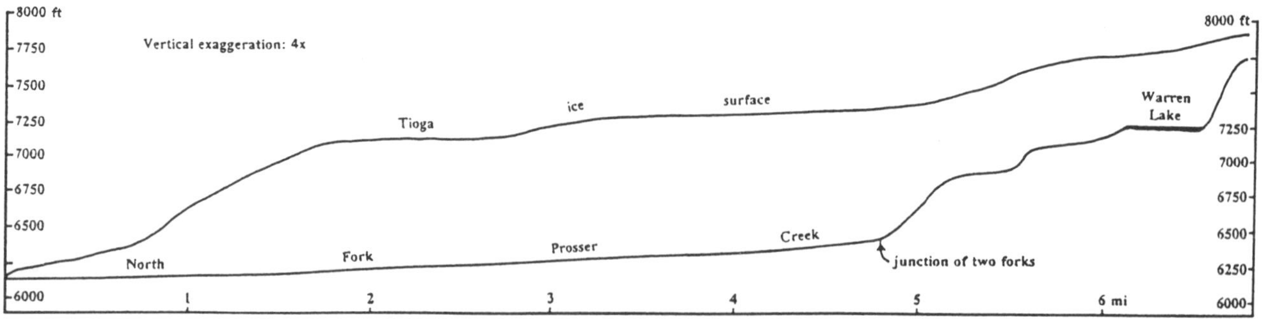


**Figure 23. Profile of Tioga ice surface in North Fork Prosser Creek canyon.**

flowing—hardly a good start for a large glacier. Furthermore, how did this ice originate? The Paradise Lake saddle is largely a shallow basin, and it is too low in elevation for an ice sheet to form. Two lines of evidence indicate the ice was much thicker: polish and striations. Here I find it very important to stress that in the entire Sierra Nevada I have found polish and striations only where Tioga glaciers once covered the land. I have found none in areas supposedly covered by Tahoe or pre-Tahoe glaciers. Locations where past geologists have identified presumably Tahoe and pre-Tahoe polish and striations all were covered with Tioga glaciers. So until someone can find bona fide Tahoe and/or pre-Tahoe polish and striations, they remain evidence of Tioga glaciers only.

Polish and striations are abundant on the barren bedrock surrounding the lake, but they also continue both north and south up granitic ridges. The granitic ridge ascending from the lake's northwest shore extends to 8170 feet elevation before yielding to volcanic sediments, and I found striations along this ridge up to 8140 feet. The granitic ridge ascending from the lake's northeast shore has poorer evidence, apparently because it is increasingly mafic (rich in dark minerals) upward and it is highly fractured—two rock characteristics unfavorable for preservation of polish and striations. (Even by the lake, the polish and striations on this mafic, fractured bedrock are much less than elsewhere around the lake.) The granitic ridge yields to volcanic sediments at 8290 feet

**Photo 45. Paradise Lake, from highest north-side striated site.**

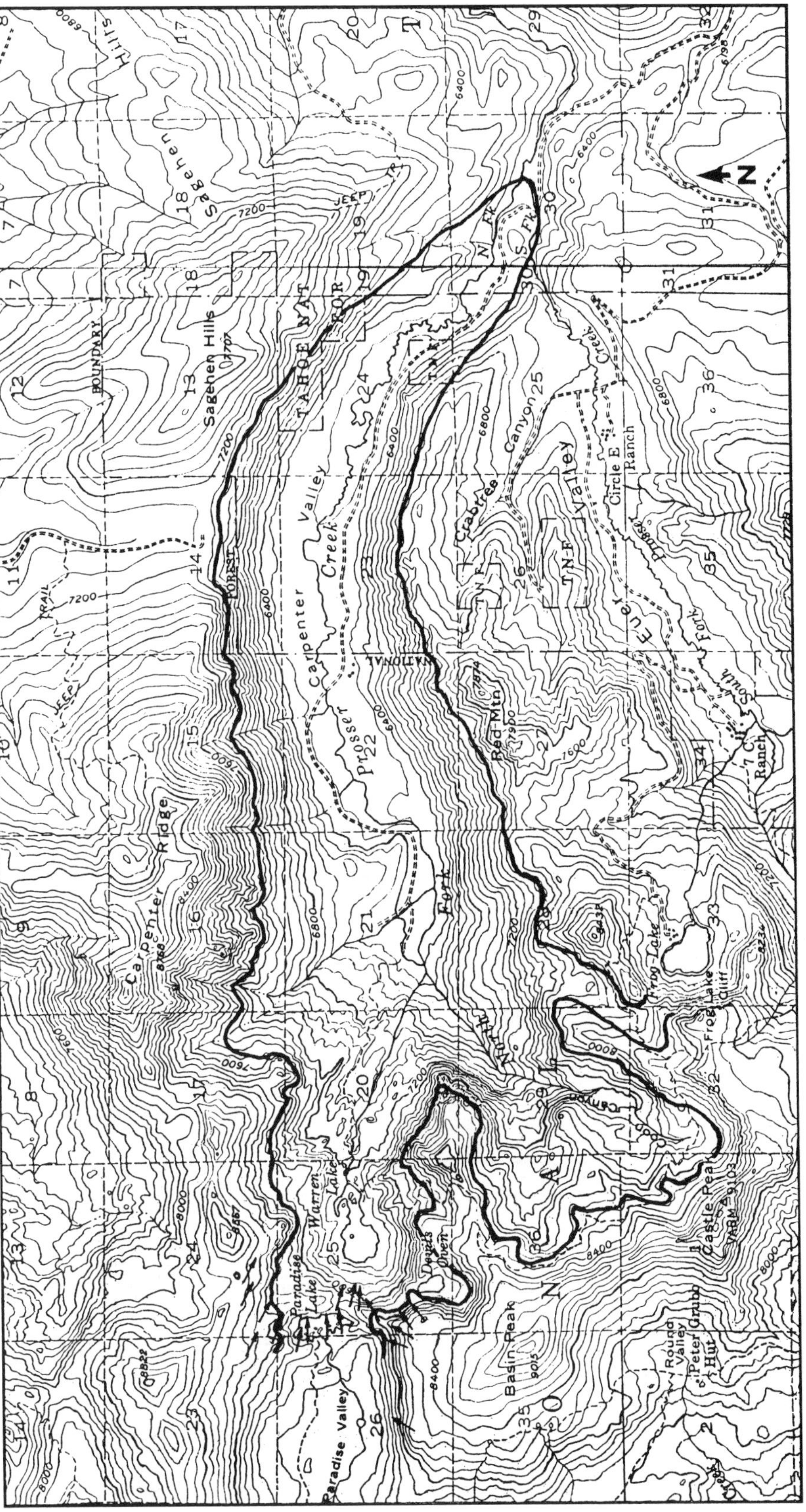


**Map 13. Extent of Tioga ice in upper North Fork Prosser Creek canyon. Arrows show direction of ice flow, based on glacial striations. Scale 1:62,500.**

elevation, this high point located about 700 feet due west of a lakelet on a broad crest saddle. All glacial ponds, lakelets, and lakes that I (like others) have dated in the Sierra Nevada have post-Tioga-maximum dates, and this shallow lakelet should be no exception. Evidence that it is post-Tioga lies in a small patch of fresh, striated polish on the high point (Photo 45). From the lake north up to here, all striations are oriented approximately parallel to the longitudinal profile of the canyon, having bearings of about 90° lower down and about 110 to 120° higher up. The highest striations, as well as those on a small granitic outcrop immediately south of the lakelet, have a bearing of 120°, and this orientation demonstrates that the lakelet's basin does not owe its origin from a glacier spilling south across the lakelet's divide. Had that occurred, the bearings would be about 180°.

The fresh, striated polish on high point, 8290 feet elevation, indicates that the ice surface of the Tioga glacier was at least 8440 feet elevation, if one uses an average figure of about 150 feet for the thickness of brittle ice riding atop a glacier's flowing ice. The elevation of the ice surface can be estimated from evidence on the canyon rim south and south-southwest of and above Paradise Lake. The granitic rim is between about 8300 and 8400 feet elevation, so the Tioga glacier should have buried it under thin ice. Striations along it, however, are not parallel to the longitudinal profile of the canyon. Rather, they have bearings of about 20 to 30°. These indicate that a glacier originating on the northern slopes of Basin Peak flowed directly downslope to the brink of the canyon to join its glacier. Consequently, the ice surface of the canyon's glacier could not have been much higher than the granitic rim, or else its easwtard motion would have altered the direction of the lower part of the descending, Basin Peak glacier. The ice surface of the descending glacier, where it flowed across the gentle benchlands just south of the canyon's brink, would have had a minimum

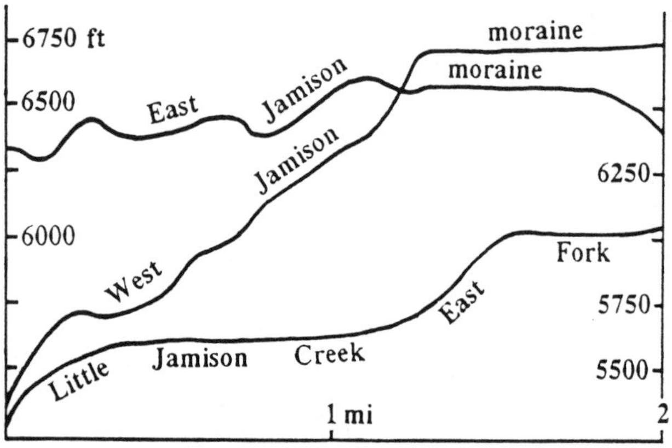


**Figure 24. Profiles of the pair of Tioga lateral-moraine crests of Little Jamison Creek canyon.**

ice-surface elevation of about 8500 feet. Any less, the descending glacier would have been too thin to flow. Without further evidence, which may be difficult to find on this area's volcanic slopes, the actual ice-surface elevation is unknown. Therefore, at present, all one can say is that the thickness of the Tioga glacier above the 7750-foot Paradise Lake saddle was a minimum of 750 feet, five times the thickness that Birkeland had determined. Colman and Pierce, in their 1992 paper, also have disagreed with Birkeland's conclusions in this area. They reassigned his Tahoe to Tioga, which, if a correct assignment, suggests that perhaps other presumably Tahoe moraines in the range actually may be Tioga. Indeed, as I show in future chapters, most of the Sierra's west-side moraines identified as Tahoe actually are Tioga.

## Lakes Basin area

Although the Sierra's east-side Donner Pass area poses significant problems in reconstructing the thickness and extent of past glaciations, it is not alone. The east-side Lakes Basin area, north of Highway 49, poses even more. This is in spite of the area's moraines having been mapped in incredible detail by Scott Mathieson for his lengthy 1981 M.S. thesis. These problems are a prelude to the problems faced in the glaciated lands of the west slopes of the range. Map 14 shows the maximum Tioga and Tahoe lateral-moraine crests in and west of the Lakes Basin area, while Figure 24 shows longitudinal profiles of the pair of Tioga lateral-moraine crests of Little Jamison Creek canyon. If the crests of that canyon's lateral moraines actually represented the Tioga ice surface when the glacier was at its maximum, then their profiles would be quite similar. However, they are very different, and furthermore, neither represents the profile of a modern glacier. The ice surface of the east-side moraine begins by actually increasing in elevation from near the cirque, then is anomalously flat, and finally starts to drop slightly, but in an undulating manner. In contrast, the west-side moraine stays level, but at a higher elevation than its counterpart, then drops very rapidly, thinning to almost nothing where this tributary canyon begins its plunge northwest to Jamison Creek canyon. Mathieson mapped no lateral moraines along this plunge, although he extrapolated a continuation of the west-side moraine down to the main canyon. The Tioga lateral moraines in this main canyon also have problems. The south-side one dies out before reaching the tributary canyon, but is extrapolated to an end moraine located at the tributary canyon's mouth. In contrast, the crest of Jamison Creek canyon's north-side moraine is an irreconcilable 1000 feet above the main canyon's floor opposite the tributary canyon's mouth.

Finally, above the two canyons' confluence, Tahoe lateral moraines are absent, which could imply that Tahoe

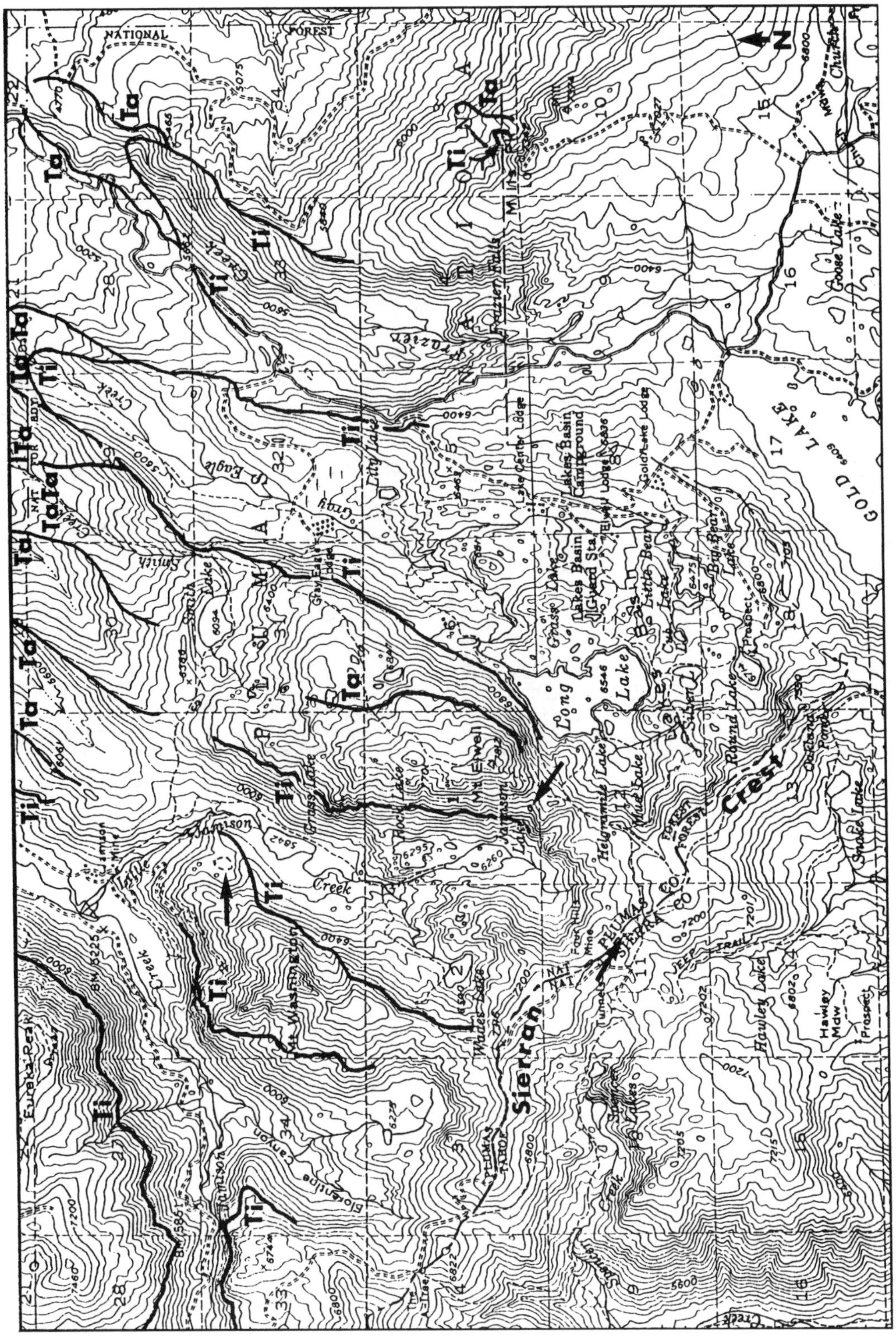


**Map 14. Tioga and Tahoe lateral-moraine crests in part of the Lakes Basin area. Scale 1:62,500.**

**Photo 46. View southeast across Long Lake, Lakes Basin, and Sierra Buttes, from Mt. Elwell. The basin was totally overridden by Tioga glaciers, so the moraines left within it do not represent the ice surface of the maximum Tioga.**

glaciers were smaller than Tioga glaciers. However, Mathieson mapped Tahoe lateral moraines in the lower part of Jamison Creek canyon, and these stand above and extend beyond the Tioga lateral moraines, indicating that the Tahoe glacier, as in all his other mapped canyons, was larger than the Tioga glacier. The absence of any Tahoe moraine or till above the confluence raises another question: how did the seasonal lake just west of the lower part of Little Jamison Creek canyon's west-side moraine originate (Map 14, oval, dashed line with arrow pointing to it)? Mathieson's mapping indicates that no glaciation occurred at the lake or on any land above the lateral moraines of either canyon. As was mentioned under the "Donner Pass area," all dated Sierran lakes formed after Tioga glaciers retreated. One might assume that this seasonal one is a rare, nearly infilled relict of a pre-Tahoe glaciation. However, such a glaciation must have been smaller than the Tioga and the Tahoe, for Mathieson, despite detailed mapping, was unable to find any pre-Tahoe till, moraines, or erratics. What the abundant evidence in the Lakes Basin area clearly shows is that here reconstruction of former ice surfaces on the basis of lateral-moraine crests is impossible. Note on Map 14 that in this drainage there also are a pair of ponds, identified with two arrows, which are well above the Tioga lateral moraines, and they require a similar explanation.

# 14

# Implausible Moraines and Implausible Glaciers

It is obvious from the last two east-side areas that lateral moraines previously identified as Tioga maximum or Tahoe maximum are not only implausible looking, but also produce implausible glacial reconstructions. This same problem occurs on the west side of the Sierra Nevada wherever such moraines have been mapped and assigned a date. Examples occur in every glaciated river drainage from the Mokelumne south to the Kern. North of the Mokelumne, only the trans-crest Lakes Basin area (North Yuba River/Middle Fork Feather River), just discussed, has been studied in detail. In addition to that area, four other glaciated drainages have been *field mapped* for glacial evidence in considerable detail: the Stanislaus (my own field work, which raised questions that ultimately led to this book), the Merced (Matthes, 1930), the San Joaquin (Matthes, 1960; Birman, 1964), and the Kern (Matthes, 1965). Elsewhere, the field mapping was rather superficial, and in too many cases it was done in the office, geologists perceiving Tioga, Tahoe, and pre-Tahoe glacial deposits on aerial photos.

There are three types of already suggested problems with this century's glacial mapping on the western slopes, and examples are presented for each. First, as was just mentioned, the crests of lateral moraines yield implausible glacial reconstructions. Second, some west-side canyons lack Tahoe lateral moraines above the Tioga lateral moraines, even though the Tahoe glaciers were concluded to have been larger. And third, the lengths of Tioga glaciers relative to the lengths of the Tahoe glaciers do not fit the 80 to 95% range found along the east slopes north to about the vicinity of Donner Pass. On the west side, however, the problem is compounded, for not only do relative lengths vary widely, but so do absolute lengths, as is shown in Table 4 for the Stanislaus, Merced, and San Joaquin river drainages. All three are known to have had extensive early (pre-Tahoe) glaciation. Something is terribly wrong with previous interpretations.

In the Middle Fork Stanislaus River drainage, the pre-Tahoe and Tahoe lengths are based on my field work. As I discuss under Chapter 20's "Stanislaus River drainage," the lengths of these two glaciers are uncertain. I feel uneasy about them; either the pre-Tahoe is too long with respect to the Tahoe, or the Tahoe is too short with respect to the pre-Tahoe. I have not found sufficient evidence to locate the farthest extent of either glacier. The Tioga length is based on Slemmons' 1953 dissertation

**Table 4. Lengths (in miles) of pre-Tahoe, Tahoe, and Tioga glaciers in three west-side canyons.**

| River drainage | pre-Tahoe | Tahoe | Tioga | Tioga/Tahoe |
|---|---|---|---|---|
| Stanislaus, Middle Fork | 42.0 | 27.2 | 4.8 | 18% |
| Merced | 36.2 | less than 26.1 | 26.1 | 105-150% |
| San Joaquin, South Fork | 61.3 | 50.6 | 30.3 | 60% |

map. Today, no one would accept this abnormally short length, but I use it because he arrived at it by using Blackwelder's criteria for east-side Tioga glaciers. Obviously, those criteria do not work.

In the Merced River drainage, I arrived at the pre-Tahoe length by measuring Matthes' reconstructed map (Professional Paper 160, Plate 39) of his "El Portal/Glacier Point" (pre-Tahoe) glaciers. For the length of the Tioga glacier I measured his reconstructed map of his "Wisconsin" (Tahoe and Tioga) glacier, which he believed ended at Bridalveil Meadow. Under the current 1980s and '90s interpretation of glaciation in the Merced River drainage, the Tioga glacier advanced to this meadow, but the Tahoe glacier was shorter. This interpretation is based on Wahrhaftig's proposed "Tahoe moraine," which drops below a Tioga moraine (Chapter 1, Figure 2). The crest of this "Tahoe moraine" (neither Tahoe nor moraine) east of and above Little Yosemite Valley descends gently west over most of its length, then drops steeply near its end. If one extrapolates the gentle part, then one might conclude that the Tahoe glacier was just slightly shorter than the Tioga glacier. But if one extrapolates the steeper part, then the Tahoe glacier was much shorter. Therefore the Tioga/Tahoe percent could range from about 105% (if it ended near the base of El Capitan) to about 150% (if it ended in western Little Yosemite Valley). Unlike anywhere else in the range, the Tioga glacier here supposedly was longer than the Tahoe.

In the South Fork San Joaquin River drainage, I arrived at the pre-Tahoe length by measuring Matthes' reconstructed map (Professional Paper 329, Plate 1) of the "pre-Wisconsin" (pre-Tahoe) glaciers. For the Tahoe length I used the Wisconsin of Matthes' reconstructed map, plus Birman's Tahoe terminus, which is about 2.4 miles longer than Matthes'. I chose Birman's terminus because: 1) his glacial map is more recent and far more detailed; 2) he describes how he arrived at its location; 3) Matthes did not separate his Wisconsin into Tahoe and Tioga; 4) Matthes' reconstructions were theory-driven and much of his mapped deposits were imaginary; and 5) Matthes' work is posthumous, so it may not represent his true intent. Likewise, I used Birman's Tioga terminus.

The Tioga/Tahoe ratio of Table 4 shows extremely high variability, and none of the ratios is within the 80-95% range of the Tioga/Tahoe glaciers along the east side of the central Sierra Nevada. This is only the first of a number of problems that have arisen due to relative dating of the west-side moraines by, logically, the same methods used for the east-side ones. What follows are additional problems.

## Mismatched moraines

The Tioga and Tahoe lateral moraines in and above Little Yosemite Valley—as interpreted by Clyde Wahrhaftig on the 1989 USGS geologic map of Yosemite National Park—illustrate the magnitude of the problem. (Little Yosemite Valley is the most controversial area in the whole range, and so will be the focus of this section on problems.) Map 15 shows Wahrhaftig's maximum Tioga and Tahoe lateral moraines in this area, while Figure 25 shows longitudinal profiles of the crests of these lateral moraines as they occur on both sides of the valley. It also shows, for comparison, Matthes' Wisconsin glacier's ice surface (from his Figure 22). As in the previous chapter, you can skip this map (and others that follow) and still comprehend the following arguments. You do, however, need to understand the figure's profiles.

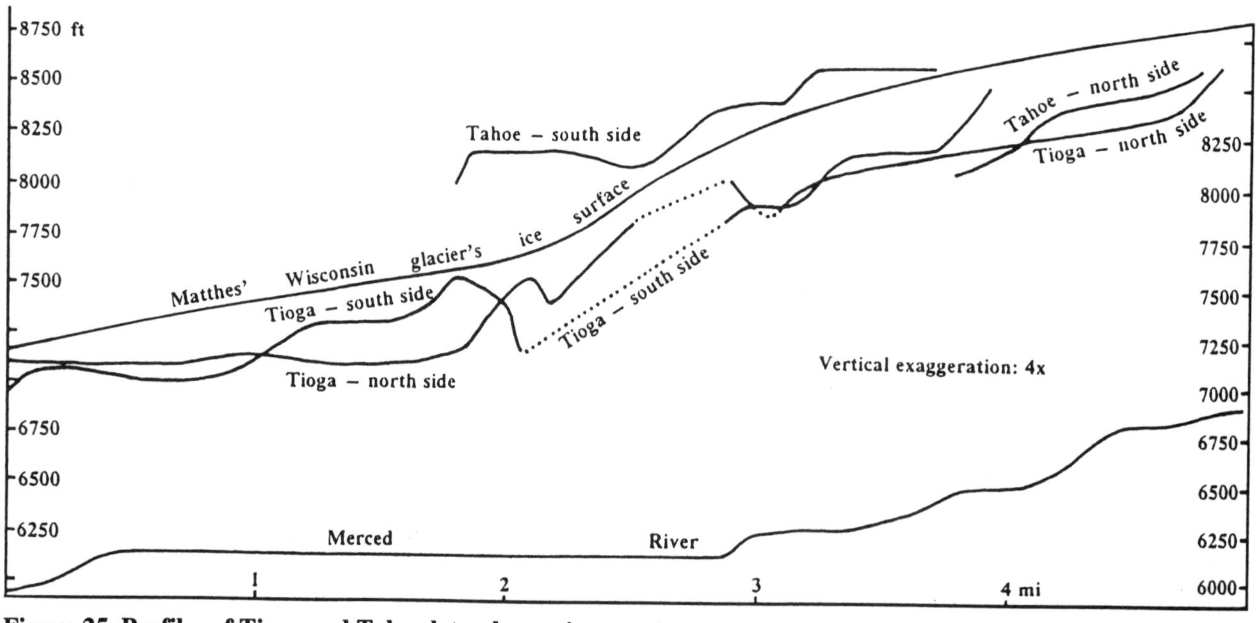


**Figure 25. Profiles of Tioga and Tahoe lateral-moraine crests on north and south sides of Little Yosemite Valley.**

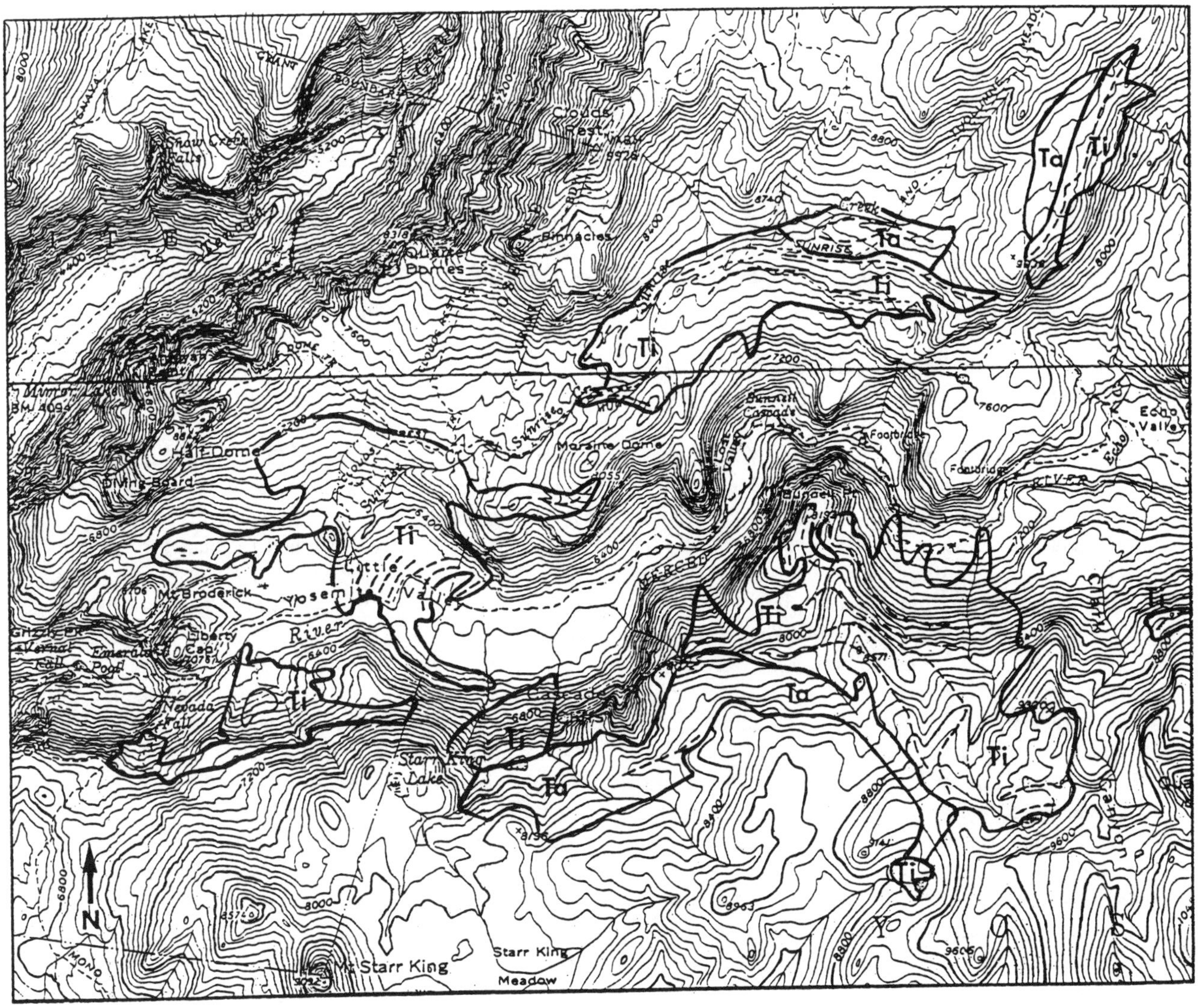


**Map 15. Tioga and Tahoe lateral moraines of Little Yosemite Valley. Scale 1:62,500.**

On the north side, there is only one short Tahoe lateral moraine (right side of figure), whose western end appears to descend beneath the longer, narrow-crested Tioga lateral moraine. This implies that the Tahoe glacier diminished faster down-canyon than did the Tioga glacier, and therefore, it was shorter. In a 1990 letter to me, Wahrhaftig wrote:

> I found the same relation between Tahoe and Tioga moraines on the Tuolumne River and in Illilouette Creek basin that we find on the East slope of the Sierra, in the San Joaquin basin, and elsewhere. When I found the Tahoe lateral moraine on the north side of Little Yosemite Valley disappearing downstream beneath the Tioga lateral moraine, on Sunrise Creek east of Clouds Rest, I realized that the reason we don't see Tahoe moraines in Yosemite Valley is because *for some unknown cause,* the Tioga glaciers actually extended beyond the Tahoe glaciers in Yosemite Valley. The Bridalveil moraines at the west end of the valley have all the weathering characteristics of Tioga, not Tahoe moraines. (Italics mine; underline his.)

Before discussing the Little Yosemite Valley situation, I should point out that the moraines on the Tuolumne River that he mentions *appear* to have the same relations (nesting, relative length, morphology) as those on the east side. One could even argue that they deserve to be the type-locality Tioga and Tahoe west-side lateral moraines. However, under closer examination (as discussed later), they too are just as enigmatic as are the ones in Little Yosemite Valley.

Also I should point out that although Clyde Wahrhaftig, until he died in 1994, was universally acclaimed as *the* Sierran glaciation and geomorphology expert. He appears to have gotten this recognition due to lack of completion in the two fields. More than anything else, he compiled the previous glacial studies of others. While in Yosemite Valley with me in June 1993, he stated that he had done very little actual glacial field work, and that he never visited the crucial area of the Starr King bench above Little Yosemite Valley or any of the San Joaquin's glaciated lands. Huber too alluded to this in the 1990 *Yosemite Centennial Symposium Proceedings*: "With the aid of modern aerial photographs, unavailable to Matthes, Wahrhaftig was able to improve on Matthes' mapping and also to extend map coverage to the entire park for the Tioga glaciation." By performing minimal field work, one runs the risk of missing crucial evidence, and Wahrhaftig in Little Yosemite Valley missed most of it.

Of the four Little Yosemite Valley profiles, only the north-side Tahoe lateral moraine appears to make sense in that it mimics the profile of the valley's Merced River. This moraine is the diving one Wahrhaftig used to conclude that the Tahoe glacier was shorter than the Tioga. However, the profile of the south-side Tahoe lateral moraine presents a contradictory picture, since not only does it stay above the south-side Tioga lateral moraine, but the elevation difference between the two south-side crests, which is minimal up-canyon, increases to about 650 feet by about 2 miles down-canyon. This implies that the Tioga glacier was diminishing much more rapidly than the Tahoe, and consequently the Tahoe must have advanced significantly farther than the Tioga. Evidence for this exists west beyond Yosemite Valley proper, high above the Merced River gorge, where the Cascade Creek bridge till (as discussed later) has an assortment of subsurface boulders, exposed in a road cut, that most closely resemble ones in recognized west-side Tahoe moraines as Wahrhaftig so defined them back in 1984. An extrapolation of the Tahoe glacier down-canyon from the Cascade Creek site indicates that it could have advanced south entirely through Merced Gorge.

There are other notable problems with the four moraine profiles. First, a projection of the north-side Tahoe moraine westward beneath the north-side Tioga shows that down-canyon in the vicinity of where the south-side Tahoe moraine dies out, the difference in elevation between the two Tahoe crests would have been 600 to 900 feet. The valley width here is about 2 miles, and so in cross profile the ice surface of the Tahoe glacier would have dipped north at a 5½% to 8½% gradient. Such ice surfaces do not exist. Second, the two Tioga moraines have short stretches that increase substantially in elevation down-canyon. These stretches are too great to be attributed to random, local heavy accumulations of till. Indeed, field inspection shows no more thickness than elsewhere. Rather, on each side of Little Yosemite Valley proper, the north and south bedrock rims locally increase down-canyon, and the till lies on these upslope stretches. Third, the two undulating Tioga moraines indicate that the glacier thinned in a believable fashion through the first 60% of the profile, but then it stayed relatively level, especially if one considers only the north-side moraine. The last 20% of that profile is taken from Matthes, who did not address why the glacier diminished rapidly and then maintained essentially constant thickness.

Be aware that in his Figure 22 he smoothed the longitudinal profile of his Wisconsin (Tahoe/Tioga) glacier's ice surface, thereby ignoring the undulations of the highest lateral moraines mapped on his Plate 29. In like manner, in the Tuolumne River drainage from O'Shaugnessy Dam (Hetch Hetchy Reservoir) down to Mather, Wahrhaftig, in the geologic map of Yosemite National Park, mapped the south-side's highest Tioga lateral moraines as nearly level to gently descending for about 3 miles, then as steeply descending about 1000 feet over the next 2 miles. (My field work confirms the mapping as correct. It is the ages assigned to these moraines that I dispute.) But in his glacial reconstruction of the park's Tioga glaciers, Wahrhaftig (for a 1987 map drafted by Tau Rho Alpha) averaged out the nearly level and steeply descending stretches to produce a very believable reconstruction—one at odds with the evidence on the geologic map. Both the park's geologic map (USGS Map I-1874) and the park's Tioga-glaciers map (USGS Map I-1885) should be available at Yosemite Valley's Visitor Center. If you are good with maps, purchase a copy of each, and compare the Tioga evidence in Little Yosemite Valley and in the Hetch Hetchy area with the reconstruction for each area. Very interesting.

Problems appear when one reconstructs a past glacier based on mapped evidence. In Map 16 I have reconstructed Wahrhaftig's Tahoe moraine from Merced Lake to the west end of Little Yosemite Valley. The first problem, which is more obvious in Figure 25 (which covers only part of this map's length), is that there is no correlation between drops along the canyon floor and drops along the glacier's ice surface. Where the ice surface drops most dramatically, between 8000 and 7200 feet elevation, the floor of Little Yosemite Valley is flat. Inexplicable ice thinning. Second, ice contours should be more or less perpendicular to the glacier's direction of flow. Generally, they are not. And finally, some basal shear values indicate a flawed reconstruction. I divided the glacier into four lengths: low-gradient, 8800-8600 feet; intermediate-gradient, 8600-8000 feet; high-gradient, 8000-7200 feet; and low gradient, 7200-6800 feet. Using the average elevation to determine the average thickness of each length, I arrived at the following respective basal shear values: 0.77, 3.10, 7.28, 0.93. The first and last are permissible values (between 0.5 and 1.5 bars), the two in between definitely are not. It is little wonder that Wahrhaftig did not publish a map of his Tahoe glacier.

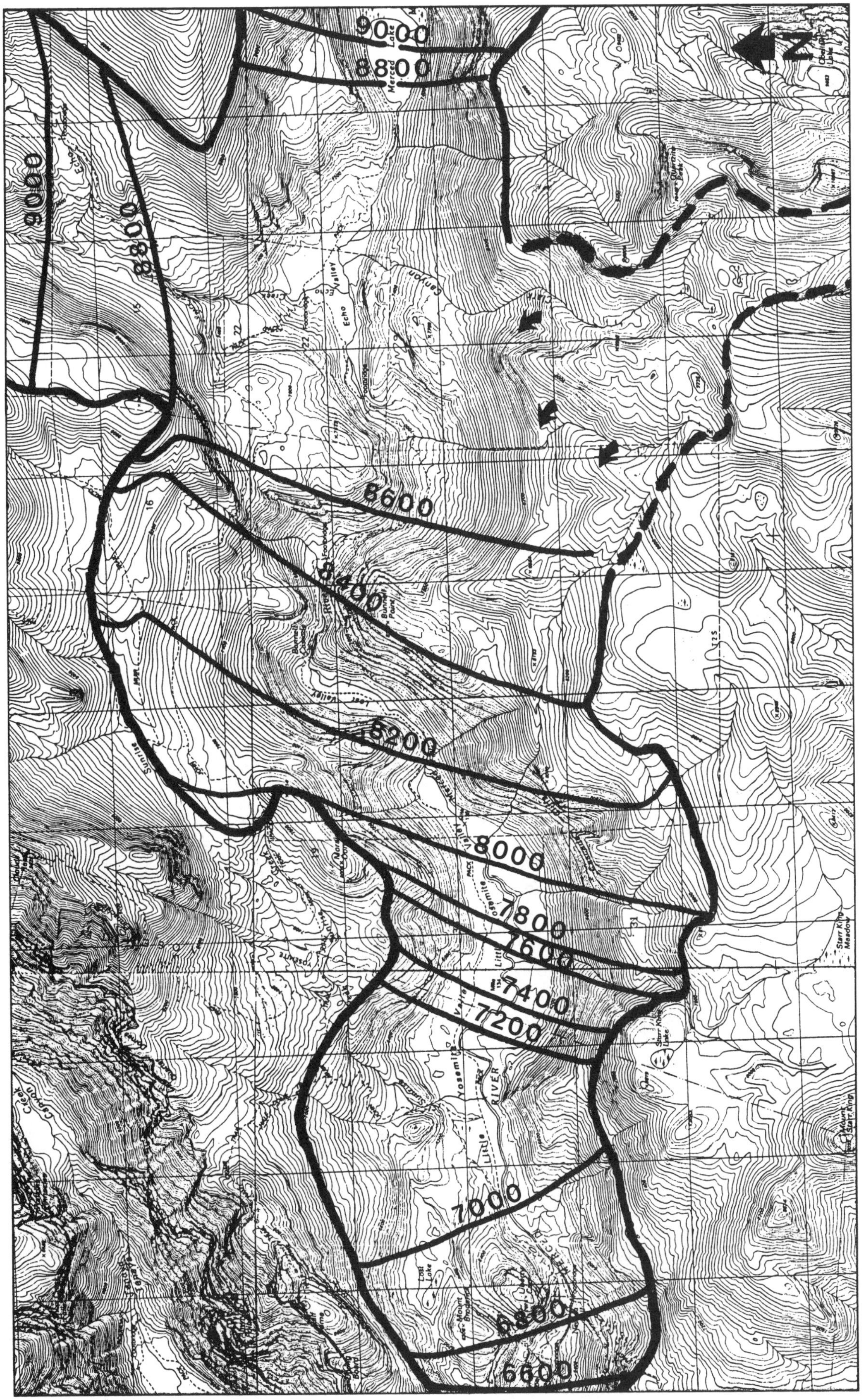


**Map 16. Wahrhaftig's Tahoe glacier from Merced Lake to Nevada Fall. Scale 1:48,000.**

## Two relative-dating methods to date glacial deposits

The foremost problem with reconstructing the ice surfaces of former glaciers, such as those that flowed through Little Yosemite Valley, is that all no moraines, till, and erratics on the west side of the Sierra can be absolutely dated, so relative dating has been used. The universally used standard method has been one of relative position of glacial deposits and the morphology, or shape, of each. For the west side of the Sierra, this method divides glacial deposits into three categories, as follows.

The foremost problem with reconstructing the ice surfaces of former glaciers, such as those that flowed through Little Yosemite Valley, is that no moraines, till, and erratics on the west side of the Sierra can be absolutely dated, so relative dating has been used. The universally used standard method has been one of relative position of glacial deposits and the morphology, or shape, of each. For the west side of the Sierra, this method divides glacial deposits into three categories, as follows.

*Tioga:* moraine crests are sharp and rugged and have abundant fresh boulders; lateral moraines lie just below and inboard of Tahoe lateral moraines, and its terminal moraine lies behind the Tahoe terminal moraine.

*Tahoe:* moraine crests are low, broad, and smooth and have fewer boulders, these somewhat weathered.

*Pre-Tahoe:* moraines are absent; isolated erratics and possibly some sparse deposits (patches of till) exist beyond the Tahoe's moraines. In contrast, on the east side pre-Tahoe (Sherwin) moraines exist, but the lesser precipitation there—as Blackwelder observed back in 1931—has resulted in slower weathering and erosion, hence their preservation.

The great benefit of this relative-dating method is that Tioga and Tahoe moraines seem to be quite evident on topographic maps and aerial photographs, and so *field work is unnecessary*. As I have said, Yosemite National Park's Tioga, Tahoe, and pre-Tahoe moraines, depicted on the park's geologic map, were a compilation of former work supplemented by modern aerial photographs plus minor field reconnaissance. In like manner, in preparation for mapping glacial evidence in the Stanislaus River drainage, I examined aerial photographs and identified what I perceived as Tioga and Tahoe lateral moraines. Subsequent field work revealed that most moraines were bedrock ridges, some with till, some without. For the few moraines that existed, I had misidentified their age. This method simply does not work.

Another method was inferred by Blackwelder in 1931 and Wahrhaftig in 1984: the Tioga and Tahoe moraines and till can be differentiated by the amount of weathering of their *subsurface* boulders. Tioga moraines should have no weathered subsurface boulders. Actually, a partly weathered, although still coherent, boulder from a Tahoe moraine could be incorporated into a Tioga glacier and then deposited, so an occasional weathered subsurface boulder may be found. In contrast, Tahoe moraines have many, which often are little more than assemblages of poorly cohesive mineral grains. The disadvantage of this method is that it requires exhausting, resourceful, and occasionally dangerous field work, since one must examine subsurface boulders, often in remote places and sometimes on unstable slopes. Digging a pit in a moraine can be extremely time consuming, even though thoroughly weathered subsurface boulders can lie within a foot of the surface. Since you don't know exactly where, you may have to excavate several cubic yards of deposits. Consequently, manmade road cuts and excavations are the most desirable sites, although most moraines and till don't have any. In the wilderness, where excavations are not allowed and roads are absent, trail erosion can expose subsurface boulders (Photo 47). Natural subsurface exposures in moraines and till can be generated in four

**Photo 47. Fresh, formerly subsurface, Tioga boulders exposed along a trail. This spot is just below and west of the Cottonwood Creek moraine, located between Camp Mather and Hetch Hetchy Reservoir.**

**Photo 48. Fresh boulders exposed after a fire consumed a tree's trunk and roots. Not Tioga boulders, these *blocky, homogeneous* ones are the result of subsurface exfoliation of granitic bedrock. The former tree's prying, enlarging roots aided exfoliation.**

ways: erosion by a creek flowing either across a deposit or along its base; mass wasting such as gullying or soil creep; uprooting of trees by strong winds; and consumption by fire of entire trees, including subsurface base and major roots (Photo 48). I used all of these different exposures, manmade and natural, to separate Tioga deposits from Tahoe ones.

## Pinecrest Lake area: subsurface boulders of Tioga and Tahoe moraines

Clyde Wahrhaftig, quoted earlier under Chapter 12's section on Tahoe glaciation, suggested that the east moraine of Cascade Lake, along Highway 89 in the southwestern part of the Lake Tahoe Basin, is a good site to see fresh younger Tioga till that rests on weathered Tahoe till. Actually, a better site lies closer to Yosemite National Park: the prominent Dodge Ridge Road cut above the Pioneer Trail Group Campground (Map 17). This site in the Stanislaus River drainage, east of Highway 108 and above the south shore of Pinecrest Lake, has the advantage of far less traffic than found along maddeningly busy, shoulderless Highway 89. At it, from where the road switchbacks at a guard rail, a low cut through a moraine extends about 75 feet southeast, and all its subsurface boulders are fresh, and hence, by Blackwelder's and Wahrhaftig's criteria, it is Tioga (Photo 49). From it a higher cut curves south, and the first part is about 210 feet long. This 25-foot-high cut proved to be too steep and unstable to permit firsthand examination of the boulders (Photo 50). On a visual, rough examination, I estimated them to be about ¼ fresh, ½ intermediate (partly decomposed), and ¼ grussified (decomposed). By Blackwelder's and Wahrhaftig's criteria, it is Tahoe. An equally high cut, about 140 feet long, is less steep, and with caution, one can reach virtually all its boulders (Photo 51). In this Dodge Ridge cut I examined 54 boulders that were a foot or greater in their greatest dimension. All essentially are granodiorite, and I classified each according to: 1) degree of weathering—fresh (solid surface; sharp ring from hammer), grussified (weathered surface and interior; dull thud from hammer), and intermediate; 2) boulder size—1 foot to 6+ feet diameters, at ½-foot intervals; 3) crystal size—coarse (4-5 mm), medium (2-3 mm), and fine (1 mm); and 4) proximity to the moraine's surface. Additionally, I examined 24 boulders in a more-recent Water Tank cut (Photo 52), which is smaller, and ranges in height from about 6 to 12 feet. This site (also shown on Map 17), lies just below and west of the spot from where the climbing road begins to curve southeast.

In the Dodge Ridge Road cut, 31% of the subsurface boulders were fresh, 26% were intermediate, and 43% were grussified. In the Water Tank cut, 4% were fresh, 42% were intermediate, and 54% were grussified. Both cuts, by Blackwelder's and Wahrhaftig's criteria, are Tahoe, and for the two cuts combined, 23% were fresh, 31% were intermediate, and 46% were grussified. The specifics for each cut, as well as for the two cuts combined, are presented as Table A in the Appendix. Of interest is how the distribution of fresh, partly weathered,

**Photo 49. Tioga moraine, Dodge Ridge Road cut.**

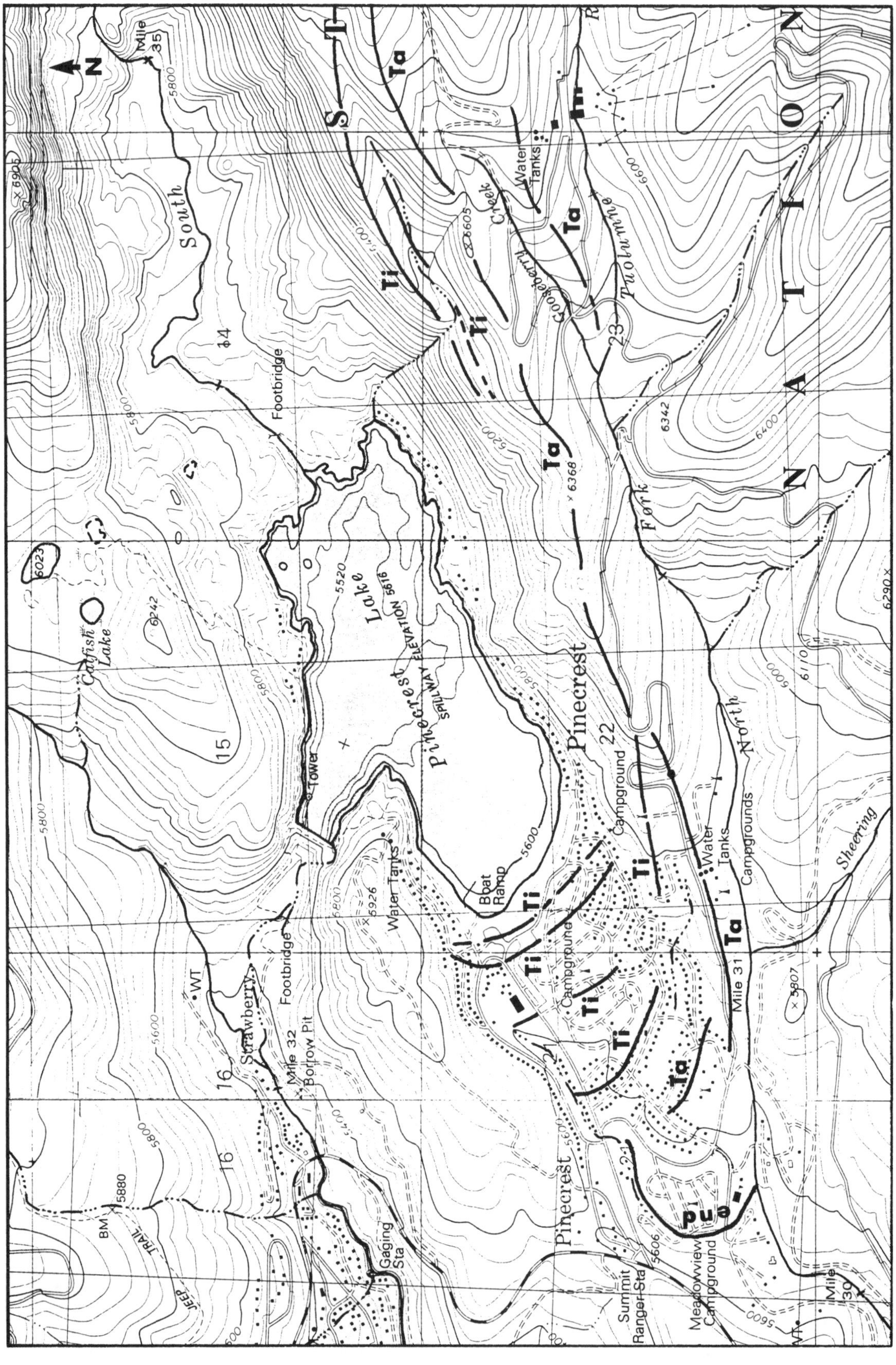


**Map 17. Tioga and Tahoe moraine crests in the Pinecrest Lake area. Scale 1:24,000.**

**Photo 50. North part of Tahoe moraine, Dodge Ridge Road cut.**

**Photo 51. South part of Tahoe moraine, Dodge Ridge Road cut.**

and grussified boulders varies *significantly* between the two cuts, *despite a separating distance of only some 120 feet.* They are on opposite sides of Dodge Ridge Road and are obviously part of the same Tahoe moraine. As just stated, the road cut (A) has 31% fresh boulders, the water-tank cut (B) 4%. The difference can be explained if the subsurface boulders of the water-tank cut were exposed to more ground water and hence to more chemical weathering.

Had these two cuts' weathering proportions appeared in two moraines, each located, say, in a different canyon, you could reasonably conclude that one moraine, having 31% of fresh boulders, must be significantly younger than the other moraine, having only 4%. Therefore, you might conclude that the first moraine was late Tahoe (or even the discredited Tenaya), and that the second one was early Tahoe. But as the two cuts show, you would be wrong. Consequently, one cannot attempt to date moraines based

**Photo 52. Water-tank cut, immediately west of Dodge Ridge Road cut. Rudy Goldstein and the author's daughter, Mary Anne, provide scale.**

on the *relative percentages* of weathered and/or partly weathered subsurface boulders. Their presence in fair numbers indicates a Tahoe age; their absence or virtual absence indicates a Tioga age. The high variability in percentages of weathered and/or partly weathered subsurface boulders is similar to the high variability in distribution of surface boulders, a method much used in earlier studies, but shown to be unreliable. For example, in the lateral moraines above the southeast corner of Pinecrest Lake, the boulder density varies so greatly over short distances that one can conclude at one crest site a moraine must be Tioga, whereas at another crest site 100 yards away it must be Tahoe!

Another observation is based on a comment by Wahrhaftig, who was in the field with me (and others) in June 1993 at the controversial Cascade Creek bridge till, above and west of Yosemite Valley. He had suggested that subsurface boulders would weather at varying rates, depending on the size of their mineral grains. However, the boulders I examined showed that there is no pattern (see Appendix, Table A). Not only is weathering independent of mineral-grain size, but also of boulder size. Additionally, I had expected that the percentage of weathered boulders would increase with depth, but the two cuts strongly indicate that fresh, intermediate, and grussified boulders occur randomly; weathering does not vary with depth. This field evidence has important implications, for Wahrhaftig believed that the Cascade Creek bridge till was pre-Tahoe. He asserted that its subsurface boulders had weathered abnormally and mysteriously slowly, unlike any other pre-Tahoe deposit—another riddle in the rocks.

## Reinterpretation of ages of glacial deposits

The reason that correlation of glacial ages from canyon to canyon on the west side of the Sierran crest has been so plagued with problems becomes obvious in the following cross profile, Figure 26. This is a cross profile of a typical west-side Sierran canyon, one quite similar to that of upper Little Yosemite Valley near the equilibrium-line altitude of its Tioga glacier. (Above the "ELA," as it is often called, the ice surface is concave—lower in the center than at the edges—while below the ELA it is convex—higher in the center than at the edges. Only at the ELA is a cross section of the ice surface relatively level.) Furthermore, the distribution of moraines and erratics in the figure are typical of that in west-side canyons, and the distribution is almost identical to that in upper Little Yosemite Valley.

Two interpretations of the cross section's erratics and lateral moraines are possible. The standard one, based on east-side moraines (where the glaciers largely were unconfined and field relations are quite obvious), suggests that three general-age groups exist. From youngest to oldest—and smallest to largest—they are Tioga, Tahoe, and pre-Tahoe. On the right side of the canyon in the figure, the standard interpretation shows a recessional moraine near the canyon floor, a sharp-crested Tioga moraine at the rim of the inner canyon, a subdued-crest Tahoe moraine behind it, and isolated erratics of former pre-Tahoe moraines scattered across the bench. The lake on the bench would have appeared after the pre-Tahoe glaciers left the bench, presumably about 800,000 years ago (terminal Sherwin). However, the left side of the

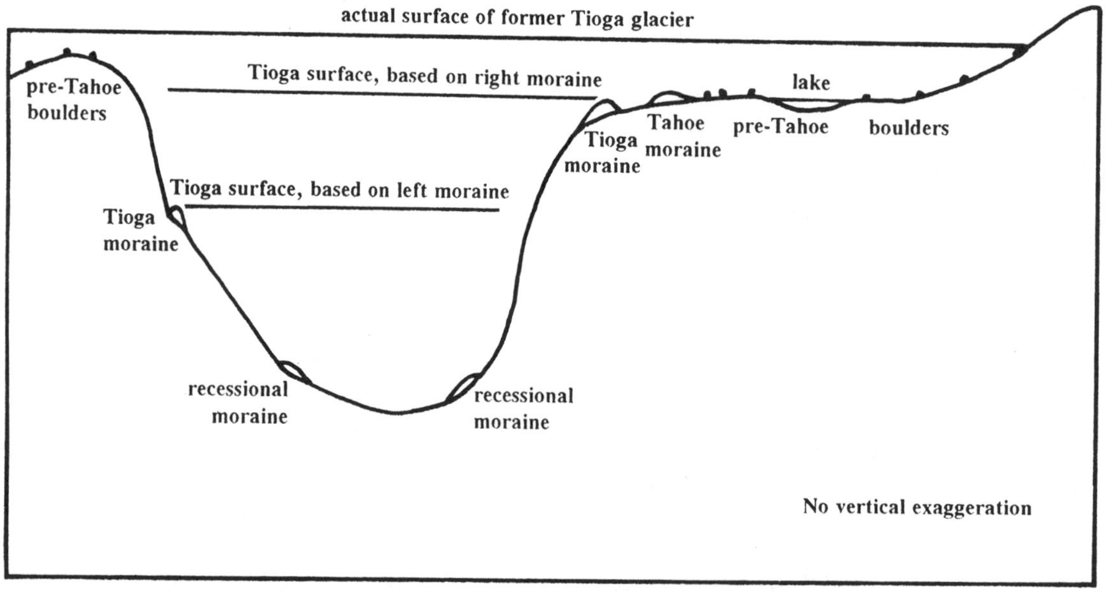


**Figure 26. Cross profile of a typical west-side Sierran canyon, showing the distribution pattern of glacial evidence. "Actual surface of former Tioga glacier" is arbitrarily a straight line (cross section located at equilibrium-line altitude).**

canyon presents a very different, contradictory interpretation. It shows a matching recessional moraine near the canyon floor, a sharp-crested Tioga moraine on a sloping bench, and only pre-Tahoe erratics above it. The left-side Tioga moraine is too low with respect to the right-side one, and a left-side Tahoe moraine is missing.

My interpretation is that the moraines and virtually all the erratics are all late-Tioga, and that the ice surface of the glacier, rather than being pre-Tahoe, is Tioga. Under this interpretation, as the glacier grew in its early stages, it eventually topped the rim of the inner canyon and lapped onto the bench. Eventually it grew large enough to advance across the bench and bury it. Note that only the farthest-right, highest erratic (at the surface of the glacier) truly represents the thickness of the glacier at its maximum. Had this bench been bordered by two high ridges, one up-canyon and one down-canyon from it, then the ice would have become trapped, and would have been stagnant. This was the situation down in Yosemite Valley. As the glacier advanced down-valley, it flowed laterally into the valley's recesses (broad side canyons), not only burying and protecting the bedrock from erosion, but also burying the overlying talus. As is discussed later, the valley has a surplus of talus, not the deficit that every published account from Whitney's onward concluded. Accurately estimating the amount of talus produced over time is crucial for reconstructing the valley's geomorphic evolution.

In Figure 26 I have assumed that the bench is not confined, so the glacier can flow slowly down-canyon, thus over time excavating a shallow lake basin out of fractured or weathered bedrock. As I mentioned in Chapter 11, the velocity of a glacier is proportional to the third power of its ice-surface gradient and to the fourth power of its thickness. In this figure the ice-surface gradient over both thin ice and thick ice would be about the same, so the velocity varies only with the thickness. Since in this figure the ice above the canyon's floor is about six times thicker than that above the bench, the relative velocity would be $6^4$, or 1296 times faster than across the bench. If one assumes an even distribution of debris atop the glacier, then the inner canyon part is transporting 1296 times as much debris per unit time as is the bench part. On the left side of the inner canyon, the rim (which is similar to Moraine Dome above Little Yosemite Valley) is buried under thinner ice. In this example (as at that dome) the ice is so thin that it is essentially the upper 150 feet of brittle, non-flowing ice. Consequently, there is little or no transport of debris here, and the highest fresh erratics, which would be very sparse due, are close to the actual surface of the former Tioga glacier.

In the glacier's waning stages the ice surface would have lowered more or less uniformly, and the left side (a domelike ridge), with its very sparse deposits, would have been first exposed. On the bench the glacier would have gradually retreated and left the sparse deposits it had been transporting. As wasting away continued, the glacier's lowering ice surface approached the inner canyon's rim. As the glacier fluctuated in size, this edge would have fluctuated laterally, leading to the creation of a broad, low moraine. With more waning the glacier's edge eventually would have retreated immediately beyond the inner canyon's rim. The velocity of the glacier just beyond the rim would have been considerably greater than it was on the bench, and so considerably greater deposits would have been dropped off the glacier's edge onto the rim. This creates a relatively high moraine. The moraine is narrow because the edge of the glacier, now confined within the rims of the inner canyon, cannot fluctuate laterally. The ice surface of the glacier lowers over time, leaving the Tioga moraine on the narrow bench midway up the left side of the canyon, and ultimately leaving the moraines on the floor. An examination of the subsurface boulders, where they have become exposed in the moraines, would show that *all have fresh boulders*, and so all are Tioga, or more correctly with respect to the field relations, late-Tioga.

The following parts in this chapter present evidence that the second interpretation of Figure 26 is correct, and that on west-side drainages, the complex granitic topography dictated the pattern of the deposition of moraines, till, and erratics. In contrast, the first interpretation conflicts with much of the field evidence. The first interpretation has worked fairly well in many areas, such as in the east-side drainages, where glaciers were confined within canyons. But it does not work well on west-side ones, where glaciers overflowed trunk canyons, benches, side canyons, and even major drainage divides.

## Diachronous lateral moraines: Little Prather Meadow and Mt. Reba

Under my interpretation of the glacial history in the cross profile of Figure 26, the moraines are all late-Tioga, but the highest one is oldest while the lowest one is youngest. Such moraines, when viewed in a longitudinal profile, are seen to be diachronous (time transgressive), that is, some parts were deposited before other parts. Figure 27 shows the relative ages of parts of a lateral moraine deposited on an irregular bedrock rim of an inner canyon. This profile is similar to a (diachronous) lateral moraine extending down-canyon from Moraine Dome toward Little Yosemite Valley. The ice surface of the Tioga maximum is above the rim at time T-0. The glacier is transporting deposits along its surface. Most of these are transported directly down the inner canyon, but a small amount is transported laterally by thinner ice that flows slowly across the benchlands above the rim of the inner canyon. During deglaciation, deposition first occurs at time T-1 atop the major knoll. With more deglaciation, the ice surface lowers, and at time T-2 deposition occurs below

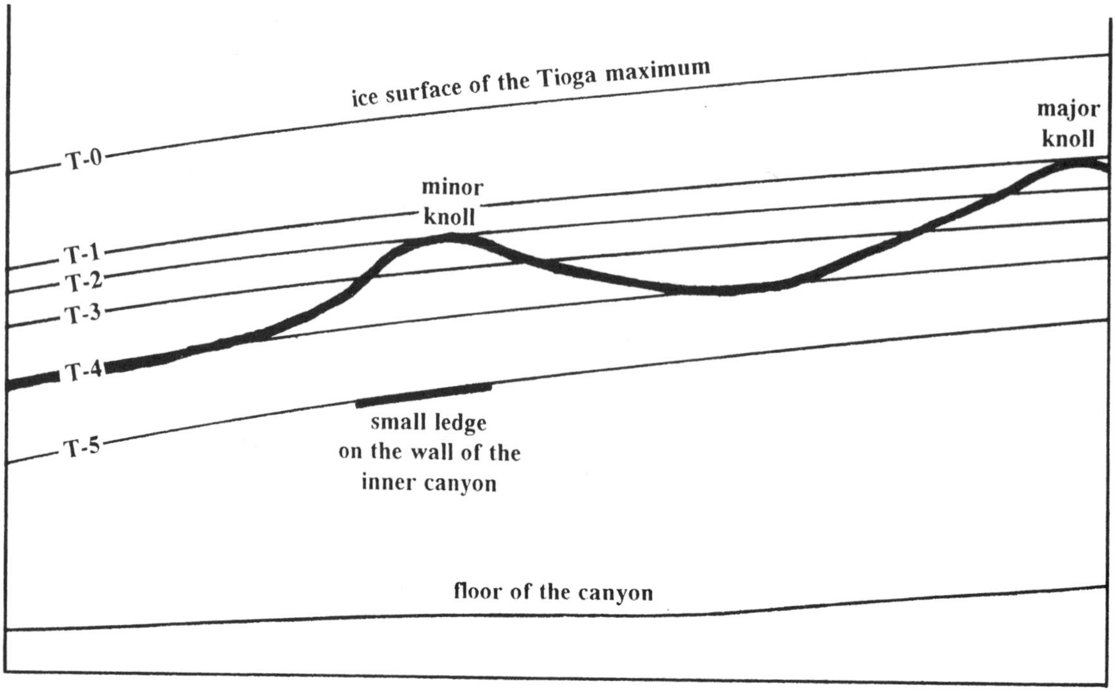


**Figure 27. Longitudinal profile of a lateral moraine on an irregular rim of an inner canyon.**

the T-1 site and concurrently atop the minor knoll. At time T-3 it occurs on the lower slopes of the major knoll and on middle slopes on both sides of the minor knoll. At time T-4 it occurs on the saddle between the two knolls as well as along the rim segment extending west from the minor knoll. Also at this time the ice on the lands above the rim of the inner canyon finally disappears, dropping the last of its sparse deposits. Finally at T-5 (and later), the glacier is confined to the inner canyon, which in this example is too steep-walled to support any deposits except where a small ledge might exist.

Field evidence for this diachronous deglacial pattern exists in west-side canyons, especially in and above Little Yosemite Valley, and its evidence there soon will be presented. However, insight into this pattern first came to me not in Yosemite National Park but rather from two areas in the glaciated lands of the Stanislaus and Mokelumne river drainages, north of the park. The first one is centered on the hanging side canyon of Little Prather Meadow in the North Fork Stanislaus River drainage. This locality is shown on Map 18, and the thick, curving line across the topography separates numerous erratics to the north from scarce erratics to the south. Photo 53 shows this boundary as viewed west, from slopes east of and above Little Prather Meadow. My field mapping of erratics, till, polish, striations, and moraines in this river drainage, combined with my late-Tioga radiometric (carbon 14) dates for ponds and lakes within it, establishes a Tioga maximum ice surface above the floor of the Little Prather Meadow side canyon, and nearby striations indicate that the glacier flowed west across the side canyon, not up or down it. Had the edge of the Tioga glacier been along the line, a lateral moraine would have been constructed, but none exists. Furthermore, had the line represented the edge of the glacier, then in the unglaciated upper part of this side canyon the mass wasting of its loose, volcanic slopes should have collected at the edge of the glacier, perhaps in an ice-dammed pond. However, glacial, alluvial, stream, and lake sediments are absent. The most logical conclusion is that the upper part of the side canyon was filled with stagnant ice. If so, in parts of the glaciated Sierra Nevada ice that was slowly moving or stagnant would have left very few erratics and very little till. Furthermore, a boundary between abundant and sparse glacial deposits does not necessarily represent the edge of a former glacier.

The second area contains three severely hanging canyons in the vicinity of the Mt. Reba Ski Area: Horse Valley, Snow Valley, and Grouse Valley (Map 19). Each canyon has a steeply descending lateral moraine (Chapter 3, Photo 7; and Photo 54, below) that is believed to have formed from deposits left by the Tioga-age Mokelumne glacier when it was at its maximum. Therefore, each moraine serves as a gauge to the local elevation of that glacier's ice surface. A maintenance jeep road descends Horse Valley before swinging over to Snow Valley, and this provides rather easy access to glacial polish and striations within each canyon and to the moraines on the brink of each side canyon. My reconnaissance of the ridge that separates the Mokelumne River and North Fork Stanislaus River drainages provided abundant evidence—

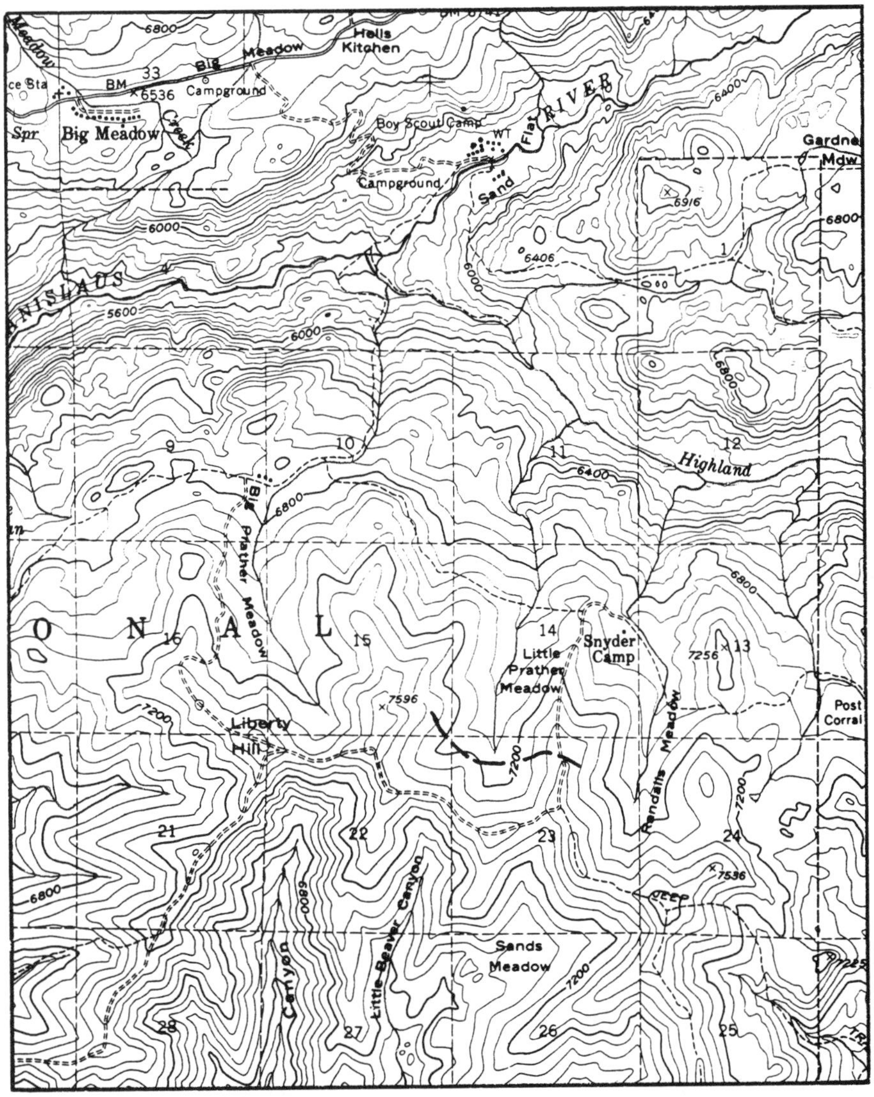


**Map 18. Little Prather Meadow and vicinity. Heavy line separates numerous erratics to the north from scarce erratics to the south. Scale 1:62,500.**

in the form of fresh erratics, fresh polish, and fresh striations—that the Mokelumne glacier spilled massively over a low divide to flow into the North Fork Stanislaus River drainage to form the North Fork Stanislaus glacier (Map 20). Part of this glacier then overflowed the river drainage to spill back into the Mokelumne River drainage. This occurred at the head of each of these three side canyons. Within Horse Valley are fresh striations on the granitic bedrock that confirm that a fairly massive Tioga-age glacier flowed down this canyon to unite with the Mokelumne glacier. On the lower bedrock ridge between Horse and Snow valleys there is a jeep-road cut through glacial deposits, and the subsurface boulders clearly are fresh (Photo 55), indicating that these deposits are Tioga or late-Tioga in age.

If this union occurred, as the fresh striations suggest, then how are the moraines explained? In one unpublished anonymous reconstruction, a geologist suggested that the Horse Valley glacier stopped just short of the Mokelumne glacier, and consequently the moraine at the head of the hanging canyon is an end moraine. However, end moraines curve, but the Horse Valley moraine is straight. Furthermore and more importantly, at the mouth of each hanging canyon there is only one moraine, and each is located on the northeast (up-canyon) side. Where is the other half of the hypothesized end moraine? In contrast, the geologist interprets the Snow Valley moraine as the Mokelumne glacier's south-side lateral moraine. Interestingly, on the north side of the Mokelumne River canyon opposite the mouth of Snow Valley, he placed the ice surface about 650 feet higher, and this elevation produces a laterally sloping ice-surface problem similar to that of Wahrhaftig's Tahoe moraines in Little Yosemite Valley. Another problem is that if the Snow Valley moraine is a Tioga-age lateral moraine, as the high density of its fresh boulders suggests, then where is the Tahoe-age lateral moraine? Above each should be a parallel Tahoe-age lateral moraine, but none of these three side canyons has two sets of lateral moraines.

These problems disappear if the three moraines are diachronous, lateral moraines of the late-Tioga Mokelumne glacier, formed in the manner shown in Figure 27. Under my interpretation, as the Mokelumne glacier began to wane, the overflow into the North Fork Stanislaus River drainage decreased tremendously, which caused its large glacier to catastrophically fail. This conclusion is based on the magnitude of the Tioga-age glacier overflow at the head of this drainage, and how much it would have been reduced by a relatively modest reduction in the thickness of the Mokelumne glacier. With the demise of the North Fork Stanislaus glacier, its overflow into Horse, Snow, and Grouse valleys ceased, and the glaciers flowing down them disappeared. The waning Mokelumne glacier, whose ice surface at this time was about 300+ feet above the mouths of these side canyons, was on the order of 2000 feet thick, perhaps about 1000 feet thinner than at its maximum. Nevertheless, being this thick, it flowed at a relatively high velocity. Ice passing the mouth of each canyon had insufficient time to slowly flow briefly up each canyon. Had they done so, they would have left terminal moraines facing up-canyon.

As the ice surface lowered to the low-angle slopes that make up the cross profile of each canyon's mouth, the Mokelumne glacier deposited a diachronous lateral moraine on each side of each canyon's stream. It started deposition near the sides and ended it down at the lowest point, beside each side canyon's creek. The absence of a lateral moraine on the down-canyon half of the mouth of each side canyon can be explained by the action of their creeks. As each creek approached the mouth, it was deflected along the edge of the glacier. So no sooner was glacial debris deposited than the creek transported it away. The creeks were destroying their down-canyon, time-transgressive moraines as rapidly as they were being created.

In a June 1993 letter, Wahrhaftig took issue with my interpretation of the Mokelumne's "diachronous moraines" as interpreted and as applied to the lateral moraines of Little Yosemite Valley. He said that what I overlooked was that the Mokelumne glacier would dam the creek of each side canyon, consequently a lake would form in each. So he concluded that "there is no reason to believe that these prominent lateral moraines do not mark

**Photo 53. South edge of field of erratics above Little Prather Meadow.**

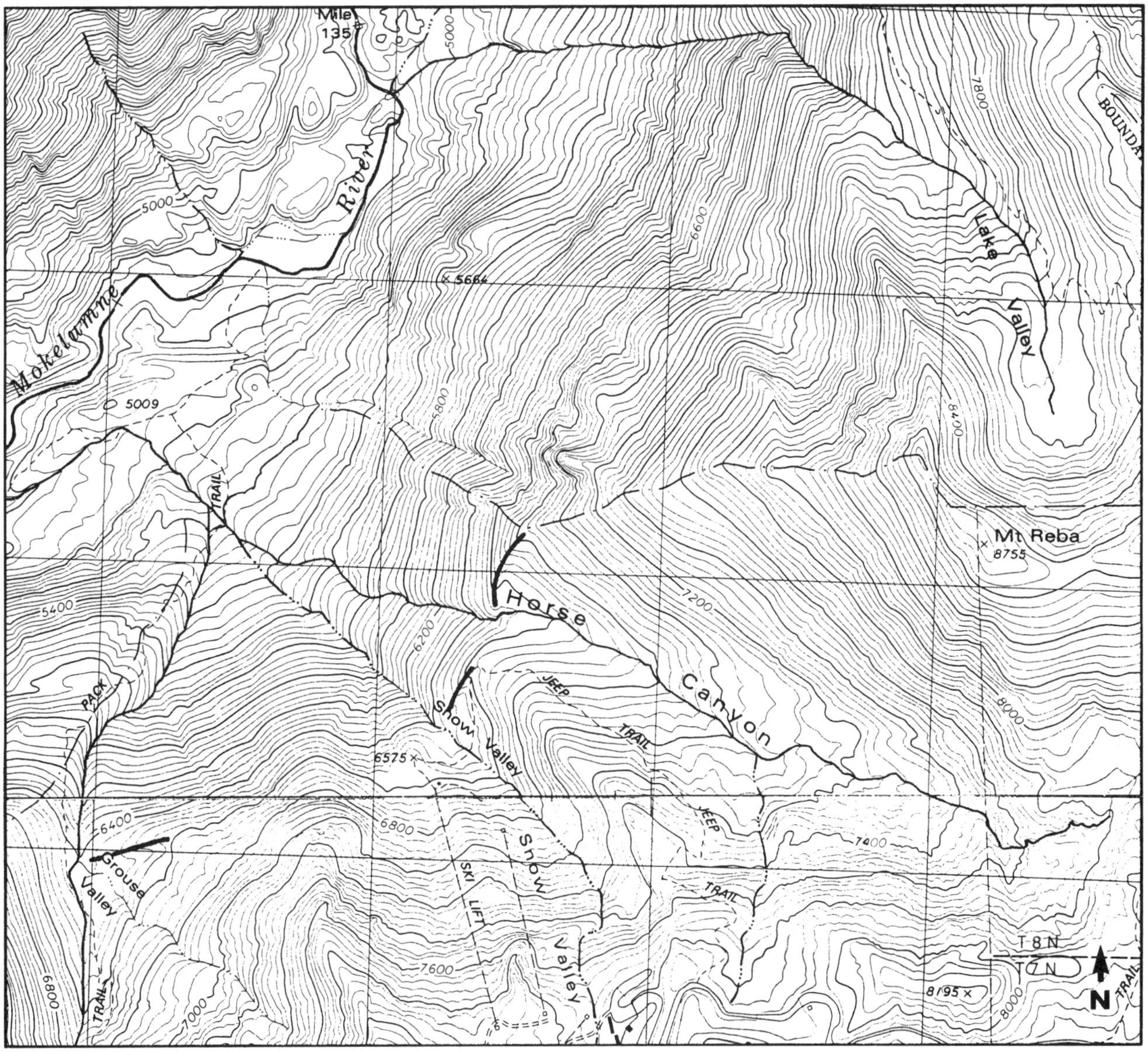


**Map 19. Lateral-moraine crestss on the brink of Horse, Snow, and Grouse valleys. Scale 1:24,000.**

the outer limits of the Tioga glaciation, as Matthes and we say they do." Wahrhaftig had not visited any of the Mokelumne's side canyons, so his letter incorrectly implies that each one had a *pair* of moraines rather than just one moraine on the eastern (up-canyon) side. Furthermore, a lake would have left sediments, but despite two trips to Horse and Snow valleys, I found none, which would have been unavoidably obvious. Hence, no lake existed. In Chapter 16 I present evidence of two large glaciers flowing past the mouths of tributaries (Clark Fork Stanislaus River and Illilouette Creek), and in each case only stream sediments accumulated. The streams, rather than being dammed behind the side of a glacier, flowed alongside it.

Finally, if Wahrhaftig is taken literally—the prominent lateral moraines mark the outer limits of the Tioga glaciation—then the highest outer moraines are very representative of the ice surface at the glacial maximum. But as I demonstrated earlier in this chapter, under "Mismatched moraines," the highest Tioga and Tahoe moraines yield unrealistic, imaginary ice surfaces and impossible basal shear values. On his Tioga glaciation map of Yosemite National Park, Wahrhaftig was able to produce a *believable* ice surface for the glacier through Little Yosemite Valley only by *ignoring* his prominent lateral moraines. His own criterion failed him.

## The Hetch Hetchy area's moraines and the significance of Laurel Lake

In previous sections I have suggested that there are two distinct methods to relative-date lateral moraines. The traditional method is based mostly on the relative position

**Photo 54. Aerial view south toward Horse, Snow, and Grouse valleys. Horse Valley is in the left part of the photo and has an obvious jeep road descending its western (right) side. A conspicuous moraine descends along the eastern half of the side canyon's brink. The jeep road heads west (right) into a group of trees at the lower part of Snow Valley. Just below the road is a conspicuous moraine along the eastern half of that side canyon's brink. Right of a major rockfall is conspicuous moraine along the eastern half of Grouse Valley's brink.**

of deposits. Inner lateral moraines are Tioga, outer ones are Tahoe, and deposits beyond them are pre-Tahoe. Furthermore, deposit morphology is important. The inner lateral moraines are narrow, quite bouldery, and sharp-crested; the outer lateral moraines are broader, less bouldery, and subdued; and the deposits beyond them are but sparsely scattered, large erratics. This method has a crucial, implicit assumption: deposits were left in similar concentrations everywhere, despite extremely variable rates of flow between the thick canyon ice and the thin uplands ice. This assumption has been made in complete disregard of the law of glacial physics which states that the surface velocity is proportional to the fourth power of the ice thickness, so thin, slow ice would transport and leave much fewer deposits than thick, fast ice. With assumed similar concentrations of deposits everywhere, the initial Tahoe and pre-Tahoe moraines would have resembled Tioga moraines. Then with much time the Tahoe moraines were weathered and eroded to their present, subdued condition, and with much grater time the pre-Tahoe moraines were weathered and eroded until only large, resistant erratics remained.

My method is based primarily on the degree of weathering of subsurface boulders. Tioga deposits (moraines and till) have fresh boulders, Tahoe deposits have varying amounts of fresh, partly weathered, and grussified boulders, and pre-Tahoe deposits do not have continuous cover, but rather exist as isolated erratics above the highest Tahoe deposits. Furthermore, the morphology of moraines is due not so much to weathering and erosion but far more to pattern of deposition, which is controlled by the morphology of the landscape. Thus thin ice moving very slowly over a broad, upland surface will have relatively little debris atop it when it finally melts—in accord with the just mentioned law of glacial physics. Along the rim of an inner canyon, the edge of a waning glacier will oscillate slightly, leaving more conspicuous deposits, which the first method interprets as a Tahoe moraine. Then when the ice surface of the waning glacier finally lowers to the rim, the glacier's edge becomes

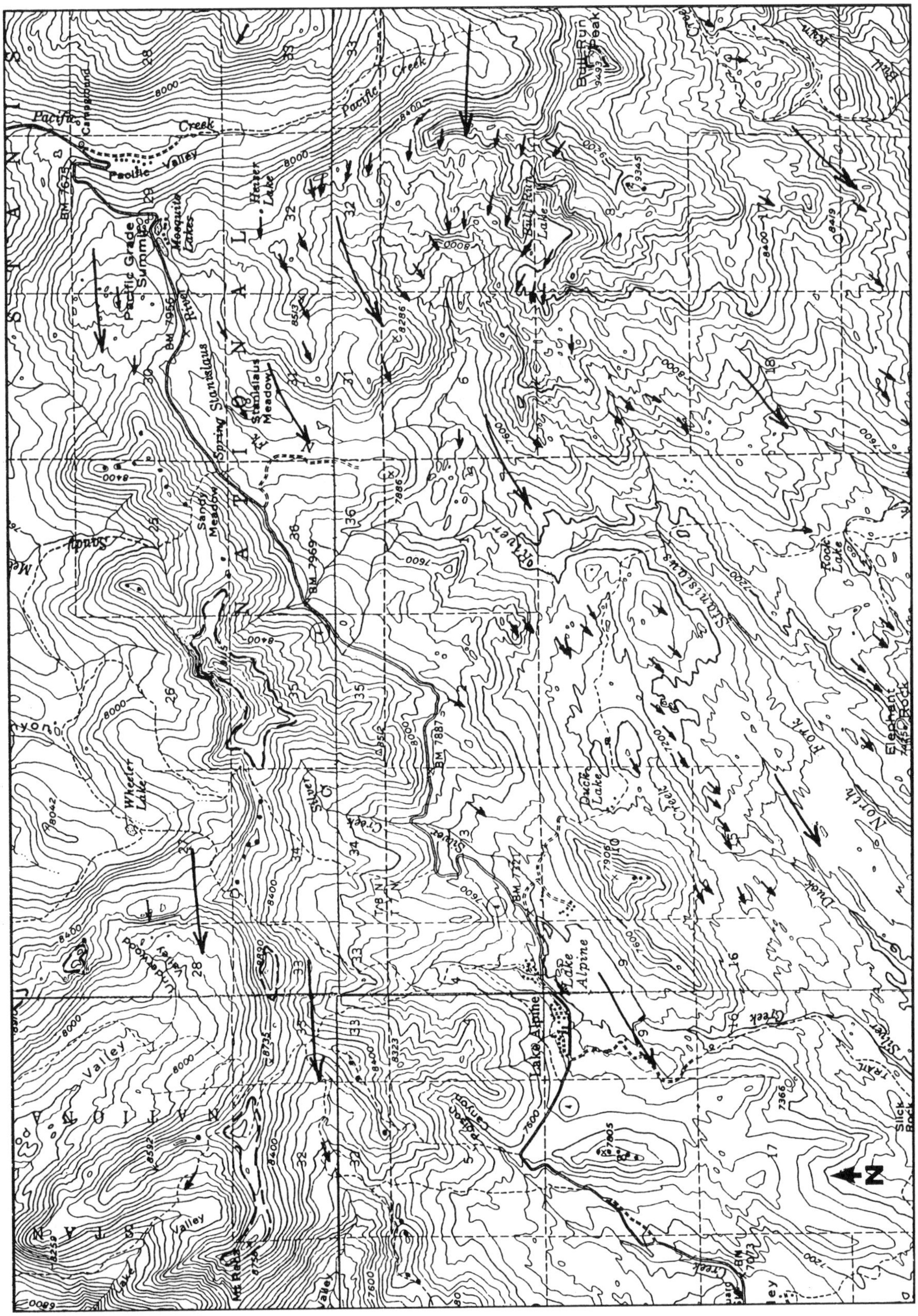

Map 20. Tioga-glacier evidence in the upper Mokelumne and Stanislaus river drainages. Dots on ridges are fresh, usually large (3+ feet) erratics; short arrows are bearings of striations on fresh polish; long arrows are inferred ice flow at Tioga maximum. Scale 1:62,500.

**Photo 55. Road cut through fresh (Tioga) deposits on lower bedrock ridge between Horse and Snow valleys. Ken Ng provides scale.**

fixed, and because it cannot oscillate, it leaves a narrow ridge of deposits, which the first method interprets as a Tioga moraine. Both methods predict the same morphology, but under my method, virtually all of the deposits are late-Tioga. Relatively few are Tioga maximum, Tahoe, or pre-Tahoe. The Hetch Hetchy area and relatively nearby Laurel Lake are an ideal place to examine the merits of each method.

As in Little Yosemite Valley, in the Hetch Hetchy area the lateral moraines at first appear to represent maximum ice surfaces, but on scrutiny, problems arise. Map 21 shows the Tioga, Tahoe, and pre-Tahoe moraines as Wahrhaftig interpreted them for the park's geologic map. His mapped lateral-moraine crests closely conform to those of a more detailed map of the Lake Eleanor 15' quadrangle, produced in 1987 by Dodge and Calk, and this indicates general acceptance in their relative dating. However, Figure 28, which shows the profiles of the highest Tioga and Tahoe lateral moraines along the southeast side of the Tuolumne River canyon, raises two issues (of *many* in this area). First, neither profile is smooth; rather, each drops steeply, levels out, drops steeply again, and levels out again, despite the profile of the Tuolumne River being relatively uniform. Today's glaciers don't have this pattern. Second, for unconfined glaciers on the east side of the Sierra Nevada, the crests of the Tioga and Tahoe lateral moraines gradually diverged from each other down-canyon. But as the profile shows, whereas they first diverge, they later converge. Note that Wahrhaftig's profile of the Tioga ice surface (the thick line, which I derived from the glacier map drafted by Tau Rho Alpha) does not match the profile of the crest of the highest lateral moraines, despite his emphatic statement, above, that the elevations of a moraine's crest are essentially synonymous with those of the glacier's ice surface. The discrepancy between the two is too great to be attributed to cartographic errors by Tau Rho Alpha. As in Matthes' and Wahrhaftig's reconstructions of glaciers through Little Yosemite Valley, the field evidence here has been generalized to produce a realistic (although undocumented) ice surface.

Under my interpretation the Hetch-Hetchy-Mather area's moraines are developed best along the *rim* of the Tuolumne River's canyon, where ridges and benches give way to steep slopes. On the southeast side both the "Tioga" and the "Tahoe" lateral moraines of Wahrhaftig begin as deposits on a southwest-descending bedrock ridge. The "Tahoe" then is deposited along the northwest

see inset at left

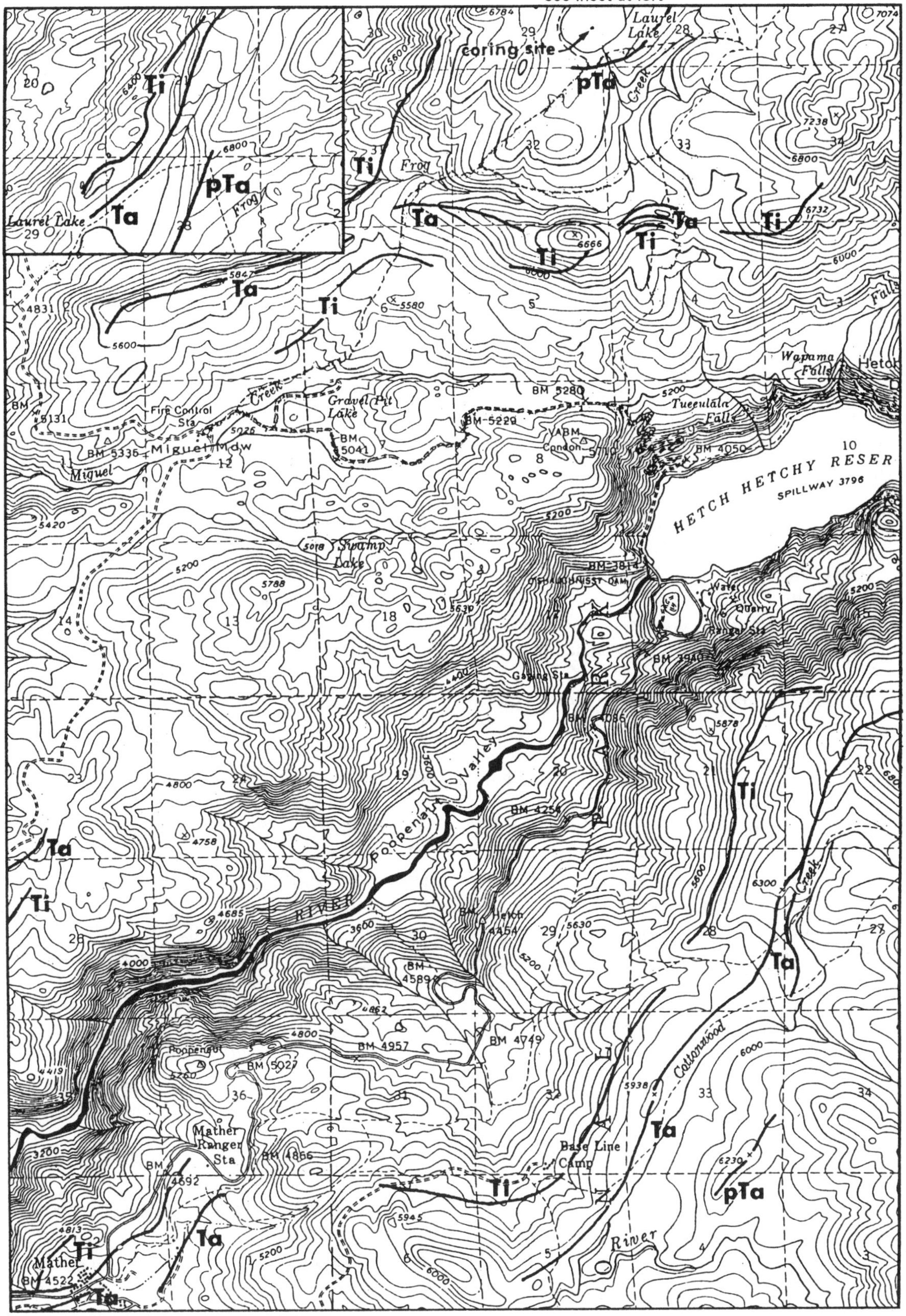


Map 21. Tioga, Tahoe, and pre-Tahoe lateral-moraine crests of the Hetch Hetchy area. Scale 1:62,500.

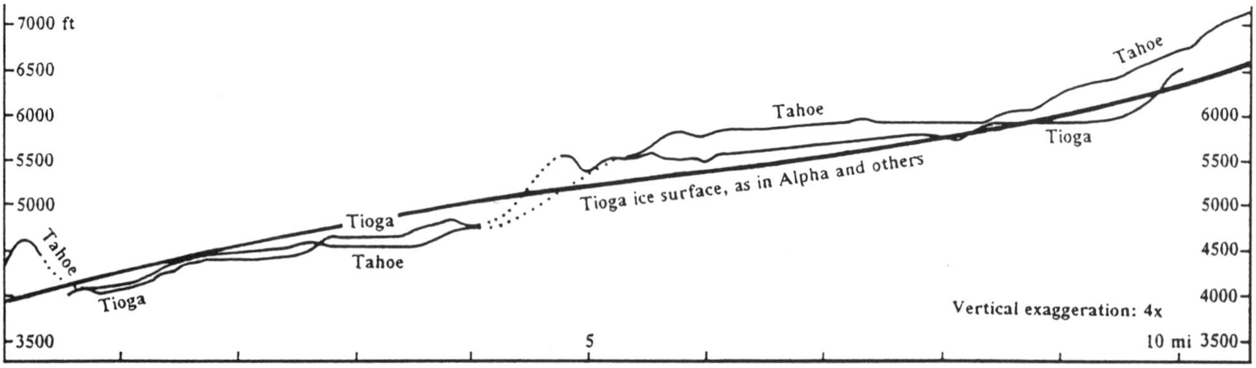


**Figure 28. Profiles of highest Tioga and Tahoe lateral moraines along southeast side of Tuolumne River canyon.**

edge of the nearly flat Cottonwood Creek bench. This moraine is quite thick, not just a veneer of till on a buried ridge, as is the usual case. Evidence for this exists on its northwest slopes just southwest of where the trail descends toward Base Line Camp. On those slopes is a conspicuous spring line, fed by flow from Cottonwood Creek beneath the moraine. Consequently, the creek has a tendency to dry up in this vicinity. This proves that there is no buried bedrock ridge lying just below the moraine's surface. Along the descending trail, patches of erosion expose hundreds of formerly subsurface boulders, and all are fresh, making the moraine either Tioga or late-Tioga according to Wahrhaftig's subsurface weathering criteria (Photo 47).

Below the "Tahoe," the "Tioga" lateral moraines exist locally on varied, till-mantled slopes where there are linear, narrow bedrock benches amenable for thick deposits. Lack of detail on the small-scale park geologic map hides the fact that some of these moraines rise *upslope* in the down-canyon direction—by as much as 200 feet along two moraine crests that are shown on the more detailed Lake Eleanor geologic map. Beyond the Cottonwood Creek bench the rim of the Tuolumne River's canyon makes a steep, major drop to the Mather bench, and the glacial evidence drops with it. The north edge of the Mather bench is nearly level, but it does undulate, as does the "lateral moraine" (veneer of till) atop it. Of significance is the 4813-foot knoll just north of Camp Mather (lower left corner of map). This linear knoll had *fresh* erratics up to its summit (Photo 56), indicating that the Tioga glacier was at least as high. But if it was, then it would have been about 300 feet thick at the knoll's west base, and consequently it would have been thick enough to flow across the Mather bench. Wahrhaftig mapped glacial deposits there as Tahoe, although where subsurface boulders were exposed, I could find only fresh (i.e., Tioga) boulders. The north edge of the lengthy Mather bench, with a lateral moraine atop it, eventually descends to a low point, near bench mark 4264. Then, according to Dodge and Calk (1987), the crest of the moraine dies out at 4400 feet elevation—yet another upslope rise in the down-canyon direction. Under my interpretation, the lengthy deposit, like the Cottonwood Creek bench "Tahoe," is a late-Tioga diachronous lateral moraine, based on the observation that the subsurface boulders are fresh. The surface boulders too are fresh, although this is not as reliable an indicator, since I (like others from Blackwelder, 1931, onward) have observed fresh spalling of boulders in burned areas, and this spalling keeps the sides of the boulders fresh looking. As in Figure 26, the late-Tioga deposits rapidly diminish in quantity with distance away from the edge of the Mather bench.

To test which method of relative-age dating of glacial moraines is correct, you can date the age of a lake, which lies on uplands above the Tuolumne River gorge, occupying a relative position similar to the lake shown in Figure 26, which is pre-Tahoe by the standard interpretation, and post-Tioga (less than 16,000 years) by mine. Such a lake exists: Laurel Lake (top of the map), which lies about 3.4 miles due north of the trailhead at the (O'Shaughnessy) dam that backs up Hetch Hetchy Reservoir. However, to reach the lake by trail is a strenuous 8.4-mile hike with some 3200 feet total elevation gain, one way. My lake-coring trip was on August 20, 1994, and was done as a day hike. My self-designed backpackable drilling platform, which breaks down to manageable pieces, combined with the necessary drilling equipment and packs to carry everything, amounts to a total weight of about 250 pounds, or to about an average of 50 pounds for each of my five-man party (Photo 57; also see Photo 1). Temperatures on the mostly shadeless ascent reached 90°F.

For this lake to be a valid test site, the moraines must have been dated properly with respect to the standard method. (Map 21 shows the moraines, north of the reservoir, as they are dated according to Wahrhaftig in Huber *et al.*) The trail momentarily tops out on a bench at a spot about 0.6 mile east of a prominent 6666-foot knoll. The edge of the bench has a conspicuous lateral moraine that is higher and narrower than one to the north of it (a linear pond, near the south edge of Section 33, lies between

**Photo 56. Large erratic atop summit 4813 north of Camp Mather. These erratics are as fresh as those found on the universally recognized late-Tioga end moraines of Yosemite Valley. Former geology student Peter Bernard provides scale.**

**Photo 57. The Laurel Lake coring party atop O'Shaughnessy Dam. Left to right: author, Greg Schaffer, Rudy Goldstein, Doug Sornberger, Jeff Middlebrook.**

them). This makes them, by the standard method, respectively, Tioga and Tahoe, and Wahrhaftig assigned them these dates. Beyond them, moraine crests are absent, so any glacial deposits must be, by the standard method, pre-Tahoe. Laurel Lake lies beyond these deposits, and glacial deposits there must also be pre-Tahoe. Note in the map's inset that a broad, and therefore Tahoe, moraine extended southward to Laurel Lake's north shore. This moraine is on the southeast ridge of the Eleanor Creek canyon. A Tioga moraine lies within the canyon, just north of and below the Tahoe. Therefore, under Wahrhaftig's interpretation, a Tahoe glacier lapped up to the north shore but never traversed the lake's basin. Consequently, the last glacier to occupy the basin must have been pre-Tahoe. According to the standard method, all the glacial deposits along the trail to the lake have been dated properly.

However, there is reason to believe that the conspicuous lateral moraine east of the prominent knoll is not a true moraine, but rather is in essence a veneer of fresh till on a linear bedrock ridge. If this moraine had been deposited on a flat bench, then the linear pond backed up behind it would have seeped through it, forming a spring line along the base of the moraine (as along the Cottonwood Creek bench). Also, the pond would have disappeared shortly after the snowpack had melted. However, no seep or telltale water-loving vegetation existed, and the shallow pond had dropped 2 feet by early August 1994—a drought year. (This is a typical amount for all bedrock-lined Sierran lakes of similar elevation. In contrast, an unnamed, trailside lakelet trapped behind a lateral moraine on the east rim of Tanglefoot Canyon—north and above Salt Springs Reservoir, in the Mokelumne River drainage—had dropped by 5 feet by *mid-June* of the same drought year.) Both the "Tioga" and "Tahoe" lateral moraines along the way to Laurel Lake contain only fresh subsurface boulders, indicating that both are Tioga or late-Tioga. Both ascend westward partly up the prominent knoll rather than descend gradually down-canyon, and both have now exposed, formerly subsurface fresh boulders. Thus by my interpretation they must be late-Tioga diachronous lateral moraines (Photo 58). Finally, the "pre-Tahoe moraine" enclosing Laurel Lake is, like the "Tioga" lateral moraine, essentially a veneer of fresh till on a linear bedrock ridge. The lake remains high throughout the summer and the "moraine" lacks a spring

**Photo 58. Fresh rocks exposed at toppled, burned trunk in "Tahoe" moraine. This site is approximately 700 feet west of the west end of the linear pond.**

**Photo 59. Assembling backpackable drilling platform on Laurel Lake.**

line. The lake's outlet creek has eroded away some of the till, exposing formerly subsurface boulders, and all are fresh (i.e., Tioga).

In the Stanislaus River drainage of the western Sierra Nevada, all *non-seasonal* lakes dated by me or by others in Cal Berkeley's Geography Department have about 10 to 30 feet of organic sediments lying above the glacial flour, which is a gray, gravely layer between these sediments and the surface bedrock of the lake basin. The glacial flour is ground up rock and gravel left by the former glacier. All the lakes have basal organic-rich sediments that indicate all lakes formed within the last 16,000 to 13,000 years. A pre-Tahoe (Sherwin) lake should have about 500 to 1500 feet of sediments, assuming a constant sedimentation rate over time, and assuming no increase due to influx of sediments from Tahoe or Tioga glaciers. (Sedimentation rates vary with size; the smallest lakes receiving about 2 feet per 1000 years; most lakes receiving about 1 foot per 1000 years, and the largest, such as Tenaya Lake, receiving about ½ foot per 1000 years.) Laurel Lake, of about average size of the lakes I've cored, has a maximum depth of about 45 feet—too deep for our coring gear—so the team had to core in shallower water, although still near the center of the lake (which is deepest toward its north end). In 18.7 feet of water we penetrated 14.0 feet of sediments to reach the glacial flour (Photos 59 and 60). (These *thin* sediments *alone* confirm the Tioga age of the lake.) The carbon-14 date on the basal organic sediments is 10,800 B.P., which corrected for true age, indicates they were deposited about 13,000 years ago (11,000 years B.C.).

But a date on basal sediments from the deepest part of a lake can be several thousand years older than one from a shallower part of it. Laurel Lake's date is comparable to one from Rock Lake, in the North Fork Stanislaus River drainage, where coring was done in 6.7 feet of water and through 10.7 feet of sediments: 10,590 B.P. The age of that lake increased significantly to 12,860 B.P. in slightly deeper water—8.7 feet—and in thicker—12.8 feet—sediments (the maximum depth is about 9.2 feet). Rock Lake lies 11 to 12 miles up-canyon from the drainage's lowest lakes—Lake Moran and Paradise Lake—which according to their carbon-14 dates became deglaciated slightly earlier. Likewise, since Laurel Lake lies up-canyon from Swamp Lake, which according to Susan Smith and R. Scott Anderson became deglaciated at about 15,000 years ago, it too should be slightly younger. Therefore the age of Laurel Lake is late-Tioga. The method of dating moraines by degree of weathering of subsurface boulders is supported; the standard method, based largely on moraine morphology, is rejected.

Because the standard method has been used exclusively since the early 1900s, all the glacial mapping west of the Sierra Nevada crest, which in more cases than not yields implausible glacial reconstructions, needs to be reassessed. The subsurface-boulder method indicates that most (possibly as much as 70% to 90%) of the Sierra's west-side lateral moraines have incorrect age assignments, which would explain why the creation of believable glacial reconstructions under the standard method has been impossible. Indeed, past ice surfaces are easier to construct when moraines are few, as in the Stanislaus River drainage, where I had to rely on the highest erratics on dozens of ridges and summits to reconstruct the ice surface. In that drainage, reconstruction problems developed only at the North Fork's few "moraines" (till on sloping ridges), where at first I forced the evidence to conform to the standard method. Fortunately, the Middle Fork drainage essentially lacks moraines, thus simplifying its glacial reconstruction, and the South Fork has unconfined moraines, which yield true glacial reconstructions. For such reconstructions the ideal glaciated canyon would have gentle slopes that allow the deposition and preservation of lateral moraines, and would not have recesses or benches, where differentiation between glacial stages can be difficult. The lower part of Lee Vining Creek canyon is close to this ideal. In contrast, the least ideal glaciated canyon for reconstructions would have steep sides that do not preserve lateral moraines, and would have recesses and benches occupied by glacial ice.

In the entire Sierra Nevada, Little Yosemite Valley (and Yosemite Valley below it) is likely the farthest removed from the ideal conditions for glacial reconstruction, due to seemingly contradictory glacial evidence in the former and to precious little glacial evidence in the latter. Yet it is precisely there that the most important glacial work was done, by François E. Matthes, and then his methods were used by others to identify and relative-date most of the remaining glacial deposits west of the Sierran crest. I will address glacial evidence in Little Yosemite Valley and Yosemite Valley first, and in the greatest detail, then will address it in other western drainages. For each area I will present: 1) the present description of the glacial evidence (extents and ages); 2) where possible, on what evidence or logic that interpretation was reached; 3) my description of the evidence; and 4) my interpretation of the evidence. In every glaciated drainage except the Kaweah River—the only one with glacial evidence that can be correctly interpreted by the standard method—and I explain why), I show that the standard method yields implausible ice surfaces that are at odds with glacial physics, while the subsurface-boulder method yields plausible ones in agreement with glacial physics. Examples of implausible ice surfaces include: 1) sudden, major (300+ foot) drops; 2) significant rise in the ice-surface elevations in the down-canyon direction; 3) ice-surface elevations on one side of a canyon at considerable variance with those on the opposite side; and 4) stepped ice-surface profiles—glacier ice remaining quite uniform in thickness for a considerable length, then rapidly thinning before resuming a quite uniform thickness for another considerable length.

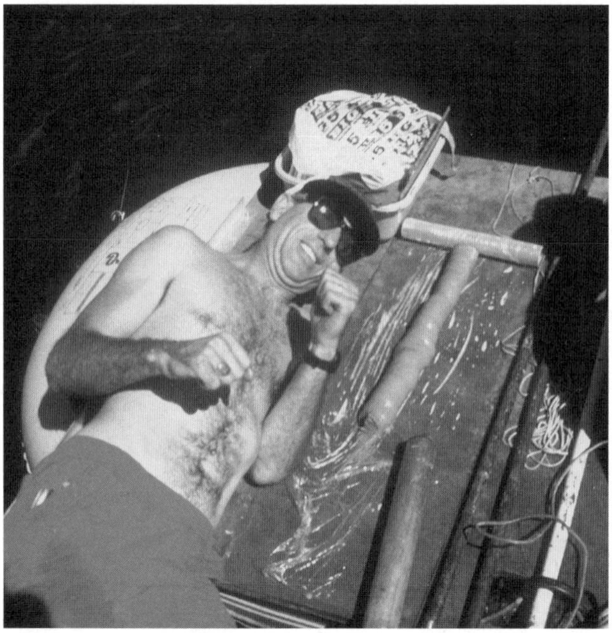

**Photo 60. Rudy Goldstein gives "thumb's up" upon reaching glacial flour.**

# 15

# Little Yosemite Valley's Glacial History

The field diary of François Matthes indicates that in 1905 he spent about 40 days surveying the lands in and about Little Yosemite Valley. Back then, he already was aware of what glacial evidence looked like, having observed it previously in the Sierra Nevada, the Rocky Mountains, and the Cascade Range. While surveying in the valley, he could not help but notice the evidence. Therefore, when he returned to the valley in 1913 to officially examine its glacial evidence, he knew where to look, and completed his examination in about 18 days. It is safe to say that he saw more glacial evidence than anyone else. Therefore, it first seemed surprising to me that he said relatively little about it. He left out a lot of evidence that he must have seen. This probably was not an oversight, since most of the evidence raises problems with his interpretations.

The field evidence is so abundant in and about Little Yosemite Valley that I have parceled it into five smaller, more manageable areas. First I cover the north half of the area, east to west (down-canyon), then the south half. Each area is centered around one or more prominent geographic features. Finally, evidence for determining the thickness of the Tioga, Tahoe, and pre-Tahoe glaciers is used to reconstruct the profiles of their ice surfaces through Little Yosemite Valley. In chapters following this one, evidence from Tenaya Canyon, Illilouette Creek valley, Yosemite Valley, and Merced Gorge is used to reconstruct the profiles there. Then in Chapter 20 I apply the analysis of Yosemite's Merced River drainage to local, enigmatic areas in seven other west-side drainages, demonstrating that this new method of analysis works where the standard one generally failed. Only in the Kaweah River drainage has the standard method worked, and I explain why it worked only there.

Chapter 15, on Little Yosemite Valley's glacial history, is necessarily complex, and perhaps most readers will gloss over it. You may get the gist of the evidence from this chapter's photographs and their captions and from the summary paragraph at the end of this chapter. (This also applies to chapters 16, 17, 18, and 20.) The details must be presented so that serious geoscientists know exactly where the evidence lies. If they are skeptical about my interpretations (and the best scientists are very skeptical), then by using the maps in this chapter and others, they can visit the sites and can personally evaluate the evidence. I highly recommend they do so. Matthes explained very little of the evidence he saw, and Wahrhaftig, who rode on horseback along a trail through the valley (and thereby saw very little evidence), never wrote how he delineated and relative-dated his glacial deposits. In contrast, I am presenting evidence from all my sites. No evidence is omitted. This is important, because in the past, geologists who did geomorphic and glacial studies in Yosemite and in the Sierra Nevada did not address evidence conflicting with their interpretations.

## Sunrise Creek

The "Sun" on the right side of the top (north) edge of Map 22 is Sunrise Mountain, a mile-long nondescript, undulating ridge. From near its south end a subordinate ridge drops south 1.3 miles in stepped fashion to a minor 9030-foot summit along the north edge of the Merced River's inner canyon. My brief field reconnaissance along Sunrise Mountain leads me to concur generally with Matthes' reconstruction of the Wisconsin (Tioga/Tahoe) and pre-Wisconsin (pre-Tahoe) glaciers on and below that mountain—with one exception. On Map 22 at site 1 is a minor saddle with a seasonal pond, which is about 40 yards in diameter, and which is scarcely 3 feet deep when full. Matthes' Plate 39 shows the saddle as unglaciated, yet the pond's origin would be hard to explain without glaciation. I have observed shallow hollows in the unglaciated southern Sierra, but the largest were only a

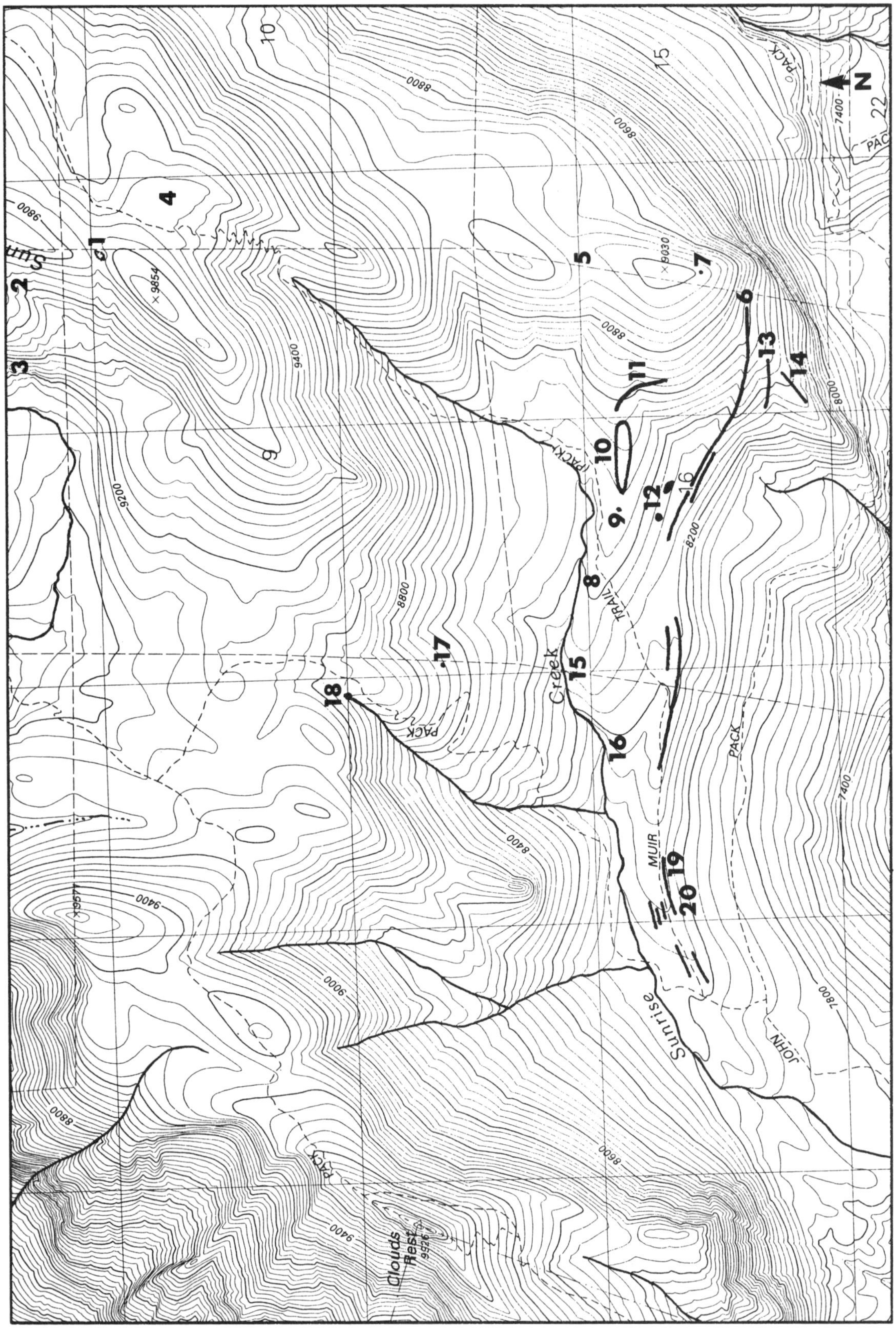


**Map 22. Little Yosemite Valley area: Sunrise Creek. Scale 1:24,000.**

few yards in diameter and several inches deep. Furthermore, they had developed on vegetation-free, level surfaces, apparently from water temporarily residing atop them, year after year, allowing chemical weathering to free mineral grains, which then were swept away by winds. The seasonal pond at site 1 is in a very different environment. It lies in a small basin, not on a flat. Furthermore, the basin is forested, not exposed, making wind deflation impossible. Finally, the bedrock is fresh, similar to that of other areas traversed by Tioga glaciers. Therefore, a Tioga glacier probably flowed across this minor saddle.

But did the glacier flow west or east across it? I suggest that flow across the saddle was eastward. With a cirque glacier originating just off the map—above site 2—its ice surface would have been barely above the saddle, and so only stagnant ice would have covered it. This accounts for the absence of polish and striations, but not for the absence of erratics. The reason erratics are not found is that the cirque glacier originated on gentle, stable slopes that before glaciation, as after, were producing only gruss, not rockfall. In contrast, at site 3 (and several others off the map) rockfall is and was produced from several unstable cliffs, and abundant, large erratics lie downslope from them. An argument against westward flow by the Echo Creek glacier across the shallow saddle is that if it had deposited only stagnant ice there, it still would have left erratics slightly lower on the upper part of the subordinate ridge, site 4, because unlike the local cirque glacier, this glacier flowed through a drainage very prone to rockfall. Absence of erratics implies no glaciation there whatsoever, as Matthes has so mapped.

From site 4 the first half of the subordinate ridge drops about 700 feet in elevation before essentially leveling out. Matthes shows this southern half as glaciated in pre-Tahoe time, whereas Wahrhaftig on the park's geologic map shows it glaciated in Tahoe time. At site 5, a broad, shallow saddle, the glacial evidence consists only of about two dozen large (6 to 16 foot) erratics. One might argue that the Tahoe glacier buried this saddle, but that by covering it only with stagnant ice, the flow was extremely slow and so few erratics and no till were deposited. However, glaciers leave erratics of all sizes, not merely large ones. In my extensive glacial mapping of the Stanislaus River drainage, I found only large erratics beyond the limits of Tioga and Tahoe glaciers, in the realm of the pre-Tahoe glaciers. Apparently pre-Tahoe erratics less than about 3 to 5 feet in diameter have weathered completely away, given the antiquity of their deposition, and hence usually are absent. Consequently, their absence at site 5 indicates that the glaciation was pre-Tahoe.

A conspicuous lateral moraine pinches out at its upper end at site 6, which is at 8620 feet. Evidence from most of the remaining sites suggests that this is a late-Tioga moraine; hence the maximum-Tioga ice surface at this site could have been higher, though not by much. An extrapolation of the moraine up-canyon shows that the Tioga glacier's ice surface would have been approximately at the break in slope between gentle slopes above and steep ones below. If the Tioga glacier's ice surface had been above site 6, then it would have lapped onto the gentle slopes. However, the ice would have been thin and its motion minor, so it would have left few erratics and till. The ice surface of the Tahoe glacier could have been slightly above it, and likewise it would have left few erratics or till. As was mentioned above, neither topped site 5, and so the ice surfaces of the Tioga and Tahoe glaciers have an upper constraint. It limits the Tioga and Tahoe ice surfaces in the vicinity of site 6 to about 8800 feet. This accords with the crest of a moraine east of and below site 5 as shown on the park's geologic map.

Site 7 is relatively inconsequential except that here is perhaps the most dramatic, large, 12-foot-wide erratic, precariously perched on a pedestal that has a maximum height of 10 feet (Photo 61). Matthes, especially at Moraine Dome (in the next section), used the height of pedestals to determine the amount of post-glacial (El Portal/pre-Tahoe) denudation, that is, how much the bedrock has been weathered and eroded down since a glacier left the erratic. But as I demonstrate for the dome's famous erratic, pedestal height can be a poor indicator of actual post-glacial denudation.

At site 8 the John Muir Trail crosses the crest of Wahrhaftig's diving Tahoe-age lateral moraine, and in that vicinity this broad, smooth ridge *almost* appears to be a classic moraine of Tahoe age. However, the boulders here are numerous and fresh, and this suggests a Tioga age (Photo 62). But if you leave the trail, this perception changes by walking east along the crest to site 9, where

**Photo 61. Large erratic balanced on pedestal. Erratic is Cathedral Peak granodiorite, bedrock is Half Dome granodiorite. Broad, glaciated summit of Bunnell Point is in the right background, and unglaciated Mt. Starr King rises above it.**

Photo 62. Trail view west down the "diving Tahoe-age lateral moraine."

boulders are far fewer. Indeed, there stands a 10-foot-long, coarse-grained erratic with two deep solution holes—a general sign of lengthy weathering. Farther east, in the zone of site 10, erratics are nearly absent, and one could interpret this part of the ridge as pre-Tahoe. Just beyond this zone the ridge steepens, and the curving line, site 11, marks the upper limit of erratics. If this line represents the ice surface of a glacier, then its extrapolation southeast indicates that at site 6 the ice surface was at about 8720 feet. This is close to the site 5 constraint for the ice surfaces of the Tioga and Tahoe glaciers.

The true nature of the "diving Tahoe-age lateral moraine" is revealed by a thorough examination of it. West of the John Muir Trail the till along the crest becomes increasingly thin until at last it becomes discontinuous to expose the west-diving nose of a bedrock ridge. Two exposures of bedrock (the considerably smaller one is shown in Photo 63), exist along its south slopes at site 12. By the larger, eastern one are six 3-foot-diameter exhumed (formerly buried) erratics, all fresh. Immediately below this exposure, just above the base of a narrow gully, is a short line of boulders that comprise a lateral

Photo 63. Thin, eroding till exposes bedrock on diving lateral moraine.

Photo 64. Partly exhumed, 10-foot-long fresh erratic on eroding slope.

moraine. This line of boulders is not nearly so prominent as the lengthy Tioga-age lateral moraine, which has a smaller lateral moraine immediately south of and below it. This may be a late-Tioga, diachronous, recessional lateral moraine, since it follows bedrock, first along a descending ridge then along the edge of a nearly level bench. Below this moraine and to the east are two smaller, similar moraines at sites 13 and 14. The lower end of the site 13 moraine actually descends below the lower end of the site 14 one, which is a technically impossible condition under the nested-moraines relative-dating paradigm, which requires the innermost crest to be the lowest. As with the main lateral moraine, their origins are readily explained as deposits left along bedrock ridges by the lowering ice surface of the waning glacier.

The solution-pocketed erratic at site 9 argues for a Tahoe age for the glacial deposits on the bedrock ridge. However, virtually all the other erratics appear fresh and argue for a Tioga age. (Fires can exfoliate the surfaces of small erratics and the *sides* of large ones, keeping these surfaces fresh. However, the *tops* of large erratics are quite protected from fires, and so should develop solution pockets over time.) Therefore, the site 9 erratic could be a partly weathered though coherent erratic from a former glaciation that later fell atop a Tioga glacier and was transported here. That almost all the erratics on the bedrock ridge are not pocketed argues against a Tahoe age. Furthermore, at sites 15 and 16 (Photo 64) along the north side of the bedrock ridge, the steepness of the slopes has resulted in partial erosion of till. All exhumed erratics I observed there not only looked fresh but also produced a solid ring when struck with a hammer, so likely they are Tioga. Finally, along the lower slopes above the bank of Sunrise Creek in the vicinity of site 17 is a zone of abundant boulders that extends about 250 feet upslope. These are mapped by Wahrhaftig as Tahoe, but the boulders, like those along the creek and at site 15, are fresh. Because the hundred-plus surface boulders and dozens of exhumed subsurface boulders that I examined on this descending bedrock ridge all were fresh, the veneer of deposits on it must be Tioga.

To determine the elevation of the Tioga ice surface in this vicinity, I hiked north up slopes to find the highest erratic. Scattered boulders lie along the ascent, and the highest one encountered is at about 8690 feet, site 17. It is about 4 feet in length and width but, more importantly, is composed of Cathedral Peak granodiorite, which is readily identified because of its large potassium-feldspar phenocrysts (see Photo 61). Since all the bedrock that occurs in, above, and well beyond Little Yosemite Valley is medium-grained Half Dome granodiorite, the boulder not only is an erratic, but also must have had a source area in or east of the Echo Creek drainage, where Cathedral Peak granodiorite occurs. This erratic marks the minimum elevation for the ice surface of either the Tioga or the Tahoe glacier. It probably was not left by a pre-Tahoe glacier for two reasons. First, the slopes here are about 18°, and over some 800,000 years of post-Sherwin glacial time, repeated earthquakes should have caused all erratics to eventually slip downslope. This process is readily observed at Bear Gulch, in Pinnacles National Monument, in the inner Coast Ranges of central California. There, a similarly steep slope dips north to the brink of the gulch, and large blocks from a thick, fractured, pyroclastic flow have broken free, slowly sliding down the slope and over the brink into the steep-walled gulch, apparently moving only when an earthquake occurs on the San Andreas Fault, which lies about 5 miles to the east. Due to the nearness of this fault to Pinnacles, the rate of sliding is much faster than would be expected in the Little Yosemite Valley area, which is about 17 miles from the main fault along the west edge of the Mono Basin.

A second reason why the site-17 erratic probably is not pre-Tahoe is that, as was mentioned just above, pre-Tahoe erratics of this size generally weather completely away. (At Turtleback Dome, west of Yosemite Valley, there are giant pre-Tahoe erratics on its summit area, but none on its long, north slopes, which are just as steep as the slopes

**Photo 65. Erratics left by the Tioga-age Sunrise Mountain glacier.**

**Photo 66. View east along a Tioga-age lateral moraine. Compare with Photo 59, the "diving Tahoe-age lateral moraine."**

in the site-17 vicinity.) On the slope above the erratic are boulders, but they could have been derived from local bedrock. Genuine erratics finally appear where the slope angle finally abates at about 8900 feet. During the Tioga glaciation, the Sunrise Mountain glacier extended at least as far southwest as site 18 (Photo 65), spilled hundreds of boulders down a ravine below it, and that glacier also would have left erratics on the gentle slopes above 8900 feet.

Midway between this site and Clouds Rest to the west is a steep, south-flowing drainage that Matthes' map (Plate 29) indicates was occupied by a Wisconsin-age cirque glacier, which descended to about 8400 feet. Although I did not field check it, I doubt its existence. If a cirque glacier had existed there, then one should have existed in the drainage descending south-southwest from site 4 (the one that has the John Muir Trail), since the head of a glacier there would have been about 400 feet higher. I found no Tahoe or Tioga glacial evidence in that area, and neither Matthes nor Wahrhaftig, mapped a Wisconsin or a Tioga glacier there.

Sites 19 and 20 are along the western part of a multi-crested lateral moraine of universally recognized Tioga age, though I differ from others in specifying it as late-Tioga diachronous. Photo 66 is taken at site 19, which is on the crest of the highest of several nested moraines. The evidence here is virtually identical to that along the crest of the "diving Tahoe-age lateral moraine" at site 8. Both have broad, gentle crests and abundant, fresh erratics. The only lateral moraines in this Sunrise Creek area that have sharp crests are the ones found along the rim of the inner canyon, such as at site 20 and south of site 12. The evidence here supports my interpretation of the glacial deposits in Figure 26 (cross section of a typical west-side Sierran canyon): moraine morphology is dictated mostly by location and slope angle, not necessarily by moraine age.

## Moraine Dome

Like most of the Sierra's so-called domes, glaciated Moraine Dome is not even close to being a true dome, but is simply a high point (barely) on a smoothed ridge. In contrast, unglaciated Mt. Starr King, southwest across Little Yosemite Valley, is a true dome. The summit area of Moraine Dome (Map 23) is significant because it has both protruding dikes and erratics on pedestals, and Matthes used these to measure how much the bedrock surface had lowered since the last glacier blanketed the summit. The glaciation there he considered El Portal (pre-Tahoe), based on weathered bedrock that lacked till, polish, and striations, on pedestals 3 feet high, and on dikes up to 8 feet high. (Actually, only a few dikes are more than 6 inches high; most are 2 inches or less.) Presumably with the demise of the El Portal-age glacier, which according to Matthes' detailed glacial map (Plate 29) buried the summit under about 500 feet of ice, erratics lay on a glacially smoothed bedrock surface, and dikes, if they had protruded at all, had been planed down to that surface.

One can visit any spot in the Sierra Nevada where *thick* Tioga-age glaciers flowed through a canyon and verify these relations. Consequently the heights of pedestals and dikes for Matthes seem to represent the minimum amount of denudation of the dome's summit area. One can observe in the field, as Matthes did, that an erratic can fall off its pedestal. Just northeast of the actual summit is one such example, an 8-foot-long, fallen erratic (Photo 67). Matthes suggested that a fallen erratic could then begin to develop another pedestal beneath it. This would explain in part why erratics left after the same glaciation have pedestals of varying heights (from essentially zero up to 3+ feet)—they have been on pedestals for varying amounts of time—and why Moraine Dome's pedestals—some with erratics, some without—are much lower than the highest dike. A caveat must be given about erratics on pedestals: they are rare. I have seen hundreds of pre-Tahoe erratics, and the great majority have no discernible pedestal. This suggests precious little post-glacial weathering of bedrock surrounding the pedestal has occurred, so pedestals did not form. Had most of these erratics fallen off their pedestals, then pedestals with

**Photo 67. An 8-foot-long, fallen erratic on Moraine Dome's summit area. Actual summit is to the right. Mt. Starr King is in background. A discontinuous veneer of gruss covers the summit area.**

**Photo 68. The dipping 8-foot-high dike on Moraine Dome's summit.**

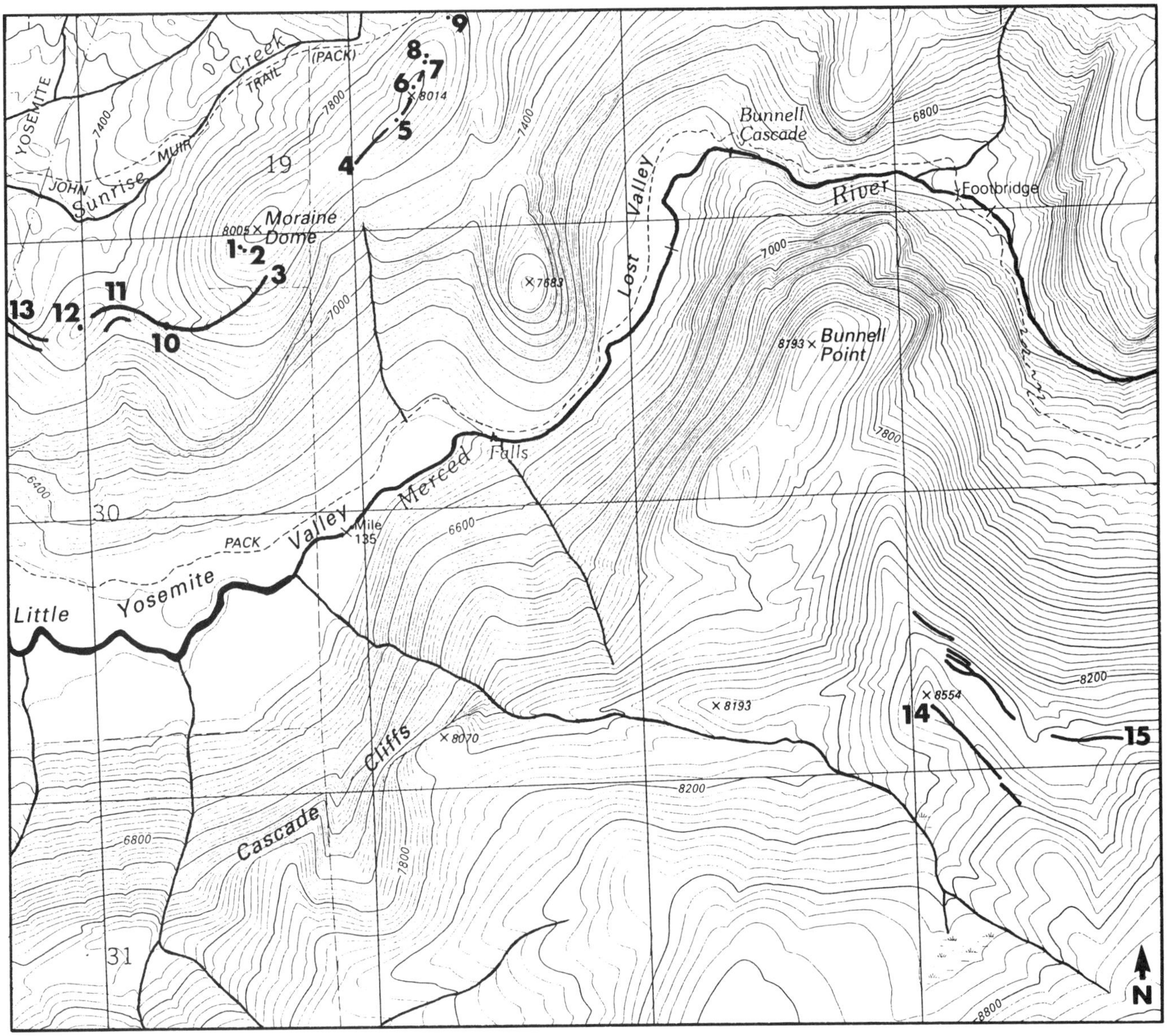


**Map 23. Little Yosemite Valley area: Moraine Dome. Scale 1:24,000.**

adjacent, toppled erratics would abound. Erratics usually do not fall off and roll away, rather they tilt downward and stay in place, as did the erratic in Photo 67.

Matthes used the 8-foot-high, southwest dipping dike (Map 23, site 1; Photo 68) as a minimum estimate of denudation, noting that its top was not a remnant of the glaciated surface. Still, he estimated that less than one foot of weathering had occurred, so the amount of post-glacial weathering would have been about 9 feet. In contrast, the giant, flat-based, pre-Tahoe erratics atop Turtleback Dome quite obviously have not fallen off their pedestals, and these average only about a few inches. Why the discrepancy? My explanation, based on evidence presented later, is that recently deglaciated surface bedrock is depressurized, and so tends to exfoliate at a faster rate than normal. I present evidence immediately below that Moraine Dome was covered by Tahoe and Tioga glaciers, and that it would have experienced enhanced exfoliation after each one. Turtleback Dome, in contrast, was not covered by these glaciers. Under this interpretation the dike cannot be used as a gauge of post pre-Tahoe denudation.

The famous perched erratic just below the summit of Moraine Dome (site 2) is a special case, and it is one of the key elements in unraveling the actual glacial history of the dome. The following photos document not only the extremely high rate of denudation at this site, but also the cause. Photo 69, taken in 1908 (reproduced from Matthes, 1930, Plate 37B), shows G. K. Gilbert standing beside this large erratic composed of Cathedral Peak granodiorite. Note the dense stand of scrubby huckleberry oaks in the foreground. Photo 70, taken in 1966, shows

**Photo 69. Moraine Dome's perched erratic, 1908.**

**Photo 71. Moraine Dome's perched erratic, 1993.**

**Photo 70. Moraine Dome's perched erratic, 1966.**

me atop the erratic. Only one living oak remains, although the ones in the background are healthy. The broken bedrock that Gilbert stood on now is quite visible. Photo 71, taken in 1993, shows that all the foreground oaks have died, although the ones in the background still are healthy, despite several years of drought. The pedestal, now revealed by a lack of vegetation, is seen to be composed of three exfoliating shells. (In Photo 69, Gilbert is standing on the second shell, which is in the process of splitting into two parallel shells.) These three shells have all slipped, the lower one about 10 inches down a 25% gradient from the intact, matching exfoliating shells to the right. The broken-off shells to the left are ready to topple off the basal shell. The root of the problem is, well, just that—a root! A Jeffrey pine has taken advantage of a fracture across the dome's exfoliating shells, and has sent a long root down along it (Photos 72 and 73). The exfoliating shells downslope from the fracture will eventually slide, root or no root, but the growing root is accentuating the rate of movement. Because the root does not extend beyond the pedestal, on its opposite (far) side the exfoliation shells and huckleberry oaks atop them are unaffected.

Therefore, like the 8-foot-high dike, the pedestal is not a gauge of post pre-Tahoe denudation, since major denudation has occurred locally here in just under a century. Indeed, that mass wasting is occurring at such a rapid rate suggests that the erratic probably did not rest here for some 800,000 years before facing sudden denudation and imminent toppling. The erratic likely has been perched here for a much shorter period of time, that is, since the Tioga glaciation. Had it resided for so long a time, or even since the Tahoe glaciation, it should have developed significant weathering pits, as the fallen erratic (Photo 67) should have, but neither did. Additionally, one piece of evidence for a Tioga age is the evidence that is *missing,* but should have been present if Matthes' interpretation were correct. The sharp-crested Tioga-age lateral moraine that descends steeply from Moraine Dome dies out at its upper end, site 3, simply because northeast of this point the slopes are too steep (>35°) to allow deposits to remain. However, it should resume on the gentle ridgecrest slopes northeast of and below the dome's summit. But at site 4, which marks the southwesternmost point of uncontested Tioga till, a sharp-crested lateral moraine should have developed over thousands of years of deposition, but is

**Photo 72. Jeffrey pine and Moraine Dome's perched erratic. A long root has extended from it along a fracture across the dome's exfoliating shells. Note in the left below the tree the upper part of the dome's sharp-crested Tioga-age lateral moraine.**

absent. Instead, all that exists is thin, discontinuous till, which is all one would expect to exist when stagnant ice of a waning glacier melted away, leaving behind its debris. If sites 4 through 9 were covered by a Tioga glacier, while those atop Moraine Dome were covered by an ancient pre-Tahoe glacier, we would expect to see notable differences. An examination of sites 5 through 9 shows that there are none.

**Photo 73. View from perched erratic toward base of Jeffrey pine. The exfoliation shells to the right are breaking apart from the intact shells to the left.**

**Photo 74. Weathered bedrock exfoliating in area of Tioga glaciation.**

At site 5, directly atop the ridgecrest, is a 5 by 4 by 1 foot high "pedestal"—really an exfoliation shell—that has become detached from its base in post-glacial time, much as the pedestal of the famous erratic has. At site 6, which is immediately northeast of the ridgecrest's highest point (at 8014 feet, about 9 feet above Moraine Dome's "unglaciated" summit), the till is locally absent, so the bedrock subjected to Tioga glaciation is exposed (Photo 74). Except for vegetation and pine needles, the scene is almost identical to that of Moraine Dome's summit. There is a thin veneer of gruss except where *weathered* bedrock protrudes above it. There also are exfoliating shells 6 to 10 inches thick, similar to ones on the dome. Where the ridgecrest descends northeast at a gradient comparable to that in the vicinity of the famous erratic, features similar to it exist among the large, fresh erratics. At site 7 (Photo 75) is an intact "pedestal," 3 feet high—essentially as high as the one supporting the famous erratic, but nevertheless it has managed to escape being planed down by the Tioga glacier known to have flowed across this ridgecrest. At site 8 is a close analog (Photo 76) to the

**Photo 75. Protruding bedrock "pedestal" in area of Tioga glaciation. Note giant erratic in background.**

**Photo 76. Large erratic on slipping exfoliation shell in area of Tioga glaciation.**

**Photo 77. Protruding dike in area of Tioga glaciation.**

famous erratic and its pedestal, a 6-foot-long erratic on an exfoliation shell that is slipping away, as are adjacent ones. And finally at site 9 is a dike (Photo 77) that protrudes up to 1.5 feet above the adjacent bedrock (almost all of the dikes atop Moraine Dome are lower than this one).

As Matthes and Wahrhaftig stated, the Tioga glacier spilled across the ridgecrest to leave the deposits to the northwest of it, more or less as shown on their maps. Along the part of the ridgecrest that has undisputed Tioga deposits, the difference in elevation between their highest elevation (on the summit) and their lowest (on the northeast saddle just beyond the north edge of Map 23) is about 224 feet. But the small overflow lobe of the Tioga glacier quite obviously had more than 0 feet of thickness where it traversed the summit, the maximum thickness of the lobe would have been greater. If we use an average, if arbitrary, figure of 150 feet for the minimum thickness of ice required to flow across the ridgecrest summit, then the ice-surface elevation there would be about 8165 feet. Because the dike is about 175 feet below the ridgecrest's summit, it then would lay under ice with a minimum thickness of 325 feet. However, since the precise ice-surface elevation at the ridgecrest summit is unknown, the minimum thickness of the ice over the dike is uncertain. The ice surface there, lying down-canyon from the ridgecrest summit, could have been slightly lower, so the ice could have been a few feet (or even a few tens of feet?) thinner.

The foregoing evidence indicates that there is only one significant difference between the glacial evidence on Moraine Dome and that on the ridgecrest: the dome lacks the sparse till and abundant erratics. Both have fresh erratics, protruding dikes, protruding bedrock, and weathered bedrock. My interpretation of the glacial evidence of both areas is that it was left by the *waning* Tioga-age glacier. During the maximum-Tioga glaciation, the ice surface was about 8400+ feet, this figure based on collective evidence of highest Tioga deposits in and above Little Yosemite Valley. The upper 150 feet of ice was brittle, so the thickness of the flowing ice across the dome and ridgecrest was rather thin, about 230+ feet. The glacier's velocity here was minimal, as was erosion. Neither dikes nor bedrock were appreciably worn down. One of the best examples of non-erosion of solution pockets by a Tioga glacier is also the most accessible: a low knoll beside the Vikingsholm parking lot, adjacent to Highway 89 above the southwest end of Emerald Bay, Lake Tahoe. The knoll's summit area has pockets on the order of 1 foot deep, despite formerly lying beneath 200 to 400 feet of unquestionably Tioga-age ice (Photo 78).

As the glacier waned, the base of the brittle ice grounded on Moraine Dome, and this stopped the flow of glacial ice immediately up-canyon from it, that is, on the ridgecrest. There the ice surface was a bit higher, and the ice was able, for a relatively short time, to spill over the ridgecrest and advance the short distance down to Sunrise Creek. The glacier continued to wane, and stagnant ice in the Sunrise Creek area began to melt, leaving in place its relatively sparse debris when the glacier ice here finally melted completely. With continued lowering of the ice surface the glacier began to create Moraine Dome's namesake, the sharp-crested, diachronous lateral moraine. Deposition began at site 3, while lower down (e.g., sites 10-13) the glacier still continued to flow laterally north into the Sunrise Creek area. As the ice surface of the waning glacier continued to fall, deposits were left increasingly lower on the bedrock ridgecrest. When deposition occurred at site 10 (Photo 79), it also occurred at site 12 (a small summit on the bedrock ridgecrest), since both became ice-free at the same time. With an additional drop of 80 feet, the saddle east of the small summit became ice-free and received deposits, as did contemporary site 13. Continued dropping led to deposition of the lateral moraine west of site 13.

There are two lines of evidence to support a late-Tioga age for the glacial deposits on the dome and the ridge-

**Photo 78. A foot-deep solution pocket on a knoll above Emerald Bay.**

crest. First, if a glacier had spilled across the ridgecrest throughout the Tioga glaciation—that is, for thousands of years—it would have deposited an enormous amount of sediments on the northwest slopes. But what exists are sediments so thin that they can be easily missed if one is not searching for them—they pale in comparison to the volume of sediments left by the Sierra's short-lived Little Ice Age glaciers. These thin sediments are not the remnants of an originally thick accumulation, for Sunrise Creek lacks the discharge to transport the boulders currently choking its stream bed. Second, as was just mentioned, collective evidence of highest Tioga deposits in and above Little Yosemite Valley indicates a higher maximum-Tioga ice surface, rather than along this ridgecrest, where Matthes and Wahrhaftig have put it. Bunnell Point, which lies southeast across the Merced River canyon from the ridgecrest, is littered with abundant, fresh erratics left by the Tioga glacier. Its highest point is 8193 feet elevation, and even if the ice atop it was stagnant, and therefore only 150 feet thick, the ice surface there would have been at about 8343 feet elevation, which is 329 feet higher than the highest point of the ridgecrest.

About 0.8 mile south-southeast from Bunnell Point's highest elevation is a 8554-foot summit, site 14. From this a sharp-crested diachronous moraine—low in the middle and high on each end—extends southeast. It could have been deposited by a glacier flowing northwest from the north cirque of Mt. Clark, located about 1.5 miles southwest of the corner of Map 20, by the west-flowing Merced River glacier, or by both. I suggest that it was deposited by the late-Tioga Merced River glacier.

**Photo 79. Moraine Dome's sharp-crested lateral moraine. Photo is from just above site 10, and viewed toward the saddle (site 11) and the small summit of bedrock ridgecrest (site 12), hidden behind trees in center of photo. Liberty Cap is in the left background.**

Because that suggestion is somewhat speculative (since I did not think to look for erratics of Cathedral Peak granodiorite while in the field), I use only the sharp-crested moraine extending west from site 15 as evidence for the *minimum* height of the maximum-Tioga Merced River glacier. The orientation of this stretch of lateral moraine indicates that the glacier depositing it flowed west (i.e., the Merced River glacier); it could not have been deposited by the northwest-flowing Mt. Clark glacier.

That the lateral moraine exists at all indicates that the Mt. Clark glacier had retreated above its union with the Merced Canyon glacier, exposing bedrock rim to deposition. Thus the lateral moraine was deposited after Tioga deglaciation had begun, and so does not even represent the ice surface during the maximum Tioga. Beyond a washout at a gully this late-Tioga lateral moraine continues northwest beneath the 8554-foot summit, ending at an elevation of 8420 feet. This is about 400 feet above the ridgecrest extending northeast from Moraine Dome, across the canyon. The ice surface on that ridgecrest should have been at about this elevation, or higher, since after all this is a late-Tioga ice surface, not the maximum. The evidence presented so far, as well as that presented in the following sections on Little Yosemite Valley, together indicate that the Tioga glacier had an ice surface higher than one that extended along Moraine Dome's ridgecrest. Furthermore, when all the evidence is assembled (as shown in Figure 29 and on Map 32), that higher ice surface is seen to have a smooth, slightly descending down-canyon profile, not one that fluctuates in perfect synchroneity with the rim of the Merced River's inner canyon. The higher ice surface suggests that Moraine Dome was last buried under Tioga (and presumably Tahoe) ice, not just under El Portal (pre-Tahoe) ice.

## Half Dome and Quarter Domes

Along the north slopes of Little Yosemite Valley proper, Matthes laboriously mapped the miserable, dense-brush crests of about 30 subparallel lateral moraines, to which he assigned a Wisconsin age. Wahrhaftig assigned them a Tioga age. The Half Dome Trail crosses the highest moraine (Map 24, site 1) at about 7200 feet, and above that the ground appears to be unglaciated. Fortunately for geoscientists but not for hikers, the heavily used trail is somewhat eroded—erosion has exhumed hundreds of formerly buried boulders between 7200 and 7450 feet elevation. Boulders composed of Cathedral Peak granodiorite leave no doubt that in this area of Half Dome granodiorite bedrock the thin veneer of soil really is a thin veneer of glacial deposits. To these, Matthes assigned an El Portal age, Wahrhaftig a pre-Tahoe (El Portal equivalent) age. However, because *all* of these exhumed boulders I inspected were fresh, they are Tioga age. Under this interpretation, the thin veneer was left during the maximum Tioga glaciation by relatively sluggish ice, whereas the assemblage of lateral moraines below it was left during the waning late-Tioga glaciation. This pattern is very similar to deposits located between Laurel Lake and Hetch Hetchy Reservoir, discussed in the previous chapter. On the trail up to Laurel Lake, you first climb up through an assemblage of lateral moraines, then above the "Tahoe" moraine encounter only fresh exhumed boulders. All deposits are from Tioga glaciers, and the same should hold true on the Half Dome Trail.

The trail turns west and climbs to the top of a sparsely forested ridge, near whose west end Matthes mapped an erratic (site 2). It indicates that at least one glacier—assumed by Matthes to be El Portal (pre-Tahoe)—reached this elevation, 7890 feet. However, evidence on the Starr King bench, across Little Yosemite Valley to the southwest, indicates that both the Tahoe and Tioga ice surfaces were at least this high, and so this erratic, not on a pedestal, could be as young as Tioga. If either glacier flowed parallel to the ridge, it would have left a conspicuous moraine atop it, but none exists. However, if either spilled across the ridge—as I suggest—it would have left only the sparse debris it was carrying when the glacier there melted.

The trail then climbs via short switchbacks to an open, somewhat broad ridge (site 3) that is sufficiently large to hold at least a few erratics. That none exist *suggests* (but does not prove) it was not glaciated. The slopes between this ridge and the glaciated one below are too steep to preserve glacial evidence, and so the highest ice surface of a former glacier might be constrained between 7950 and 8350 feet. Matthes put the El Portal (pre-Tahoe) ice surface at about 8200 feet, a compromise value that accords well with the highest glacial deposits on the Starr King bench. Tahoe and Tioga glaciers were almost as high, as determined by deposits on that bench as well as on the Quarter Domes.

But above that bench, on an 8300-foot saddle northwest of Mt. Starr King (mentioned later) are two erratics, which indicate that the higher Half Dome ridge, situated across Little Yosemite Valley from this site, should have been glaciated. Although this 8200-foot ridge is large enough to hold at least a few erratics, the pre-Tahoe ice there, with a surface at about 8300 feet elevation, would have been brittle and stagnant, so erratics need not have been deposited. Therefore, the weight of the evidence suggests that it was under pre-Tahoe ice. Supporting evidence exists at the summit of Inspiration Point, near Highway 4 in the upper North Fork Stanislaus River drainage. This clearly was under deep Tioga ice, based on abundant evidence on a higher ridge to its northwest. However, despite being five times the size of Half Dome's 8200-foot ridge, it has only one definable erratic, on its southeast corner. Without that solitary erratic one could conclude—erroneously—that this broad, rubbly volcanic summit, which also lacks polish and striations, was never glaciated.

Along an ascent northeast up slopes toward the Quarter Domes, you see very little glacial evidence on account of dense brush. However, at low elevations, at least two erratics exist. A 15-foot-high one at site 4 is on a slight pedestal, and it has several knobs protruding from it, these indicating an inch or so of weathering. A Tahoe age assignment might be appropriate for it except that it is in a local wind gap that experiences severe winter weather, which causes intense physical weathering. Such weathering on high ridges is extremely common, so polish and striations left by Tioga glaciers are seldom preserved. For example, along the Sierran crest from Sonora Pass north to Ebbetts Pass, a distance of about 25 miles, I found only one tiny patch of polished, striated rock—on a small, resistant dike. Along *most* of the high ridges in the entire glaciated area of the Stanislaus River drainage I could not find any fresh surfaces.

The granodiorite bedrock at site 4 is quite weathered and is exfoliating, so the slight pedestal, like the much taller ones atop Moraine Dome, does not signify an ancient age for the erratic. A 5 foot-high one at site 5, which is more protected, is not on any pedestal, and it is fresh Cathedral Peak granodiorite. It thus appears to be Tioga age, which makes the erratic down at site 4 also Tioga age. Lower Quarter Dome (site 6) has erratics that could be interpreted as Tioga or Tahoe, depending on your perception of the amount of weathering. They contrast with those on Upper Quarter Dome (site 7), some of which are on low pedestals and are very highly pitted and potholed. These appear to be pre-Tahoe, but again, given their site's exposure to severe winter weather, they could be severely weathered Tahoe erratics.

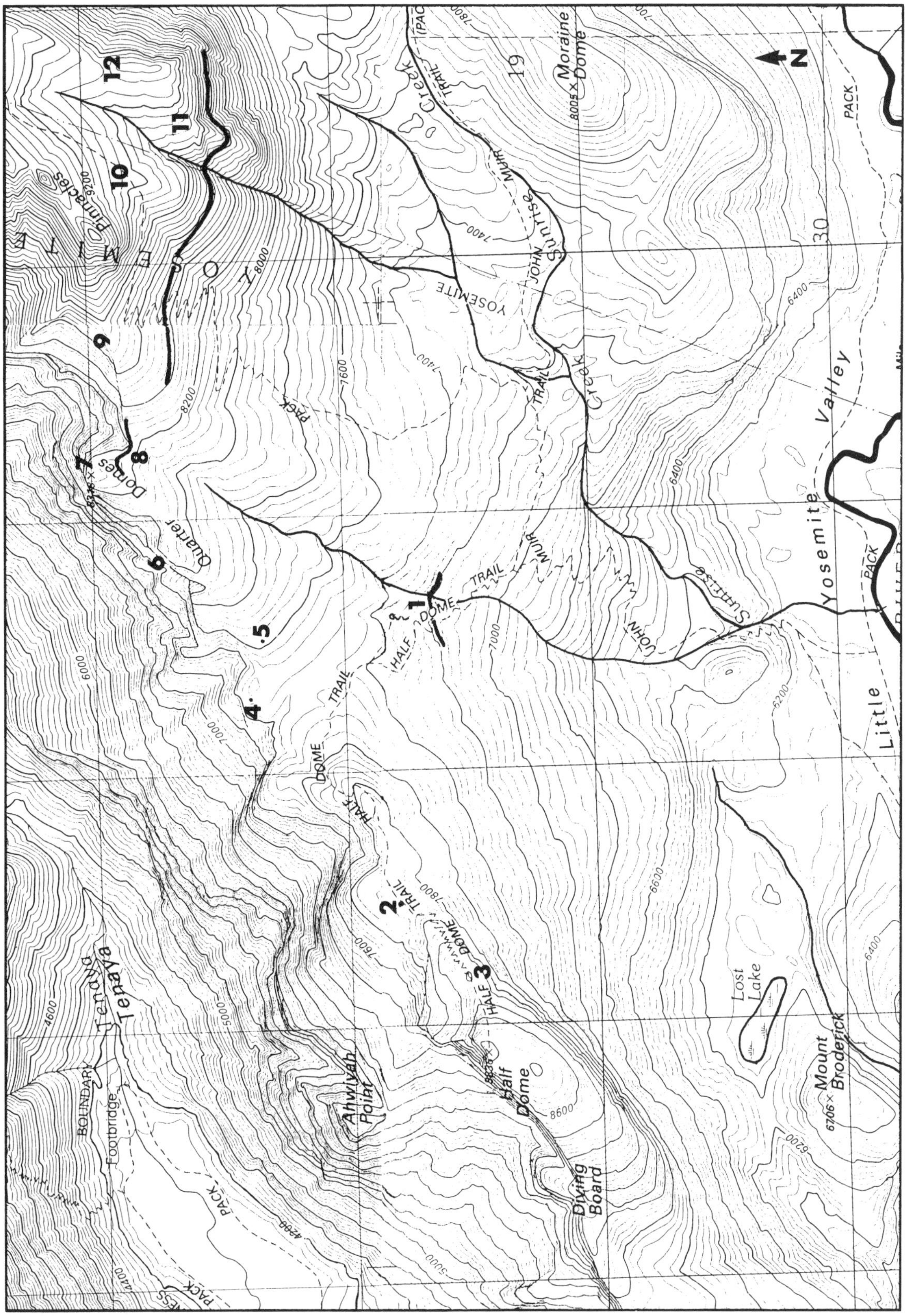


Map 24. Little Yosemite Valley area: Half Dome and Quarter Domes. Scale 1:24,000.

**Photo 80. Moraine descending to saddle southeast of Upper Quarter Dome.**

Evidence for a Tahoe age exists near Upper Quarter Dome, at site 8, where a lateral moraine descends west to a saddle (Photo 80), then ascends up the south slopes of Upper Quarter Dome (Photo 81) before dying out. This short moraine dies out above due to steepness of the crest and below due to the broad flat surface, on which the moraine gets spread out into discontinuous till. At the saddle the north side of the moraine, perched atop the head of a treacherously steep gully, is eroding away, and roughly 10-20% of the exhumed boulders are moderately to well weathered (Photo 82). On the basis of subsurface boulders of the Tioga and Tahoe moraines above Pinecrest Lake, the ones here would be judged to be between Tioga and Tahoe in age. There are too many weathered boulders to be classified as Tioga. The relatively low percentage of weathered ones can be attributed to a relatively thin, gravely matrix, a virtually nonexistent soil, and a lack of heavy snow accumulation, all of which results in a deficit of subsurface water compared with that of the Pinecrest Lake Tahoe-age moraines. Consequently, reduced subsurface weathering is to be expected, and the

**Photo 81. Moraine ascending from saddle northwest up Upper Quarter Dome.**

lateral moraine is very likely Tahoe in age. However, because it descends to and ascends from the saddle, it is diachronous, and so is late Tahoe.

Also of note here are three other points. First, because the east upper end of the moraine is at least as high as the summit of Upper Quarter Dome, the erratics on that summit would have been left at some time during the Tahoe glaciation. Erratics left by one or more pre-Tahoe glaciers would have been removed by the Tahoe glacier. Second, lateral moraines exist alongside glaciers, so the *waning* Tahoe glacier was flowing west past the saddle, not spilling north over the brink into Tenaya Canyon. However, from Upper Quarter Dome the canyon's rim drops moderately southwest, and the glacier, being thicker there, would have spilled into the canyon, as the slightly

**Photo 82. Exhumed boulders at saddle of the Upper Quarter Dome moraine.**

smaller Tioga glacier also would have. If they did so, they left minimal glacial deposits when they grounded and stagnated, dropping their sparse load. This is seen in the Stanislaus River drainage where major Tioga-glacier spillovers at Ebbetts Pass, Boulder Lake canyon, Mosquito Lakes, and the Dardanelles all lack moraines.

A third point is that during the Tahoe maximum, the glacier would have spilled north across Upper Quarter Dome's summit. The thicknesses of the Tahoe and the pre-Tahoe are constrained by lack of glacial evidence on gentle slopes and level areas (sites 9 through 12) east of the Quarter Domes and below the Pinnacles. The line on the map, which is significantly below these sites, is Matthes' ice surface for the El Portal (pre-Tahoe) glaciation. However, this is a better indicator of the Tahoe ice surface, since deposits left on moderately steep slopes above it would have been preserved, but none have been found. In contrast, deposits left there much longer ago by pre-Tahoe glaciers would have been eroded away. This conclusion is based on the absence of such deposits on the moderate slopes below Strawberry Peak ridge and below Dodge Ridge (both in the Stanislaus River drainage), each ridge having dozens of pre-Tahoe erratics. Therefore, the elevation of the Tahoe glacier's ice surface in the vicinity of the Upper Quarter Dome saddle would have been about 8400 feet, and the Tioga's, if the Lower Quarter Dome's erratics are Tioga, about 8200 feet.

This interpretation also agrees with that in the Stanislaus River drainage, where the crests of lateral moraines left by the Tioga and Tahoe glaciers do not begin to drop below pre-Tahoe deposits until at about 8000 feet elevation. In contrast, Matthes and Wahrhaftig placed the Wisconsin (Tioga/Tahoe) ice surface well below the pre-Tahoe, down at about 7200 feet. They indicated it was about 1100 feet thick, about half of what I propose. Both geologists further indicate that it was much thicker—by about 820 feet—immediately east of Moraine Dome. No mechanisms can account for this significant loss of thickness—from 1920 feet to 1100 feet—in only 1.0 mile, and its ice-surface gradient of 15½% results in an impossibly high basal shear value, as discussed earlier.

## Diving Board, Liberty Cap, and Panorama Cliff

The southeast side of Half Dome is too steep to hold any glacial deposits. Furthermore, it seems to lack polish that might indicate a minimum Tioga-glacier thickness. The brushy bedrock slopes along the base of this side also lack any evidence that can be used to determine ice thickness, although Liberty Cap has some. Matthes placed the ice surface of the Wisconsin (Tioga/Tahoe) glacier there at about 7000 feet, which is just below its 7076-foot summit. His elevation is based on what he judged was the highest Wisconsin lateral moraine on the south slopes of Little Yosemite Valley. Less-defined moraines above it he judged as El Portal (pre-Tahoe). Matthes climbed to Liberty Cap's summit twice in 1905, while surveying for his Yosemite Valley topographic map. However, these ascents were years before his glacial field work, so likely he wasn't looking for glacial evidence, which exists. In contrast, he climbed Half Dome's Diving Board in 1922, during his glacial field work, but did not draw its moraines nor its prominent erratics on his glacial map (Plate 29). Abundant lodgement till was plastered by the glacier on the lower northeast (up-canyon) slopes of Liberty Cap between about 6300 and 6500 feet (Map 25, site 1), but not on the steeper upper slopes. Matthes' mapped till up to about 6800 feet, but it is only rockfall. Above 6900 feet the slopes are gentle enough to hold moraines. Had a Tahoe or Tioga glacier scraped alongside this upper part of Liberty Cap, it would have left prominent moraines, but not even sparse till exists.

Superficially Liberty Cap's summit appears not to have been glaciated during the Tioga glaciation. Solution pockets exist, and most are about 2 inches deep, although one is 8 inches deep (Photo 83). However, as is the case at Moraine Dome and the ridge extending northeast from it, the Tioga and Tahoe glaciers overriding the summit would have been thin, somewhat stagnant, barely eroding ice. Supporting evidence for thin ice occurs on slopes to the north (site 5) and south (sites 11 and 12). The few erratics on Liberty Cap's summit area not only are fresh, but lack pedestals, and so are likely Tioga. Furthermore, at 7010 feet (site 2), there are fresh, if faint, glacial striations, bearing 240°, on a small patch of polished bedrock. To produce polish and striations, the ice had to flow. Since approximately the upper 150 feet of glacier ice is brittle, the glacier here to be at least this thick. If it was only 50 feet thicker (producing an ice surface of 7210 feet), then the actual summit (at 7076 feet) would have been under about 134 feet of ice, which would have been stagnant, explaining the non-erosion of the solution pockets. (Like other summits, this one is exposed to extreme cold-season physical weathering.) The Tioga ice surface above Liberty Cap then would have been on the order of 7200 feet, or slightly higher.

A Tioga glacier with this ice surface would have buried the 6706-foot summit of Mt. Broderick, just northwest of Liberty Cap, under about 500 feet of ice, and so it has a smoother-looking surface. Matthes' Plate 29 indicates no glacial deposits on its summit, which is much larger than Liberty Cap's, yet it is strewn with erratics, patches of till, and one recessional end moraine (site 3, Photo 84). Obviously it was not under stagnant ice. Although lower and less imposing than Liberty Cap, Mt. Broderick is more difficult and dangerous to climb, and Matthes didn't climb it. (I've done it in tennis shoes, but I recommend climbing shoes.) He would have viewed the evidence from Liberty Cap, but as mentioned above, he climbed that years before he began his glacial field work.

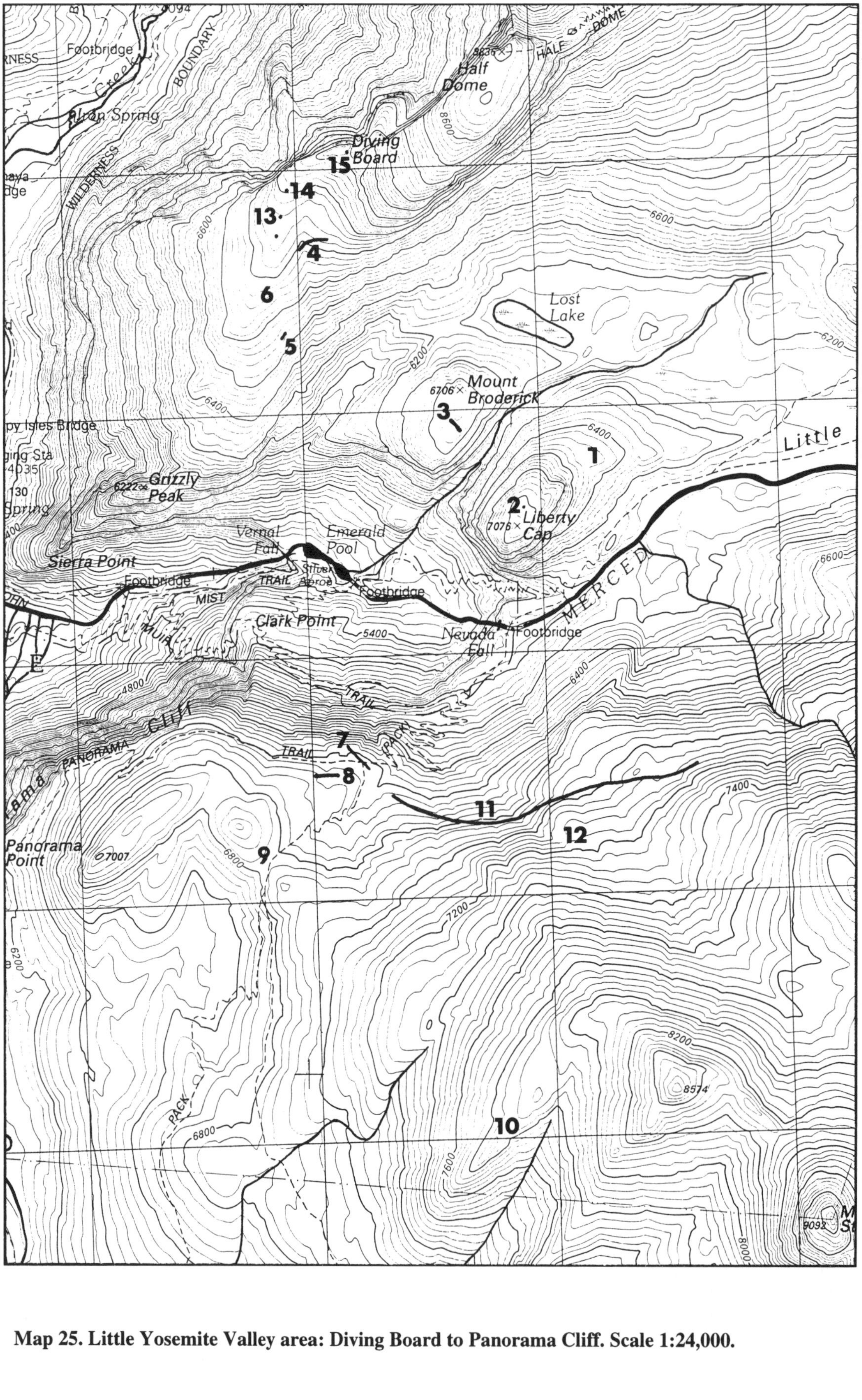


Map 25. Little Yosemite Valley area: Diving Board to Panorama Cliff. Scale 1:24,000.

**Photo 83. Shallow solution pockets on Liberty Cap's summit. Half Dome is in background.**

Evidence on slopes south of and below the Diving Board sets limits on all three glaciations. These slopes are reached by starting on a surprisingly good use trail that first strikes west along the north slopes of Liberty Cap's long east rib, skirts between Mt. Broderick's base and Lost Lake, and then dies out on the saddle just west of and above the lake. A ducked cross-country route continues onward, and there are several possibilities up to the Diving Board, easier ones being farther to the west (especially up along a swath of conifers—see Photo 84). If you pick a proper ascent route, you will find yourself on gentler slopes leading up to the Diving Board. When you descend from it, chances are good that you may not be able to find the route you ascended, and instead may encounter drop-offs. In short, only experienced mountaineers should visit this fine area of glacial evidence.

On the gentler slopes, two minor, short lateral moraines exist, one at site 4, the other at site 5. The east end of the one at site 4 is buried under rockfall from the southwest face of Half Dome, while the west end is actively eroding away. Of about 100 boulders exposed there, I found only two that were grussified. Since this site is in a gully, perhaps these two may be older erratics originally left higher up but then redeposited here during an extreme downpour, such as in the great storm of 1867 that James Hutchings had described. Therefore it appears to be Tioga rather than Tahoe, although the latter cannot be ruled out. The smaller lateral moraine lower down at site 5 is more subdued, and therefore characteristic of Tahoe under

**Photo 84. Mt. Broderick's summit, viewed from Liberty Cap's summit. A recessional end moraine rests roughly midway between the upper and lower ends of the summit area. The slopes below the Diving Board are in the background, and the best ascent to them is up along the swath of conifers.**

**Photo 85. Eroding moraine at 6700 feet elevation, south of the Diving Board.**

**Photo 86. Tioga-age lateral moraine at the junction of the Glacier Point-Panorama and Mono Meadow trails southwest of and above Nevada Fall.**

conventional age assignments (Photo 85). Resting on a moderate slope, it is actively eroding away, and of over 100 exhumed boulders, I found none that were grussified, so it must be Tioga. If the gradient of the site 4 moraine is projected down-canyon, it ends up high above the site 5 moraine, say at site 6, which is devoid of deposits. However, in this vicinity the glacier was plunging into a deep inner canyon, so its ice surface also would have plunged. From this I conclude that the moraines at sites 4 and 5 represent the ice surface of the Tioga glacier. The Tahoe glacier here must have been slightly thinner than the Tioga glacier. *Remember this,* for this relation occurs at several other sites that are mentioned in the following chapters on glacial evidence in the Yosemite area. Clyde Wahrhaftig had proposed a similar relation with his Tahoe moraine diving beneath a Tioga one farther up-canyon (Map 22, site 8), but his moraine is a bedrock ridge that does not match up with his mapped Tahoe moraines south of and above Little Yosemite Valley.

The best evidence for the ice surface of the Tahoe glacier being below that of the Tioga glacier lies to the south, at the Glacier Point-Panorama and Mono Meadow trails junction, site 7. Here a fresh, unquestionable Tioga lateral moraine diagonals northwest across the junction (Photo 86), and I could find no moraine above it. Had a Tahoe glacier been higher, it would have left a lateral moraine. Just west of this site, Matthes mapped a west-trending moraine (site 8), but it is a bedrock ridge. On lands west from about the diagonal line descending from summit 8574 in the southeast part of Map 25, Matthes mapped about two dozen lateral moraines. No moraine exists; rather, his "moraines" are merely *gruss* (granitic gravel) atop nearly level or gently sloping ridges and linear benches. They are most plentiful along the broad ridge surmounted by the Mono Meadow Trail, at site 9, but all that exists is a thin layer of gravel no different from gravel on unglaciated slopes. Matthes also mapped three erratics on a prominent ridge (site 10), but I found only weathered bedrock and blocks derived from it.

If the moraines at sites 4 and 5 represent the ice surface of the Tioga glacier that was steeply plunging down into the chasm west of Mt. Broderick and Liberty Cap, then the Tioga-age lateral moraine at the junction of the Glacier Point-Panorama and Mono Meadow trails (site 7) should do the same. Matthes sort of indicated this, as

**Photo 87. Boulders being exhumed on a minor ridge. View north toward Half Dome. Fires such as the one here can keep the surface of lowly boulders fresh through heat-induced exfoliation, so only formerly buried boulders can be used to determine the relative age of the deposit.**

shown by a line identified as site 11, which represents his highest Wisconsin lateral moraine. His actual line differs somewhat, for it follows a linear, 50-foot-high ridge (the moraine) that appears only on *his* Yosemite Valley topographic map. No such ridge appears on the two later USGS Yosemite Valley topographic maps, nor could I find it in the field (and it would have been impossible to miss). Instead, I found coarse- and fine-grained boulders mantling a minor ridge at site 12, some of them being exhumed, and all of them fresh, indicating a Tioga age (Photo 87). From this site, at about 7150 feet, the descent to the moraine at the trail junction, at about 6000 feet, is similar to the one between the two moraines (sites 4 and 5) on slopes below the Diving Board.

Above sites 4 and 5 are several erratics. Two prominent ones are at sites 13 (Photo 88) and 14, these being, in their greatest dimension, respectively 20 feet and 17 feet. No small ones were found along the ridge northeast up to the Diving Board, which is the typical situation for pre-Tahoe erratics throughout the Stanislaus River drainage. Neither erratic was pitted and neither was on a pedestal, but this too is the usual case for pre-Tahoe erratics throughout that drainage. However, an interesting erratic exists at site 15, just 30 feet below the knife-edge ridgecrest of the Diving Board (Photo 89), and 100 feet west of the notch separating this ridgecrest from Half Dome. Being only 10 inches across, it is anomalously small. However, it is composed of Cathedral Peak granodiorite, which has large phenocrysts, and over time these do weather out, so this erratic could be all that remains of a once large boulder. At 7440 feet elevation, it indicates that at least one pre-Tahoe glacier was on the order of 400 feet above the buried upper end of the moraine at site 4. Matthes places the ice surface on the southwestern face of Half Dome at a considerably higher elevation, 8100 feet. No evidence exists there; rather, it is correlated with his imaginary evidence (gruss) below the aforementioned summit 8574.

**Photo 88. Giant erratic below the Diving Board.**

**Photo 89. The Diving Board; view west through Yosemite Valley.**

## Starr King bench

This bench (Map 26) was particularly important for Matthes (1930, p. 62), for on it are well defined "old moraines" that exist because "conditions evidently were particularly favorable for their preservation"; consequently here "the student may best learn to recognize these older moraines by their constituent materials as well as by their indistinct forms." However, the preserving conditions are not at all evident, and Matthes suggests none. The well-defined lateral moraines are generally on exposed, descending bedrock ridges, where extreme physical weathering (not preserving) conditions prevail during the colder, wetter months. And one moraine just north of Starr King Meadow is on gentle slopes that are water-saturated most of the year, and therefore subject to above average chemical weathering. Because his "El Portal" moraines are so well preserved here, in contrast to all other sites, they are the exception, yet Matthes has made them the rule—in essence, the type locality. Additionally, Matthes correlates one moraine just north of Starr King Meadow with the one at the saddle below Upper Quarter Dome. But as mentioned earlier, that is Tahoe, and so by correlation this one also should be Tahoe. Moraines inboard of it may be Tioga. The evidence indicates this is the case.

The low knolls on the Starr King bench have weathered summits that appear very ancient, and thus not glaciated since pre-Tahoe time. But both likely were buried under thin, essentially stagnant Tioga ice, as were the summits of Moraine Dome and Liberty Cap. The low knoll southwest of Starr King Lake (Helen Lake on Matthes' Yosemite Valley topographic map) has on its 8052-foot summit, site 1, shallow solution pockets similar to those on Liberty Cap's summit (compare the two summits, shown respectively in Photos 90 and 83). In like manner the low knoll northwest of Starr King Lake has on its 8019-foot summit, site 2, gruss and exfoliated slabs similar to those on bedrock exposed on the uncontested Tioga-age glaciated ridge northeast of Moraine Dome (compare the two areas, shown respectively in Photos 91 and 74).

Starr King Lake's existence also argues for a Tioga glaciation in this area, for no Sierran lake, as dated by carbon 14 in its lowest sediments, is older than 16,000 years, i.e., late Tioga. In a June 1993 letter, Clyde Wahrhaftig, who admitted to not visiting the lake, took issue with this conclusion, stating:

**Photo 90. Shallow solution pockets atop summit 8052, southwest of Starr King Lake. Half Dome, to the northwest, is in the background.**

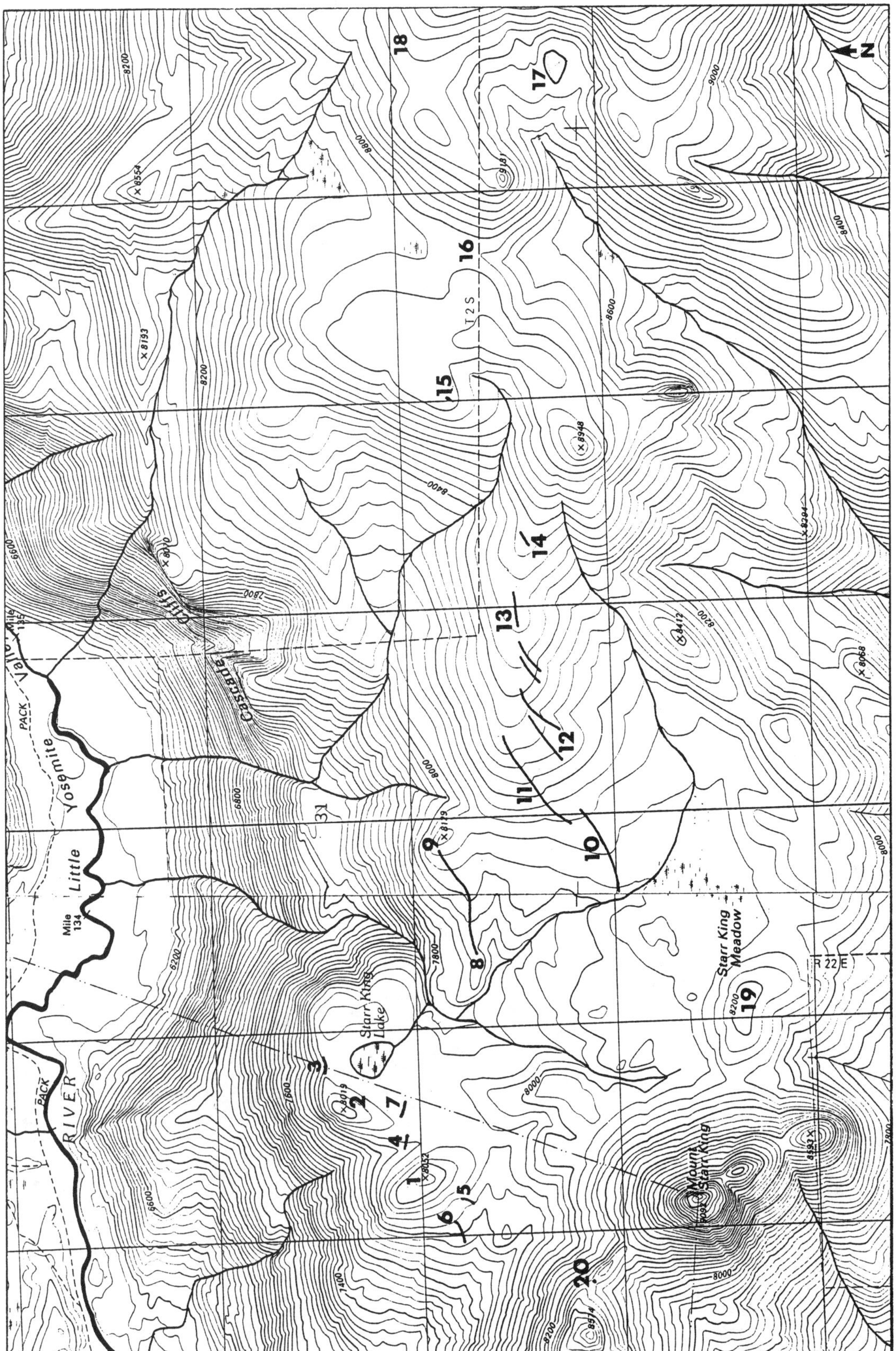


**Map 26. Little Yosemite Valley area: Starr King bench. Scale 1:24,000.**

> I am told that Starr King Lake is so shallow you could just about wade across it, except for the possible danger of being swallowed up by the marsh vegetation on its floor. I suspect that if it does not have a depositional dam of some sort, it is probably a deflation basin created in a thick body of gruss that underlay the bench on which the lake lies.

It was indeed shallow, about 30 inches deep in the center, when I visited it on July 15, 1993, during a drought year, and undoubtedly it would dry up well before the rainy season began. (At high-stand, the lake would be about 3 feet deep.) I was not swallowed up by the vegetation, which was minor. I used a hollow PVC pipe to probe the sediments, and found them of uniform thickness (Photo 92). Essentially, about 4 inches of soft, organic muck lies atop 4 inches of more cohesive organic-rich sediments, which in turn lies atop glacial flour. Glacial flour and gruss both are gray, but the flour does not feel gritty like grains of gruss, but rather, true to its name, feels smooth, having been ground fine through glacial processes. The thin layer of organic-rich sediments above the flour is typical of those of about a dozen other seasonal or occasionally seasonal glacial lakes and ponds I've cored in the Sierra Nevada. Most were in the North Fork Stanislaus River drainage. One, Harden Lake, which dries up only in drought years, is in west-central Yosemite National Park. In these the sediment thickness above the glacial flour ranges from about 4 inches up to about 1 foot. Unfortunately, being thin, these sediments are contaminated with modern carbon 14, and so cannot be accurately dated. For example, two unquestionably post-Tioga seasonal ponds that I dated in the North Fork Stanislaus River drainage up-canyon from Lake Moran yielded carbon-14 dates of 560 A.D. and 730 B.C.

**Photo 91. Gruss and exfoliated slabs atop summit 8019, northwest of Starr King Lake. Half Dome, to the northwest, is in the background.**

**Photo 92. Author probing sediments in Starr King Lake. Mt. Clark is in the background. Photo by Ken Ng.**

Starr King Lake is not dammed behind a depositional dam, as Wahrhaftig speculated, so is it a deflation basin? The answer is no, because glacial flour—not gruss—is present. Furthermore, deflation hollows require two conditions to evolve, which are not met here. First, to create a hollow on a bench, strong winds must be able to blow away some of the gruss that is exposed after whatever ephemeral water that accumulates evaporates. However, the lake is surrounded entirely by a dense forest, so the winds here are weak. (I cored the lake on a *very* windy afternoon.) Consequently, the gruss cannot be blown out of the basin. Second, the gruss has to be exposed in the first place, but after the water evaporates, it will still be buried under the vegetation and its underlying organic-rich sediments. Finally, given that the Sierra Nevada has dozens, if not hundreds, of granitic benches suitable for deflation basins, they should abound. However, despite hiking thousands of miles in the range, I have found none.

Matthes mapped quite a number of moraines on, below, and above the Starr King bench, and although I did not attempt to verify the existence of all of them, I checked a number of them. In essence, his lateral moraines do exist, while his till—indistinguishable from soil and decomposed bedrock in unglaciated areas—is highly questionable. In particular, El Portal (pre-Tahoe) till on the steep slopes below Mt. Starr King and its northwestern outlier, summit 8574 would have long ago been removed by mass wasting, including soil creep. One sees here that soil and underlying detritus on bedrock slopes are actively creeping downslope, as they are on the lateral moraines constructed on such slopes. One such moraine exists at site 3 (Photo 93), at the top of a steep slope immediately north of a shallow saddle east of summit 8019, and its exhumed boulders are fresh. Surprisingly this moraine was not mapped by Matthes. Because there is exfoliating, mass-wasting, coarse-grained granitic bedrock immediately above it, this deposit could be considered colluvium (products of mass wasting). However, the fine-grained granitic boulders in it require a non-local—i.e., glacially transported—origin.

The bouldery moraine at site 4, in the gully between summits 8019 and 8052, also has fresh, exhumed boulders, as does the one at site 5 (Photo 94), which is very actively mass wasting due to the steepness of the slope (35% gradient). The moraine at site 6, southwest of summit 8052, is particularly interesting, because it has two characteristics of classic Tioga moraines: hummocky surface and abundant, fresh boulders (compare this moraine, shown in Photo 95, with the uncontested Tioga-age lateral moraine shown in Photo 66). Nevertheless, Matthes classified it (and others here) as El Portal, apparently because they are well above the highest conspicuous (Wisconsin) lateral moraines along the north slopes of Little Yosemite Valley. Along the valley's south slopes below the brink of the Starr King bench, Matthes'

**Photo 93. Exhumed boulders in moraine east of summit 8019.**

**Photo 94. Exhumed boulders in moraine south of summit 8052.**

**Photo 95. Very bouldery moraine southwest of summit 8052. View is southwest down along its crest.**

Wisconsin ice surface parallels the highest north-slope moraine. I cannot overstress that it is an *arbitrary* ice surface; on these south slopes the lateral moraines above his ice surface are indistinguishable from those below it.

Lateral moraines on the Starr King bench are a different matter, for any glacier, which would be slowly flowing across the bench, would be unconfined by topography and hence its edge would be free to oscillate. Hence its lateral moraine of sparse sediments would be spread over a linear zone rather than concentrated along a line. Matthes mapped four of these long, variably defined lateral moraines. The first is shown descending past sites 9 and 8 and appearing at site 7. Matthes drew in a stretch of moraine between sites 8 and 7, but the terrain there is so uniform that I could not detect even a hint of a moraine. From site 9 (Photo 96), at an elevation of about 8100 feet, this moraine descends a broad ridge. The moraine in that vicinity has a slight convexity to it, which is due to the shape of the moraine itself and not to a veneer of till atop a convex ridge (which actually is quite flat). However, farther down at site 8 (Photo 97) the moraine is little more than a veneer of till atop a curved ridge. The moraine has erratics, but none exhibit deep solution pockets. Because this moraine conforms to a low bedrock ridge as it drops west to a creek bed and then, if Matthes is correct, climbs up about 80 feet to a flat bench, it is diachronous.

The lateral moraine to the south of it, which apparently is responsible for the creation of the bog known as Starr King Meadow, has a similar morphology: curved above and flattening below. The boulders on it are deeply weathered, as shown in Photo 98 (site 10). This indicates a Tahoe or pre-Tahoe age, although the latter can be discounted since there are far too many boulders less than 3 feet across and the matrix is too abundant. Unfortunately I was unable to find any exhumed boulders at either this moraine or the one north of it (sites 8 and 9). On the basis of surface boulders, I provisionally designate the north one Tioga and the south one Tahoe.

Northeast of the provisionally Tahoe moraine is another one (site 11), which would be of the same age. Both appear to be diachronous. The site 11 moraine descends a crest, then dies out where that crest dies out, while the site 10 moraine begins nearby on a slightly curved, lower-gradient crest and follows it down to its end. The till-free "moraine" above it (site 13) is a swath of boulders where it plunges southwest over the crest, but east along it the moraine is merely a collection of boulders up to an elevation of about 8500 feet. These range from very small up to 7 feet. Their abundance plus their small size discount a pre-Tahoe age, and so this assemblage is provisionally Tahoe. At site 14 is the crest's highest erratic, and several lie between it and the till-free "moraine." It too may be Tahoe, but then, it may be pre-Tahoe.

This erratic, at 8570 feet, stands not only well above Starr King Meadow (by some 530 feet), but also above the drainage divide south of the meadow. Therefore the glacier—Tahoe or pre-Tahoe—that left it would have

**Photo 96. View southwest down a lateral moraine, from site 9.**

**Photo 97. View southwest down same lateral moraine, from site 8.**

overflowed the drainage divide. The ice surface of that glacier decreased westward, as does the drainage divide (from about 8400 feet to about 8200 feet), so the thickness of the ice overflow would have been relatively constant along it. The ice surface at site 14 is governed by the absence of any glacial evidence east of it on a broad, flat bench, site 15, which is at about 8810 feet elevation. Consequently, one can conclude that ice thickness along the drainage divide was relatively minor, less than 300 feet, so ice flow across it was slow and essentially nonerosive, as atop the Tioga-age glaciated ridge northeast of Moraine Dome. At site 15, the southwest edge of the flat

**Photo 98. View northeast up a lateral moraine, from site 10.**

bench, stands a 15-foot-high, upward curving dike, which like the one atop Moraine Dome cannot be used to calculate a denudation rate.

Whereas Matthes had mapped pre-Tahoe moraines and till south of sites 10 through 14, Wahrhaftig did not. He mapped the moraines of sites 10 through 13 (but not 14) as pre-Tahoe. If we apply his assumption that the ice surface of a glacier closely parallels the crest of the highest lateral moraine left by that glacier, then his pre-Tahoe glacier would have plunged—for no logical reason—about 500 feet on its descent to the north edge of Starr King Meadow. Finally, the ice surface of the pre-Tahoe glacier would have to have risen in order to deposit two erratics at site 20, some 300 feet above Starr King Meadow. This clearly is impossible. Like his Tioga and Tahoe ice surfaces, Wahrhaftig's pre-Tahoe ice surface defies logic. No glacier today has such ice surfaces.

East of the flat bench is a broad, shallow gap (site 16) in the drainage divide, which contains a few erratics. I believe these were left by a northward overflow from a small glacier that emanated from a shallow bowl holding an unnamed lake, site 17, and that this glacier was responsible for the origin of the boggy meadow just north of and below the gap. But without reconnaissance in its drainage, these conclusions are tentative. East of and above the shallow gap there is no evidence of glaciation until at site 18, which marks the west edge of an impressive, northwest-descending rock glacier originating from Mt. Clark's north cirque.

The aforementioned drainage-divide ridge ends at the base of Mt. Starr King, and immediately east of it is a broad, flat-topped knoll, site 19, with about a dozen large erratics on it. These do not appear to be very weathered, which suggests a Tahoe age, and the glacier that left the many erratics at site 13 could have left the ones here. Additionally, because the elevation of the Tahoe glacier's ice surface in the vicinity of the Upper Quarter Dome saddle would have been about 8400 feet, as mentioned earlier, the ice surface here would have been only slightly lower, and so the glacier might have spilled across the drainage divide. On the other hand, the erratics are few in number and all are large, which suggests a pre-Tahoe age, as does one perched erratic (Photo 99). Its 30-inch pedestal appears to have formed over a long period of time, not over a few centuries, as may be the case for the Moraine Dome pedestal. The knoll lacks till, as do the low gaps at its east and west bases, but absence at an overflow site can occur even in areas of Tioga-ice overflow. (Matthes' map indicates morainal material at both, but only gruss exists.)

Extensive field work on the Starr King bench may resolve the age of the two well-developed moraines (between Starr King Lake and Starr King Meadow) and of the erratics along the drainage divide, which I *very* tentatively call pre-Tahoe. There is one more important site, 20, where on the saddle between Mt. Starr King and summit 8574 are two erratics that Matthes identified as Mt. Clark granite. By modern terminology, this is the leucogranite of Mt. Clark. I did not visit this site, and given that much of his glacial evidence does not exist, the two erratics here could be suspect. However, one would not mistake leucogranite for the local bedrock, Half Dome granodiorite, so the boulders must be of glacial origin. The age is unknown, and I, like Matthes and Wahrhaftig, believe it is pre-Tahoe. (Matthes assigned it to his Glacier Point stage, which in recent years has been dropped.) At about 8300 feet elevation, it seems too high to be Tahoe, based on the somewhat speculative Tahoe evidence on the Starr King bench and the drainage divide. At this elevation it is higher than Matthes' highest ice surface on Half Dome.

## Longitudinal profiles of Little Yosemite Valley glaciers

Some of the evidence in the preceding sections sets limits in several locations for the minimum or maximum elevation of the ice surface of the Tioga, Tahoe, and pre-Tahoe glaciers that flowed through Little Yosemite Valley, and with these limits I have reconstructed their longitudinal profiles in Figure 29. Some limits are poor and were not considered. Two examples of poor limits are: 1) the unglaciated summit of Half Dome stood high above the glaciers, providing poor control; and 2) most of the Tioga evidence is for the late-Tioga stage, hence underestimating the true elevation of the Tioga ice surface. The best limits are those of sites 5, 6, 11, and 17 in the Sunrise Creek section, sites 14 and 15 in the Moraine Dome section, sites 6 to 12 in the Half Dome and Quarter Domes section, sites 4, 5, 7, 12, and 15 in the Diving Board, Liberty Cap, and Panorama Cliff section, and sites 1, 2, 9, 13, 14, 15, and 20 in the Starr King bench section. At the west end of the profile the pre-Tahoe ice surface is

**Photo 99. Perched erratic on the east end of the flat-topped knoll between Mt. Starr King and Starr King Meadow.**

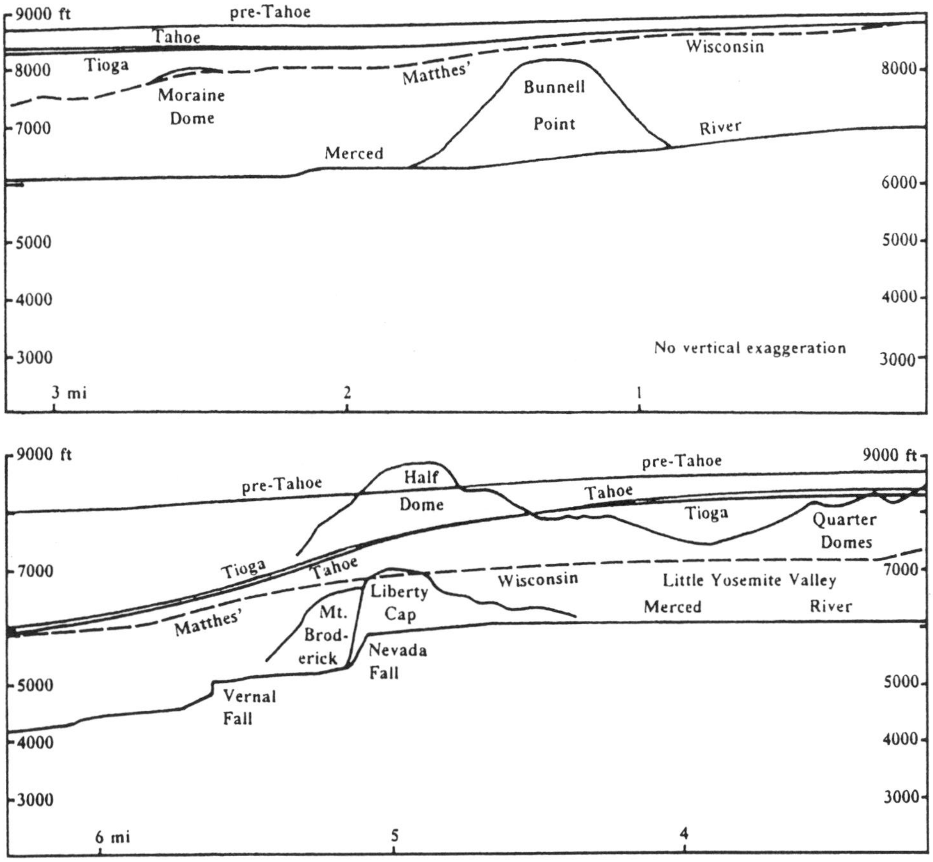


**Figure 29. Longitudinal profiles of Little Yosemite Valley glaciers.**

limited by the highest erratics below the summit of Sentinel Dome, and the Tahoe and Tioga ice surfaces by the glacial deposits just above the brink of Illilouette Fall. Both are discussed in the next chapter.

The elevation at each site was projected perpendicular to a straight line having a bearing of 262° and passing through the highest point of Liberty Cap. This line of "best fit" was determined by taking the meandering course of the inner gorge of the Merced River and progressively straightening it out. Projections to this line result in a believable Tioga ice surface that does not conflict with any strong limit until the vicinity of Liberty Cap and Mt. Broderick. If the Tioga glacier plunged over them perpendicular to their nearly vertical southwest (lee) faces, then they plunged along a bearing of about 222°. The one striation atop Liberty Cap shows a flow along a bearing of 240°. Consequently, site 5 below the Diving Board and site 7 at the trail junction on the opposite side of the valley yield slightly discordant results if projected perpendicular to the straight line bearing 262°. When they are adjusted to the change in direction where the glacier plunged over Liberty Cap and Mt. Broderick (240°), they produce a believable Tioga ice surface.

My pre-Tahoe ice surface is essentially identical to Matthes' El Portal, which is not shown on the profile. However, my Tioga and Tahoe ice surfaces are very different from his Wisconsin, hence this is shown for purposes of comparison. Wahrhaftig's Tioga ice surface is identical to Matthes' Wisconsin. Both show a 40 percent decrease in ice thickness from the summit of Moraine Dome to where the ice surface becomes horizontal, about ¾ mile down-canyon. This massive decrease occurs despite a nearly horizontal canyon floor here and the occurrence of similar cross-sectional morphology and area along this stretch. Wahrhaftig's Tahoe ice surface is not shown because, as stated above, the mapped Tahoe moraines on one side of Little Yosemite Valley are highly discordant with those on the opposite side, and thus result in two mutually irreconcilable ice surfaces (as first discussed in Chapter 1).

Whereas in many Sierran canyons the glaciers had an unobstructed flow through them, they did not in the Merced River drainage between Nevada Fall and Echo Valley. Bunnell Point in particular was a major north-projecting monolith that stood directly in the way of the glaciers, and most of the ice was forced to make a broad meander around its north slopes (Map 23). Ice then was forced around the south slopes of a slightly smaller monolith that projected southward into the canyon. Locally, such as along the north base of Bunnell Point, the canyon is so narrow that constructing a trail through this stretch was quite difficult. Past glaciers, despite being on the order of some 2000 feet thick in this vicinity, barely widened the canyon. They certainly were poor erosive agents. Nor did they significantly deepen the canyon. Extending east about ¼ mile from Bunnell Cascade (Photo 100) is an inner gorge similar to those in Tenaya Canyon and in the Clark Fork and Middle Fork Stanislaus River canyons. In Chapter 22 I present evidence for the antiquity of these inner canyons and for the inability of glaciers to erode down through resistant bedrock. About 4 miles down-canyon, in the vicinity of Nevada Fall, much of the ice was forced to flow between the sides of Liberty Cap and of Mt. Broderick. That these features still stand, despite lying squarely in the path of mammoth glaciers, is additional testimony to the resistance of monoliths to glacial erosion.

What the glacial evidence in this lengthy chapter demonstrates is that when the moraines, till, and erratics are interpreted by the criteria I developed in Chapter 14, then realistic ice surfaces are produced for the Tioga, Tahoe, and pre-Tahoe glaciers. In contrast, Matthes was able to produce a realistic ice surface only for his El Portal (pre-Tahoe) glacier. And Wahrhaftig was *unable* to produce a realistic ice surface for *any* of the three glaciers. No glacier in the world today has an ice surface like his. Nevertheless, the USGS geologists and the Yosemite National Park personnel uncritically have accepted his interpretations.

**Photo 100. Bunnell Cascade, beneath the north face of Bunnell Point. Author's brother Greg, in this 1964 photo, provides scale.**

# 16

# Illilouette Creek Valley's Glacial History

In John Muir's sixth study in the Sierra, the "Formation of soils," he attributed the origin of all of California's deposits (his "soils") to glacial deposition, and thus all of California had been glaciated. In like manner, Matthes gave a special interpretation to gruss. Granitic bedrock decomposes to gruss, and this can be found from low elevations near the western edge of the Sierra Nevada up to gently sloped summits of high peaks such as Mt. Hoffmann, in the center of Yosemite National Park. Gruss at neither site has a glacial origin. Matthes acknowledged the gruss on unglaciated uplands, but where he believed ancient (El Portal/Glacier Point) glaciation had occurred—for example, on the Starr King bench—he interpreted it as "morainal material." But this is just gruss on level or gently sloping surfaces. One of the best examples of this is along site 1 on Map 27. As precisely mapped by Matthes on his Plate 29 of glacial deposits, this "moraine" is 1880 feet long, but it is only a thin veneer of gruss on the bedrock lip of a hanging valley (Photo 26, Chapter 8).

Many of Matthes' mapped "lateral moraines" descend southwest toward the floor of lower Illilouette Creek valley, indicating that the El Portal glacier from Little Yosemite Valley overflowed into it. In his Professional Paper 160 he stated (p. 63):

> The older moraines indicate, further, that at the time of maximum glaciation the Illilouette Glacier and the opposing lobe met and united, forming a broad sea of ice that eventually rose high enough to find an outlet westward into the Bridalveil Basin. Later, when the Illilouette Glacier and the lobe melted apart, there was formed between them a temporary lake—ancient Lake Illilouette it may be called. The gravel and sand deposited in this body of water remain to attest its former presence.

In the largest of the pre-Tahoe glaciations, lower Illilouette Creek valley must have been under considerable ice, for the highest erratic on the east slope of Sentinel Dome, site 2 (upper left edge of map), is at about 7930 feet, which is about 2100 feet above the brink of Illilouette Fall. Nevertheless, no evidence of a temporary lake exists. Photo 27 (Chapter 8) was taken at Matthes' site of his best lake sediments, his Plate 32A. As elsewhere in the lower part of Illilouette Creek valley, this spot, site 3 (lower right corner of map), is a granitic bench with a thin, discontinuous veneer of gruss. This bears no resemblance to true lake sediments, as in Photo 101. The south side of the bench, site 4, which is composed of slopes above the north bank of Illilouette Creek, also lacks lake sediments and has instead glacial-outwash deposits (Photo 102). These slopes are eroding, so the stream-rounded cobbles are exhumed. Their fresh nature indicates a Tioga age.

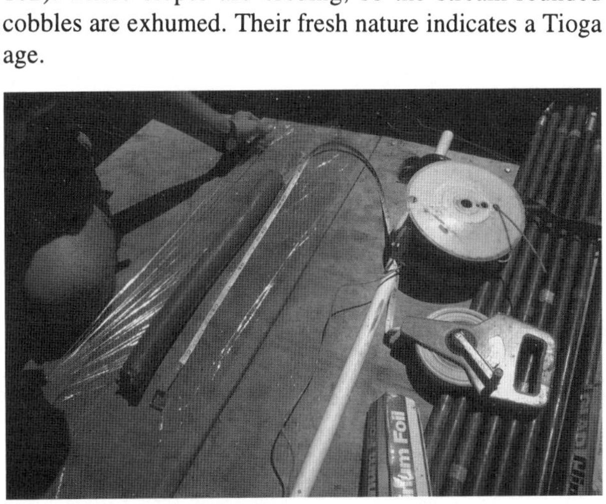

**Photo 101. Wrapping up lower sediments from Summit Lake, North Fork Stanislaus River drainage. These are typical of Sierran lakes: dark brown, organic-rich sediments transitioning to basal, gray glacial flour.**

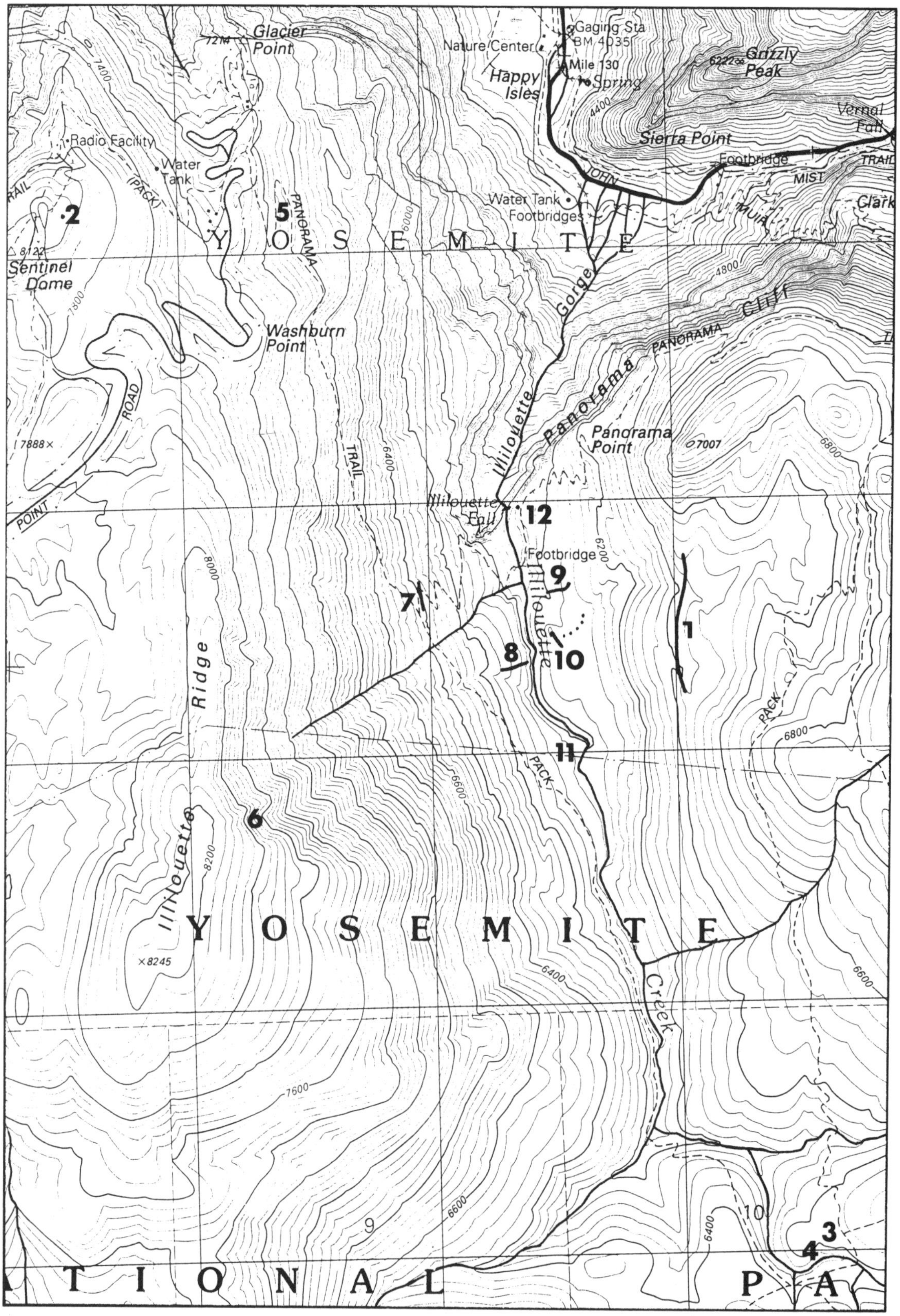


Map 27. Lower Illilouette Creek valley. Scale 1:24,000.

**Photo 102. Cobbles in glacial-outwash deposits.**

Both west and north from this vicinity I could find neither lake sediments nor pre-Tahoe moraines or erratics. I am unaware of any Sierran lake sediments indicating that a glacier advanced *up* a side canyon and impounded a lake. There is one Sierran site where a glacier should have spilled into a side canyon or at least should have blocked its mouth and impounded a lake, but it did not. This site is at the confluence of the Clark Fork with the Middle Fork Stanislaus River (Dardanelles Cone 15' quadrangle). My observations of glacial erratics and one striation along the Sierran crest between Boulder Peak and summit 10379 (Sonora Pass 15' quadrangle) indicate that during the Tioga glaciation the East Carson River glacier spilled west into the Clark Fork drainage in a swath that was about 2.5 miles wide, greatly augmenting that canyon's glacier. As the Tioga glaciation waned, the East Carson River glacier's overflow rapidly diminished, and the volume of ice in the upper Clark Fork drainage decreased dramatically. (This decrease was similar to that of the waning Mokelumne glacier's overflow into the North Fork Stanislaus River drainage, mentioned in Chapter 14.) The Middle Fork glacier was not fed by any major overflow, so it did not experience a similar substantial reduction. There is no evidence that it advanced up the Clark Fork canyon. Rather, stream sediments were deposited in this vicinity (where the Fence Creek Campground is situated on level ground). These are best viewed 1/3 mile up the Clark Fork Road past its junction with Road 6N06, which is just past the Clark Fork bridge (Photo 103). There the bedded sediments have abundant cobbles, which argues against a lake-sediment interpretation. In a north-side Clark Fork Road cut between bridges across the Middle Fork and the Clark Fork the stream sediments are in direct contact with glacial deposits (Photo 104), indicating that these sediments were accumulating along the edge of a glacier that was not spilling up-canyon. Instead, the edge of the glacier was fixed and the Clark Fork Stanislaus River must have been flowing west along its edge.

In lower Illilouette Creek valley (the area shown on Map 27) Matthes' map has only El Portal (pre-Tahoe) glacial evidence. However, where I checked his considerable evidence it was mostly nonexistent, being either gruss or colluvium, and the rest was misinterpreted with regard to nature and/or age of the deposit. Most of his mapped El Portal (pre-Tahoe) deposits, especially on slopes below Illilouette Ridge, lie on slopes on nearly impenetrable brush. I was fortunate during the summers

**Photo 103. Bedded stream sediments in Clark Fork Road cut.**

**Photo 104. Contact of stream sediments with glacial deposits, north-side cut between Middle Fork and Clark Fork bridges.**

**Photo 105. View down avalanche chute lying below Illilouette Ridge. Rudy Goldstein provides scale.**

of 1992 and '93 in that a major fire had swept through the area, removing virtually all of the brush, exposing the ground, and revealing the following.

At site 5, high on the Panorama Trail, there is a broad, formerly well-forested, shallow bowl between Washburn and Glacier points. There Matthes' map indicates extensive morainal material in sheets from about 7700 feet down to about 6750 feet, but only thin soil and underlying detritus exist.

The head of the avalanche chute, site 6, is depicted (reasonably enough) as an El Portal (pre-Tahoe) cirque glacier. However, the bedrock looks fresh enough to have been glaciated during the Tioga (Photo 105). The top of the northeast-facing headwall is at about 8160 feet, which is an elevation virtually identical to the top of a cirque headwall 3.9 miles to the south, on the northeast-facing slopes of Horizon Ridge, where Matthes mapped a Wisconsin-age cirque glacier. To be consistent then, one would conclude that the head of the avalanche chute is also a Wisconsin-age cirque. When I examined this site, I quickly found five patches of sublinear striations on feldspar dikes between 7700 and 7800 feet. The key word here is "sublinear." Glacial striations should be straight, not slightly meandering. Consequently, the striations could have been imparted by slowly moving boulders, with or without a snowfield, or by boulders in a slowly moving cirque glacier. The striations are somewhat similar to almost equally small ones found near the upper limits of Tioga-glaciated bedrock in the Paradise Lake area, north of Donner Pass (Chapter 13). Lack of moraines argues against a cirque glacier. Still, a snowfield forms every winter, and rocky debris breaking from the wall above should accumulate at the base of the snow, but instead it continues down the avalanche chute. Likewise, debris from a cirque glacier would have continued down the avalanche chute. Without additional field evidence, I cannot say with certainty that a Tioga-age cirque glacier existed.

At site 7, just below a trail junction, Matthes' map shows a linear deposit of morainal material, yet the colluvium there is no different from that which surrounds the postulated deposit. Till could not remain in place for hundreds of thousands of years on a slope with a 40+% gradient—this denies the existence of soil creep. And site 8, a proposed end moraine from the El Portal (pre-Tahoe) Merced glacier advancing southward, up-canyon, is no more than colluvium bordering site 6's active avalanche chute. Site 9 is another proposed end moraine from the same glacier, but I could not find any sign of a moraine crest anywhere. On this flat there are extremely abundant boulders, probably fallen from above, which are mostly sub-angular and of all sizes up to 10 feet in diameter. Beneath them, as seen in steep exposures along Illilouette Creek, is bouldery glacial outwash.

The only moraine-like deposit is at site 10. This is a linear, northwest trending, sharply defined, 300-foot-long

**Photo 106. Tioga-age moraine-like deposit in lower Illilouette Creek valley.**

deposit of several hundred large, fresh, mostly subangular boulders up to 6 feet in diameter (Photo 106). The deposit's surface gains about 3 feet of elevation to the northwest. (Matthes' map incorrectly shows the deposit starting east from the brink and then curving northward, which would have been the case *if* it had been left by a south-advancing glacier.) Outwardly it appears to be a Tioga-age moraine, and it does not appear to be a *recent* rockfall deposit for the following reasons. First, the slopes east of and above it are too low (about 350 feet) and too gentle (about 35% gradient) to generate a rockfall that would be sufficiently large and that had sufficient momentum to slide about 550 feet west across a horizontal bench. Second, the linear shape of the deposit is not typical of rockfalls deposited on broad, level benches. Third, its long axis is oriented northwest rather than east, as it should be if it had originated from eastern slopes. And fourth, while the avalanche chute high above to the southwest (site 6) could generate such a rockfall, it likely would not end atop a bench that is about 120 feet above Illilouette Creek.

However, there is evidence against site 10 being a moraine. First, a *south*-advancing Merced glacier up lower Illilouette Creek valley could not leave an end or lateral moraine with a northwest bearing. Nor could a *north*-advancing glacier have left such a narrow moraine with this bearing. The boulders of this site are too fresh to be anything but Tioga or younger. But Illilouette's Tahoe glaciers, which were slightly longer than its Tioga glaciers, advanced no farther than to within about 3 miles southeast of site 3. Second, this deposit of fresh-looking boulders resembles the fresh, Tioga-age moraine at the junction of the Glacier Point-Panorama and Mono Meadow trails (site 7 in the "Diving Board, Liberty Cap, and Panorama Cliff" area)—only with far more boulders and without matrix. All Tioga-age moraines I have examined have had sand, gravel, and cobbles mixed with boulders. Third, if a moraine from a Merced glacier, then it should contain boulders derived from bedrock of the upper, southern part of the Merced River drainage. A significant amount of the bedrock there is metavolcanic, leucogranite, and leucogranite porphyry, and since these are very different from the Half Dome granodiorite of lower Illilouette Creek valley, they should have been readily apparent. But of the 100+ boulders I examined, all appeared to be locally derived and essentially of the same rock type. Finally, if the deposit were a moraine, then the Merced glacier must have spilled south into the lower part of Illilouette Creek canyon to at least this location. Doing so, it would have dammed Illilouette Creek. As at the confluence of the Clark Fork with the Middle Fork Stanislaus River, sand-rich stream and/or lake sediments then would have accumulated and would have been preserved, but I found none. Also of relevance, the evidence at that confluence clearly indicates that the Middle Fork glacier did not advance up the Clark Fork canyon, despite absence of a lake to block it, as was also the case at the mouths of the Horse, Snow, and Grouse hanging-tributary valleys above the waning Mokelumne glacier, mentioned in Chapter 14.

This enigmatic deposit can be explained as a *Tioga-age* rockfall. It was generated from a west-slope avalanche chute, slid across the canyon's lower slopes, which were buried under Illilouette Creek's glacial outwash, and then came to rest on a bedrock bench. Rounded pebbles and small boulders on the bench demonstrate that a shallow layer of outwash had buried it, as does outwash that extends 0.3 mile upstream where, at site 11 above the west bank, it occurs up to at least 6050 feet elevation. The rockfall would have deflected the Tioga-age Illilouette Creek eastward around it, and then this creek could have eroded part of the deposit to create its present northwest orientation. Rounded pebbles and cobbles scattered among the generally subangular boulders indicate that the creek was high enough to deposit them.

At the rockfall site the edge of the bench is steep-sided, but to the north it is less steep, and abundant, large, rounded outwash boulders remain intact on the slope, and they can be traced down-canyon to the vicinity of the brink of Illilouette Fall, which officially is at 5816 feet. Immediately east from the brink the outwash gravels (*rounded* pebbles, cobbles, boulders—Photo 107) are very conspicuous up to the Glacier Point-Panorama Trail, site 12, and along it I traced them about 370 feet northeast to an elevation of about 5950 feet. The highest gravels there are significant, since they lie well above and beyond the brink of the fall, far too high to have been deposited during modern, post-glacial floods. The only way they could have been deposited there was if the ice surface of the Tioga-age glacier had been that high, damming Illilouette Creek and causing it to drop its sediments. If the ice surface of the Tioga-age glacier had been below the brink of the fall, as Matthes mapped the Wisconsin ice surface (at about 5750 ft), then outwash sediments would not have accumulated, but rather would have been transported over the fall's brink. If the ice surface had been much higher, the glacier would have left a diachronous lateral moraine on this east side, just like the ones left on the east sides at the mouths of Horse, Snow, and Grouse hanging-tributary valleys in the Mt. Reba

**Photo 107. Glacial outwash lying just beneath the Glacier Point-Panorama Trail.**

area. Lack of a moraine suggests that the highest gravels were very close to the ice surface of the Tioga-age glacier, which would have been about 5950 to 6000 feet.

A final point to be made is the absence of any convincing Tahoe or pre-Tahoe glaciation in lower Illilouette Creek valley. One would expect a pre-Tahoe glacier to have left at least a few large erratics on the broad ridge above east slopes, but I found none. Matthes identified El Portal (pre-Tahoe) erratics on his map, but he also showed none on the ridge. Were it not for the erratics on the east slope of Sentinel Dome (site 2), and on the saddle between Mt. Starr King and summit 8574 (Map 26, site 20), the highest ice surface of the pre-Tahoe glaciers would be very speculative. The only evidence I could find of a Tahoe trunk (Merced River) glacier advancing up lower Illilouette Creek valley is at site 11, where there are some exposed boulders that are weathered through to the core (i.e., classic Tahoe). However, these are more likely just the few remnants of Tahoe outwash produced by Tahoe glaciers that existed upstream. Hence there is no real evidence of a Tahoe trunk glacier advancing up lower Illilouette Creek valley. Furthermore, because Tahoe moraines and erratics were not found at the brink of the valley, I infer that the Tahoe glacier's ice surface, as at previously mentioned sites to the east, was slightly lower than the Tioga-age glacier's ice surface. Again, this is an important relationship, since throughout the range the Tahoe glacier was a bit thicker than the Tioga. It was here too, but in Yosemite Valley and just east and northeast of it, the Tahoe's ice surface was slightly lower. Can a thicker glacier have a lower ice surface? Yes, under *special* circumstances, as we'll see in Chapter 19.

In summary, Illilouette Creek valley's glacial history is generally straightforward: Tioga and Tahoe glaciers from the highlands of the Clark Range and the Buena Vista Crest advanced to within no more than 5 miles from Illilouette Fall. The valley's lower lands in those 5 miles were glaciated only by one or more pre-Tahoe glaciers. Glacial outwash near the fall demonstrates that the Merced River's Tioga glacier was thick enough to rise slightly above the brink of Illilouette Fall, and in doing so, it dammed the creek. Local burial of the lowest ½ mile of the valley floor beneath this outwash then permitted a large, Tioga-age rockfall, descending from high slopes of Illilouette Ridge, to sweep east across the outwash and end up on a granitic bench. It may be the only *known Tioga-age* rockfall in the Sierra Nevada.

# 17

# Tenaya Canyon's Glacial History

On June 25, 1863, members of the Whitney Survey came upon Tenaya Lake, and from it they proceeded northeast to Tuolumne Meadows, noting abundant glacial evidence in this area. In the next decade John Muir documented in his second Sierran study the direction, thickness, and length of this Tenaya Canyon glacier. Like others who have come after them, all have recognized that an immense lobe of the mammoth Tuolumne glacier flowed across a low pass separating the meadows from the lake, and then continued down to Yosemite Valley. In his Professional Paper 160 François Matthes presented a very believable reconstruction of the Wisconsin glacier. This later was divided into the Tahoe and Tioga glaciers. How large was each?

In this area I limited field work to south of Snow Flat (top edge of Map 28) and west of Tenaya Creek. Lands east were not investigated (except superficially) because the broad Tioga-age glacier originating from Sunrise Mountain (mostly north of Map 22) flowed west and over the east brink of Tenaya Canyon, so I deemed it virtually impossible to find any evidence there that might show relations among the canyon's Tioga, Tahoe, and pre-Tahoe glaciers. My field work principally was done to determine whether the ice surface of the area's Tioga glacier was slightly higher than that of the Tahoe.

Olmsted Point, site 1, is an overly popular stop (post T24 along the Tioga Road) that provides tourists with an awe-inspiring view southwest down Tenaya Canyon to Clouds Rest and Half Dome and an equally inspiring one northeast to Tenaya Lake, flanked by Polly Dome on the west and Tenaya Peak on the east. Virtually all the lands seen here have been glaciated, but how thick were the past glaciers? Abundant fresh polish and striations at Olmsted Point leave no doubt that it was overridden by a Tioga-age glacier, and these exist in diminishing amounts up to within about 40 feet below a 8900-foot summit, site 2, 0.7 mile north of the point. Northwest across the canyon, at site 3, which is the same elevation as site 2, is the highest striation (bearing 187°). Abundant polish and striations lie about 20 feet lower, east of site 3. No glacial evidence is in the vicinity of summit 9206, site 4. If the Tioga-age glacier flowed sufficiently to impart polish and striations just east of and below site 3, here it must have been at least 150 feet thick, the approximate thickness of the upper layer of brittle ice. Consequently the ice surface at sites 2 and 3 must have been at least 9050 feet.

Did the Tioga-age glacier override summit 9206? Its summit is dotted with solution pockets, the deepest one, in the foreground in Photo 108, having a depth of about 9 inches. This argues either for no glaciation in a very long time (the conventional interpretation) or for a covering of thin, essentially stagnant ice that did not affect the older surface, as I have proposed at sites in the Little Yosemite Valley area and as was definitely the case at Lake Tahoe's Emerald Bay knoll. Site 5, midway along a linear ridge, has even more convincing evidence of ancient glaciation, where the pits are up to 15 inches deep. But none of the erratics along that ridge are on pedestals, which argues, though not convincingly, for a younger age. As on Moraine Dome's summit area, the crucial evidence at sites 4 and 5 may be the evidence that does *not* exist: had a Tioga-age glacier flowed past both sides of summit 9206 rather than overriding it, the glacier *might* have left a conspicuous lateral moraine along each side. I say "might" instead of "should" because the late-Tioga equilibrium-line altitude in this area, as it was east of and above Little Yosemite Valley, was at about 9000 feet elevation, if one judges by the highest conspicuous moraines, so Moraine Dome was below the ELA, but summit 9206 was above it. Consequently my conclusion—that the summit and the linear ridge south from it were mantled under thin Tioga ice—is somewhat tenta-

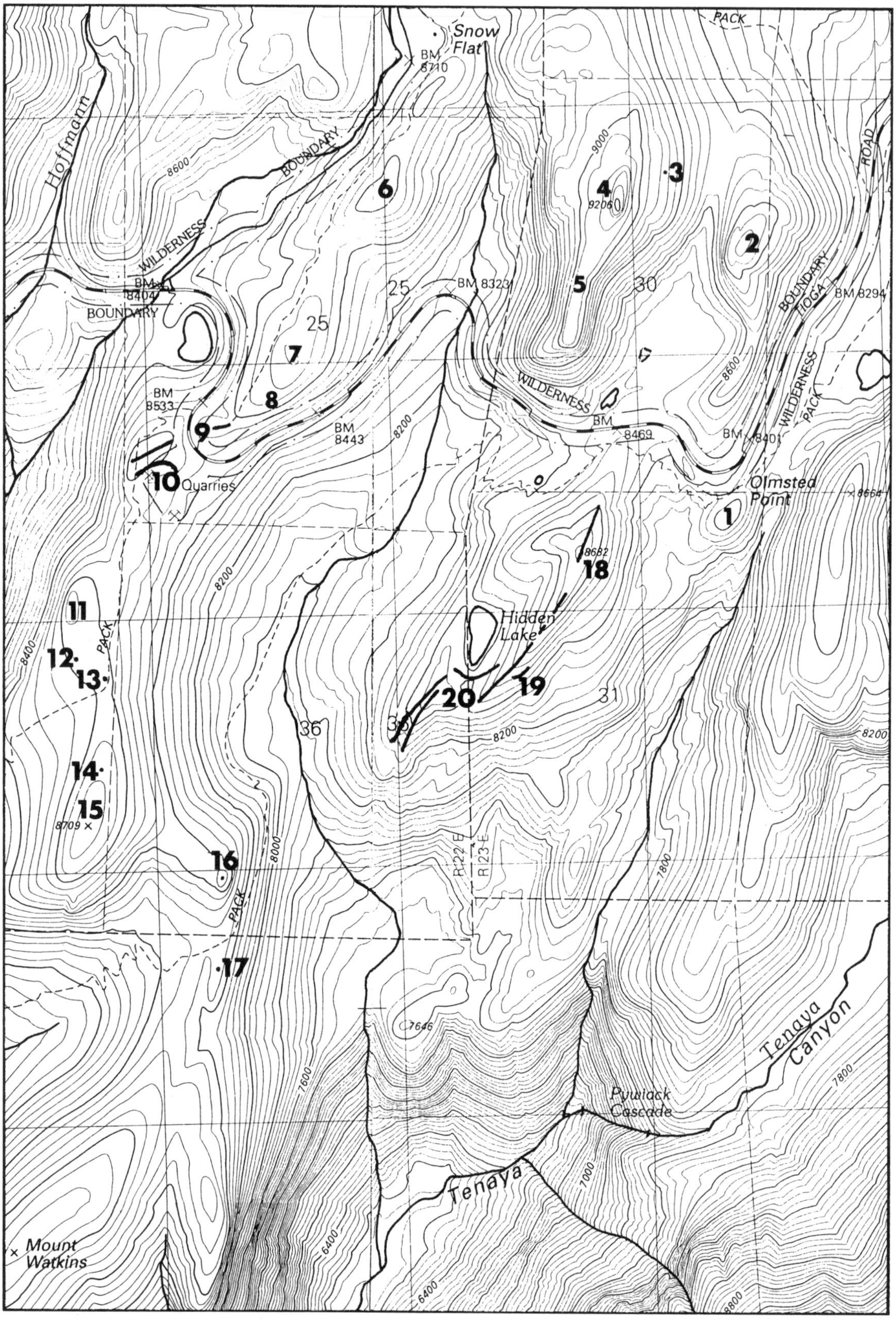


Map 28. Tenaya Canyon. Scale 1:24,000.

**Photo 108. Solution pockets on summit 9206. In the background, Clouds Rest (left) and Half Dome (right) delineate the east rim of Tenaya Canyon.**

tive. Nevertheless it is supported by the presence of almost completely barren bedrock on the summit area and on the slopes surrounding it. Had the summit not been glaciated since Tahoe or pre-Tahoe time, conspicuous patches of soil should abound in spots protected from strong winds, but none exist (Photo 109).

In contrast, Wahrhaftig reconstructed the Tioga ice surface just below summit 9206 and the linear ridge south from it. And Matthes reconstructed his Wisconsin (Tioga/Tahoe) ice surface just above summit 9206 and below the linear ridge, leaving the section from about site 5 southward ice-free. He shows this ice surface plunging steeply, at about a 26% gradient, despite the valley floor west of the ridge having only a 3½% gradient and the valley floor east of it being essentially flat. This indicates severe thinning, which would not occur near the ELA.

West of summit 9206 is a ridge that descends from site 6 to site 9, where the Tioga Road (at road post T22) cuts through the park's most easily studied Tioga-age moraine. The cut is quite deep, as shown in Photo 110, indicating that *there* it is not just a thin veneer of deposits on a bedrock ridge. However, this spot is an anomaly. The broad, nearly level summit area of site 7 is exposed bedrock with scattered erratics, and site 8 is a bedrock bench. Till does exist nearby along the descending ridge, but it must be thin. The moraine is thick at site 9 simply because it is a low point on the ridge, and being low, the Tioga Road fortuitously was constructed through it.

The curved line, site 10, marks the southern edge of the morainal deposits, beyond which are only a few isolated erratics. This is perhaps the most crucial site to determine the relative thicknesses of the Tioga, Tahoe, and pre-Tahoe glaciers. The till down to site 10 is unquestionably Tioga, but south of that curved line the undulating, broad ridge has only a few erratics. About 650 feet south of the edge of morainal deposits, I found only three erratics along the broad ridge, these at sites 12, 13, and 14 (maximum dimensions of each, respectively, 6, 20, and 6 feet). No erratics existed on the summit areas of sites 11 and 15. Was this broad ridge mantled under thin Tioga ice, or were these erratics left by an earlier glacier?

Had the broad ridge been mantled under thin Tioga ice, then when the glacier began to wane, it should have

**Photo 109. View west from site 2 toward summit 9206's barren slopes. Highest glacial polish is about halfway up the slopes. Mt. Hoffmann is in the right background.**

deposited conspicuous recessional lateral moraines along each side of it, but none exist. Furthermore, on the subordinate ridge extending southeast down to a point, site 16, there should be evidence of a moraine or at least of till. Its western part has a few large erratics, and its eastern part has dozens of them just before the high point, which has a giant one (25 feet), but nowhere is there a moraine or till—only gruss. Finally, south of and below that point is a linear ridge with a 10-foot erratic at site 17. This bedrock ridge, at about 8250 feet elevation, is well below the ELA, and had it been overrun by either Tioga or Tahoe ice, it would have had a diachronous lateral moraine deposited on it, but none exists. Furthermore, the large erratic has solution pockets up to 6 inches deep and rests on a pedestal up to 6 inches high. The moraine's absence plus the erratic's solution pockets both argue for a pre-Tahoe glaciation. And if site 17 stood above the ice surfaces of Tioga and Tahoe glaciers, then obviously the significantly higher sites from 11 through 16 also must have stood above them. That they have only large erratics (typically 6+ feet) is another indication that they were left by a pre-Tahoe glacier.

The highest elevation of the Tioga ice surface on slopes below and northeast of site 16 appears to be at about 8360 feet elevation, this based on the highest elevation of abundant boulders as well as what I interpret as late-Tioga lateral moraines northeast across the canyon at site 20 (mentioned later). This elevation is about 110 feet *above* site 17, but that site apparently escaped glaciation because in the vicinity of site 16 the floor of the adjacent tributary canyon begins a 410-foot plunge to a bedrock flat, and the ice surface of the glacier also would have plunged. Thus in this vicinity the highest possible elevation of the Tioga-age glacier is no more than about 8200 feet elevation.

My examination of evidence south of the Tioga Road in the Hidden Lake area created additional disagreement with glacial reconstructions by Matthes and Wahrhaftig. The ascent south up to summit 8682, site 18, is along a bedrock ridge with a mostly continuous cover of thin till (a down-canyon-*ascending* diachronous late-Tioga moraine). Exhumed and partly exhumed boulders of this deposit are all fresh, indicating a Tioga (or late-Tioga) age. The actual summit must have been quite close to the ice surface of the Tioga-age glacier, for this elevation is very close to that of the Tioga-age moraine between sites 10 and 11. From the summit a ridge extends west, and it is abundantly covered with fresh (i.e., Tioga) boulders. Also from the summit the diachronous late-Tioga moraine descends fairly continuously south-southwest along a bedrock ridge.

The only site that could be interpreted as Tahoe or pre-Tahoe is slightly below and just east of the ridge at site 19, where there is a narrow, 150-foot-long, gently sloping bench with a veneer of gruss. Along its east edge stands a 10-foot-high column (Photo 111) that under conventional wisdom argues for a very long period (i.e., since pre-Tahoe) of post-glacial weathering and mass wasting. However, if it lay under thin, nearly stagnant ice, erosion may have been minimal. I am not suggesting that this column has remained unchanged since the Tioga glaciation, but rather that a larger mass of rock of unknown size stood above the bench during that glaciation, and what remains today is a smaller piece of it. Neither Matthes nor Wahrhaftig mapped this site as pre-Tahoe. Matthes showed only a small area around the site 18 summit area—that part above 8640 feet on the modern Tenaya Lake 7.5' quadrangle—as being above the Wisconsin (Tioga/Tahoe) ice surface.

In contrast, Wahrhaftig, in Alpha *et al.*'s map of the park's Tioga glaciers, depicted all of this ridge as being above the Tioga ice surface. That interpretation is contradicted in his mapped field evidence on Huber *et al.*'s geologic map of the park, which clearly shows a continuous band of Tioga till starting at about 8600 feet and extending north-northeast up to site 18's 8682-foot summit and then along its descending, west-trending ridge, and from it down to and across the canyon floor.

**Photo 111. A 10-foot-high column south of summit 8682. Half Dome is in the right background, the southwest shoulder of Clouds Rest is in the left.**

Additionally, he shows Tahoe till within and below this arc, extending from the north edge of Hidden Lake southward down to about the 8000-foot elevation. There the Tahoe till lay about 200 feet below the Tioga ice surface as shown in Alpha *et al.* Besides the conflicting data, there are three additional problems with Wahrhaftig's Tahoe till. The first is that the Tioga glacier, as mapped in Huber *et al.*, was thick enough to completely overrun this older till, and yet its deposits were not eroded away. The second is that all dated glacial lakes in the Sierra are

**Photo 110. Fresh boulders exposed in a Tioga Road cut through a moraine.**

shown to have existed for 16,000 years or less, and their sediments range from about 10 to 30 feet in thickness. If Tahoe, then Hidden Lake would be the first pre-Tioga lake to have been discovered. At the post-Tioga usual sedimentation rate of about 1 foot per 1000 years for a lake this size, Hidden Lake should have accumulated about 130 feet of sediments since the end of the Tahoe glaciation about 130,000 years ago. Because the lake appeared to me to be about 30 feet deep, the total depth of its basin then should be about 160 feet. For a lake this small, this basin depth strains credulity. The third problem is that the Tahoe till south and southwest of the lake contains extremely bouldery, very fresh, sharp-crested, classic-Tioga lateral moraines (Photo 112, site 20). Because the western one rises almost to the highest point at site 20 before descending southwest, it, like the longer, higher diachronous moraine across summit 8682, must be a late-Tioga diachronous moraine.

Below the Hidden Lake area, the floor of Tenaya Canyon makes a major drop of about 1300 feet in just 0.6 mile. Of those who have commented on the geomorphic or glacial history of Yosemite Valley, only two—John Muir and I—have traversed the entire length of this canyon, and when I did so back in 1977, I saw no significant glacial evidence in this lower part. As elsewhere in the Sierra Nevada, glacial polish and striations are uncommon below about 6000 feet and are virtually absent below 5000 feet. For example, along the Merced River between Nevada Fall at the west end of Little Yosemite Valley and Happy Isles at the southeast end of Yosemite Valley, I have not seen polish and striations. Consequently, a search for them on the lower and middle slopes of Tenaya Canyon (via multiple roped ascents or descents) is likely to be unproductive. On the other hand, fresh till may exist on minor benches, and these may be discovered in a future major climbing effort. Without further evidence in Tenaya Canyon proper, one is left with the relatively unconstrained Wisconsin (Tioga/Tahoe) ice surfaces of Matthes and Wahrhaftig.

West of Tenaya Canyon and just beyond the west edge of Map 28 is the hanging tributary canyon of Snow Creek. Matthes' Plate 39 shows extensive El Portal (pre-Tahoe)

**Photo 112. View southwest along two Hidden Lake moraine crests.**

morainal material in sheets along the lower slopes, but as elsewhere, I found only colluvium. He stated (p. 79 and shown on Plate 39) that the Wisconsin (Tioga/Tahoe) glacier "ended fully 2 miles north of the brink of Tenaya Canyon", which is plausible, but I did not check that location. In contrast, Wahrhaftig in Huber *et al.* mapped Tioga till near the brink east of and above Snow Creek, indicating that the Snow Creek glacier must have joined the Tenaya Canyon glacier. In that vicinity I could not find any till, moraines, polish, or striations, and so Matthes' interpretation seems more correct. Paradoxically, Wahrhaftig in Alpha *et al.* reconstructed the Tioga-age Snow Creek glacier as ending about 2 miles north of the brink, just as Matthes had.

Finally, in this general area of Tenaya Canyon and vicinity I could not find any verifiable Tahoe moraines or till, indicating that, as in the western part of Little Yosemite Valley and in the lower part of Illilouette Creek valley, the Tioga glacier in this area had an ice surface slightly higher than that of the *thicker* Tahoe glacier. My reconstructions of the Tioga and Tahoe ice surfaces are similar to Matthes' reconstruction of his comparable Wisconsin ice surface. As in Little Yosemite Valley, the glacial evidence in the Tenaya Creek drainage, as mapped by Wahrhaftig, has serious problems.

# 18

# Yosemite Valley's Glacial History

Yosemite Valley is perhaps the best known, most studied glaciated valley in the world. About half of the 129-page text of Matthes' Professional Paper 160 is devoted solely to the valley's glacial history, and the remainder is essentially a reconstruction of the valley's topography before glaciation, done so that the changes wrought by glacial erosion could be determined quantitatively. Ironically, glacial evidence that would limit the depth and length of Tioga and Tahoe glaciers advancing through this valley is very sparse. Fresh Tioga-age glacial polish and striations apparently occur only in two locations within the valley proper: just above the base of a cliff (below Rixon's Pinnacle) situated west of and above Leidig Meadow along the north wall of the central part of the valley (site 8, west edge of Map 29), and on Ranger Rock (a.k.a. Manure Pile Buttress), situated between El Capitan and Eagle Creek (site 1, east edge of Map 30). There, polish exists on a flat bench just below the actual summit as well as just above the rock's south base, the start of several popular rock climbs. Matthes mentioned a *possible* third site, at Artist Point, which is discussed later in this lengthy chapter. From the valley floor, you can imagine seeing abundant polish high on the nearly vertical walls of the valley, but despite at least 70 different roped ascents up these walls from 1964 onward, I never found a single patch of glacial polish. What you see from below is water polish. This appears to be the case for Glacier Point Apron, the curving lower half part of the cliff system below Glacier Point, where on various roped ascents I have found only water polish. Consequently, when Huber stated in his *Geologic Story of Yosemite National Park* (p. 50) that it has glacier polish, most likely he was viewing water polish, the two being indistinguishable from a distance.

## Medial Moraine

When the Tioga glacier retreated up Yosemite Valley about 14,000 years ago (this age based on radiocarbon dates on basal sediments of Sierran lakes), it did so sporadically, leaving several recessional end moraines where its front briefly stagnated between Bridalveil and El Capitan meadows. Its front then apparently retreated several miles without any stagnation, since no moraines occur east of El Capitan meadow until at the far end of the valley. This is a typical Sierran deglacial pattern, as noted by Malcolm Clark back in 1976, and as can be inferred from Joseph Birman's detailed 1964 map of part of the San Joaquin River drainage. At the east end of Yosemite Valley is a large moraine, which is parallel with the length of the valley rather than perpendicular to it. On some valley maps, including the USGS's 1990 Half Dome 7.5' topographic quadrangle, this is identified as "Medial Moraine" (Map 29, site 1), since before Matthes did his glacial field work, it was believed to be one. This view is old, dating back to at least Whitney (1865, p. 422). Unfortunately, the name stuck, even though Matthes argued convincingly that it is not a medial moraine, since such a moraine is composed of only the deposits atop two merged glaciers (here, the Tenaya and Merced) when they finally melted away. This is quite different from lateral moraines, which can be constructed over many years of deposition. In short the thickness of the "Medial Moraine" is far to great. This easily is the valley's largest moraine, and Matthes suggested that it is an end moraine left by the Tenaya glacier, the Merced glacier, or both.

Along two lines of evidence he concluded that this *recessional* end moraine was left by a glacier from Tenaya Canyon. First, this west-extending moraine seems to

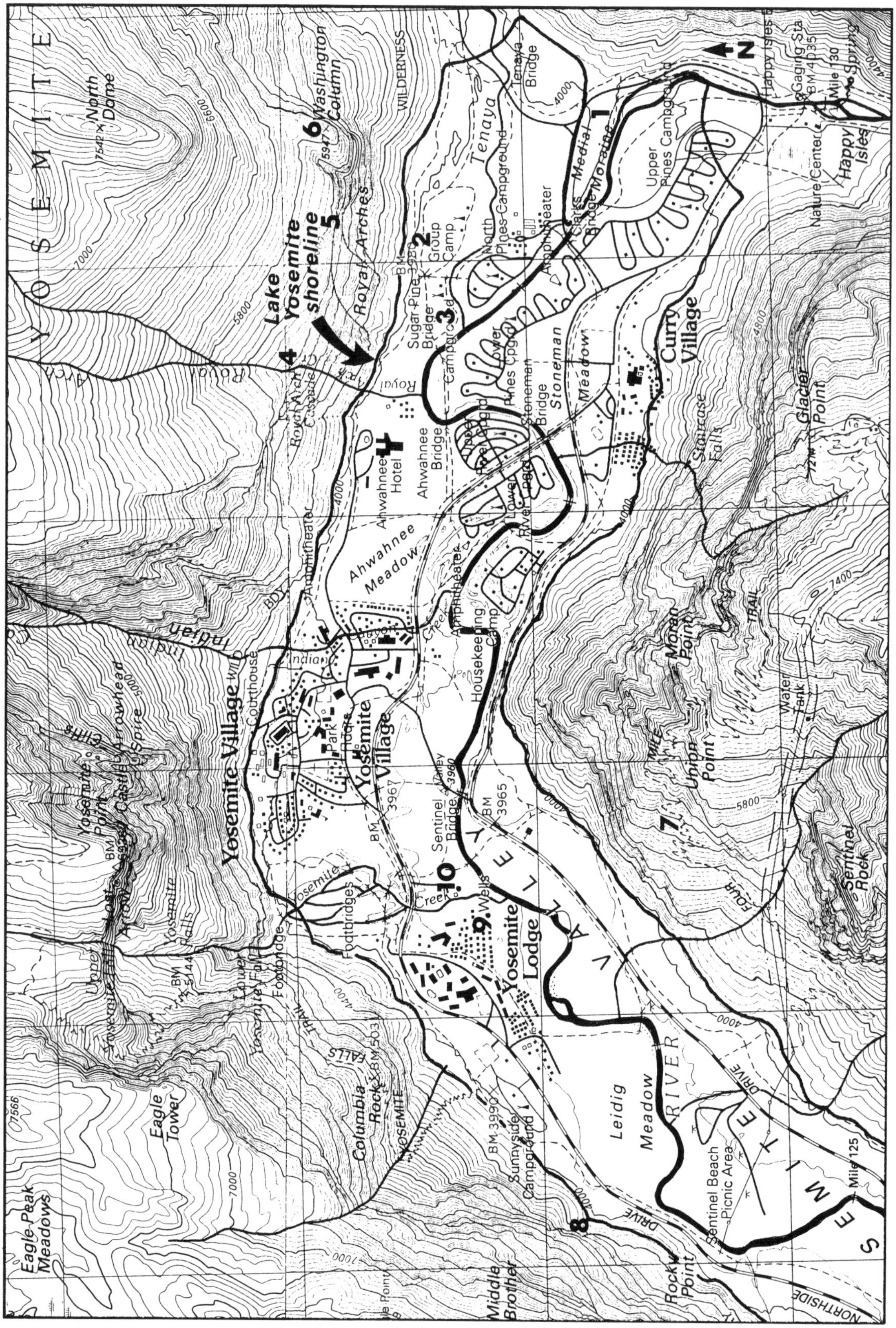


Map 29. Yosemite Valley, east part. Scale 1:24,000.

curve slightly northward (the curvature is extremely slight), so one could extrapolate the *very slight* curvature north over to the base of Royal Arches. However, because the Merced River has been undercutting the south flank of this moraine, its shape could have changed over time, so the slight curvature proves nothing. Second, at the base of the Royal Arches talus slope Matthes found what he believed to be the north end of this recessional moraine (site 2).

However, the "Medial Moraine" at its southeast end (immediately west of the paved road) has a large, white, fine-grained, crystalline leucogranite boulder (Photo 113), which, essentially devoid of dark minerals, is almost completely composed of quartz and feldspar. This boulder could not have come from above Tenaya Canyon since there is no leucogranite bedrock in that glacier's source area. Rather, this bedrock exists in the Mt. Clark area of the upper Merced River drainage, so the moraine must be the product of the Merced (Little Yosemite Valley) glacier. One might expect a western limb of the moraine to exist in the vicinity of Camp Curry, where the ground is flat and a moraine should stand above it. However, if the Merced River originally flowed through this area and then over time its channel migrated laterally northeast to its present site, as minor western channels suggest, the river would have eradicated this limb.

Matthes' minor limb at site 2 does look like an end moraine left by the Tenaya Canyon glacier. This Royal Arches deposit is located in the Group Campground, and its north edge exists at a bike path that is located near the base of a more recent talus slope. Should you wish to visit this site 2, you will find it located about 400 feet east of the bike path's junction with an old road at the northwest edge of the campground. The deposit has a rather flat crest with a bearing of about 190°, and it dies out at the north bank of Tenaya Creek near where that creek bends from northwest to west-northwest. Important to deciphering the nature of this deposit is a large rockfall deposit about 300 yards to its west, which I call the Confluence rockfall deposit (site 3), since its distal (lower) end is at the confluence of Tenaya Creek and the Merced River. This deposit has diverted Tenaya Creek, forcing it to angle southwest over to the nearby Merced River. Gerald Wieczorek *et al.*, in a comprehensive 1992 report on Yosemite Valley rockfalls, have mapped it, as I have, as a prehistoric (pre-1850) rockfall deposit.

There are two lines of reasoning that suggest this Royal Arches deposit may be a moraine, not a rockfall deposit. First, its crest is narrow and so it superficially resembles recessional end moraines. It is linear and it protrudes about 5 feet above the valley-floor sediments immediately to the east and about 15 feet above those immediately to the west. This asymmetry is similar to that of the valley's Bridalveil Meadow moraine. Second, its surface is mostly soil; few boulders are present. One might expect a

**Photo 113. Large, leucogranite boulder on "Medial Moraine."**

rockfall deposit to be very bouldery and soil-free, as the adjacent talus slopes are.

However, the deposit is not as narrow as it looks. Actuality it is quite broad, almost as wide as it is long. When formed, it could have been easily recognized as a rockfall deposit, but then it could have changed over time as its flanks were eroded away by major Tenaya Creek floods. Furthermore, sediments collecting over time behind it would have buried its eastern part, which explains why that part stands about 10 feet higher than the western part. Thus its overall morphology and asymmetrical east-west cross section can be explained equally well by mass wasting and ensuing stream processes as by glacial processes.

Three lines of evidence supporting a rockfall origin for this deposit are its lack of definable erratics, its failure to extend southward across Tenaya Creek (instead, the creek flows around it), and especially its morphological similarity to the Confluence rockfall deposit. If it had been derived from deposits of a Tenaya Canyon glacier, then some of the boulders should be of Cathedral Peak granodiorite, yet there are none. However, since there are not many boulders exposed in it, absence of Cathedral Peak granodiorite boulders does not absolutely negate the possibility of this deposit being a moraine, since the number of boulders may be too small to be a statistically valid sample.

The deposit reaches its high point at its distal end, just above a north-bank meadow, against which it ends abruptly, not continuing above the south bank of Tenaya Creek. Its absence across the creek cannot be explained by burial under an alluvial fan, like the one that apparently has buried low moraines below the base of Bridalveil Fall, since Tenaya Creek lacks such a fan. In contrast, the Merced River has produced a sizable one below Happy Isles.

Finally, while this deposit superficially resembles a recessional moraine in gross morphology, in detail it more closely resembles the Confluence rockfall deposit, which has a high distal end that northward slopes gently down toward the base of Royal Arches. In contrast, moraines generally rise toward the base of a valley wall. Unlike the large-block talus slopes accumulating at the base of Royal Arches immediately north of each, the two deposits are mostly soil-covered, the Royal Arch deposit distinctly though not significantly more so than the Confluence rockfall deposit. To become mostly soil-covered would have required much time, so I surmise that both are early Holocene in age, the Royal Arch rockfall deposit being the older of the two. In degree of soil development, they resemble the rockfall deposit below and southeast of the prow of El Capitan (Map 30, site 2; described by Matthes on p. 108 and mapped on Plate 3 by Wieczorek *et al.*). The Royal Arch rockfall deposit also resembles a similar one about 0.4 mile northeast above the Mirror Lake site, where Tenaya Creek is forced to curve around its distal end. I examined this soil-covered, rocky deposit, checking about 100 boulders, and none were composed of Cathedral Peak granodiorite. Had the deposit been a moraine, then about 25 to 50% should have been of this rock type, given the expansive outcrops of this bedrock along the glacier's route.

In the field, Wahrhaftig suggested to me that the low point near the proximal (upper) end of the Royal Arches rockfall deposit (located where the road crosses it) could be low due to erosion by a stream flowing along the side of the waning Tioga-age glacier. Such glacial streams exist, as I have mentioned earlier with regard to the three hanging tributary canyons in the Mt. Reba area, to the Clark Fork-Middle Fork confluence, and to Illilouette Creek by its fall. However, during the Tioga glaciation today's talus slope at the base of Royal Arches did not exist, and the edge of the glacier would have been along the bedrock base of Royal Arches, about 650 feet north of the low point. Furthermore, as just mentioned above, the Confluence deposit has a similar low point, and its rockfall origin is not in doubt. My interpretation of this distal-end high topography is that at each of the two sites, large blocks broke loose from Royal Arches and came to rest in their current locations, and these formed a dam behind which additional blocks accumulated immediately and possibly during later rockfalls.

Therefore there is no evidence to support the "Medial Moraine" as being derived in part from a Tenaya glacier. Evidence against that is based on the late-Tioga deglacial histories in Tenaya Canyon and the upper Merced canyon. Matthes' Wisconsin-glaciers reconstruction and Wahrhaftig's Tioga-glaciers reconstruction both demonstrate that the glacier descending Tenaya Canyon was fed mostly from overflow of the giant Tuolumne glacier. Under my reconstruction it was also fed (to a much smaller extent) by overflow from the Merced glacier between the Quarter Domes and Half Dome. Without these overflows, the only sources were from the east slopes of Mt. Hoffmann (the south slopes fed Snow Creek, not Tenaya Canyon) and the west slopes of Sunrise Mountain. This situation is analogous to the North Fork Stanislaus River glacier being fed largely from overflow of the giant Mokelumne River glacier. As was mentioned earlier, the deglaciation there caused rapid failure. Likewise, the Tenaya Canyon glacier, having had its source greatly diminished relatively early during deglaciation, would have failed similarly, as did the Clark Fork Stanislaus River glacier. At this time the Merced glacier still would have occupied part of eastern Yosemite Valley. The Merced glacier, which received only minimal overflows of the Tuolumne glacier, would have retreated more slowly, so that by the time it was depositing the "Medial Moraine," the Tenaya glacier was far up-canyon. (Matthes' Plate 39 shows four overflows, which east to west were: Tuolumne Pass, which was by far the largest, though small compared to the Tuolumne overflow down Tenaya Canyon; through the shallow saddle north of Rafferty Peak; through a similar one east of the Cockscomb; and through Cathedral Pass.)

## Brinks of Royal Arch Cascade and Royal Arches, summit of Washington Column

These sites limit the maximum elevation of the ice surface of the Tioga and Tahoe glaciers. Had the Tioga or Tahoe glaciers reached any of them, they would have left erratics, till, or a lateral moraine, and this evidence still would exist today. However, I could not find even a single erratic at any of these sites, and apparently neither could Matthes. Therefore, these two glaciers had an ice surface no higher than about 5400 feet at the brink of Royal Arch Cascade (site 4), no higher than about 5600 feet at the brink of Royal Arches (site 5), and no higher than about 5920 feet on the summit of Washington Column (site 6). Although the slopes rising from this north rim of eastern Yosemite Valley were mantled under pre-Tahoe ice and probably had at least a veneer of glacial deposits left upon deglaciation, they were too steep (about 50% gradient) to preserve any evidence for the ensuing few hundred thousand years. Were it not for several erratics well above them, which compare favorably in elevation with the highest erratics below the summit of Sentinel Dome, mentioned earlier, there would be no evidence that these slopes had ever been glaciated.

## Four Mile Trail till

The only Wisconsin (Tahoe/Tioga) till that Matthes

**Photo 114. Mass-wasted deposits along the Four Mile Trail at a small washout. Note the curving trunks of the live oaks—their attempt to compensate for ongoing soil creep.**

**Photo 115. Polished, striated bench, viewed from Ranger Rock's summit.**

mapped within Yosemite Valley proper is conveniently located on moderately steep, unstable slopes traversed in part by the Four Mile Trail. His map shows the highest Wisconsin till at about 5150 feet elevation, at site 7 (Photo 114), and El Portal (pre-Tahoe) till continuing uninterrupted above it from this elevation. Of thousands of boulders that exist along the trail and just off it, I inspected about one thousand to see if they were locally derived (same composition as bedrock) or were transported by a glacier (different composition). All boulders were locally derived. Furthermore, only bedrock, mass-wasted deposits, and soil exist. Till, as at most of Matthes' mapped sites, is completely absent. Apparently like Muir, he equated boulders to erratics and soil to till. Matthes' Wisconsin ice surface here is purely hypothetical.

## Ranger Rock

The only constraining glacial polish in the valley is on a small bench just below the summit of Ranger Rock (Map 30, site 1). This bench (Photo 115), at an elevation of about 4570 feet, provides a minimum thickness of the Tioga glacier—the polish and striations are too fresh to be Tahoe or older. (Actually, I have never seen any polish or striations on bedrock covered by Tahoe and pre-Tahoe glaciers.) To impart these features the glacier ice atop the bench must have been at least 150 feet thick, which places the minimum ice surface at 4720 feet. The glacier could have been much thicker, and Matthes placed the Wisconsin (Tioga/Tahoe) ice surface here at about 4900 feet. This is just below what appears on his topographic map (and on later ones by others) as a relatively broad ridge near Split Pinnacle, which, at about 5050 feet, if covered by the Tioga or Tahoe glacier, should have left evidence. In reality the ridge is mismapped. It is very narrow—too narrow to have preserved any evidence. Matthes mapped morainal material on a fairly steep ridge between Ranger Rock and Split Pinnacle, but I found none, and none would even be expected, given the instability of the slopes.

## Cathedral Rocks

Matthes' diary implies that he never visited the summits of the Lower, Middle, or Higher Cathedral Rocks, and probably never visited the saddles between them. Nevertheless, he mapped seven small patches of El Portal (pre-Tahoe) morainal material in this area. At the most easily accessible site—the deep notch at Higher Cathedral Rock's south base (site 7)—I found no trace of till. The only boulders there are easily explained by local mass wasting. The same applies to boulders at the Gunsight—the deep notch between Lower and Middle Cathedral Rocks (site 8, Photo 116). I also inspected the five remaining sites between these two, their deposits easily explained by mass wasting.

I also examined a dozen or so large boulders on the summits of Middle and Higher Cathedral Rocks that appeared to be different in composition from the bedrock they rested upon (Photo 117). In almost every case they were locally derived. The bedrock on each summit is very heterogeneous, bands of Rockslides diorite interlaced with El Capitan granite and Bridalveil granodiorite. In many instances a boulder need slide only a few yards down the gently sloping summit areas to rest on different bedrock. Boulder movement probably is painstakingly slow, requiring innumerable earthquakes to produce minor movement. Some are rounded and quite weathered, resembling pre-Tahoe erratics, and they could have broken from the bedrock tens or hundreds of thousands of years ago. Apparently, however, evidence of pre-Tahoe glaciation does exist on these summits. Matthes (p. 64-65) mentioned that Frank Calkins had found several boulders of Half Dome granodiorite, one boulder of Cathedral Peak granodiorite, and a fragment of limestone. Photo 118 is of one possible erratic, about 6 feet across, whose composition contrasts with that of the bedrock. Located

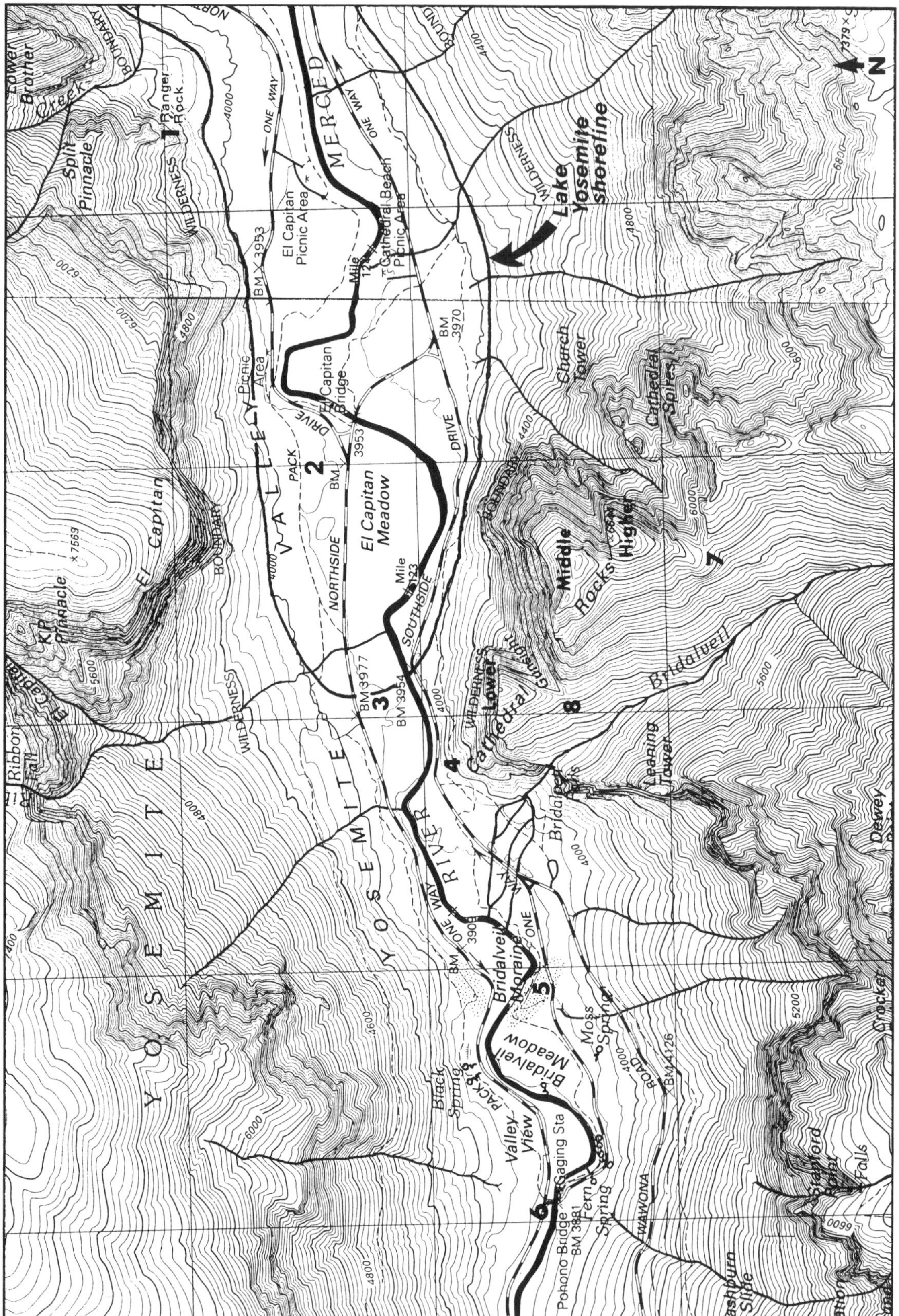


**Map 30. Yosemite Valley, west part. Scale 1:24,000.**

**Photo 116. Weathered boulders on Gunsight's notch.**

**Photo 117. A weathered boulder atop Higher Cathedral Rock. The boulder has tilted from its original orientation and has acquired a rusty rock varnish, making it appear as an erratic, but like the bedrock it is fine-grained Bridalveil granodiorite.**

**Photo 118. Possible erratic atop Middle Cathedral Rock.**

only a few yards from the actual west summit of Middle Cathedral Rock, it has not rolled very far, if it has rolled at all, so it does not appear to have a local origin.

## Upper Ribbon Creek drainage

I did not visit the upper Ribbon Creek drainage, which begins immediately above the brink of Ribbon Fall, shown along the north edge of Map 30. Matthes mapped El Portal (pre-Tahoe) moraines, till, and erratics in it, and he stated that the glacier descending the drainage was largely an overflow of the Yosemite Creek glacier. Whereas I may doubt that his early moraines existed here, as often is the case the Merced River drainage above Yosemite Valley, I must accept that at least some of his 20 mapped erratics are legitimate erratics. However, even if any of this evidence exists, it offers no constraints for the pre-Tahoe ice surface in Yosemite Valley. This is because the ice surface of this tributary glacier could have been accordant with that of Yosemite Valley's trunk glacier, *or* the tributary glacier could have cascaded down cliffs to the trunk glacier. Either interpretation is permissible, even if all of Matthes' mapped glacial features do exist.

## El Capitan Meadow moraine

This is the easternmost of several end moraines (site 3), lying immediately west of El Capitan Meadow, and hence I have given it this unofficial name. Like the "Medial Moraine," this one is controversial. It supposedly is the dam for *post-Tioga* Lake Yosemite, discussed below. Matthes considered this moraine and the one immediately downstream from it, the Bridalveil Fall moraine, to be products of a younger Wisconsin glaciation, and four others downstream from these two to be products of an older Wisconsin glaciation. In today's terminology, these moraines would be, respectively, Tioga and Tahoe. Blackwelder had considered all as Tahoe, based on surficial characteristics. However, today all are considered to be Tioga, based on unweathered subsurface boulders within them.

## Bridalveil Fall moraine

Both this moraine and the next one have been called, on different maps, the Bridalveil Moraine. To avoid confusion, I have renamed each after its adjacent local attraction, respectively, Bridalveil Fall and Bridalveil Meadow. Bridalveil Fall moraine (site 4) is a mere remnant of a moraine that is cut by the south-side valley road at a bend near the upper northeast end of a long parking area used for viewing Bridalveil Fall from the road. The cut through it exposes a matrix containing fresh—Tioga-age—boulders. Its up-canyon side is mantled by stream boulders, indicating that in the past the bed of the Merced River was about 15 feet higher—the elevation it currently has at the El Capitan Meadow moraine. Likewise, stream boulders perched on the north bank of the river between these two moraines indicate a similar height. However, these boulders could be glacial outwash rather than postglacial stream boulders deposited atop rockfall deposits in the river. Trenching may reveal the true origin of the boulders, which could be a combination of both situations.

## Bridalveil Meadow moraine

This is another controversial moraine (site 5). According to Matthes this lowest moraine marks the maximum extent of his early Wisconsin glacier, which would be Tahoe in age by modern criteria. However, the road cut exposes fresh subsurface boulders, indicating a Tioga age. My evidence from western Little Yosemite Valley, Illilouette Creek valley, and Tenaya Canyon indicates that the Tioga ice surface in these locations was higher than the Tahoe ice surface. Hence one could reasonably conclude, as Wahrhaftig did, that the Tioga glacier—*unlike any other in the entire range*—was *longer* than its preceding Tahoe glacier. (As I'll explain later, it wasn't).

If either the Tahoe or the Tioga glacier advanced only ½ mile beyond the Bridalveil Meadow moraine, that is, to the vicinity of Pohono Bridge (site 6), it would have exited Yosemite Valley proper and would have entered the upper gorge of the Merced River. At the bridge the river's meandering course gives way to a straight one that is confined both by the narrowness of the gorge and the accumulation of talus along its floor and lower slopes. Creation of any end moraine would have been unlikely, since sediments would have been swept away by the wall-to-wall river as rapidly as they were deposited. Should a minor amount of sediments actually have been preserved in spite of major flooding, they would have been buried under rockfall talus. It is not surprising, then, that no end moraines have been found in the *two* gorges below Yosemite Valley (the upper one is west-oriented; the lower, named one, Merced Gorge, is south-oriented). All agree that these unquestionably were glaciated in pre-Tahoe time, as evidence presented below indicates.

Only in the Merced River drainage has the lowest end moraine been used to mark the maximum extent of Tioga and Tahoe glaciers. Matthes (p. 55-56) stated that *by definition* the lowest end moraine in any canyon marks the maximum extent of its glacier, although he was well aware that in the Tuolumne and San Joaquin river drainages this definitely was not the case. In the Tuolumne and the South Fork San Joaquin canyons he mapped many lateral moraines on canyon rims, but mapped the lowest end moraine many miles up-canyon from the westernmost lateral moraines. Indeed, in the entire North Fork and Middle Fork San Joaquin canyons he mapped no end moraines whatsoever, although he mapped lateral moraines on their rims. Given the narrowness of each canyon's floor, no end moraines would be expected. In other major river canyons—the Mokelumne, North Fork Stanislaus, and Middle Fork Stanislaus—whose lower *glaciated* sections are gorges, no one has found any end moraines. Where end moraines exist, the lowest in each river drainage lies many miles up-canyon above the lower glaciated section, and so it does not mark maximum glacier extent. To be consistent, then, the Tioga-age Bridalveil Meadow moraine should lie many miles above the Merced's lower Tioga-age glaciated section. This is the first of *nine lines of evidence* that indicate that this moraine does not represent the maximum extent of the Tioga glacier.

Disregarding *his own mapping* of lateral and end moraines elsewhere, Matthes used this westernmost end moraine to mark the maximum extent of the Wisconsin (Tioga/Tahoe) glacier in Yosemite Valley. There is a serious problem with this proclamation, as can be seen from Table 5. Whereas the Tahoe glaciers in three river drainages descended close to elevations reached by pre-Tahoe glaciers, in the Merced River drainage the difference between Tahoe and pre-Tahoe lowest elevations is about seven times greater than the difference for the average of the three other drainages.

Clearly something is wrong with the lowest-elevations difference in the Tahoe/pre-Tahoe glaciers of the Merced River drainage. Either the lowest elevation of the pre-Tahoe glacier is much too low or the lowest elevation of the Tahoe glacier is much too high. Pre-Tahoe erratics west of Yosemite Valley support the lowest elevation of the pre-Tahoe glaciers, so the lowest elevation of the Tahoe glacier must be too high. This is the second line of evidence.

A third line of evidence that the Tioga-age Bridalveil Meadow moraine is merely recessional lies in its relative volume. If it were a terminal moraine, it would have received deposits for thousands of years—so why is it small? There is abundant talus on the valley's lower slopes east of El Capitan, and if a glacier were to advance

**Table 5. Lowest elevations of the greatest advances of pre-Tahoe glaciers and Tahoe glaciers in selected Sierra Nevada river drainages.**

| River drainage, n. to s. | pre-Tahoe | Tahoe | Difference in elevations |
|---|---|---|---|
| North Fork Stanislaus | 3600 feet[1] | 3900 feet[1] | 300 feet |
| Tuolumne | 2000 feet[2] | 2400 feet[3] | 400 feet |
| Merced | 1800 feet[4] | 3900 feet[5] | *2100 feet* |
| San Joaquin | 3000 feet[6] | 3170 feet[7] | 170 feet |
| Average difference in elevations, *excluding* the Merced drainage: | | | *290 feet* |

[1]Interpretation of my field mapping of pre-Tahoe and Tahoe glaciations.
[2]Matthes, 1930, p. 53.
[3]My extrapolations of Tahoe moraines by Dodge and Calk, 1987, and by Huber *et al.*, 1989.
[4]Matthes, 1930, Plate 39.
[5]Matthes, 1930, Plate 29. This actually is Tioga, not Tahoe, as Matthes had indicated.
[6]Matthes, 1960, Plate 1.
[7]Birman, 1964, Plate 1.

down the valley today, much of it would get entrained, producing an enormous terminal moraine. Yet the Bridalveil Meadow moraine is smaller than the short-lived, recessional "Medial (end) Moraine." In volume it is dwarfed by preserved Tioga-age end moraines along the Sierran east-side drainages such as: 1) Glen Alpine Creek (Fallen Leaf Lake, Lake Tahoe Basin), 2) West Walker River, 3) Rush Creek (Grant Lake, Mono Basin), and 4) Mammoth Creek (Long Valley). Although the terminal moraines of these drainages are huge compared to the Bridalveil Meadow moraine, all their glaciers were smaller than the Merced River's glacier. Matthes (p. 57) was aware of Bridalveil Meadow moraine's diminutive size compared to the Medial Moraine, and he therefore stated that all the lowly frontal moraines originally were much larger, deposited on the valley's bedrock floor, but that later they were mostly buried by sediments. He shows this in his reconstruction of the El Capitan Meadow moraine (his Figure 18). The original volume for this moraine, as reconstructed by Matthes, then could have been about four times larger than what is exposed today. If the partly buried Bridalveil Meadow moraine also were four times larger, it still would be too small to be a terminal moraine.

A fourth line of evidence is end-moraine morphology. During maximum glaciation the snout of a glacier fluctuates, so the end moraine is spread over a wide area and has a series of hummocky, subparallel crests. In the four numbered examples above, the longitudinal lengths across their Tioga-age terminal-moraine assemblages are, respectively, 0.7 mile, 1.0 mile, 0.7 mile, and 2.0 miles. For comparison, the length across the *single-crested* Bridalveil Meadow moraine is 0.1 mile. One could argue that the width of the valley's end moraines combined—1.0 mile—is comparable. However, as cuts through them by the Merced River show, these few moraines are discrete entities, not crests on one large moraine. This pattern—discrete entities—occurs in many east-side drainages, and it is characteristic of recessional moraines, not terminal moraines.

A fifth line of evidence is the relative amounts of talus on either side of the Bridalveil Meadow moraine. If this moraine is the Tioga's terminal one, then the talus just behind it has accumulated in the last 14,000 ice-free years. In contrast, the talus in front has accumulated since the time when the last preceding glacier advanced beyond it. According to a 1990 article by Huber, the preceding more extensive glaciation was the Sherwin, which ended about 800,000 years ago. Therefore, talus has been accumulating in front of the moraine for that many years, but behind it, the talus was removed by the Tioga glacier. If one assumes a steady rate of talus production, there should be about 57 (800,000÷14,000) times more talus in front of the moraine than behind it. But the talus in front is *less.* My interpretation is that talus on *both* sides is post-Tioga, but the talus behind the moraine is slightly higher in elevation because it rests on glacial-outwash and post-glacial sediments trapped behind the moraine. Since the cliffs above the moraine are retreating at about the same rates, the relative amounts of talus in front of and behind the moraine cannot be ascribed to differences in rock types that would lead to differential rates of rockfall.

A sixth line of evidence is based on Matthes' observation (p. 70) that El Portal (pre-Tahoe) striations on glacial polish can be found "under favorable light" in the vicinity of Artist Point (Map 31, lower right edge, at about 4700 feet elevation). I examined the Artist Point locality but found no polish or striations, despite visiting it under ideal lighting conditions. However, even under optimal conditions, one worker may not be able to verify another worker's *faint* striations; hence Matthes' evidence may actually exist. If his striated polish does exist, it probably is Tioga in age, since no reputable Tahoe or pre-Tahoe striations and/or polish exist. Other than François Matthes, only Joseph Birman claims to have found some, although, as discussed in Chapter 20, virtually all of his Tahoe and Tenaya moraines are late Tioga, so his polish and striations also would be late Tioga.

There is a discontinuous cover of debris at Artist Point, which usually is a fraction of a yard in thickness, but it is not solely of glacial origin. Most appears to be derived from mass wasting of the cliffs above, some from local weathering and mass wasting, and perhaps only a minor component from glacial deposition. Moisture staying within the debris for most of the year accelerates its weathering. Therefore, had the debris been pre-Tahoe till, it would have weathered away long ago. For example, in the Stanislaus River drainage as well as on lands above Yosemite Valley, the pre-Tahoe deposits generally amount to only large, isolated boulders up to several yards across, located atop bedrock ridges and gentle slopes of bedrock. Actual till does not exist. The slopes in the vicinity of Artist Point are too steep not to experience mass wasting, particularly soil creep, so pre-Tahoe till is an impossibility and any polish or striations could not be protected for long before being exposed to weathering.

Artist Point, located along the abandoned Wawona Road on the south side of the valley, is about 1.0 mile down-valley from and 800 feet above the Bridalveil Meadow moraine. If the ice at that locality had been thick enough to polish—i.e., thick enough to flow—then the local thickness there must have been at least 150 feet. As was noted earlier for the Horse, Snow, and Grouse valleys above the Mokelumne River and for Illilouette Creek valley, a glacier can be higher than the mouth of a hanging valley and still not flow up it. Whereas moraines were left at the mouths of Horse, Snow, and Grouse valleys, apparently none were left at the mouth of Bridalveil Creek valley (i.e., the brink of Bridalveil Fall). However, none would be expected since, in contrast to the Mokelumne glacier flowing rapidly past those mouths, the Merced glacier at the mouth of Bridalveil Creek valley was composed of stagnant, virtually till-free ice (like the ice in the Little Prather Meadow area, discussed in Chapter 14). Consequently, absence of a moraine does not prove that the ice surfaces of the Tioga and Tahoe glaciers must have been below the brink of Bridalveil Fall; they could have been well above it. (In contrast the Tioga glacier's edge skirting across the mouth of Illilouette Creek valley was rapidly moving and sediment-laden, and it would have deposited a moraine if it had been high enough.)

A devil's advocate could argue that because the Bridalveil Creek drainage had Tioga glaciers, glacial outwash would have accumulated at the ice dam at the brink of the fall, just as it did at the brink of Illilouette Fall. However, these glaciers were fewer and much smaller, and Bridalveil Creek apparently dropped the outwash when its gradient slackened almost to zero as it traversed the benchlands south of the Glacier Point Road. Neither I nor Matthes was able to find any outwash in the lower stretch of Bridalveil Creek above Bridalveil Fall, despite Tioga and Tahoe glaciers having existed about 8 miles upstream.

A seventh line of evidence for the Bridalveil Meadow moraine being recessional is based on the relative distance of the first lowest recessional moraine behind the terminal moraine. In 1976 Malcolm Clark observed that in most drainages on both sides of the Sierran crest, distinct recessional moraines are located in a zone about 60 to 80 percent of the glacier's length. If this observation (also made by others) is correct, and if the Bridalveil Meadow moraine was the Tioga-stage terminal moraine, then the Merced glacier would have been about 25.6 miles long, and its easternmost frontal moraine, the El Capitan Meadow moraine, would be at a position marking about 96% of the maximum length of the Merced glacier. The other western moraines would be at even higher percentages. This is at variance with the known end-moraine patterns. But if the Bridalveil Meadow moraine was the lowest of the recessional moraines and if it was deposited when the retreating glacier was at 80% of its maximum length, then the glacier would have advanced an additional 6.4 miles for a total length of 32.0 miles. This would put the glacier's snout at an elevation of 2670 feet and just above the lower end of Merced Gorge. At such elevation it would be 870 feet above the lower end of the pre-Tahoe glacier. This is a still a high value for Table 5, but that was for Tahoe, not Tioga, glaciers. If the Tioga glacier had been shorter than the Tahoe glacier, as in Chapter 19 I assert it was, then an advance to the lower part of Merced Gorge (and the Tahoe advancing a bit farther) would fit the pattern that occurs in other glaciated canyons.

An eighth line of evidence is the lack of lateral moraines. On the east side of the Sierra Nevada, each terminal moraine curves up-canyon from its center, merging imperceptibly with its lateral moraines. This relation also holds true on the west side, where unquestionable terminal moraines exist, such as in the Pinecrest Lake area (Chapter 14, Map 17).

Finally, there is a ninth line of evidence that Tioga and Tahoe glaciers advanced well beyond the Bridalveil Meadow moraine: the *relatively youthful* Cascade Creek bridge till, perched high above the floor of Merced Gorge and far west beyond the Bridalveil Meadow moraine.

## Cascade Creek bridge till

A long-standing tradition with roots back to at least Whitney (p. 422-423) is that the more recent glaciers advanced no farther than the Bridalveil Meadow moraine. Hence the idea was already about a half century old when Matthes began his glacial field work. If the idea is correct, any glacial evidence beyond that moraine must, by definition, be from an earlier glaciation. This view perseveres even today, even though it requires the Tioga and Tahoe glaciers to be anomalously short and their snouts anomalously high with respect to the largest pre-Tahoe glacier when compared to other sets of glaciers on both sides of the range (e.g., Table 6). Consequently Huber in his *Geologic Story of Yosemite National Park* (p. 51) says of the Cascade Creek bridge till:

> A small patch of *old, deeply weathered till* is exposed in a road cut on the Big Oak Flat Road just east of the bridge crossing Cascade Creek. This till, about 1,000 ft above the valley floor, would hardly be recognized as such without the artificial cut. These erratics and till are assigned to one or more pre-Tahoe glaciers that filled Yosemite Valley. (Italics mine.)

Mass wasting should have removed it over the few hundred thousand years that it is believed to have perched on a small bedrock ledge traversed by the road (Map 31, site 1; Photo 119). Why should this till, *a priori* assumed to be pre-Tahoe in age, exist in this precarious location when no *verifiable* pre-Tahoe till has been found in the entire west side of the Sierra Nevada? If pre-Tahoe till had survived here, it would have survived at myriad sites.

That the Cascade Creek bridge till is not old and deeply weathered can be concluded from the data presented in Table B in the Appendix. These are summed here: the ratio of fresh to intermediate to grussified boulders is 66% : 25% : 9%. For comparison the ratio for the two cuts through the Tahoe lateral moraine above Pinecrest Lake (Table A) is 23% : 31% : 46%. The Cascade Creek bridge till has nearly three times the percent of fresh, subsurface boulders and only one fifth the percent of grussified boulders as the *Tahoe-age* moraine. While in the field

**Table 6. Tahoe glaciers, distance from 1000-foot thickness to snout.**

| Drainage | 15' quadrangle(s) | Distance |
|---|---|---|
| South Fork Stanislaus River | Pinecrest & Long Barn | 2.7 miles |
| Glen Alpine Creek (Fallen Leaf Lake) | Fallen Leaf Lake | see "Note" |
| Rush Creek (Grant Lake) | Mono Craters | 3.7 miles |
| Green Creek | Matterhorn Peak & Bodie | 5.9 miles |

| Drainage | Crest-floor elevations | Terminal-moraine elevations | Distance |
|---|---|---|---|
| South Fork Stanislaus | 6720-5720 feet | 5580 feet | 2.7 miles |
| Glen Alpine | 7010-6010 feet | 6300 feet | 2.3 miles |
| | 7300-6400 feet | 6300 feet | 4.2 miles |
| Rush Creek | 8100-7100 feet | 7040 feet | 3.7 miles |
| Green Creek | 8920-7920 feet | 6800 feet | 5.9 miles |
| | Average, *excluding* Glen Alpine's 2.3 miles: | | 4.1 miles |

Note: Two interpretations, 2.3 miles and 4.2 miles, are possible. The first has a measuring site based on the maximum depth of the lake, about 365 feet, which *arbitrarily* is placed midway along the lake. Note that in this interpretation the base of this site is about 290 feet *below* the crest of the lowest terminal moraine, so the base of the glacier flowed that much upslope. The second interpretation is measured from the highest point on the east-side lateral moraine, above the south end of the lake. Note that it is only 900 feet thick. During the Tahoe glaciation, Lake Tahoe was at least 90 feet higher than its current elevation of 6229 feet, so the snout would have been bordered by lake water at least 20 feet deep. This shallow depth probably had minimal influence on melting of ice at the glacier's snout.

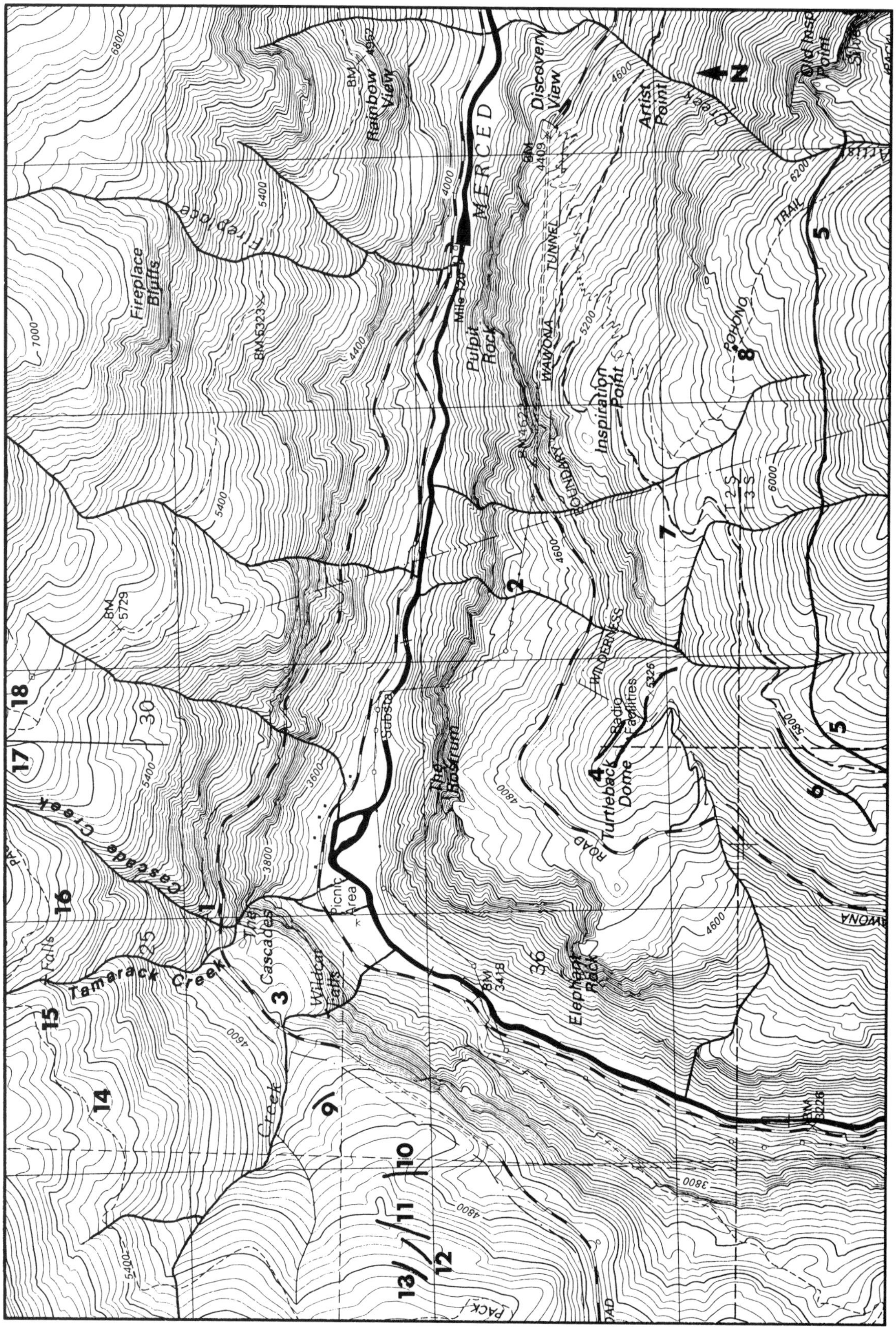


**Map 31. Merced Gorge. Scale 1:24,000.**

**Photo 119. Cascade Creek bridge till.**

with me in June 1993, Clyde Wahrhaftig speculated that this till was exceptionally preserved by a cover of talus that was impermeable to water. This argument does not hold water because, first, water percolates readily through talus slopes; and second, because the talus blocks atop the till are too few to completely cover it. Despite the abundance of fresh, subsurface boulders, the Cascade Creek bridge till nevertheless has too many partly fresh ones to be considered Tioga age.

The distribution of its grussified boulders is interesting in that all three occur in the boulder-poor lower part, a silt-and-gravel till that lacks fresh or intermediate boulders. The two parts could be interpreted as a late-Tahoe till atop an early Tahoe till were it not for the lack of a definable contact or any sign of a soil horizon. Note in Photo 119 that the left (west) side of the upper part has a greater boulder density than the right (east) side, which grades smoothly downward into the lower part. The Pinecrest Lake cuts demonstrate that extreme variation in the degree of weathering of subsurface boulders can occur over a very short distance (the road cut versus the water-tank cut), presumably due to more ground water at the lower site, and the same could apply here. The lower part of the Cascade Creek bridge till, being rich in silt and gravel, certainly would hold more ground water; hence subsurface chemical weathering should have been more intense.

Since throughout the entire Sierra the crests of Tahoe lateral moraines are not that much higher than those of the Tioga's (except near the terminus). Therefore, here, the Tioga's crest should lie just below the Cascade Creek bridge till. Unfortunately, a cliff lies below it, so no deposits would be left. Also, a cliff lies above it, so no deposits would be left there. Possibly the ice surface of the Tahoe glacier was higher that the till. If so, then one should see both Tahoe and Tioga till on gentle slopes just to the southwest, but I found none. Therefore, the Cascade Creek bridge till appears to represent the Tahoe ice surface. This would have been at about 4400 feet elevation, or about 1000 feet above Merced canyon's floor. On the Sierra's east-side slopes, where lateral moraines are 1000 feet thick (such as at the first three sites mentioned earlier—Glen Alpine Creek, West Walker River, and Rush Creek), the Tioga moraine lies only a few yards below the Tahoe moraine. Consequently, the Merced's Tioga glacier should have been about 950 feet thick.

The steep-walled, rockfall-prone Merced Gorge preserves no known evidence of Tioga or Tahoe glaciers; therefore an estimate of the distance these glaciers advanced down it must be based on well-constrained evidence of other Sierran glaciers. Because several glaciers left well-preserved moraines, the distance from a thickness of 1000 feet to the crest of farthest Tahoe end-moraine is known. Four glaciers are presented in Table 6. The South Fork Stanislaus River glacier is on the west side; the other three are on the east side. The glaciers are arranged in order of the elevation of the crest of the lateral moraine where it is 1000 feet above the floor of the valley.

Table 6 shows that the lower the crest elevation at the 1000-foot thickness, the shorter, in general, was the lower part of its Tahoe glacier. At the Cascade Creek bridge till, the elevation is far lower, about 4370 feet elevation, so one might conclude that Merced Gorge's Tahoe-age glacier would have ended in considerably less than 2.7 miles beyond the till. If so, then the glacier may have advanced only half-way through the gorge, ending at about 3000 feet elevation.

However, this pattern of decreasing length with decreasing elevation, while likely real, may not be applicable. The canyons of the four glaciers in Table 6 maintained relatively constant cross-sectional areas. In contrast, the Tahoe-age Merced glacier flowed through the broad, essentially flat-floored Yosemite Valley, then was forced to flow through the narrow, west-trending upper Merced gorge and the narrow, south-trending Merced Gorge proper. Because of this constriction, ice would have had to flow faster. An important consideration is how much the canyon's floor descends from the 1000-foot-thick site to the terminus. In Table 6 the elevation losses from this location to the crest of the lowest terminal moraine for the South Fork Stanislaus, Glen Alpine Creek (second interpretation), Rush Creek, and Green Creek glaciers are, respectively, 140 feet, 100 feet, 60 feet, and 1120 feet. Table 6 shows that the Green Creek glacier is significantly longer than the other three, which may be due in part to its overall higher elevations, and consequent cooler temperatures and slower melting rate. However, its longer length may have been due in part to the considerably steeper drop, which would have steepened its ice-surface gradient. Because the velocity of a glacier is proportional to the third power of this gradient, the Green Creek glacier should have flowed faster, allowing it to advance farther down, to lower elevations. Likewise, the Merced's Tahoe-age glacier dropped considerably. From the Bridalveil Meadow moraine down through the upper Merced gorge the elevation drops 480 feet, then through Merced Gorge proper it drops 840 feet. Both drops would cause the Merced's Tahoe-age glacier to increase its velocity, especially along the lower, more-substantial drop through the 3.5-mile-long Merced Gorge. Consequently an advance entirely through the gorge, if not slightly beyond it, is reasonable.

There are three lines of circumstantial evidence to support this. First, as was shown back in Table 5, the average difference in elevations between the snouts of pre-Tahoe and Tahoe glaciers is 290 feet. If one uses only the values from the North Fork Stanislaus and Tuolumne drainages, an average of 350 feet, then for the Merced drainage, where the pre-Tahoe descended to about 1800 feet, the Tahoe should have descended to about 2150 feet.

The second line of circumstantial evidence is from the seventh line of evidence under the Bridalveil Meadow moraine. If it was the lowest of the recessional moraines, and if it was left when the glacier was at 80% of its maximum length, then the glacier would have advanced an additional 6.4 miles for a total of 32.0 miles. This would put the glacier's snout at an elevation of 2670 feet and located just above the lower end of Merced Gorge. At such elevation it would have been 870 feet above the lower end of the pre-Tahoe glacier. If the Bridalveil Meadow moraine was closer to 60% of the glacier's maximum length, the glacier would have advanced beyond the lower end of the gorge.

The third is the distribution of major rockfalls. Perhaps the *major* difference between glaciated and unglaciated *granitic* canyons in the Sierra Nevada, which heretofore has never been acknowledged, is that glaciated ones have abundant rockfalls and resulting talus slopes, while unglaciated ones have few rockfalls and little if any lasting talus slopes. (The major difference is *not* the often stated, false generalization that glaciated canyons have U-shaped cross profiles and unglaciated canyons have V-shaped ones. As Matthes and Wahrhaftig stated—but ignored the implications thereof—both types of cross profiles occur in glaciated *and* unglaciated granitic, Sierran landscapes.) Apparently if a sufficiently thick glacier buries a granitic landscape for a lengthy period of time, it exerts a strong pressure on the bedrock surface; then when the glacier melts, which happened rapidly in the Sierra Nevada, the surface is rapidly depressurized, and the bedrock can mass waste, often by exfoliation. This proposed process is not an original idea. Others before me have observed it both in glaciated mountain canyons and in glaciated lands of North America and northern Europe covered by the great ice sheets.

Where roads cut through granitic bedrock, the thickness of concentric, exfoliating layers is revealed to be only a few yards. One of the thickest exfoliation layer I have seen is at the brink of Half Dome's summit (Photo 120), where the shells total about 40 feet thick. That they aren't thicker implies that the overlying weight is sufficient to prevent exfoliation. Since the density of granodiorite is about 2.7 and that of glacial ice is about 0.9 (water is 1.0), a thickness of ice required to generate this weight is about 120 feet, or not quite the thickness of the brittle ice comprising the upper layer of a glacier. This implies that a glacier need not be all that thick to have an effect—upon retreating—on the underlying, rapidly depressurizing rock. Consequently, even some glaciers of the Little Ice Age should have been thick enough to have produced a post-glacial response in the form of increased mass wasting. In a glaciated canyon the lowest bedrock of the canyon's walls is under greater pressure than higher up, and therefore it is more likely to fail and thus create mass wasting. If mass wasting occurs, generating a rockfall, then the overlying, unsupported bedrock can become unstable and generate another rockfall, and so on up the wall. Yosemite Valley has many examples of this process.

With the foregoing in mind, one can observe that from the upper Merced River drainage down to the floor of Yosemite Valley, as in all glaciated drainages, there has been a lot of rockfall. As the Tioga glacier thinned westward through the valley, it would have exerted less and less pressure on the valley's lower walls (its floor is thick sediments and is not an issue). Therefore, the amount of rockfall production also should have diminished westward. This is the pattern on the detailed maps by Wieczorek *et al.* (Plates 3 and 4, scale 1:12,000). Had the Tioga (and/or Tahoe) glacier stopped at the Bridalveil Meadow moraine, then canyon walls east of it (i.e., from the Cathedral Rocks and El Capitan eastward) should have had significantly more rockfall production than those west of it. Instead, the amount of rockfall production continues to diminish westward through the upper Merced gorge southward through Merced Gorge proper, and then west just beyond it. This implies that the Tioga glacier may have reached this far, down to about 2300 feet elevation. The Tahoe glacier, being slightly longer, may have advanced down to about 2200 feet elevation, that is, to within 400 feet above the lowest elevation reached by pre-Tahoe glaciers.

On the other hand, in the North Fork Stanislaus River drainage, discussed in Chapter 20, the greatest pre-Tahoe glacier appears to have advanced about 4.8 miles beyond the Tahoe glacier, although this figure could be off by a mile or so. This length suggests that perhaps the Merced's Tahoe glacier advanced—like the South Fork Stanislaus River's Tahoe glacier—only about 2.7 miles beyond the point of 1000 feet of thickness (the Cascade Creek bridge till), that is, down to about 2860 feet elevation. The midpoint between these two interpretations is the south end of Merced Gorge, at about 2600 feet—considerably above the pre-Tahoe—and there, for lack of demonstrating one interpretation superior to the other, I arbitrarily have placed the terminus of the Tahoe glacier.

Likewise, I somewhat arbitrarily have placed the terminus of the Tioga glacier about 0.6 mile above that of the Tahoe, which is at an elevation of about 2800 feet. This distance is based on the relative separation of the lowest of the Tahoe and Tioga end moraines in the South Fork Stanislaus River drainage's Pinecrest Lake area, the separation being about 0.3 mile. Since the Tahoe and Tioga glaciers of the Merced River drainage were about twice the length of those in the South Fork Stanislaus River drainage, the relative separation of the lowest of its Tahoe and Tioga end moraines also should have been twice the length, or about 0.6 mile.

**Photo 120. Exfoliating granodiorite on summit of Half Dome. Exposed part of metal rod is about 2 feet high. View is of the upper, overhanging part of its northwest face, beyond which lies Yosemite Valley.**

Two remaining issues related to the Cascade Creek bridge till need mention. The first is that if the Tahoe and Tioga glaciers were respectively about 1000 feet and 950 feet thick at the head of Merced Gorge, then up-canyon they should have lapped onto the gentle surfaces (site 2) northeast of and below the summit of Turtleback Dome. I checked the gentle surfaces at the brink of the gorge, from the tunnel west to the Rostrum and then south to Elephant Rock, but failed to find any till. But in the vicinity of site 2 the ice would have been stagnant ice, and as such, it should not have left a continuous cover of till. However, one would expect that as the glacier waned and its ice surface lowered, a diachronous lateral moraine would have been constructed along the brink. Its absence may be due to the catastrophic failure of the Tioga (and earlier) glaciers due to the rapid reduction of the Tuolumne glacier's overflow into Tenaya Canyon. The ice surface of the stagnating glacier might have descended too rapidly to deposit any till other than perhaps a few boulders. This may have been what happened when the North Fork Stanislaus River glacier rapidly deteriorated as the Mokelumne River glacier's overflow rapidly diminished. There is a paucity of moraines and till in the lower glaciated part of the North Fork Stanislaus River drainage, despite some slopes being gentle enough to hold and retain them (Photo 121). In support of this argument, which may seem contrived, I point out that on gentle bedrock slopes at site 3, southwest of and *below* the Cascade Creek bridge, one or more lateral moraines should exist, *but none do,* even though the slopes clearly had been covered by the same glacier that left the Cascade Creek bridge till.

The second issue is why the Tioga glacier was thicker in Yosemite Valley and just above it, while the Tahoe glacier was thicker in most of Little Yosemite Valley *and* in Merced Gorge. As suggested earlier at the end of the Tenaya Canyon chapter, the apparent reason for this

**Photo 121. Lower, glaciated lands of the North Fork Stanislaus River drainage. View is east toward a clearcut, seen from a ridge (at 5380 feet elevation) located immediately east of a jeep trail that descends northeast to the Ramsey site (Boards Crossing 7.5' quadrangle, Sections 22 and 23). All lands but those near the skyline were under Tioga ice, yet glacial evidence in this view, including the ridge in the foreground, is so sparse that without a detailed field study, one would conclude it never had been glaciated. Erratics lie along a minor ridge (5831 on map, and reached by Road 6N17), just left of the upper left edge of the clearcut.**

anomaly, not found anywhere else in the Sierra Nevada, lies with another anomaly, also not found anywhere else in the Sierra Nevada: Yosemite Valley's own post-glacial Lake Yosemite. Such a lake likely appeared after the end of a number of glaciations. I suggest that the last lake to form was after the retreat of the Tahoe glacier, and that it was filled in mostly with glacial outwash from early and mid-Wisconsin glaciers, which failed to reach the valley. Then the late-Wisconsin Tioga glacier advanced, but it did so across thicker sediments, and hence although it may have been slightly thinner than the Tahoe glacier, its ice surface was slightly higher because its base was higher. This higher ice surface caused the Tenaya and Merced glaciers up-canyon from it also to have higher ice surfaces for several miles. Below Yosemite Valley, both the Tioga and Tahoe glaciers flowed on a bedrock floor, and since this base was essentially the same for both, the Tahoe glacier, being thicker, would have left higher deposits. Under this interpretation there was no post-Tioga Lake Yosemite, as discussed in the next chapter.

## Turtleback Dome and environs

Matthes' glacial map of the valley shows a double-crested El Portal (pre-Tahoe) lateral moraine more or less atop a short east-west ridge known as Turtleback Dome, site 4. It further shows the longer, southern moraine connected by till that continues uninterrupted up to his El Portal ice surface, site 5. Below and west of this ice surface (and south of Turtleback Dome), his map shows a sharp-crested lateral moraine, site 6. Despite at least five trips to the summit of the dome and two to the slopes above it, I have been unable to find any of this mapped evidence; it simply does not exist.

The two moraines atop the dome are no more than local soil plus a discontinuous veneer that appears to be derived from local bedrock. Nevertheless, the dome was glaciated, since one can find an exotic boulder or two that match bedrock up-canyon. Still, overall the Turtleback Dome deposit most closely resembles one in lower Illilouette Creek valley. Located on an open east slope about 0.6 mile above Illilouette Fall, the deposit is due to slow mass wasting and long-term weathering. (Compare the deposits of Turtleback Dome, Photo 122, with those of lower Illilouette Creek valley, Photo 123.) If Turtleback Dome's deposit was the remnants of an ancient moraine, then many of its boulders should be exotic. However, I examined several hundred and concluded that at least 90% (and as much as 99+%) were locally derived. Only one boulder was a definite erratic, although some others could have been. Hence this appears to be a mass-wasted deposit with perhaps a few remnants of formerly larger erratics.

This issue is not trivial. If the deposit is the remnants of a lateral moraine, such a moraine would have been created *only* by a glacier that flowed west past the dome's

**Photo 122. Deposit atop Turtleback Dome.**

summit, leaving sediments on it. This would imply that the ice surface must have been at a similar height. Had the glacier been considerably thicker, it would have flowed southwest across the dome. Then, as it waned and the ice surface lowered, the ice over the dome eventually would have become thin and stagnant. That ice would have carried only a few erratics, not enough debris to leave a moraine. Hence, if a morainal deposit really does exist here, then the pre-Tahoe ice surface is known. On the other hand, if the deposit is the result of local mass wasting, then the very few genuine erratics in it would have been left by a thicker glacier, and the pre-Tahoe ice surface is unknown. Given the overwhelming preponderance of locally derived boulders, I conclude that the second alternative was the case.

Worth mentioning is that atop the dome near and below the west part of its summit are four large boulders, these about 10+ feet high. Two are locally derived, as are smaller boulders, some of which can be seen in the process of breaking apart from the bedrock. These boulders, like the bedrock, have mafic (black, iron-rich) inclusions. In contrast, the two large boulders that are erratics lack these inclusions, and each has an aplite (white, feldspar-rich) dike—and such dikes are absent in the local bedrock. The larger, more prominent of these two erratics is shown in Photo 124. These are important in that they show that very little denudation of bedrock has occurred since they were deposited—a very important piece of evidence elaborated on later, under a discussion of rates of Sierran denudation.

As elsewhere, Matthes' mapped till that continues uninterrupted up slopes to his El Portal ice surface is merely soil and underlying detritus derived from local bedrock. Furthermore, this material is identical both above and below his ice surface. Perhaps at one time there had been till, but the slopes simply are too steep to have held it for long. Matthes' till is suspect in that its area is essentially congruent with the area bounded by two seasonal creeklets above their junction, which lies east of

**Photo 123. False glacial deposit: weathering, mass-wasted blocks on an east slope in lower Illilouette Creek valley.**

Turtleback Dome. If a pre-Tahoe glacier had left deposits above the dome, then some erratics should have been found at site 7, the relatively flat top of a west-descending ridge. Additionally, they should have been deposited and preserved on this ridge, which has a gradient of about 25%. That Matthes mapped neither till nor erratics along it suggests, but does not prove, that it was unglaciated. If it was not glaciated, then the ice surface would have been below it, which in turn implies that at Turtleback Dome the ice surface lay just above the dome's summit and the deposit there is a remnant of a lateral moraine. However, it could have been glaciated and have left no surviving evidence, so the actual ice surface cannot be determined unambiguously at these two sites.

Matthes mapped a trailside erratic above this ridge, at site 8, which is just a few yards north of the Pohono Trail and at an elevation of 5960 feet. A large boulder indeed stands exactly where he had mapped it, and since no other large boulders exist in the vicinity, I have presumed it to be his mapped erratic. Although it looks like an erratic

**Photo 124. Largest erratic atop Turtleback Dome. Ken Ng provides scale.**

**Photo 125. Matthes' trailside "erratic" along the Pohono Trail. Rudy Goldstein stands on bedrock connected to the "erratic."**

from its north, south, and west sides, the lower part of its east side definitely is connected to adjacent bedrock (Photo 125).

Matthes also mapped lateral moraines and till in the upper Cascade Creek drainage (that part beyond the north edge of Map 31) and along the upper west slopes of Merced Gorge. In the Wildcat, Tamarack, and Cascade creek drainages I looked for evidence along the trail that traverses across them (sites 14-16), and also looked for evidence along the old road from Cascade Creek down to the start of the El Capitan Trail (sites 17 and 18). I also checked the lower part of that trail. Glacial moraines and till were absent at every site. Additionally, I checked Matthes' mapped El Portal (pre-Tahoe) deposits west of and above the Big Oak Flat Road south of Wildcat Creek, sites 9 through 13. All the sites are colluvial deposits on bedrock, although initially sites 11 and 13 appear different. Site 11 is a gently sloping surface above a cliff of El Capitan granite, and it has a veneer of small diorite clasts (not rounded boulders). However, it is very unlikely that a glacier would have left solely diorite clasts and nothing else. Furthermore, if pre-Tahoe in age, they should have disintegrated long ago, since diorite weathers rapidly. (I cannot recall seeing any pre-Tahoe diorite erratics in the Sierra Nevada, although I've seen hundreds of granitic ones.) Because the clasts are fresh and angular, I suspect that their source is local, just a few yards away, under shallow colluvium. Site 13 is a slope that also has a lot of diorite clasts in its soil, yet the only *prominent* outcrops are El Capitan granite. However, near the slope's lower south end I found a low outcrop that over its length of 16 feet rapidly changes from coarse granite to fine granite to coarse diorite. I suggest that since diorite weathers rapidly, it is less likely to form outcrops, hence it mostly lies under colluvium rather than protruding from it. At none of these sites did I see any exotic boulders. Had any boulders been deposited by a glacier, then I should have found at least a few Cathedral Peak granodiorite boulders, one of which occurs in Turtleback Dome's deposit.

Matthes (p. 66) mentioned that another ancient moraine diverted Grouse Creek, which lies about 3 miles south of Turtleback Dome (and lies south of Map 31). I did not visit this site, but it is mapped by Huber *et al.* solely as a low bedrock ridge of El Capitan granite that is in contact with granodiorite of Arch Rock at its west end, and with tonalite of the Gateway along its south base. Mapping these contacts would have been possible only with detailed field work, which would have resulted in the discovery of a moraine, had it existed. I take its absence on their map as absence in the field. Therefore, it appears that the only pre-Tahoe evidence in the area covered by Map 31 is that atop Turtleback Dome, which lies about 650+ feet below Matthes' unconstrained, purely hypothetical (although logical and possibly correct) ice surface.

## Longitudinal profiles of the Yosemite Valley glaciers

As in and above Little Yosemite Valley, some of the evidence along the rims of Yosemite Valley and Merced Gorge sets limits for the minimum or maximum elevations of the ice surfaces of the Tioga, Tahoe, and pre-Tahoe glaciers. However, unlike those higher areas, Yosemite Valley and the Merced Gorge have pitifully few constraints. The Cascade Creek bridge till, above the head of Merced Gorge, tightly limits the ice-surface elevation of the Tahoe glacier, suggesting a thickness there of about 1000 feet, and if the Tioga glacier had been slightly shorter, as is the case in all other Sierran canyons, then here it would have been slightly thinner, about 950 feet thick. At the other end of the valley the Tioga and Tahoe ice surfaces are constrained by the absence of glacial sediments, polish, and striations along the north rim from the brink of Royal Arch Cascade east to the summit of Washington Column. Therefore, their ice surfaces must have been lower, and their thicknesses would have been about 1400 feet or less. Presumably they would not be much less, if at the head of Merced Gorge, about 9 miles

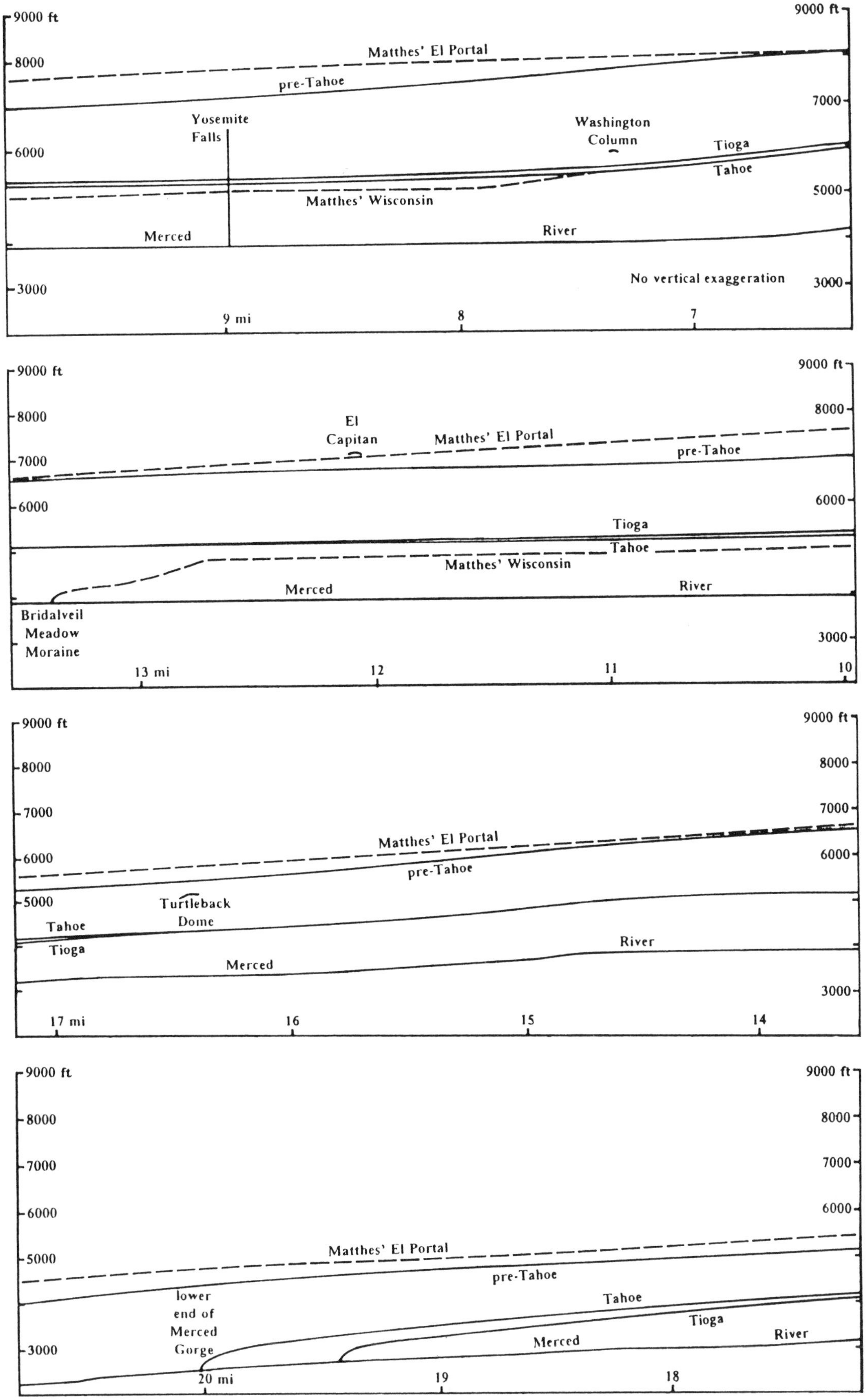


**Figure 30. Longitudinal profiles of Yosemite Valley glaciers.**

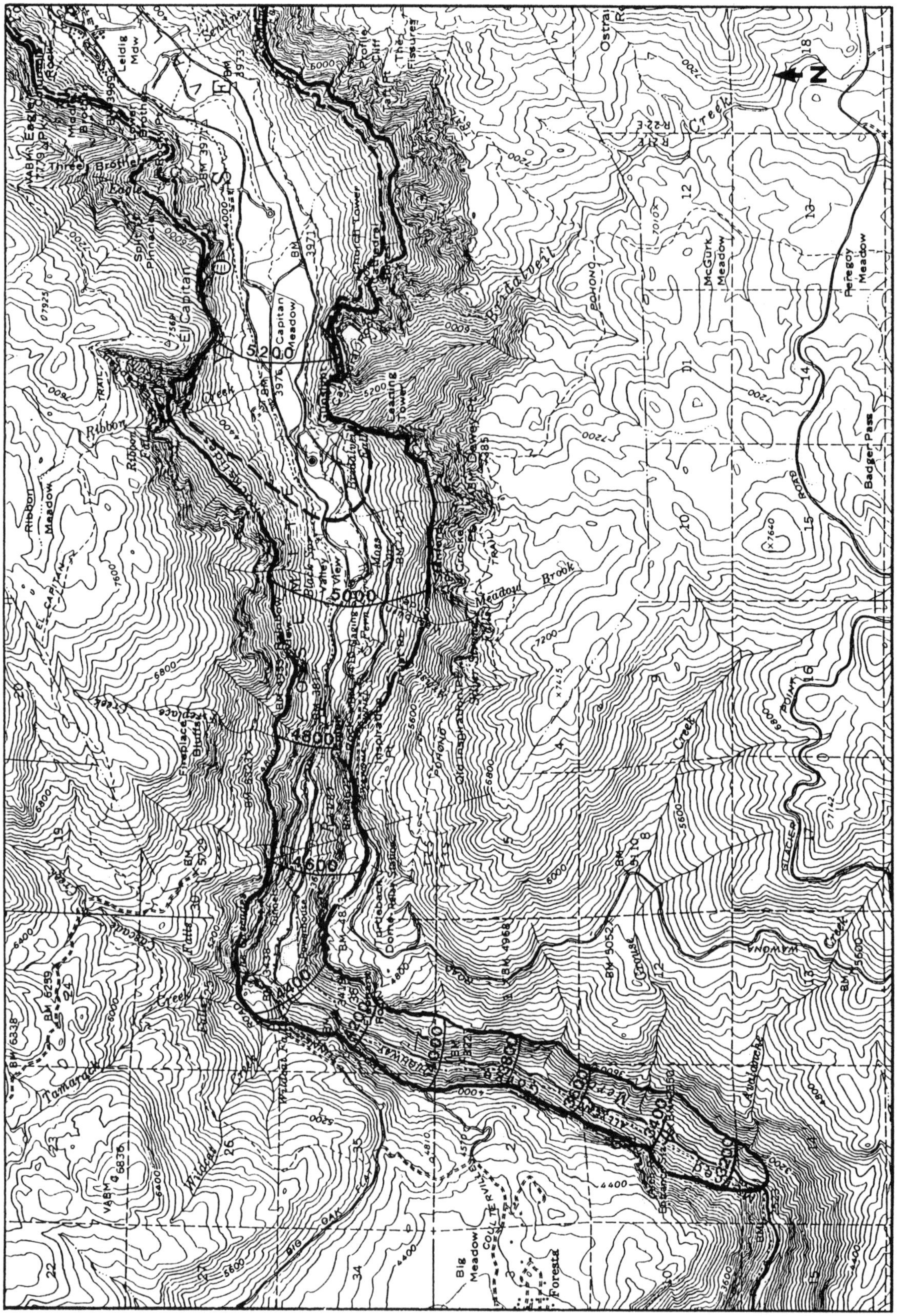


**Map 32A. Glaciers of the Merced River drainage, western Yosemite Valley and Merced Gorge. My Tahoe glaciers are solid lines; Matthes' Wisconsin glaciers, dashed lines. Scale 1:62,500.**

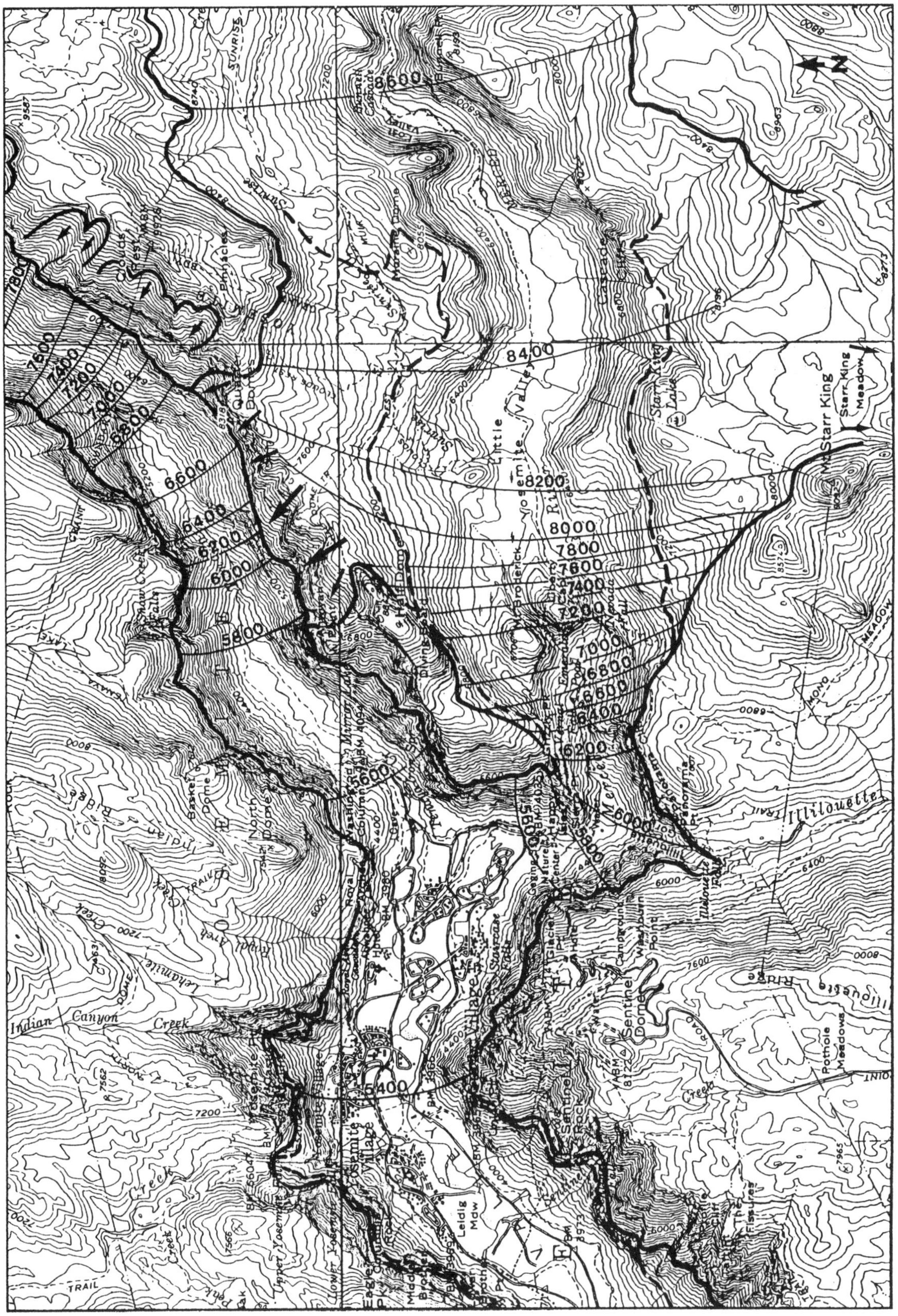


**Map 32B. Glaciers of the Merced River drainage, eastern Yosemite Valley, Little Yosemite Valley, and lower Tenaya Canyon.**

down-canyon, they were about 1000 feet thick. The pre-Tahoe glaciers are even more poorly constrained. Above the east part of the valley the summit of Sentinel Dome lacks glacial evidence so probably was unglaciated, but several erratics occur near its east base, putting the ice surface at about 7940 feet elevation. Above the head of Merced Gorge, glacial deposits atop Turtleback Dome set a minimum limit for the ice surface at about 5280 feet elevation, but it could have been considerably higher, perhaps even at Matthes' hypothetical 5900 feet elevation.

Yosemite Valley meanders more than does Little Yosemite Valley, and to construct a profile, I projected glacier elevations—both Matthes' and mine—perpendicular to the appropriate straight line drawn through each relatively straight part of the valley and gorges below it. The straight-line segments I drew were: 1) junction of Illilouette Creek (midpoint between several distributaries) with Merced River north-northwest to Clark Bridge; 2) Clark Bridge west-northwest to BM (bench mark) 3967, east of Yosemite Lodge; 3) BM 3967 southwest to BM 3970, east of Cathedral Picnic Area; 4) BM 3970 west to BM 3976, west of El Capitan Meadow; 5) BM 3976 west-southwest to Pohono Bridge; 6) Pohono Bridge west to BM 3838, at head of Merced Gorge; 7) BM 3838 south to the bottom of Merced Gorge, where the river curves from south to west; and 8) west to edge of El Capitan 7.5' quadrangle. Had I drawn a sinuous line through the valley and projected glacier elevations perpendicular to it, the resulting profile, Figure 30, would have been different, but not significantly so.

My pre-Tahoe ice surface diverges from Matthes' El Portal, dropping lower and then being flatter through the valley. This interpretation is based on the glacier flowing along the valley's bedrock floor, which seismic evidence mentioned previously shows was at a much lower elevation than was the one envisioned by Matthes. The two ice-surface reconstructions are the same around Pohono Bridge, but then mine diverges below Matthes'. Except for the upper part of the two profiles, where they are congruent in the vicinity of Sentinel Dome, there is not a single constraint to support either profile. Both are rational, and others could be drawn.

Matthes' Wisconsin ice surface drops almost at a constant gradient from just west of Vernal Fall down to the part of the valley opposite Royal Arch Cascade. It then decreases very slightly in thickness through the valley to the vicinity of Ribbon Fall. There it is about 800 feet thick, and it then thins rapidly to its snout in only about 1 mile. This rate of thinning is about three times those of east-Sierran Tahoe and Tioga glaciers, whose rates of thinning can be accurately measured by the lateral moraines they left. If Matthes' ice surface through the valley is essentially correct, then from the vicinity of Ribbon Fall, his glacier would have thinned to its snout in about 3 miles, that is, roughly midway through the west-trending upper Merced gorge. By ending his Wisconsin glacier at the Bridalveil Meadow moraine, Matthes caused it to inexplicably thin catastrophically. Because the ice surface of the glacier through the valley is virtually unconstrained, Matthes could have drawn it significantly lower, thereby alleviating the problem. However, then he would have had to explain how two massive glaciers, joining at the east end of the valley, could have produced such an anemic trunk glacier through the valley.

In contrast, my Tahoe and Tioga ice surfaces are higher than Matthes' Wisconsin ice surface by about 300 to 400 feet, and they parallel his ice surface westward to the vicinity of Ribbon Fall. But then they continue to thin at a gradually increasing rate through the upper Merced gorge and at a faster rate through Merced Gorge proper. The lower parts of the profiles of the two glaciers are similar to ones of equal length that flowed down east-Sierran slopes. As explained above, the Tioga ice surface is higher than the Tahoe ice surface in Yosemite Valley and east just above it, but the opposite is true above and below this length. My glacier profiles permit a reconstruction of the valley's former glaciers, and I show my Tahoe along with Matthes' Wisconsin on Map 32. Neither my pre-Tahoe nor Matthes' El Portal is shown, since both are similar. My Tioga is just above the Tahoe in Yosemite Valley and just below it in Merced Gorge and in most of Little Yosemite Valley.

In sum, the glaciers passing through Yosemite Valley were similar to other Sierran glaciers. In relatively short drainages on the east side of the range, the Tahoe glaciers always were a bit longer and thicker than their Tioga glaciers. In the much longer Merced River drainage, which includes Yosemite Valley, the same holds true: the valley's Tahoe glacier was a bit longer and thicker than its Tioga glacier. This relation also is true for some of the drainage's tributaries that had short glaciers that failed to reach the trunk glacier, notably in the Illilouette, Yosemite, and Bridalveil creek drainages. Where the glacial history of Yosemite Valley differs from all other glaciated canyons is that the ice surface of the slightly larger Tahoe glacier was slightly lower than that of the Tioga glacier. How can a larger glacier be thinner? To solve that riddle, we need to investigate the former, thoroughly misunderstood Lake Yosemite, which is so significant, it deserves its own chapter.

# 19

# Lake Yosemite, its Myth and Reality

The beautiful Lake Yosemite that supposedly formed after the *Tioga* glacier's retreat (Photo 126 plus Maps 29 and 30) is one of the Sierra's most enduring, cherished, and defended *myths*. The concept of a deep, post-glacial lake dammed behind the El Capitan moraine dates back to at least Whitney (p. 423), but was promulgated best by Matthes. Superficially it seems to be logical, since the valley floor behind the moraine is very flat and erratic-free, and at the moraine the river's nearly flat, meandering course gives way to a relatively straight reach of rapids. Interestingly, no one has proposed a beautiful Lake Hetch Hetchy for that valley in the Tuolumne River drainage. Its western part is flat-floored and apparently erratic-free, and like Yosemite Valley it was subjected to occasional flooding, as shown in an 1880s photo (This photo of *wall-to-wall* flooding in western Hetch Hetchy Valley is reproduced in the Yosemite Natural History Association's "Yosemite," 1984, v. 47, no. 17, p. 4. For a photo of this floor when not in flood, see "Yosemite," 1986, v. 48, no. 2, p. 5. Neither photo shows erratics.) In other words, both floors are floodplains that have developed behind local base levels. The only difference is that in Yosemite Valley the local base level is a moraine, and in Hetch Hetchy Valley it is not. One has to wonder, if the El Capitan Meadow moraine had not existed, would a Lake Yosemite have been proposed to explain the floodplain?

The Merced River's local base level in Yosemite Valley is quite modern. Early residents and tourists found the valley swampy and mosquito-ridden, so in 1879, under the direction of Galen Clark, the large blocks clogging the river's channel in the vicinity of El Capitan Meadow moraine were blasted apart, which deepened the channel by about 4½ feet. This lowered the valley's water table,

**Photo 126. Post-glacial Lake Yosemite, painted by Herbert A. Collins, Sr., and Herbert A. Collins, Jr., under the direction François E. Matthes.**

making the valley floor more suitable for residents and tourists. The river, which had meandered widely and changeably across the valley, became entrenched. However, this general vicinity has experienced prehistoric rockfalls, and if a future, large rockfall were to occur on the north face of Lower Cathedral Rock, just downstream from the moraine, a new dam across the river would become established. Perhaps in the past large rockfalls created local base levels.

Whereas geologists (including Matthes) apparently have been oblivious to potential rockfalls (and their consequences), rock climbers have been warned against them. My dog-eared copy of Steve Roper's climbing guide states that the rock's north-face route "because of loose, decomposed, and dirty rock is considered the most unenjoyable and dangerous of the Grade VI's [multiday climbs] in the Valley." Lower Cathedral Rock has several large roofs from which gigantic blocks of rock have broken loose. Consequently, since the last deglaciation the Merced River likely has been meandering across a floodplain maintained by a local base level which may have been always at the moraine, *or* perhaps at some times may have been at rockfall debris from the north face of Lower Cathedral Rock. One such rockfall deposit lies less than 300 feet down-canyon, along the south bank of the river. The river there is about 5 feet lower than at the moraine, and if a future rockfall deposit about 6+ feet were to be spread across the river, the deposit would be thick enough to become the new local base level.

Matthes had considered that the only way a large lake could have been maintained over time was by being dammed behind a sill—a mass of unfractured bedrock—that lay about 60 to 70 feet below the crest of the El Capitan Meadow moraine. Over the last three decades I have visited about 1000 lakes in the Sierra Nevada and the Cascade Range, and have found 99+% of them dammed by bedrock, although many of them superficially appear to be dammed by a moraine or by till. A case in point is the park's Tenaya Lake, one of the largest natural lakes in the Sierra Nevada. It appears to have a moraine above its lower (southwest) end and morainal boulders forming a local base level at its outlet. However, close inspection reveals that the moraine is little more than a veneer of glacial deposits on a gently curved, low, bedrock ridge, and that beneath the creek's boulders is bedrock. Matthes must have been aware that during the late-spring major snowmelt and heavy runoff, the Merced River had sufficient power to remove loose sediments around and under large blocks, causing them to be undermined and therefore lowering the local base level. However, if a bedrock sill had lain just beneath the large blocks, then after loose sediments were removed, these would be concentrated in a compact mass above the sill, forming a bedrock dam. Hence the sill was absolutely necessary for Lake Yosemite's existence. However, seismic surveys, performed in 1935 and 1937 by Cal Tech's Beno Gutenberg and John Buwalda (but not published until 1956, after Matthes' death), indicated that the bedrock floor at the site of the El Capitan Meadow moraine is about 1000 feet below the moraine.

Although Matthes' diary (September 17-18, 1937) mentions that he had observed Gutenberg and Buwalda setting off blasts on the valley floor and that he had discussed with Gutenberg the geophysical method of obtaining the thickness of sediments, he did not accept the results. Robert Sharp, who was an acquaintance of Matthes and one of the Cal Tech co-authors of the 1956 paper, wrote a letter to me in January 1989, which in part said: "Matthes, of course, did not like the seismic data as it differed greatly from his interpretations, but François, bless him, really had no data, just speculation." My notes of a March 1992 phone conversation with Sharp indicated that Matthes not only *did not like* the seismic data but rather that he *never accepted* the seismic results. Whereas the seismic evidence merely negated the dam needed for the creation and maintenance of post-glacial Lake Yosemite, it completely contradicted his reconstruction of the valley's geomorphic evolution. In short, it repudiated his method of analysis, which was based on the principles of his mentor, Prof. William Morris Davis. Using Davis' method of analysis, Matthes drew a hypothetical, but necessary, descending ridge extending from Half Dome westward into eastern Yosemite Valley (Figures 15-17). But the seismic data indicated that the valley's deepest basin, with up to 2000 feet of sediments, lay there. It is impossible for glaciers not only to completely erode away a resistant ridge that has stood for tens of millions of years but also to excavate some 3000 feet beneath its preglacial base (Figure 18). Therefore, more likely for this reason, rather than for the relatively minor matter of the lake's existence, Matthes did not accept the seismic data. Was that data correct, or were the sediments much thinner, as Matthes had supposed?

Two exploratory water wells, one drilled in 1971-72 and the other in 1974-75, went through respectively 1015 feet and 970 feet of sediment. The first one was terminated before reaching bedrock, while the second, slightly shallower one was terminated in granite, which may be either a large boulder or bedrock. These were drilled respectively on the west and east sides of Yosemite Creek (Map 29, sites 9 and 10), and here Gutenberg *et al.*'s Figure 10 indicates a depth of sediments of about 1350 feet. Therefore, the seismic surveys are supported, while Matthes' cross profiles of Yosemite Valley are refuted, since they show sediments being about 300 feet thick here (his Figure 24) and about 180 feet thick in El Capitan Meadow (his Figure 27), behind the moraine.

In a June 1993 letter, Prof. Clyde Wahrhaftig defended Lake Yosemite's existence, and in his own words his strongest argument is the record from these two wells and a third, shallower one:

> But there is now another, more compelling piece of evidence for a post-glacial lake. I have copies of logs of three [out of four] wells that have been drilled into the fill beneath the valley floor for ground-water [exploration]. This fill (of till and/or lake sediments) was shown through seismic studies by Gutenberg, Buwalda, and Sharp (GSA Bull, v. 67, p. 1051, 1956) to extend to a maximum depth of about 2,000 feet. Two of the wells encountered only sand and silt to a depth of more than 400 feet, well below the altitude of the bedrock lip of this glacially-excavated basin, before encountering gravel, cobbles, boulders, mixed with what the drillers called "glacial rock flour". The third well, located close to Yosemite Creek not far from the base of the falls, encountered gravel from 8 to 12 feet, before passing into sand and silt to a depth of slightly more than 400 feet, at which it passed into gravel, cobbles, sand, glacial rock flour, etc. Two of the wells were drilled to depths of about 1,000 feet, and the third to over 600 feet, in material such as gravel, sand, silt, and boulders, until they were stopped either by a large boulder or granitic bedrock. To me these drillers logs, together with the complete lack of glacial boulders anywhere in the river banks or on the valley floor west of the "medial moraine", indicate that the valley was occupied by a post-glacial lake.

But are the upper 400+ feet *post-Tioga* lake sediments? For that matter, are they even *lake* sediments? The sediment logs of the two deepest wells are reproduced as Table D in the Appendix, and although the wells are only about 520 feet apart, they have significant differences. First, the east site has seven layers of glacial flour, the west site none. The "glacial flour" of the east site clearly is a judgment call, being silt or sand of the west site. The silt could be from a Merced glacier or from the Merced River during deglacial and interglacial times. Tenaya Lake lies about 8 miles above the east end of Yosemite Valley, and once the retreating, giant Tioga-age Tenaya glacier, reached this vicinity, it would have deposited all sediments in the large lake's basin. These sediments, however, are only about 10 feet thick. It is unbelievable that the retreating late-Tioga Tenaya and Merced glaciers could have generated sufficient glacial flour to fill in Lake Yosemite by 394 feet—the thickness of the uppermost layer of glacial flour. It is even more unbelievable when one realizes that when the Tenaya glacier retreated to above Tenaya Lake and the Merced glacier to above Merced Lake, these lakes would be receiving the sediments, not Lake Yosemite. These lakes, much smaller than Lake Yosemite, were not filled in with sediments.

A second problem with Wahrhaftig's interpretation is that while he mentions that two of the three wells encountered only sand and silt to a depth of more than 400 feet, he ignores the east site presented in Table D, which contains cobbles and boulders. Also, in the sediments from the two additional, shallower wells there were pebbles, cobbles, and rock fragments occurring intermittently from about 216 feet down to their ends at 255 feet and 259 feet. In post-Tioga time both the Merced River and Tenaya Creek would have transported these to the hypothetical lake, but upon reaching it, their velocities would have dropped to zero and the sediments likewise would have been dropped. They could not have been transported 2 miles through the lake to the well sites. However, Gutenberg *et al.* recognized three principal sedimentary layers, the upper one about 400 feet thick

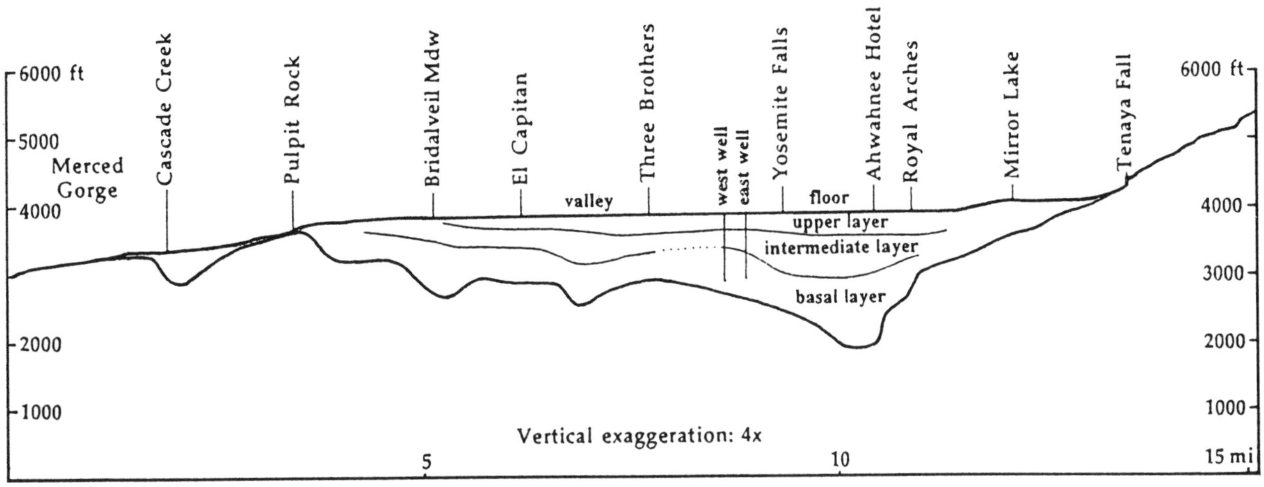


**Figure 31. Three principal sedimentary layers in Yosemite Valley.**

(Figure 31). Therefore, the upper layer could represent post-Tioga sedimentation, the intermediate one post-Tahoe, and the basal one post-pre-Tahoe.

In contrast, Robert Sharp, who wrote the geological interpretation, interpreted these respectively as post-Tahoe, post-El Portal, and post-Glacier Point or older. (Recall from Chapter 12 that the Glacier Point is not a valid glaciation, although glaciations certainly existed before the El Portal/Sherwin glaciation.) Very importantly, the upper layer extends not only to the El Capitan Meadow moraine but beyond it, to the 40-foot-lower Bridalveil Meadow moraine. If so, then the lake that was filled by these sediments was dammed by the Bridalveil Meadow moraine, *certainly not* by the El Capitan Meadow moraine, and the uppermost lake sediments would lie about 40 feet beneath El Capitan Meadow. But were there only three layers, and were they accurately delineated? Perhaps not, for in 1989 Sharp wrote me about his work in Gutenberg and others. His "geological interpretation of layering within the fill is rank speculation, just a shot in the dark." Consequently any attempt at using the seismic study's stratigraphy to argue for the age, extent, and depth of a lake is laden with uncertainty.

Given that the two well sites are far from any stream, then if a deep lake existed, the only sediments expected in their vicinity would be organic-rich clayey sediment as found in existing Sierran lakes. In these lakes silt, sand, and gravel occur increasingly closer to shore, not 1400 feet from it, which would have been the case for Matthes' post-glacial Lake Yosemite. The sediments at the sites, rather than being lake sediments, appear to be stream sediments, not unlike those deposited before the river-rock blasting in 1879, back when the Merced River had meanders that migrated unconfined across the valley floor. Not only did the river's course change with time, but in many years it flooded the valley floor. James Mason Hutchings, in his tourist's guide, described the largest historic flood known:

> On December 23, 1867, after a snow fall of about three feet, a heavy down-pour of rain set in, and incessantly continued for ten successive days; when every little hollow had its own particular water-fall, or cascade, throughout the entire circumference of the Valley; each rivulet became a foaming torrent, and every stream a thundering cataract. The whole meadow land of the Valley was covered by a surging and impetuous flood to an average depth of nine feet. Bridges were swept away, and everything floatable was carried off.
>
> Immense quantities of talus were washed down upon the Valley during this storm,—more than at any time for scores, if not hundreds, of years, judging from the low talus ridges, and the timber growth upon them.

Actually, documentation of historic and prehistoric mass wasting in Yosemite Valley by Wieczorek *et al.* shows that debris flows have been much more common than what Hutchings suggests. Historic records show that notable floods since 1867 and through 1991 occurred in 1890, 1919, 1937, 1950, 1955, 1963, 1964, 1980, 1982, and 1997, which amounts to about seven or eight per century. If that rate is extrapolated over the approximately 14,000 years the valley has been glacier-free, then about *1000 major floods* have occurred. Every time, a temporary Lake Yosemite would have formed on the floodplain behind the El Capitan Meadow moraine, and silt ("glacial flour") would have been deposited. As the lake diminished, the Merced River and its tributaries would have deposited sand, gravel, and cobbles. More numerous, nondestructive floods also would have occurred, as they have in the historic past, and in addition there would have been myriad localized torrents. For example, one historic cloudburst occurred over Tenaya Canyon on June 28, 1977, and in the canyon caused a flash flood, generated debris flows, and took two lives—and this in the *driest* year on record! One local phenomenon is the development of a tremendous snow cone at the base of Upper Yosemite Fall in years of high precipitation. Then if it gets saturated with abundant water (due to high snowmelt and runoff in hot weather), the snow cone moves *en masse,* flowing over Lower Yosemite Fall and entraining boulders at its base and transporting them downstream. By this process boulders can reach the vicinity of the two well sites along Yosemite Creek.

In the past this process could have contributed gravel and boulders to these sites, as could the Merced River at times when it was meandering across them. In short, the stream processes operating and observed in historic time, when projected over post-Tioga time are sufficient by themselves to explain the development of the valley's floodplain. Lake Yosemite is an unnecessary hypothesis. Furthermore, as Table D and my analysis shows, the sediments do not resemble those found in Sierran lakes. After the last glaciation Yosemite Valley's floor may have held a large marsh, but definitely not the 300-foot-deep lake envisioned by Matthes and Wahrhaftig. During the thousands of years of flooding and deposition in the valley, whatever erratics the late-Tioga glacier left would have been buried under sediments—hence their absence. Large erratics exist in glaciated valleys that lack wall-to-wall flooding, such as lower Lee Vining Creek canyon, above Mono Lake. They are absent where wall-to-wall flooding occurs, such as Lyell Canyon, above Tuolumne Meadows. Therefore, lack of surface erratics does not prove the existence of a deep lake.

Nor does complete lack of erratics anywhere in the Merced River's banks give support, since along most of the river's course through the valley it is above what would have been the surface of the till left by the Tioga glacier; its banks expose only post-glacial floodplain sediments. Furthermore, "complete lack of glacial boulders anywhere in the river banks or on the valley floor" is an overstatement by Wahrhaftig. Beginning in the 1880s, large granitic boulders called rip-rap revetments were placed to cover entire sections of river bank subject to stream erosion. The total length of this protective covering—which hides the river banks (and possible erratics) from view—was 14,518 feet. All or virtually all of this would have been laid before Wahrhaftig began his research in the park; consequently he cannot validate his claim.

Matthes proposed not only that huge Lake Yosemite was completely filled in post-glacial time, but also that some 14 to 16 feet of valley-floor sediments had been removed probably as the result of a rather sudden breaching of the moraine dam to its current level, and then renewed meandering widened the modern floodplain at the expense of the old one. If this were the case, then some erratics might be expected in the river banks. However, his only evidence for this older floodplain is a "particularly prominent terrace" along Tenaya Creek near the head of the valley. This is exposed along the north bank of the creek just downstream from Tenaya Bridge. The sediments exposed in the bank are mostly coarse sand and certainly are not the fine organic-rich silts and clays typically found in Sierran lakes. Furthermore, the terrace, while nearly level in its western part, dips gently to the south and southwest in its eastern part. Finally, the terrace is anomalously absent along the creek's south bank. It therefore appears to be the distal end of a former alluvial fan that now is mostly buried under the massive Mirror Lake rockfall. Hence there is no evidence of a higher valley floor.

Eliot Blackwelder appears to have been the only person before me, back in 1977, to question—albeit indirectly—the existence of a deep post-glacial lake. Apparently like me he wondered why Tenaya Lake and Merced Lake, both miles up-canyon from the proposed Lake Yosemite, were not filled with sediments while much larger Lake Yosemite was. For this reason and others he concluded that the last Tenaya and Merced glaciers to enter the valley were of Tahoe age. Hence the lake had been receiving sediments since then, which essentially is what I suggested at the end of the "Cascade Creek bridge till" section in the previous chapter. Nevertheless, I, like other modern investigators, have concluded that the fresh subsurface boulders in the moraines argue for Tioga glaciers advancing through the valley, in contrast to Blackwelder's conclusion that they did not.

Once the Tenaya glacier retreated to the Tenaya Lake basin and the Merced glacier to the Merced Lake basin, all further deposits left by them would have been transported downstream no farther than to each lake. In addition, the time from when the two glaciers retreated from the valley back to these two basins must have been relatively short—on the order of 1000 to 3000 years at most, if not considerably shorter—based on dates of post-glacial lakes in the Sierra Nevada and elsewhere. In contrast, Tenaya and Merced lakes, like most of the Sierran lakes, have been receiving sediments for about 13,000 to 14,000 years, and are far from being filled in, so how was Lake Yosemite filled in, especially in so short a time?

Ironically, Matthes used 80-foot-deep Merced Lake and 86-foot-deep Washburn Lake just upstream as evidence for the infilling of Lake Yosemite. He suggested that each lake had already been reduced to two-thirds its original length by post-glacial sedimentation. A common belief today is that relatively flat lands immediately above a glacial lake are lake fill, even though back in 1975 Spencer Wood proved that post-Tioga Sierran meadows never were former lake basins. Old myths die hard.

Matthes did not address 180-foot-deep Tenaya Lake, which he himself had sounded in 26 places, on July 17, 1919. This lake has a surface area more than twice that of Merced and Washburn combined and has twice their depths, so it should be a volume at least four times that of the two lakes combined. Unlike those two, it has been reduced by only one fourth of its original length, if—and this is a big if—the relatively flat lands immediately above it are interpreted as deposits in a former lake basin. And the sediments within the lake are not hundreds of feet thick, as theory would have us believe, but instead are only about half as thick as those in a typical (much smaller) Sierran lake. In 1995 Eric Edlund and I, investigating the nature of the protruding tree trunks in the lake, reached glacial flour after coring through only 6 feet of sediment (Photo 127). Drowned trees? In 1994 Prof. Scott Stine proposed that the lake dried up and a forest grew there, then it was flooded as the lake filled once again after a centuries-long, *extreme* drought, which by my calculations had an average annual precipitation of no more than 4 inches per year. Most of California would have had an inch or less, and almost all of the state's vegetation should have died. Sediment records around the state indicate this did not happen. In all the sediments Eric Edlund and I have extracted, we have found no evidence that any of the relatively shallow lakes we cored dried up. What I suggest happened at Tenaya Lake is that a major east-side earthquake generated massive rockfalls from the high cliffs above the lake, which swept trees into the lake. They eventually became waterlogged and sank, heavy end first, some of them lodging in the deep lake's sediments, and some not, just lying flat. Almost all Sierran lakes are

**Photo 127. Eric Edlund beside a tree stump in Tenaya Lake.**

too shallow and/or lack high cliffs to allow this process to happen, which is why upright trunks are so rare. However, *east-side lakes,* including Tahoe, which lie in the Sierra's rain shadow, do lower their surfaces in droughts, and in the past trees grew on their exposed floors.

Returning to the matter at hand, large lakes, Wahrhaftig, like Matthes and Blackwelder, was aware that Merced, Washburn, and Tenaya lakes exist, while Lake Yosemite does not. In his aforementioned letter he proposed a possible mechanism to explain why Lake Yosemite was filled in while the three others were not:

> Exactly where all the sediment came from to fill the lake, in view of the fact that there are deep glacial lakes upstream from Yosemite Valley, *is a mystery to me,* unless it came from glacial meltwater while the glacier fronts were retreating back to Tenaya and Merced Lakes. (Italics mine.)

This proposed process, however, destroys his hypothesized *miles-long,* deep lake, since as the lake was being formed by the retreat of the glacier's front, it was also being filled in with glacial sediments. So at any given time during the glacier's retreat, only a short length of lake existed. Be aware that these sediments would have abounded in pebbles, cobbles, and boulders, but the wells' sediment logs show mostly silt and sand.

If the lake was not being destroyed almost as rapidly as it was being created, but rather was destroyed mainly through post-glacial deposits of the Merced River and Tenaya Creek, as Matthes suggested, one can test this hypothesis by determining the annual rate of sediment transport into the valley and extrapolating it over post-glacial time. This I did in 1976 and then in 1977 published my results, but in summary form only, which did not explain my methods. In essence I calculated the volume of sediments (lake and overlying fill) in Gutenberg *et al.*'s upper layer and divided this amount by my calculated average historic rate of sediment transport into the valley to arrive at the time required. This volume depicted by Gutenberg *et al.* is greater than that proposed by Matthes, for not only is their proposed lake depth greater, but their lake also extends west to the Bridalveil Meadow moraine. I used their study rather than Matthes, since it was based on data, not on speculation. Using the bedrock and discontinuity contours drawn by Gutenberg and others, I calculated the volume of the sediments (lake and overlying fill) between the upper discontinuity and the valley floor at 0.218 cubic mile. This does not sound like much, so I'll convert it to cubic yards: about 1,190,000,000 (1.19 *billion*) cubic yards. Despite the enormity of today's Tenaya Lake, about 27 *million* cubic yards, its volume of water pales in comparison with that of Lake Yosemite. By my calculations, that lake's volume was about 43½ times the volume of Tenaya Lake. At the current rate Tenaya Lake is being filled with sediments, about ½ foot per 1000 years, some 200,000 years will be required to fill it. How much time would be required to fill Lake Yosemite, given today's rate of sediment transport by the Merced River and its tributaries?

At the time of my 1976 study, sediment-transport data for the Merced River were sparse at best. Measurements existed for a total of only 18 days during the water years of 1970-71 and 1972-73, these taken at the Merced River's Happy Isles stream-gauging station in the east end of the valley. I graphed these against the stream discharges for those 18 days, the points producing a good linear relationship on log-log paper (tons per day vs. cubic feet per second). Next I examined the daily discharge for the latest 14 years available at the time (1961-1974), which were the most detailed in the 59-year record. Using the graph, I then determined the daily sediment transport for several of these years to arrive at their annual sediment transport. I graphed these against annual discharge, the points producing a good linear relationship on log-log paper (tons per year vs. acre-feet per year). The annual discharge was quite variable during these 14 years, ranging from a low in the 1960-61 year of 114,200 acre-feet to a high in the water year of 1968-69 of 427,100 acre-feet. The sediment transports for these two years, based on the summation of daily amounts derived from the second graph, were 509 and 8033 tons respectively. The average amounts for this 14-year span were 263,400 acre-feet and 3002 tons of sediment per year. Since the average of the 59-year discharge record was slightly lower—249,200 acre-feet per year—I used this value to extrapolate back in time, and the corresponding sediment transport was about 2500 tons per year.

However, the 2500-ton figure was only for the Merced River at Happy Isles, whereas sediments are also transported into the valley by a number of creeks. All are insignificant except for Tenaya Creek, Yosemite Creek, and Bridalveil Creek. Since Bridalveil Creek leaps via Bridalveil Fall into the valley at a point just downstream from the west end of Matthes' and Wahrhaftig's Lake Yosemite, it did not contribute sediments to the lake. Although no reliable hydrologic data exist for the two remaining creeks, their sediment transport can be estimated. At the far west end of Yosemite Valley the Pohono Bridge stream-gauging station records the total discharge from the valley. The discharge at the Happy Isles station is about 60% that of the Pohono Bridge station, so the remaining discharge must be from other streams. Assuming that discharge was directly related to drainage-basin area, I estimated that Tenaya Creek contributed about 18% of the total discharge, Yosemite Creek about 8%, and the others (mostly Bridalveil Creek) the remainder. Using the second graph I determined that the annual sediment transport for Tenaya and Yosemite creeks combined was about 330 tons per year, giving a total annual sediment input into Yosemite Valley of 2830 tons per year. By estimating the density of the sediments, I arrived at an annual volume, and calculated that about 330,000 years would have been required to fill in Lake Yosemite and deposit sediments above it; slightly more than 3,000,000 years would have been required to fill in the entire basin, from bedrock floor to the valley's floor.

This analysis had two underlying assumptions. The first was that essentially all sediments were trapped behind the El Capitan Meadow moraine. This situation was likely while the lake would have existed, but once the lake was filled in, most of the transported sediments should have been carried by the Merced River through the valley, not deposited in it. Consequently, stream deposits would accumulated only when the river flooded its banks and left silt and sand on the river's floodplain (Photo 128). The second assumption is that the annual sediment-transport rate has been constant in post-glacial time. But as Stine repeatedly states, this time has had long periods of droughts (although not as severe as he would have them be). During them the discharge of the Merced River and its tributaries would have been less, and the sediment-transport rate would have been much less. Therefore the amount of post-glacial sediment deposition actually would have been lower—and the required lake-filling time longer—than I had calculated. Note that although during the short-lived Little Ice Age, glacial and periglacial environments could have produced greater amounts of sediments, almost all of these would have been trapped in lakes upstream from the valley, and so the sedimentation rate in the valley would not have been greatly affected.

Just how unlikely the concept of a deep Lake Yosemite is can be demonstrated by calculating the amount of fill that would have been deposited in it over the last 14,000 years of post-glacial time. The lake, as depicted on Maps 30 and 31 (and based on Matthes' final reconstruction—his earliest one was significantly smaller), has an area of about 9.90 million square yards. At an annual sedimentation rate of 2830 tons, about 1656 cubic yards would have been deposited each year. This assumes a 2.7 average density of sediment particles and a 25% porosity. The porosity figure is based on average ones for Miocene through Pleistocene sediments in G. Edward Manger's Table 1, these 22 sediments having an average of 23%. Because younger sediments tend to be more porous, I rounded the figure to 25%. Then Lake Yosemite, coming into existence about 14,000 years ago, would have

**Photo 128. The Merced River about to spill across its floodplain. North Dome rises above Royal Arches.**

accumulated sediments averaging 7 feet in thickness—about a foot more than Eric and I found at Tenaya Lake.

However, there is s faulty assumption that I used in my analysis (before my lake-coring days). Most of Tenaya lake's sediments, like those in other Sierran lakes, are derived not from stream transport, but rather from windblown silt, dust, and pollen. Therefore, this windblown amount needs to be added in. By using Tenaya Lake's sediment thickness—6 feet—as a proxy, then in 14,000 years the Lake Yosemite basin would have received an average of about 13 feet of sediments.

Another consideration is that the 18 days of sediment-transport data at the Happy Isles stream-gauging station failed to measure bed load—the amount of debris rolling along the bottom of the stream. How significant was it? Table D shows the vertical composition of sediments at two sites. If one is *very* generous and ascribes *all* sand and coarser material to bed load, then for the west well about 32.8% of the sediments was transported as bed load and for the east well about 58.4%, resulting in an average bed load of 45.6% and average suspended load of 54.4%. The total volume of annual sediments was about 3044 cubic yards, 1656 cubic yards in suspension and 1388 cubic yards along the bed. Then over 14,000 years the Lake Yosemite basin would have received an average thickness of about 13 feet. Add in the windblown sediments and you get a total of 19 feet. Both Matthes' Yosemite Valley cross profiles and Gutenberg *et al.*'s seismic data indicate a lake with an average depth of about 200 feet, so a *post-Tioga* Lake Yosemite would have lost only about 10% of its volume to infill.

The surface of the lake, marked by the elevation of the Merced River at the El Capitan Meadow moraine, would have been at about 3940 feet. At the east end of the valley the sediments are about at 3980 feet elevation, or about 40 feet higher than at the supposed damsite. If the average thickness of the stream deposits above that site is half the value, or 20 feet, then the amount of total sedimentation over the last 14,000 years would account for virtually all of these stream deposits—if one makes the absolutely unwarranted assumption that all of the sediments were trapped behind the damsite. There would be no leftover sediments for even a cubic yard of lake fill. Therefore, the reason Tenaya, Merced, and Washburn lakes exist today whereas Lake Yosemite does not is because that lake, even as a body averaging even a few yards deep, never existed after the Tioga glaciation. At best, a *marshy* lake would have existed.

As was mentioned before, a lake would appear after the end of each earlier glaciation, and the last lake probably would have been filled between the Tahoe and Tioga glaciations. The time between these two included three cool periods, the Eowisconsin, early Wisconsin, and middle Wisconsin (oxygen-isotope stages 5d-5a, 4, 3—see Table 3). If each period produced glaciers that were only about one-third to one-half the length of the Tioga and Tahoe glaciers, then the glaciers would have advanced to just beyond Tenaya and Merced lakes, and their prodigious amounts of sediments would have been transported to Yosemite Valley. The time span between the Tahoe and Tioga glaciations, as shown in Table 3, is 97,000 years, or about 6.9 times as long as the 14,000-year post-glacial existence of the floor of Yosemite Valley. At the historic rate calculated above (suspended *plus* bed), an average of about 89 feet of sediments could have been deposited during this time span. But during glacial times the rate would have been much higher. Therefore, if the average rate over the 97,000-year period had been double, then a lake of the magnitude envisioned by Matthes, by Wahrhaftig, or even by Gutenberg *et al.* would have been filled in before the onset of the Tioga glaciation.

## Summary of the Merced River drainage

Under my reinterpretation of generally known glacial evidence in Little Yosemite Valley, lower Illilouette Creek valley, Tenaya Canyon, Yosemite Valley, and Merced Gorge, problems with past reinterpretations, especially those raised by Wahrhaftig, disappear. These problems he repeatedly called "mysteries;" I call them riddles, which need a solution. Along with their solutions these are the following.

First, the Tahoe moraine diving beneath a Tioga moraine east of and above Little Yosemite Valley indicates that the Tahoe glacier was shorter than the following Tioga glacier. However, the diving moraine is merely a till-veneered descending bedrock ridge. The evidence as I have interpreted it indicates that the Tahoe glacier, like all other Sierran Tahoe glaciers, was thicker, and therefore longer, than the Tioga glacier.

Second, the diving moraine as well as the lack of Tahoe moraines above Tioga ones at the western end of Little Yosemite Valley (and elsewhere) indicate that the Tioga glacier was thicker and therefore longer. The evidence as I have interpreted it indicates that only in Yosemite Valley and along parts of the Merced and Tenaya glaciers just up-canyon from it was the Tahoe ice surface lower than the Tioga ice surface. Elsewhere above and below this extent the Tahoe ice surface was, as with all other Sierran Tahoe glaciers, higher than the Tioga ice surface. The reason for the anomaly in and just above Yosemite Valley is that the base of the Tahoe glacier was lower (by perhaps 100 to 300 feet) than was the base of the Tioga glacier; hence its ice surface was lower, despite the Tahoe glacier being slightly thicker than the Tioga glacier. One can generalize that when erosion by the earlier glaciers finally yielded to deposition later ones, this transformation presumably occurring after the Sherwin glaciation (as discussed later), the base for each glacier became successively higher with each additional glaciation. One can also generalize that during the early glaciations each succeeding glacier eroded the very weathered bedrock of the valley's floor deeper than had the previous one.

And third, the Cascade Creek bridge till is pre-Tahoe. However, the degree of weathering of subsurface boulders demonstrates that its age can be no older than Tahoe. This indicates that the Tahoe glacier here was about 1000 feet thick, and a glacier of that magnitude would have extended down to about the lower end of Merced Gorge. Doing so, the glacier—and the slightly smaller Tioga glacier—would have advanced down to elevations comparable to those of the Tuolumne and Stanislaus river drainages. Furthermore, the relative lengths of the Merced's Tahoe and Tioga glaciers, compared to the length of its greatest pre-Tahoe glacier, would have been similar to those of Tuolumne's and Stanislaus' Tahoe and Tioga glaciers compared to their greatest pre-Tahoe glaciers. Hence Divine Providence is not required, although it is for Wahrhaftig's interpretations. Wahrhaftig accepted his "mysteries" as facts to be religiously believed. On three occasions I requested that he propose a logical explanation and/or testable hypothesis for each of them, and on each occasion he refused to do so, would not consider any other alternative explanation, and said I should accept his pronouncements as fact. This is bad science.

My reconstructions of Tioga, Tahoe, and pre-Tahoe glaciers in the Merced River drainage indicate that their relative thicknesses and lengths fit the same pattern as the Tioga, Tahoe, and pre-Tahoe glaciers in the Tuolumne River drainage. One might expect all of the Sierra's western drainages to have a similar pattern, and as shown in the next chapter—on glaciation in drainages from the Mokelumne River south to the Kaweah River—they generally do, with one notable exception. Whereas Tahoe and Tioga glaciers appear to have been shorter than pre-Tahoe ones in the southern Sierra Nevada, the opposite appears to have been true in the northern half. In the Yuba River and Feather River drainages—that is, from about Donner Pass northward—there is no evidence of pre-Tahoe glaciers. In the next drainage south, the Mokelumne, pre-Tahoe glaciers appear to have been slightly larger than later ones, judging by erratics along the divide separating the Mokelumne and North Fork Stanislaus river drainages. As was mentioned earlier, the glaciers of the Mokelumne River drainage spilled over into the North Fork Stanislaus River drainage, filling its upper part to the divide and spilling back into the

Mokelumne River drainage via Horse, Snow, and Grouse valleys. However, erratics continue southwest along the ridge for about 2.4 miles and then west along the Flagpole Point ridge (Section 16) for another 1.2 miles. The glacier leaving all these erratics would have been sufficiently thick to flow across these two ridges and at least part way down into the Mokelumne's hanging tributary canyons along this stretch. These canyons, from east to west, are Corral Hollow, Bear Trap Basin, Jelmini Basin, and Mattley Meadow. Bear Trap Basin, which would have received the greatest overflow, should have had the most abundant glacial evidence. The generalized geological map for this area by Wagner *et al.* does not show glacial deposits in any of the canyons. And in a brief reconnaissance I found no moraines, which should have been prevalent and readily observable in Bear Trap Basin—if left by a Tioga or Tahoe glacier. From their absence I surmise that they were left by a larger pre-Tahoe glacier and in ensuing time, as elsewhere in the western Sierra Nevada, were weathered and eroded away. To the south, in the Middle Fork Stanislaus River drainage, the evidence is unambiguous that a pre-Tahoe glacier was considerably larger than the Tioga and Tahoe glaciers, and this pattern continues southward.

A plausible explanation for larger pre-Tahoe glaciers in the southern part of the range but not in the northern part is that the Coast Ranges of central California (south of San Francisco Bay) experienced major uplift just after the Sherwin glaciation, and so a significant rain shadow developed. Therefore, ensuing glaciers (e.g., Tahoe and Tioga) were reduced. However, north of the bay the northern Coast Ranges began uplift about 3.4 million years ago, and it may have tapered off by the Sherwin glaciation. If so, then post-Sherwin glaciers would not have been significantly affected. Evidence that uplift of the North Coast Ranges has been minor is that their marine terraces are fewer and lower than in coastal lands south of them.

# 20

# The Sierra's Other Glaciated River Canyons

"The Blind Men and the Elephant," Chapter 6, was about Sierran geologists who between the times of Josiah Whitney and François Matthes briefly visited Yosemite Valley, and then greatly overextrapolated their meager data to reach unwarranted, faulty conclusions with regard to the valley's formation. That was true not only in Yosemite Valley during the late 1800s, but also in virtually all studies on the uplift and glaciation of the Sierra Nevada from Whitney's time to the 1990s. Every geoscientist would investigate a specific river drainage (or sometimes just a small part of one), and from the evidence gleaned would extrapolate the conclusions to the entire range. By not checking other drainages, one did not have to address conflicting evidence. Consequently, over time, various drainages, examined by different geoscientists, acquired their own uplift and glacial histories. By the mid-1960s, when I was more interested in climbing the walls of Yosemite Valley than anything else, a sufficient number of discordant histories had accumulated that some geoscientist should have become alarmed. No one did. If you read Paul Bateman and Clyde Wahrhaftig's lengthy, comprehensive chapter, "Geology of the Sierra Nevada," in the Bulletin 190 of the California Division of Mines and Geology, you see no evidence that anything about uplift, glaciation, and landscape evolution was awry.

It was the calm before the storm. After the bulletin's 1966 publication, the newly created branch of geology—plate tectonics—quickly became accepted, and there arose additional conflicting uplift scenarios based on plate tectonics or its implications. The ultimate conflict occurred in 1996, as I mentioned back in Chapter 3, when two papers were published by different members of the Southern Sierra Nevada Continental Dynamics Working Group. Using the same data but different modeling, one paper supported major late Cenozoic uplift, while the other supported major late Cenozoic subsidence. If the latter conclusion were correct, then the Sierra Nevada would have been high enough until at least 20 million years ago to have supported glaciers at various times before then. This interpretation opens a whole can of worms on just how much glaciation has occurred in the Sierra Nevada.

But one needs neither theories nor models to solve the riddles in the rocks. As I have shown in Chapters 15 through 19, field evidence works just fine. At least in the Merced River drainage. What about the other drainages? If there is merit to my interpretations, then there should be field evidence for similar-size past glaciers in the other glaciated drainages. I would be remiss if I derived only the glacial history of the Merced River drainage and then categorically applied it to the other glacial drainages. So here is some of the evidence for similar glacial histories in those drainages. As in the previous chapters, some of the glacial history that I present is necessarily complex, especially for the Stanislaus River drainage, and perhaps most readers will gloss over this chapter. As in previous chapters, you may get the gist of the evidence from this chapter's photographs and their captions and from the "Conclusion of Glacial Evidence," end of this chapter, which sums up the evidence in all of the western drainages, including the Merced. For the sake of serious geoscientists, however, I feel I must present some specific evidence in these western drainages, which they can visit and evaluate for themselves.

## Mokelumne River drainage

Because the North Fork Stanislaus River glaciers were fed by overflow from the upper Mokelumne River glacier, I investigated glacial evidence there in considerable detail. Sufficient erratics, polish, and striations exist to reconstruct the Tioga ice surface, and I found no evidence

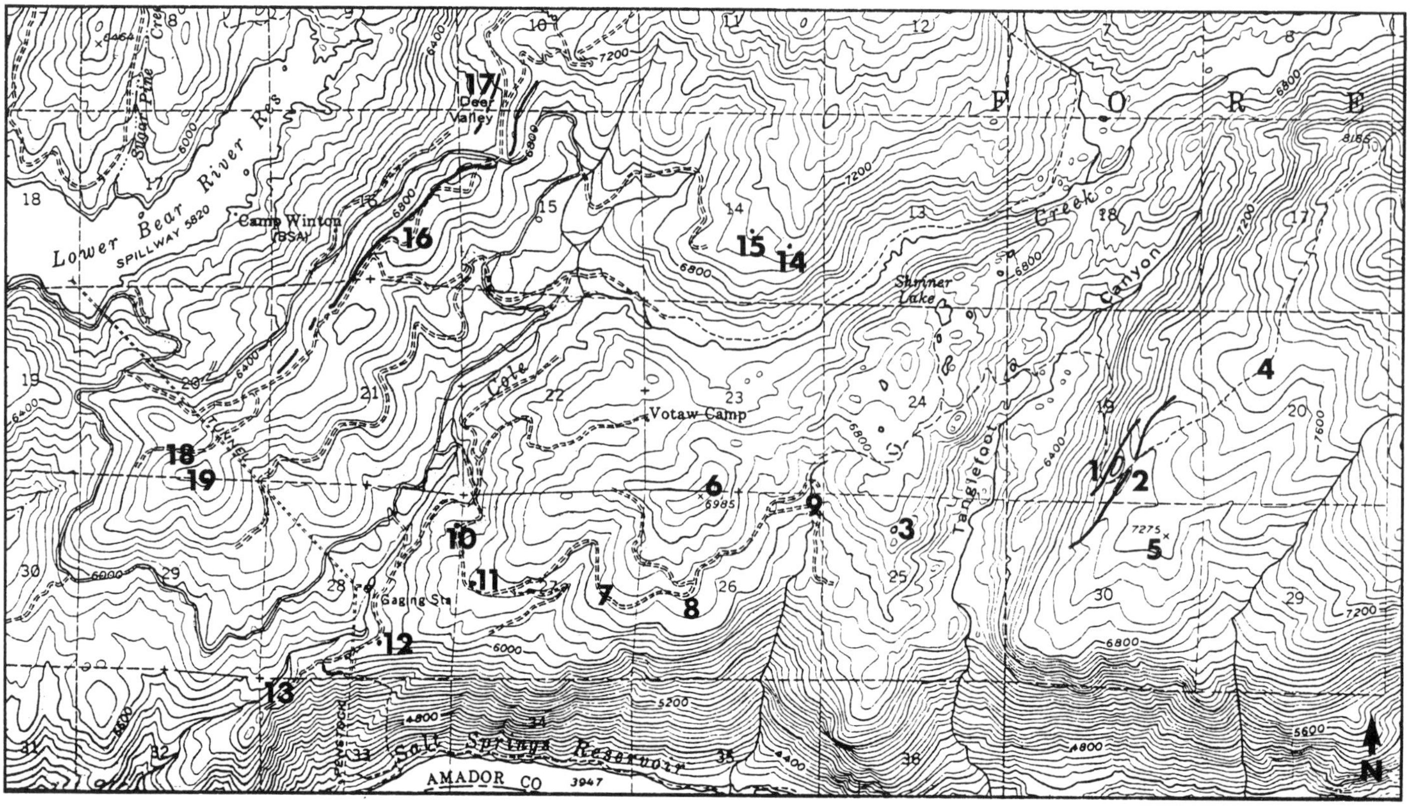


**Map 33. Glacial evidence in the Bear River-Tanglefoot Canyon area. Scale 1:62,500.**

to indicate that the Tahoe and pre-Tahoe ice surfaces were any higher. Because this area is well above the equilibrium-line altitude, no moraines exist, and hence, there is no controversy in interpretations of ice surfaces. (Unlike in major drainages from the Middle Fork Stanislaus River south, in this one and the adjacent North Fork Stanislaus River drainage the heads of canyons were too low to support Little Ice Age glaciers; consequently none of the cirques have Holocene-age moraines.) To the west below the equilibrium-line altitude are two areas with moraines and controversies. The first area is the Mt. Reba area, which has diachronous lateral moraines at the mouths of three tributary, hanging canyons high above the south slopes of the main canyon. Those moraines were discussed in Chapter 14. The second, larger area is a few miles west of the first one, above the north slopes of the main canyon. It includes the hanging tributary canyons of Tanglefoot and Cole creeks and the nearly accordant tributary canyon of Bear River. In this area prominent lateral moraines exist, which are shown on Map 33 (Silver Lake 15' quadrangle).

Roads are quite plentiful in this area, making it ideal for investigation. Furthermore, Bear River Lake Resort, situated about 2 miles in along Forest Service Road 8 from Highway 88, has a campground, lodge units, restaurant, and boat rentals, making it an attractive base camp for field work in this area.

Only the two conspicuous lateral moraines above the east side of Tanglefoot Canyon require a moderate effort (about a 4-mile walk) to reach. Under conventional interpretation the lower, inner moraine would be classified as Tioga (site 1), and the upper, outer one as Tahoe (site 2). Sandwiched between them is one of the Sierra's few truly moraine-dammed lakes, which quickly loses water after the local snowpack melts (Photo 129). (Despite viewing hundreds of Sierran lakes, I can recall only one other leaky lake: Yosemite's Harden Lake.) Corresponding moraines above the west side of the canyon are absent, although till is locally abundant. Of particular interest on that side is a seasonal lakelet at site 3, which stands at about the limit of the Tahoe ice surface, making it—suspiciously—Tahoe in age. Actually, there is no reason to believe either lateral moraine is anything but diachronous late-Tioga. They exist because they are along a sharply defined, two-stepped canyon rim. They pinch out at their upper limit not because the equilibrium-line altitude was located there but because the topography transforms from a sharp rim to a broad, rounded slope. They dive steeply at their lower ends (implying that neither glacier reached the trunk glacier) simply because the bedrock surfaces on which the moraines were deposited dive steeply. That these glaciers were larger can be concluded from the many erratics that exist on a generally broad ridge that extends from about site 4 southwest past summit 7275, site 5. Photo 130 is a view east-northeast from that vicinity, which shows erratics in various degrees of burial in a matrix of pebbles, gruss, and clay. Because I did not see any exhumed erratics, I cannot say if they were left by a Tioga or a Tahoe glacier, although I favor the former since I did not see weathering pits on them. The erratics are not pre-Tahoe because they are too abundant, they are too small (most less than a yard across), they are too fresh, and they are partly buried in soil and/or colluvium (almost all of the hundreds of pre-Tahoe erratics I've seen lie directly on bedrock).

The Tioga and Tahoe glaciers, even as conventionally interpreted by the two lateral moraines, must have buried the broad, irregular, granitic ridge between Tanglefoot and Cole Creek canyons, but how far southwest along that ridge did they extend? Evidence for the magnitude of the Tioga glacier is in the fresh erratics that extend southwest to summit 6985, site 6 (summit 7018—33 feet higher—on the Bear River Reservoir 7.5' quadrangle), which is above the ice surface derived from the lower ends of the two lateral moraines at sites 1 and 2. From that summit erratics exist west down its main ridge and also southwest down a secondary ridge. On that ridge at site 7 is a cut through till where Road 8N14 bends from south to east, and it exposes subsurface granitic boulders that are mostly weathered, i.e., Tahoe. At site 8, where the road turns from east to north, is a broad ridge that lacks erratics and till. In an alcove, site 9, there are abundant fresh erratics and some till. These observations indicate that both the Tioga and Tahoe glaciers were thick enough to cover the ridge between Tanglefoot and Cole Creek canyons, but not thick enough to spill any significant till and/or erratics down the south slopes of summit 6985. Additional road cuts at sites 10, 11, and 12 also expose weathered subsurface granitic boulders, indicating that the Tahoe glacier must have reached the Mokelumne's trunk glacier. The linear ridge, site 13, appears on the map (and in aerial photos) as a lateral moraine but instead is a bedrock ridge. Therefore the ice surfaces of the Tahoe and Tioga trunk glaciers were not this high, for had they been, they should have left a lateral moraine on the ridge. In my brief, incomplete examination of it, I did not find any erratics (although a few might exist). I interpret this lack (or almost lack) of erratics as due to the Tahoe glacier spilling across the bedrock ridge. Elsewhere in the western Sierra Nevada where glaciers spilled across ridges, only a few erratics typically were left. In my brief reconnaissance I was unable to determine the extent of the Tioga glacier in the Cole Creek drainage, but based on large and abundant erratics, it appears to have advanced to within about 0.3 mile of site 12.

**Photo 130. Erratics on broad summit 7275. Erratic in left foreground is volcanic; all others in view are granitic. View is east-southeast.**

**Photo 129. Moraine-dammed lake above Tanglefoot Canyon. Note abundant boulders in both of the lateral moraines. Also note that lateral moraines are absent on the west rim.**

About 3 miles upstream, at site 14, there is a bedrock ridge that is profusely mantled with erratics, indicating that the ice surface of a glacier there was at least this high—7170 feet. The erratics appear to be of Tioga age, and this conclusion is supported by evidence at site 15, just west and slightly lower (7130 feet), where a logging operation has exposed fresh, angular—i.e., Tioga—subsurface erratics (Photo 131). These sites, along with site 6

**Photo 131. Subsurface erratics exposed in logging operation, Cole Creek canyon.**

**Photo 132. Erratics exposed in logging operation, Bear River canyon. View is west, taken from a shallow saddle on the divide.**

south across Cole Creek canyon, indicate that the Tioga glacier was sufficiently thick to spill west across a divide into the Bear River canyon, and the slightly thicker Tahoe glacier would have done so also. This conclusion, however, conflicts with a conventional interpretation of two lateral moraines located high on the east slopes of the Bear River canyon, mostly just below this divide. These are a prominent, lengthy upper one, site 16, and a more diffuse lower one, best exposed at Deer Valley, site 17. As with the set of moraines in Tanglefoot Canyon, this set traditionally would be interpreted as Tahoe and Tioga, and as accurate indicators of their maximum ice surfaces. However, at the saddle just southeast of Deer Valley the "Tahoe" moraine *ascends* 100 feet west up the divide, making it suspiciously diachronous. Additionally, how could these *southwest*-trending moraines have been deposited if the Tioga-age Cole Creek glacier was thick enough to spill *west* across the divide?

Fresh erratics exist along the divide, and the most telling ones are at sites 18 and 19. Site 18 (Photo 132) is a large clearcut that exposes numerous erratics, while site 19, only about 30 feet above the top of the clearcut, is a broad volcanic summit with only several granitic erratics. Had the Tioga-age Cole Creek glacier left the erratics at these two sites, it would have left abundant evidence that it had advanced the entire length of Cole Creek canyon, which was not the case. The logical explanation consistent with all the field evidence is that the erratics along the divide, and especially at sites 18 and 19, were left by the larger Tioga-age Bear River glacier, which spilled east across the divide into Cole Creek canyon. Then as the glacier waned, it left diachronous lateral moraines where the topography was suitable, i.e., at sites 16 and 17.

In conclusion, just as with the Tioga and Tahoe glaciers in the Merced River drainage, the ones in the tributary canyons here were much larger than the sets of lateral moraines suggest. (The extents of the Tioga, Tahoe, and pre-Tahoe trunk glaciers of the steep-walled Mokelumne River canyon are unknown.) Furthermore, as in that drainage they blanketed most of the terrain, not merely occupying individual canyons up to their rims. Without the existence of till, erratics, polish, or lakes, reconstruction of these glaciers would be impossible, for they have left no topographic signature. In other words, glacial erosion has been too minimal to be detected by morphology alone. This statement appears to be true in all glaciated drainages of the western Sierra Nevada except for the Kaweah, mentioned below, where highly jointed bedrock led to some glacial excavation.

## Stanislaus River drainage

Before my 1990-91 100-day glacial reconnaissance here, the only descriptions and mapping of Tioga and Tahoe deposits in this drainage were by David B. Slemmons, in 1953, for the Sonora Pass 15' quadrangle, which includes the upper parts of the Middle Fork and Clark Fork drainages. His few Tioga glaciers were restricted to cirques except for several originating in the Leavitt Peak-Sonora Pass environs and coalescing to form a glacier, about 3 miles long, which flowed halfway down Deadman Creek. His Tahoe glaciers were much longer. At the time he did his field work, he had only two Sierran studies to rely on, Matthes' and Blackwelder's. Matthes had not differentiated between Tioga and Tahoe, and so provided no clue to their relative lengths in the Merced drainage. In contrast, Blackwelder suggested that the Tioga glacier advanced only to Nevada Fall, high above Yosemite Valley, while the Tahoe glacier advanced much farther, descending through the valley to the Bridalveil Meadow moraine. Since the cirques in the Middle Fork and Clark Fork drainages are at lower elevations than those in the Merced drainage, one could conjecture they would have had smaller glaciers, and hence Slemmons' mapping of Tioga glaciers would appear to be logical. But in actuality, we know today that his Tioga glaciers are post-Tioga (i.e., Holocene), occurring in the last 10,000 years, and his Tahoe glaciers are Tioga.

My mapping of glacial evidence in the Stanislaus River drainage was relatively straightforward except in the lower parts of the Middle and North forks. In the South Fork drainage the Tioga and Tahoe glaciers advanced down a canyon to relatively gentle slopes about the Pinecrest Lake area, where lateral and end moraines are exceptionally well preserved (Chapter 14, Map 17). In contrast, in the Middle Fork drainage sufficient evidence demonstrates that the ice surface of the Tioga and Tahoe glaciers descended into the inner canyon in the vicinity of Donnell Lake, and in that canyon evidence is almost nil. Tioga erratics lie on gentle slopes about 2 miles below that reservoir's dam. Beyond that obvious site, reached by the dam's rocky service road, the canyon is too steep-walled to preserve any readily viewed evidence. Given that in the vicinity of the reservoir the Tioga and Tahoe glaciers were about 1600 feet thick, I expected to find evidence of them about 9 miles down-canyon from the dam, in the vicinity of the northeast end of another reservoir, Beardsley Lake. In a brief examination of some benches in that area, I found no evidence of glaciation, which implies that Tioga and Tahoe glaciers did not reach the reservoir. However, the benches above it were deeply buried under pre-Tahoe ice, as evidence on the two rims indicates, and yet within the canyon I found no evidence for pre-Tahoe glaciers.

A thorough examination of benches may reveal that Tioga and Tahoe glaciers advanced beyond the reservoir. Until such a study is undertaken, the actual extents of these glaciers in the Middle Fork drainage are uncertain. Both definitely advanced farther than one might imply from Huber's preliminary geologic map of the Dardanelles Cone 15' quadrangle, which has an enormous, crescentic end moraine terminating at the confluence of the Clark and Middle forks and extending about 2 miles up each canyon. Much of the Clark Fork half is composed of stream sediments deposited behind the Tioga Middle Fork glacier dam (as was mentioned in Chapter 16), while much of the Middle Fork half is a veneer of till atop bedrock, and is better described as a ground moraine; no recognizable end-moraine crest exists. If past logic had been applied to this deposit, which contains the westernmost morainal material, then the Tioga glacier that left it ended here. But as I have just said, both the Tioga and Tahoe glaciers advanced considerably farther. About 4.0 miles west of the confluence, Bergson Lake lies on a bench at the brink of the deep inner canyon stretch above Donnell Lake. With my backpackable drilling platform, several of us at U.C., Berkeley, extracted a core and got a basal-sediment radiocarbon date of 13,660 B.P. This clearly indicates (as do adjacent fresh polish and striations) that the lake's sediments were deposited just after the Tioga glacier had retreated. Actually, the North Fork's glacier overflowed southeast through the broad gap between the Dardanelles and Whittakers Dardanelles, scouring the lake's basin (Chapter 22's Map 51). However, I found ample striations on fresh polish to demonstrate that this glacier merged with the Middle Fork's glacier, showing that ice flow curved from southeast to southwest as it approached the cliffs above the inner gorge now partly flooded by Donnell Lake. That curvature would have been possible only if a massive Middle Fork glacier was deflecting the North Fork's ice, and till, polish, and striations in varying amounts on both sides of the gorge indicate that the Middle Fork Tioga glacier's thickness was about 1600 feet. This evidence demonstrates the danger of equating the lowest moraine with the terminus of glacier, as was done in Yosemite Valley.

The extent of the largest pre-Tahoe glacier also is uncertain, but it was long. Abundant erratics on ridges above the canyon rim west of Beardsley Lake continue to a point where, 6 miles west of the dam, they end and show that there the pre-Tahoe ice surface dropped below the canyon's north rim. Today this point is about 2400 feet above the canyon's floor. The actual depth of the canyon back then is unknown, since the amount of canyon deepening after pre-Tahoe glaciations is unknown. To avoid the uncertainty of erosion, one can use as a proxy the length of the pre-Tahoe Tuolumne glacier beyond its 2400-foot thickness. The pre-Tahoe Tuolumne glacier was about this thick in the Mather area, based on my observations and on possible pre-Tahoe deposits of the Middle Tuolumne River shown on the park's geologic map. Matthes stated that the older glaciers descended to an elevation of about 2000 feet, which is a distance of about 8.4 miles beyond the Mather area. When this distance is applied to the pre-Tahoe Middle Fork Stanislaus glacier, its snout would be at an elevation of about 1420 feet, this spot being about 2.7 miles up-canyon from the river's confluence with the North Fork. Not only does this point seem unrealistically low, it is also unrealistically too far beyond the Tioga glacier. In the Tuolumne River drainage the Tioga glacier may have ended at about 2400 feet elevation, according to Wahrhaftig, and if so then its pre-Tahoe glacier, ending at about 2000 feet elevation, would have extended only about 3.3 miles beyond it. Consequently, at present the extent of the pre-Tahoe Middle Fork Stanislaus glacier is just as uncertain as are the extents of its Tioga and Tahoe glaciers.

Evidence for extents of glaciers in the North Fork drainage is somewhat better, although still open to some interpretation despite the presence of erratics, till, and moraines. As was mentioned in Chapter 14, the Tioga-age trunk glacier just barely spilled south across the saddle at the head of the hanging side canyon of Little Prather

Meadow. It also might have been sufficiently thick to spill south across the saddle at the head of the hanging side canyon of Big Prather Meadow, this saddle being site 1 at the right edge of Map 34 (Big Meadow and Blue Mountain 15' quadrangles). Immediately west of it is a low summit, Liberty Hill, and it plus the ridge extending southwest from it are free of erratics. This implies, but does not prove, that it was never glaciated. Given that pre-Tahoe glaciers in the Sierra were considerably larger than later ones, Liberty Hill should have been glaciated, and erratics may be absent simply because of the relatively small size of its summit area.

The thin line extending southwest from the west side of Liberty Hill, site 2, down to site 3, on a ridge, and then southwest down slopes to Little Rattlesnake Creek at site 4, marks the upper limit of the North Fork's erratics. The upper part of this line, where it traverses across Big Rattlesnake Creek's north and south tributary drainages, most likely separates flowing ice that was transporting erratics from stagnant ice that was not. The actual ice surface along this stretch, as in the Little Prather Meadow area, where a similar situation existed, probably was slightly higher, and perhaps it had a relatively constant ice-surface gradient. The lower 1.2 miles of the thin line from about the 6800-foot contour southwest to site 3 likely represents the true ice surface, since the glacier was flowing along the side of the ridge and not trapping stagnant ice. Because no glacial evidence exists above, one could reasonably infer that the line from site 2 to site 3 approximately represents the ice surface of a pre-Tahoe glacier. However, as is seen in the Middle Fork drainage, pre-Tahoe erratics are preserved only on divides and on gentle slopes; they would not have been preserved on these moderately steep slopes. Consequently, the erratics must be Tioga or Tahoe.

From the west slope of Liberty Hill, site 2, erratics occur on the ridge that descends first west and then southwest down almost to Big Rattlesnake Creek, which is trapped behind a legitimate (not veneer on bedrock) lateral moraine, site 5. Somewhat paralleling this ridge on its north side is a discontinuous set of lateral moraines, beginning west of and above Big Prather Meadow on a ridge, site 6, and ending at the rim of the North Fork's canyon on a minor ridge, site 7. Just south of that ridge (and along the west side of Road 6N17, which heads up to Lake Moran) is a pitted, granitic outcrop that appears not to have been glaciated in a long time. Conventional interpretation then would be that the moraines from sites 6 to 7 represent the Tioga maximum, and the main ridge from site 2 down to the lateral moraine at site 5 represents the Tahoe maximum.

However, the moraine that descends from site 6 and then *ascends* to another ridge about ½ mile west of it must be a diachronous late-Tioga moraine, so the Tioga's maximum ice surface would have been slightly higher, that is, along the ridge extending west from site 2. Therefore, the site 5 lateral moraine should be Tioga. Midway along the length of this moraine Road 6N17 cuts through it, and all of its subsurface boulders, as well as nearby ones churned up in a local logging operation, are fresh, also suggesting a Tioga age. Finally, at site 8, about 1.2 miles east of the moraine (and about 0.3 mile up Road 6N90 from its junction with Road 6N91), a cut exposes several very fresh subsurface boulders, the largest one about 5 feet in diameter (Photo 133). These are likely Tioga, and this suggests first that the ice surface of the Tioga glacier was close to that of the Tahoe (presumably the line just above this site), and second that the site 5 moraine is late Tioga. The same would apply to a broad linear ridge just south of it, at site 9, which superficially resembles a moraine, but as a road cut shows is thin till atop a volcanic ridge. This and other sites like it demonstrate that without road cuts, one is likely to interpret properly oriented, morphologically correct ridges as lateral moraines, when most of them are not. For example, the site 10 ridge, descending southwest from the site 9 ridge, appears on aerial photos (as well as on topographic maps) as a prominent lateral moraine, but dirty bushwhacking down along it shows you that it is granitic bedrock with a thin, mostly continuous layer of soil and a few scattered erratics. Because so many difficult-to-reach west-Sierran ridges like this one have not been visited but rather have been assessed by USGS geologists as lateral moraines on the basis of air-photo analysis, their geologic maps (GQ series) have become a treasure trove of glacial misinformation, doing more harm than good.

The long, thick line that weaves in and out of canyons as it descends from site 2 down to near the Road 5N02 bridge across the Stanislaus River, site 11, is my reconstruction of the edge of the Tahoe glacier. It overrode the ridge extending west from Liberty Hill as well as most of the ridge separating Big and Little Rattlesnake Creek

**Photo 133. Fresh erratics at 6570 feet elevation along Road 6N90.**

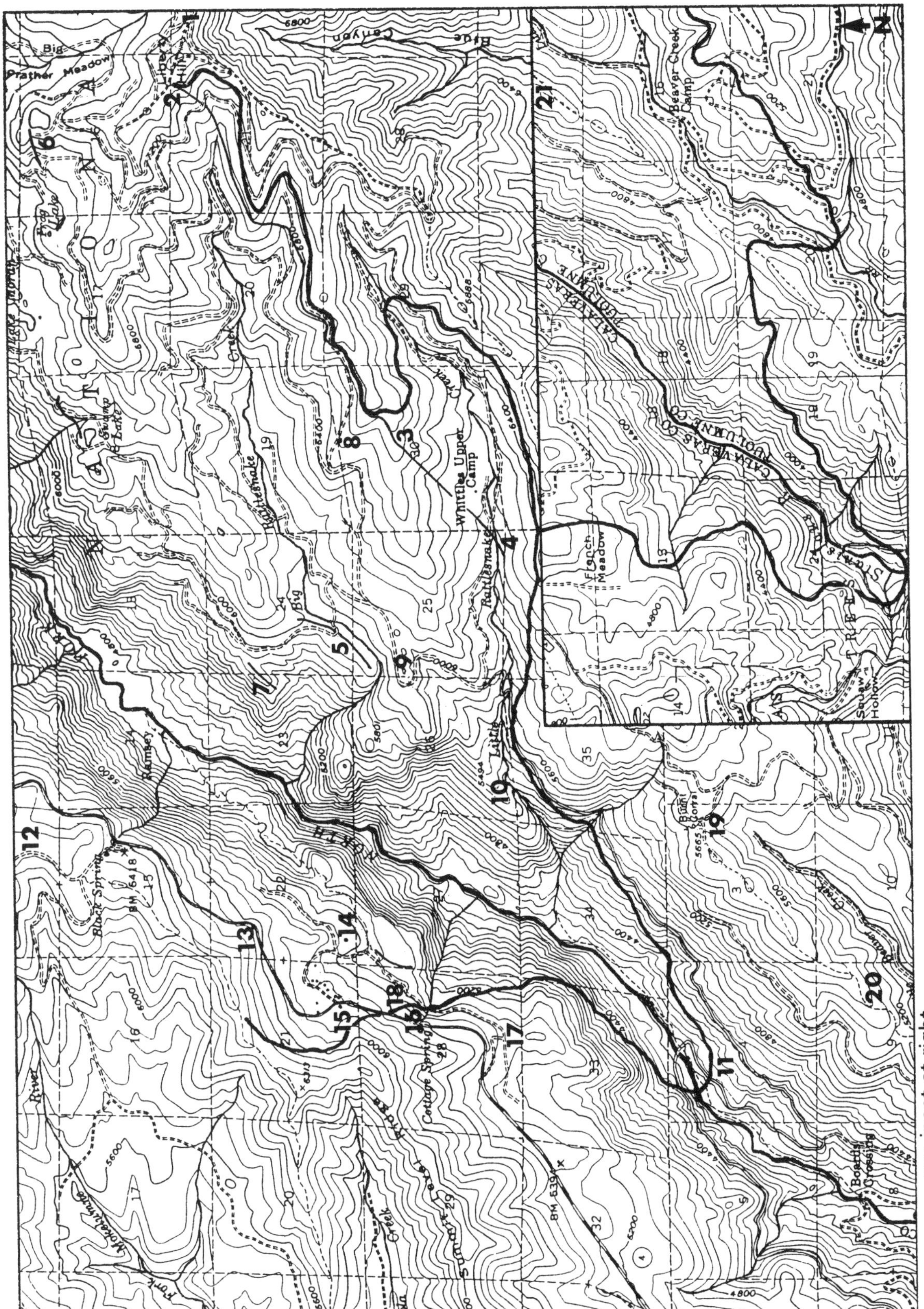


Map 34. Glacial evidence in the North Fork Stanislaus River drainage. Scale 1:62,500.

drainages. In the North Fork Stanislaus River drainage (as in the upper Mokelumne River drainage), the Tioga, Tahoe, and pre-Tahoe glaciers overrode (*rather than eroded*) ridges that were more or less perpendicular to the direction of ice flow. Between Lake Alpine and Big Meadow these glaciers overrode a half dozen ridges, then lower down they overrode the ridges bounding Randalls, Little Prather, and Big Prather meadows. As in those meadows, the glaciers in the Big and Little Rattlesnake Creek drainages filled their upper parts with stagnant ice.

As in the Merced River drainage, the Tioga glacier appears to have been just slightly smaller, perhaps ending at about 4000 feet elevation rather than at about 3900 feet. This interpretation shows that the Tahoe extended about 1.7 miles beyond the location where it was about 1000 feet thick, which is a shorter distance than those of the glaciers in Table 6. A limit on the maximum size of Tahoe and Tioga glaciers exists with Boards Crossing Road 5N14 (which does *not* bridge the river at Boards Crossing). A part of it climbs northeast up the canyon's southeastern slopes, which being about 20° steep would have held and preserved Tahoe and Tioga till. I found none, which indicates that the glaciers were below the road. Additionally, I found no glacial evidence along the generally narrow floor of the canyon, but as along the narrow gorge west of Yosemite Valley, none was to be expected.

Across the canyon along the Highway 4 route is additional evidence that supports the evidence just presented. At site 12, near the map's north edge, a glacier spilled erratics west across the divide, and because they are fresh, the glacier was likely Tioga. Exactly where the Tioga glacier (or the Tahoe glacier, for that matter) left the divide, I was unable to determine, since lateral moraines are absent. A Highway 4 cut at site 13 (6060 feet) exposes fresh and partly fresh subsurface erratics. A Tioga till should have only fresh subsurface erratics, while a Tahoe one should have at least a fair number of grussified ones in addition to fresh and partly fresh ones. Hence the age of the cut is uncertain, since its the characteristics of its boulders are intermediate between Tioga and Tahoe. I interpret the cut as a Tioga till with excessive amount of subsurface weathering due to the high amount of local ground-water flow (Photo 134). From this site a bedrock ridge, discontinuously mantled with till, descends south, and it has been locally bulldozed lower down, at site 14 (5650 feet), exposing only fresh subsurface boulders (Photo 135), which suggests a Tioga age. Whereas one could argue that the lower site is Tioga and the upper one Tahoe, the entire ridge deposits may be Tioga, since there is a continuous distribution of fresh-appearing, pit-free surface boulders between the two sites.

**Photo 134. Tioga (?) till along Highway 4 cut, 6060 feet elevation. Note spring line in volcanic stratum.**

**Photo 135. Bulldozed subsurface boulders at 5650 feet elevation.**

About 1.0 mile southwest down Highway 4 at a road junction is another cut, site 15, which has mostly partly weathered subsurface erratics (Photo 136), and hence may be Tahoe. Between this site and the highway's Cottage Spring Picnic Area, site 16, 0.6 mile to the south, evidence of Tahoe or Tioga glaciation disappears. Likewise, evidence is absent at site 17, the gentle slopes south of where Highway 4 curves west. Road 6N82 descends quickly to Road 6N80, which traverses site 17 and then traverses north. Not until site 18, just past a notable gully, does glacial evidence reappear. From here northeast ¼ mile over to a larger, broader, less steep gully, a couple of cuts plus some disrupted till display grussified subsurface boulders likely of Tahoe age. If the Tioga glacier was thick enough to leave till along the ridge from sites 13 to 14, then it was thick enough to cover sites 15 and 18, and perhaps should have eroded away Tahoe till. Two interpretations are possible. The first is that all the till is from the same glacial period, presumably Tahoe, since grussified boulders are present. However, if all the till were Tahoe, then the ice surface of the Tioga must have been lower down. This presents two problems: the ice surface of the Tioga glacier would be anomalously low with respect to that of the Tahoe glacier; and worse, its ice surface would have been considerably lower than the Tioga ice surface across the canyon. Therefore a second interpretation seems more likely: the Tioga glacier did not erode the Tahoe till. Instead, as it grew toward its maximum and spilled over the lip of the North Fork's canyon, it blanketed the gentler slopes of this half-bowl with relatively stagnant, non-erosive ice.

**Photo 136. Tahoe (?) till along Highway 4 cut, 5750 feet elevation.**

This reconstruction of the Tioga and Tahoe glaciers fits all the evidence seen on both sides of the canyon. Under conventional interpretation—based on lateral moraines, real or imaginary, perceived to represent glacial maxima—the ice surfaces of both glaciers would have been lower, the Tioga based on sites 7 or 5 and the Tahoe on sites 5, 9, or 10. This would make the two glaciers about 1 mile shorter than I have reconstructed them, and would make the Big and Little Rattlesnake Creek basins essentially unglaciated by them. The erratics in the basins then would be pre-Tahoe, which would have been preserved in place on slopes for hundreds of thousands of years despite significant soil creep and subsurface weathering.

Evidence for a pre-Tahoe glaciation larger than the Tahoe exists on uplands down-canyon. Way back in 1898, Henry Turner and Leslie Ransome mapped what they interpreted as deposits of early glaciers along the ridge southwest from the Burnt Corral saddle, site 19. Along that volcanic ridge one encounters dozens of granitic boulders, these becoming less common by site 20, the lower extent of their locally mapped deposits. However, these boulders may not be erratics. First, all the granitic "erratics" appear to resemble coarse granitic bedrock found at site 21, at the top of the map's inset. In that vicinity are similar "erratics" that clearly are derived from the local bedrock. Second, southwest beyond the map the thin layer of volcanic deposits along the ridge pinches out, exposing the underlying granitic bedrock. The pre-Tahoe glacier should have left erratics on this descending, granitic ridge, but despite a fairly thorough search for them over the next 2 miles—including an investigation of a relatively flat topped, west-trending, lower, secondary ridge—I found none.

However, if a pre-Tahoe glacier had been thick enough to leave erratics on this ridge, then it should have left erratics at similar elevations on the other side of the canyon, which may have been the case. A few granitic boulders exist midway up Summit Level Ridge, although they may have been locally derived. One, however, does seem legitimate, resting on a relatively low-angle slope (15°) at 5440 feet on a south-descending spur ridge just beyond the west edge of the map. Furthermore, along some of the roads in the northern half of the Dorrington area (east of Highway 4 and north of Boards Crossing

Road, just off the map), there are other surface, granitic boulders that appear to be erratics, since although they lie on granitic bedrock, their composition is too varied for them to have been locally derived. Additionally, some of these are very weathered, and locally there are volcanic boulders that are best explained by glacial transport. They could be ancient boulders of a former volcanic stratum, although given the variation of the granitic boulders, the sum of the evidence points to a glacial origin. A problem with doing field work here is that this area is extensively built up, having hundreds of homes. Consequently many boulders have been shoved about, and some exotics have even been brought in for landscaping. Therefore, I did not consider as evidence any boulders that may have had a human-transported origin.

The highest probable erratics occur near the top of a gently-sloped hill encircled by Chumash Circle, in Section 7 near its west edge (just off the map). They indicate that the ice surface of the pre-Tahoe glacier here was in the order of 5250 feet elevation. If so, then some of the boulders along the lower ridge extending southwest from the Burnt Corral saddle, at site 19, may be legitimate erratics. However, southwest from site 21, the ridge was low enough that it too should have erratics, but it does not, and I suggest the following explanation. The sparse west-side evidence indicates that the pre-Tahoe ice surface was close to the height of the ridge between sites 19 and 20, and so a lateral moraine was left on it, which subsequently weathered away to leave only a few large, isolated boulders. However, the ice surface was sufficiently above the lower ridge that ice spilled across it into the Little Beaver Creek drainage. As the ice surface of a glacier dropped below the spillover site, the stagnant ice left on it left very few erratics, none of them surviving today.

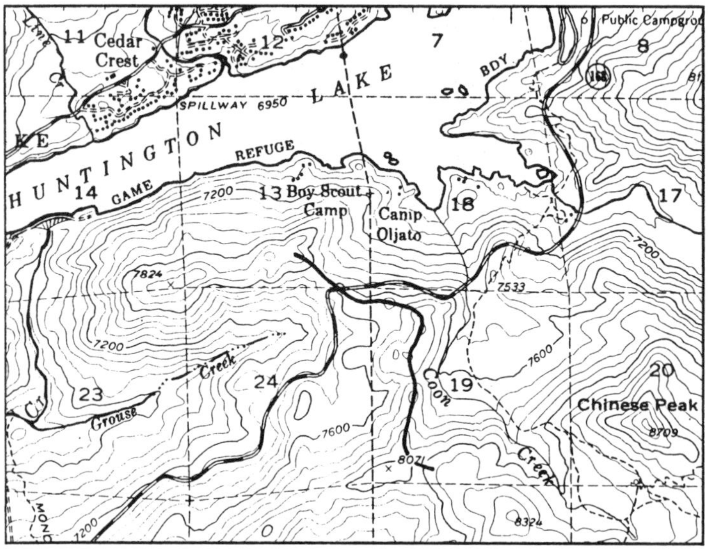


**Map 35. "Sherwin" moraine south of and above Huntington Lake. Scale 1:62,500.**

Turner and Ransome claimed to have found morainal remnants not only on this lower ridge, but also across the canyon in Squaw Hollow in what is now Oak Hollow Campground of Calaveras Big Trees State Park (map inset). ("Squaw" is politically incorrect for the late 20th century, so the hollow was renamed "Oak.") At about 4400 feet elevation, this campground stands about 800 feet above the North Fork Stanislaus River, suggesting that the glacier extended at least several miles down-canyon. However, despite a fairly thorough search of that campground plus suitable benches and slopes below it and up-canyon from it, I found no glacial evidence. In particular, I would have expected erratics on the granitic bedrock ridge in northwestern Section 25, around which the park's road makes a curving switchback. Consequently I drew the terminus of the pre-Tahoe glacier just below this ridge, at a modern elevation of about 3600 feet. A reconstruction of the lower part of the edge of the pre-Tahoe glacier appears as a thick line on the map's inset, and as drawn, it extends about 4.8 miles beyond the Tahoe glacier.

## Tuolumne River drainage

This drainage has already been discussed in Chapter 14 under "The Hetch Hetchy area's moraines and the significance of Laurel Lake," in which I presented evidence that the lateral moraines are late-Tioga, and hence the actual size of Tioga and Tahoe glaciers were greatly underestimated by Wahrhaftig and his USGS coworkers. In the *entire* Tuolumne River drainage, most of his Tahoe deposits appear to be erroneous. For example, in Chapter 14's Map 21, his only correct Tahoe deposits seem to be the easternmost part of the outer lateral moraine on Smith Peak's western slopes at about 7100 feet and the westernmost part of the Mather bench's lateral moraine where it dies out at about 4400 feet elevation, about 3 miles west of Camp Mather. The Smith Peak elevation puts the Tahoe glacier's ice surface significantly higher than the "Tahoe" moraines across the Tuolumne River canyon, which is an impossibility. But as my date of the age of Laurel Lake demonstrates, Wahrhaftig's Tahoe and pre-Tahoe moraines actually must be Tioga or late-Tioga, so his reconstruction of the Tioga glaciers in the entire western lands, as drawn by Tau Rho Alpha, has serious errors. As in the Merced, Mokelumne, and Stanislaus river drainages, using real or perceived lateral moraines to reconstruct former ice surfaces results in greatly underestimating the sizes of past glaciers.

## San Joaquin River drainage

Underestimating the sizes of past glaciers from real or perceived lateral moraines is particularly true in the San Joaquin River drainage, where moraines were mapped in considerable detail by François Matthes and then with extreme accuracy by Joseph Birman. Birman's published

**Photo 137. "Sherwin" moraine south of and above Huntington Lake. Compare the weathered boulders in this roadcut with those in cuts above Pinecrest Lake, shown in Photos 50-52.**

work is based on his Ph.D. dissertation, completed in 1957, and it was done without any knowledge of Matthes' then unpublished and unknown field work in the basin. Matthes' field work was in 1921 and 1923, and in September of 1923 he began writing up his results, then visited the area again in 1927. With his Yosemite project, Professional Paper 160, finally out of the way in 1930, he resumed work on his San Joaquin project in June. Over the years he worked intermittently on it, but did not produce a final draft before he died in 1948. Fritiof Fryxell assembled Matthes' field notes, writings, illustrations, and photographs to produce Professional Paper 329, published in 1960. Birman's field work was in 1950, 1951, and 1953, and finally was published in 1964.

Because Birman never saw Matthes' work, he independently derived an interpretation that can be compared with it. The mapped moraines in the two studies are quite similar, despite Matthes mapping on the poorer-quality early 1900s 1:125,000-scale 30' quadrangles, while Birman mapped in greater detail on 1950s 1:62,500-scale 15' quadrangles. Not only are their moraines quite similar, but also the ages they assigned them. Matthes' El Portal moraines, higher Wisconsin moraines, and lower Wisconsin moraines are equivalent respectively to Birman's Sherwin, Tahoe, and Tenaya-Tioga moraines. Perhaps this is not surprising, since Birman studied Matthes' Professional Paper 160 on Yosemite Valley and examined his glacial type localities, and therefore interpreted the evidence under the same set of assumptions. Since this area has the most studied moraines on the west side of the Sierra Nevada, it is an ideal site to test the validity of the assumptions used to date them.

The best Sherwin moraine, as acknowledged and mapped by Birman, is the one cut by Highway 168 above the southeast shore of Huntington Lake (Map 35, Huntington Lake 15' quadrangle; Photo 137). Birman stated:

> Half a mile south of Huntington Lake, a cut exposed what from very obscure exposures had been interpreted as the outermost left-lateral moraine of the Sherwin-age Big Creek glacier....
>
> The deposit is part of a moraine that was traced with difficulty along the slope south of Huntington Lake. Throughout most of its extent it veneers the crest and sides of a bedrock ridge; the morainal form is clear only near the southeast end of Huntington Lake.
>
> The highway cut made near the western end of the mapped portion of the ridge shows that the deposit is a lateral moraine plastered along the north side and crest of the bedrock ridge.

Since other pre-Tahoe/El Portal/Sherwin moraines on the west side of the Sierra Nevada have turned out to be Tahoe or younger, is this moraine also younger? Its morphology resembles Clyde Wahrhaftig's "diving Tahoe-age lateral moraine" east of and above Little Yosemite Valley, being no more than a veneer on a descending bedrock ridge. Fortunately the highway cut exposes over 100 subsurface boulders equal to or greater than 1 foot in diameter, and in Table C in the Appendix I present a breakdown of their characteristics. In Table 7, I present its summary plus those of subsurface boulders from the two cuts in the Pinecrest Lake area and in the cut by the Cascade Creek bridge. The most important difference is that the presumably Sherwin moraine has only about one fifth as many grussified boulders as do the two cuts in the Pinecrest Lake area, even though the Sherwin is supposed to be about five times older than the Tahoe. Clearly, it is a Tahoe moraine, and a legitimate pre-Tahoe moraine on the west side of the Sierra Nevada has yet to be found. Like others before and after him, Birman based his interpretation on relative positions of nested moraines and on moraine morphology (as depicted in my Figure 26). His outermost ones were Sherwin, the next ones—often along the rim of an inner canyon—were Tahoe, and the ones lower down were the (invalid) Tenaya (actually, late Tioga), and finally the Tioga.

Since Birman's type-locality Sherwin moraine is Tahoe, then his diagnostic Tahoe moraines, judged by him to be considerably younger, must be Tioga. As I have demonstrated for Little Yosemite Valley and elsewhere, lateral moraines were constructed best along the rims of inner canyons, and they are late-Tioga, diachronous moraines. This is readily apparent in examining a section of the south rim of the inner canyon of the South Fork San Joaquin River (Map 36, Kaiser Peak 15' quadrangle). The "Tahoe" lateral moraine traversing westward along the rim has an undulating nature—hardly a true reflection of an ice surface. The Corbett Lake Trail switchbacks about two dozen times up to the crest of this moraine, and the switchbacks expose fresh—Tioga—subsurface boulders (Photo 138). Likewise the crest contains abundant, fresh, Tioga surface boulders (Photo 139). South of the crest, the surface boulders rapidly diminish, but nevertheless are fresh, Tioga boulders (Photo 140). This distribution of Tioga surface boulders agrees with my interpretation of a hypothetical cross profile of a typical west-side Sierran canyon, depicted earlier in Figure 26.

Also a problem with Birman's mapping is the pattern of glaciation in the hanging tributary canyons. Note that all

**Photo 138. Exhumed, fresh boulders along the Corbett Lake Trail.**

**Photo 139. Crest of lateral moraine about 150 yards east of the Corbett Lake Trail.**

"Tahoe" lateral moraines are along prominent ridges, and that usually the profile of each canyon's east ridge does not match that of its west ridge. In contrast, along the east slopes of the Sierra Nevada, the lateral moraines on one side of a canyon closely match those on the other side. Additionally, why were some of the tributary canyons glaciated during Tahoe time while others were not? In particular there are two broad "Sherwin" canyons that should have been glaciated—one between Bolsillo Creek and East Fork Camp Creek, and the other between the

**Table 7. Comparison of subsurface weathering in four cuts.**

| Location of cut | Boulders: fresh | intermediate | grussified |
|---|---|---|---|
| Highway 168 above Huntington Lake | 33% | 58% | 9% |
| Pinecrest Lake: Dodge Ridge Road | 31% | 26% | 43% |
| Pinecrest Lake: Dodge Ridge water tank | 4% | 42% | 54% |
| Yosemite Valley: Cascade Creek bridge | 66% | 25% | 9% |

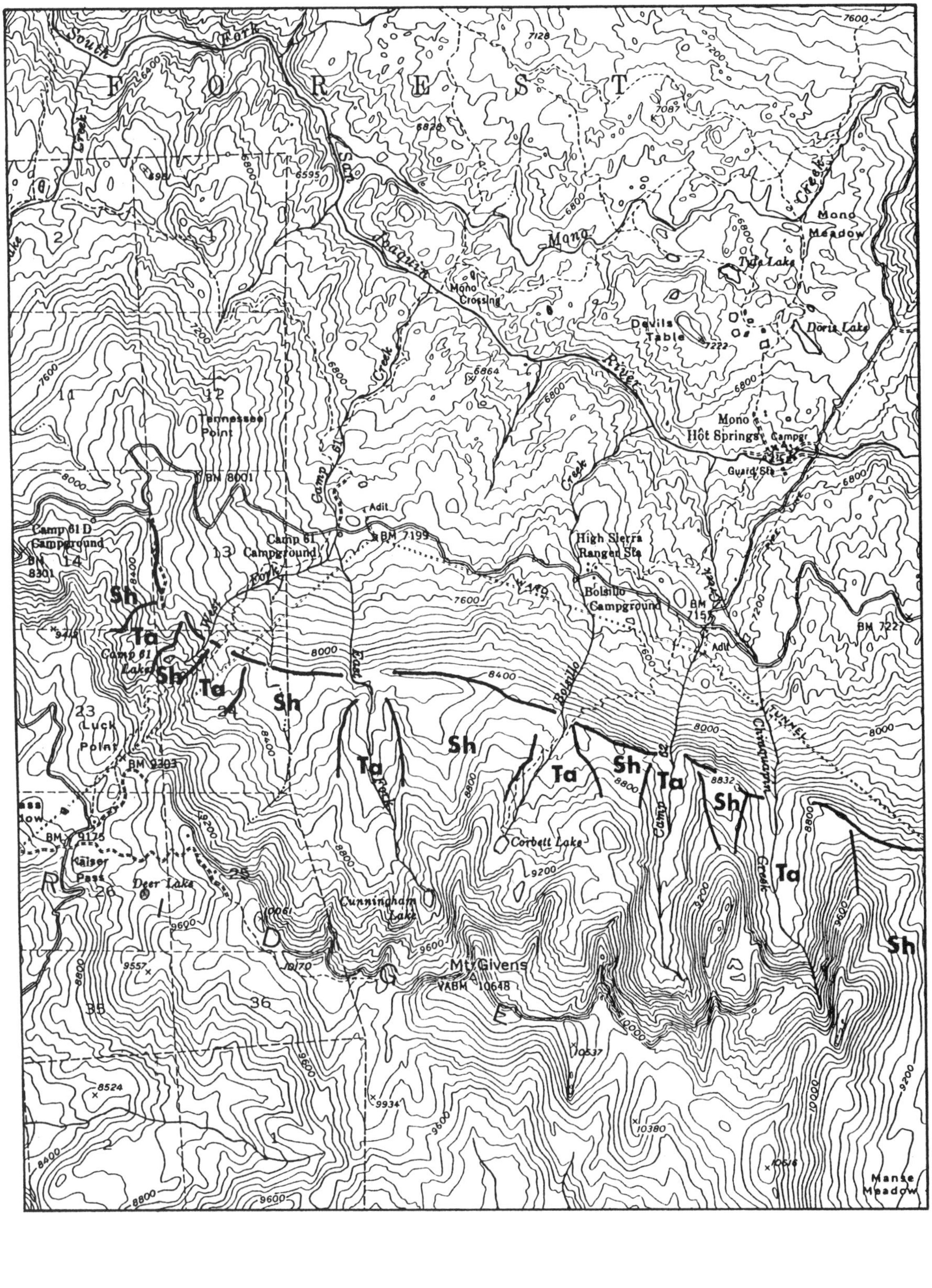


Map 36. Moraine crests of the southern South Fork San Joaquin River drainage. Ta = areas of Tahoe glaciation; Sh = areas of Sherwin glaciation. The easternmost rim moraine, just beyond Birman's area, is from Matthes' Plate 1. Scale 1:62,500.

**Photo 140. Fresh boulders south of the crest of the lateral moraine.**

East and West forks. Finally, the small West Fork canyon is mapped as not glaciated since the Sherwin, but it contains Camp 61 Lake. However, no lake dated in the glaciated Sierra Nevada has an age older than late-Tioga.

For Yosemite National Park, Wahrhaftig assigned a Tioga age to lateral moraines along the rims of the inner canyons and a Tahoe age to those on the benches or gentle slopes just above them. By sheer coincidence, then, it would appear that the ice surfaces of the Tioga glaciers just happened to coincide with the elevations of the rims. In contrast, Birman assigned a Tahoe age to lateral moraines along the rim of the inner canyon of the South Fork San Joaquin River *and* also a Tahoe age to those paralleling them on benches or gentle slopes just above them. And by sheer coincidence the ice surfaces of the Tahoe glaciers just happened to coincide with the elevations of the rims. Why should there be one pattern in the Tuolumne and Merced river drainages, and a different pattern one drainage south of the Merced, in the San Joaquin River drainage?

Like the south rim of the South Fork, its north rim also has been mapped with Tahoe-age lateral moraines along it. Are they Tahoe? I inspected them from the vicinity of the Lake Edison dam northwestward about 10 miles to two unnamed lakelets on a broad, volcanic bench (Map 37, Kaiser Peak 15' quadrangle).

About 2 miles west of the lake's resort, the Onion Spring Meadow jeep road traverses a flat bench with a conspicuous lateral moraine along its brink (Map 37, site 1; Photo 141). Boulders on it are abundant and they lack the pitting that is common on Tahoe boulders. This vicinity is particularly important because Birman's "Tahoe" moraines here form a band about 0.6 mile wide. As one

**Photo 141. Lateral moraine about 2 miles west of Lake Edison. Ken Ng provides scale.**

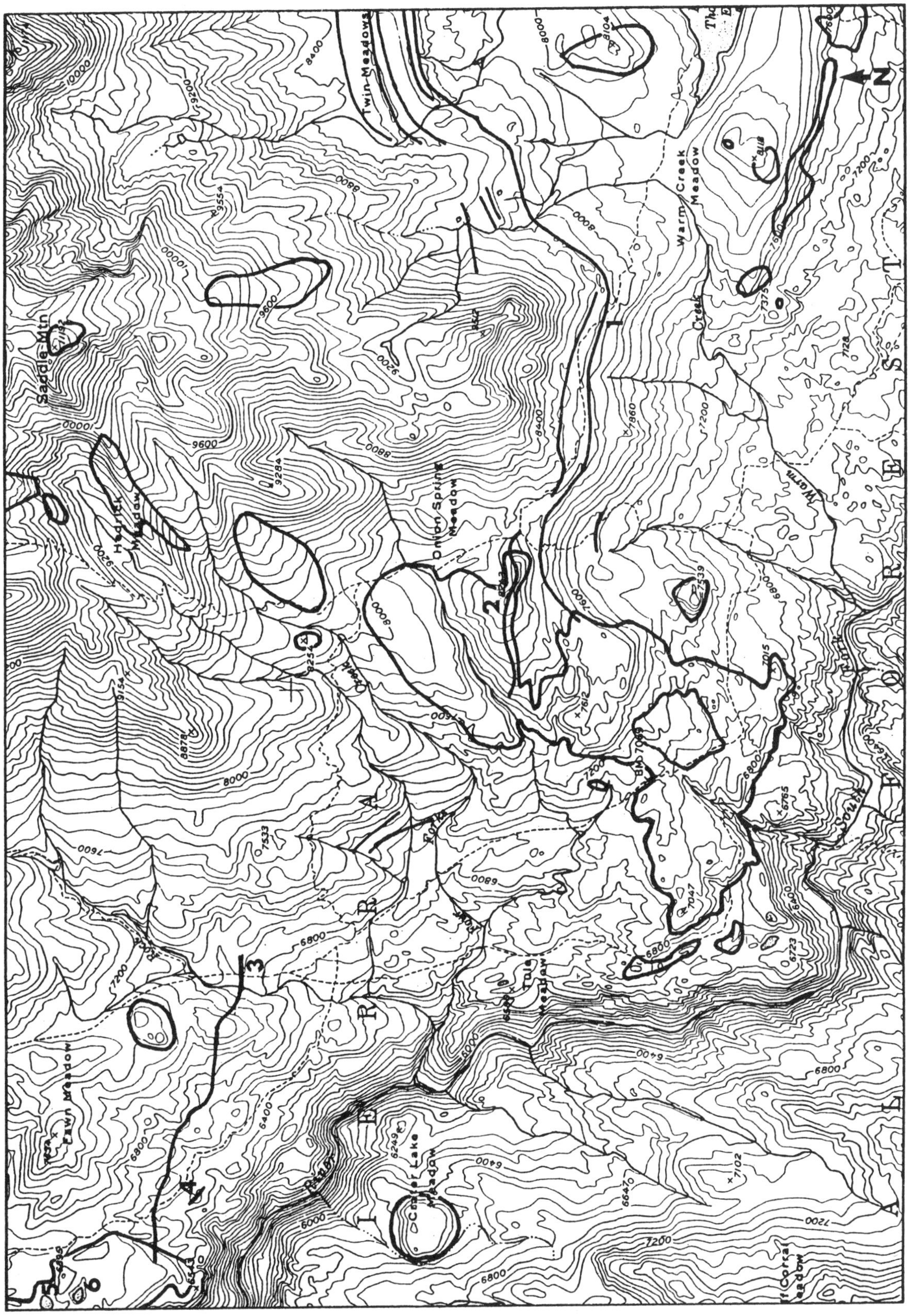


**Map 37. Moraine crests of the northern South Fork San Joaquin River drainage. Dashed line marks the highest extent of Birman's Tahoe-age deposits and moraines. Enclosed areas are trachybasalt flows or flow remnants, mentioned later under "Previous Methods of Estimating Uplift," method 9. Scale 1:62,500.**

discovers here and on other benches and gentle slopes traversed by the jeep road, the crests of the lateral moraines are sharpest where the rim of the inner gorge is best defined, and the crests rapidly become more subdued with distance away from the rim. This is the same pattern seen elsewhere in the western Sierra Nevada, but whereas other investigators assigned a Tioga age to those on the rim, a Tahoe age to those just behind them, and a pre-Tahoe age to deposits beyond, Birman assigned a Tahoe age to all these deposits, regardless of their morphology. As in other drainages, the great majority of Birman's Tahoe moraines are Tioga, and most of these are late-Tioga. Perhaps future field work may reveal that a few of the highest lateral moraines are Tahoe. Between Lake Edison and Onion Spring Meadow there are steep slopes above the canyon's benches and gentle slopes, and glacial deposits would not have survived on them, in part because of their steepness and in part due to very little source material available for deposition, since Tioga and Tahoe ice above the rim was relatively thin (about 300 feet) and hence slow-flowing.

The highest extents of Birman's Tahoe-age deposits and moraines makes an irregular line that raises questions. This line indicates that the eastern part of the Tahoe ice surface is nearly constant in elevation, then it descends west to a low divide south of Onion Spring Meadow before *ascending* briefly west from it to spill onto a short ridge's minor saddle, site 2. It then descends west again down a ridge to Onion Spring Meadow's creek. The total net descent of the ice surface in this stretch is about 1000 feet in about 2 miles, whereas in the adjacent eastern stretch of 3 miles to the east edge of the map, the descent is negligible. Over the combined stretches the San Joaquin River, in contrast, descends quite uniformly by a mere 400 feet. The profiles of the glacier's ice surface and of the river are similar to those in Little Yosemite Valley, and as was explained for that area, the reconstructed, undulating ice surface here is best interpreted as an assemblage of diachronous late-Tioga lateral moraines and adjacent deposits.

From Onion Spring Meadow's creek the ice surface descends northwest to Rock Creek, crossing it about 330 feet south of a conspicuous lateral moraine. Since it is above Birman's highest Tahoe moraine, its age should be Sherwin (pre-Tahoe), yet it is very fresh. At site 3, where a south-descending trail crosses the moraine (Photo 142), the moraine is slightly eroded, exposing boulders that once lay just beneath its surface. *All* inspected were fresh; consequently, its age should be Tioga.

About 2 miles west of Rock Creek lies a granitic bench capped with a relatively thin flow of basaltic lava, which Bateman *et al.* in 1971 dated at about 3.5 million years. Just southeast of it stands a low bedrock knoll, which is capped by a northwest-oriented lateral moraine (site 4). According to Birman, it would be one of the uppermost

**Photo 142. Fresh subsurface boulders exposed in presumably old moraine.**

Tahoe moraines, but its morphology and composition indicate otherwise. It lies along the longitudinal axis of the knoll, rising from the southeast end, topping the high point, and descending to its northwest end, making it diachronous. The boulders are fresh and abundant—i.e., Tioga.

On the flow-capped bench are two shallow lake basins, sites 5 and 6, which according to Birman have not been glaciated since Sherwin time. The southern lake is the shallower one, about 3 feet deep, while the northern one is about 5 feet deep. The shallower lake, having the same depth as Starr King Lake, like it will completely dry up in all but the wettest years. The deeper lake, having a depth closer to that of Harden Lake, mentioned in Chapter 15 under the discussion of Starr King Lake, would completely dry up in dry years, but may not always do so in normal years. Not surprisingly, because its floor stays wetter longer, less oxidation occurs, and the lake has thicker sediments, although still thin and characteristic of seasonal lakes like it. Coring the sediments with a PVC pipe (Photo 143), I reached a resistant layer—bedrock or gravel—at 22 inches—a thickness similar to Harden Lake, which is dammed by a Tioga moraine. As there, the sediments here are too thin to obtain a meaningful radiocarbon date, due to abundance of roots of modern plants. For example, in 1990 I cored about a dozen such shallow ponds and lakelets, and one lake (Duck) in the North Fork Stanislaus River drainage, and I dated three of them. In order of decreasing thickness of sediments, these three are: Duck Lake (northwest part of Spicer Meadow Reservoir 7.5' quadrangle); unnamed lakelet (near center of same quadrangle and located between Wilderness and Highland creeks); and unnamed pond on right-hand side of Spicer Meadow Reservoir road at a point 0.3 mile northeast of its start from Highway 4 (west-central part of Tamarack 7.5' quadrangle). Respectively the depths to glacial flour were 24 inches, 14 inches, and 10 inches. Respectively the carbon-14 dates were: 9340, 2680, and

**Photo 143. Coring the deeper, northern lake on a basaltic lava. The author is standing on submerged log, and is coring in 56 inches of water. Photo by Ken Ng.**

1390 years B.P. All three bodies lie considerably updrainage from deep-sediment lakes, such as Lake Moran, and these have been cored by me and others and have been dated on the order of 12,000 to 15,000 years. But just as field evidence negated Wahrhaftig's arguments for a pre-Tahoe age of Starr King Lake, the same evidence applies here too, suggesting a late-Tioga, rather than a Sherwin, age for the two lakes.

In my superficial reconnaissance of the lake's lands, I found, unsurprisingly, no erratics, for here the ice would have been thin and essentially stagnant, so few erratics would have been left. Additionally, because the area is volcanic, relatively thick soils have developed, which would bury smaller erratics. Finally, these fertile soils have dense stands of trees growing on them, making the search for erratics more difficult and time-consuming. One might assume that no erratics actually exist, and the lakes had either a post-Sherwin or a non-glacial (lava-basin) origin. But under this hypothesis, one then is at a loss to explain why neither lake basin has been completely filled in, since each would have had hundreds of thousands of years of sedimentation. Therefore, unless evidence can be provided to the contrary, the area most likely lay under thin ice of a Tioga glacier, and the two lakes are post-Tioga in age.

Additionally, with regard to the ages of Birman's moraines, his map shows that the Tioga glacier in Mono Creek canyon—one of the most classic "glacially eroded" canyons in the Sierra Nevada (e.g., Photo 144)—advanced no farther than the westesternmost end moraine in western Vermillion Valley, which now lies beneath Lake Edison. Thus the glacier ended at about 7500 feet elevation, anomalously high compared to the termini elevations of other similar-sized Tioga glaciers. Furthermore, under this interpretation it is the Sierra Nevada's *only* major tributary glacier originating on the Sierran crest that failed to reach its trunk glacier. As in Yosemite Valley, where the lowest end moraine of the Merced River drainage was arbitrarily designated as the terminus of the Tioga glacier, the same artificial designation was applied here, resulting in similar problems.

**Photo 144. Red and White Mountain at head of Mono Creek canyon. Dan Harbert in foreground.**

Finally, Matthes' glacial mapping suffers from similar problems. Both show the South Fork and Middle-North Forks Wisconsin glaciers coalescing near Rattlesnake Lake. Matthes' ice surface of the Middle-North Forks glacier dropped 1000 feet (from 7600 to 6600 feet) in elevation in *2 miles,* while the canyon floor dropped only 200 feet. Thus the glacier *thinned by 800 feet* (from 3000 to 2200 feet thick), yielding an impossibly high basal shear value of 6.6. Given that just at the upper end of this stretch the glaciers of the Middle and North forks merge to combine their masses, the last thing one would expect is drastic thinning. In contrast, the South Fork glacier dropped 1400 feet in *6 miles,* while the floor dropped 1300 feet. Thus this glacier *thinned by only 100 feet.* I urge all skeptics to examine Professional Paper 329's Plate 1, which shows Matthes' glaciers. Down a canyon, his Wisconsin-age lateral-moraine crests fluctuate with the bedrock rim; on opposing rims of a a canyon, they often do not match in elevation or gradient. The map shows literally dozens of discrepancies. No modern glacier has an ice surface like the one represented by his highest crests.

## Kings River drainage—Roaring River

Unfortunately, the Kings River drainage, despite having been extensively glaciated in its mid- and upper elevations, lacks a published glacial account. However, glacial deposits of its Roaring River tributary drainage, in southwestern Kings Canyon National Park, were drawn in considerable detail on the geologic map by James Moore and Thomas Sisson for the Triple Divide Peak 15' quadrangle, and moraine crests are shown on Map 38.

On the Triple Divide Peak 15' quadrangle, the lateral moraines of tributary canyons that comprise the Roaring River drainage are so prominent that they make an excellent moraine-identification lab exercise for undergraduate earth-science courses (Map 38). (See California Division of Mines & Geology, Bulletin 190, page 162 photograph of these moraines. This photograph also has been printed in more recent earth-science books.) Moore and Sisson appear to have identified the canyons' lateral moraines by designating the inner sets Tioga, the outer ones Tahoe. Doing so, they identify the Tioga and Tahoe glaciers as having advanced only to, or just beyond, the mouths of each canyon, leaving broad, capacious Sugarloaf Valley unglaciated. Because this drainage contains the best assemblage of lateral moraines west of the Sierran crest, it is an excellent one to test the standard relative-dating method. The map has 16 sites at which problems exist, for example: Tioga and Tahoe moraines on one side of a canyon being higher than those on the opposite side; or a Tahoe moraine stands on one side, but not one on the opposite side. Have the moraines been properly mapped?

Lacking time to examine all of them—quite a few days of field work—I instead made a rather strenuous two-day backpack trip to examine the most crucial evidence—the Roaring River moraines. I hiked east into the Sugarloaf Creek drainage, past the Sugarloaf and through Sugarloaf Valley, then into the Roaring River drainage and along the moraines between sites 1 and 2 and along the paired moraines of site 7. Along this trek I made three principal findings. First, evidence of a pre-Tahoe glaciation does exist. Second, Sugarloaf Valley was not glaciated by Tahoe and Tioga glaciers, although several of them approached or even reached the southern edge of it. And third, the ages of some of the lateral moraines indeed are incorrect.

The evidence of pre-Tahoe glaciation exists in the form of several large boulders on the lower part of a smooth bedrock ridge extending southwest from the Sugarloaf. This very resistant ridge is slowly exfoliating, and the exfoliation shells range in thickness from several inches up to 2 feet. Several large boulders are 3 to 6 feet thick, more than the thickest exfoliation shell, so they could not have been derived locally. Furthermore, the bedrock is rich in mafic inclusions, so the boulders also should be. However, the largest one I saw (Photo 145) has only two small mafic inclusions on it, when—if it were locally derived—should have had one or two dozen. Note that this location, near the "w" in Sugarloaf Meadow on Map 38, is only about a mile beyond the Tahoe and Tioga moraines of South Fork Sugarloaf Creek. This is to be expected, given that at least one pre-Tahoe glacier was somewhat longer than later glaciers. How long, we cannot say. Perhaps an intense scouring of bedrock benches on the north half of Sugarloaf Valley may reveal more pre-Tahoe erratics.

The second finding, that Sugarloaf Valley was not glaciated by Tahoe and Tioga glaciers, although they left till and outwash along its southern edge, is important because it indicates that in this drainage the glaciers could be reasonably constructed with regard to length. Absence of moraines in the valley demonstrate that these glaciers

**Photo 145. A six-foot-high boulder—probably an erratic—on Sugarloaf's southwest ridge.**

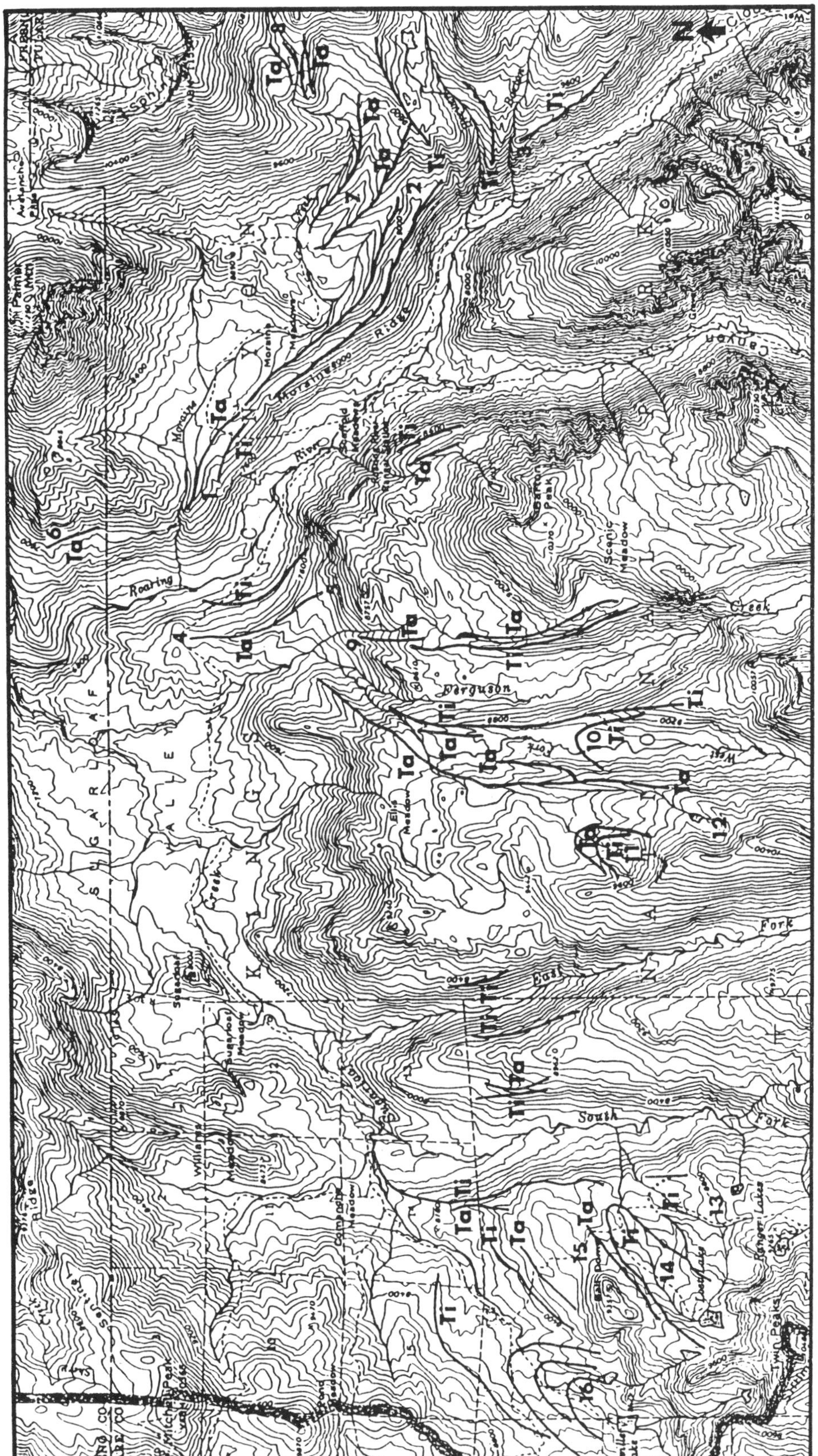

**Map 38. Tioga and Tahoe moraine crests of the Roaring River drainage, Kings Canyon National Park. Scale about 1:80,000.**

were about as long as one would infer from the mapped evidence.

But were they as thick as one would infer from the mapped evidence? The third finding indicates some glaciers were as thick, while some were thicker than their moraine crests indicate. Between sites 1 and 2 the crest of the mapped Tioga-age moraine is (to my surprise) Tioga in age. For as you struggle up the overly steep trail from Scaffold Meadows (which was not the site of a hanging, but was named for a scaffold on which early sheepmen stored their provisions out of reach from bears), you can see literally over a thousand formerly subsurface boulders now exposed by the cut of the trail. All I saw were fresh, making the moraine decidedly Tioga. Furthermore, when I examined the myriad boulders along the crest of the moraine, all were fresh. One moraine crest to the east presents a similar, but significantly different, story. There, trail cuts on it as well as eroding slopes expose abundant, decomposing subsurface boulders. Furthermore, many of the boulders along the crest appear weathered, and many have deep weathering pits. There is little doubt that it is a Tahoe moraine. But it is a diachronous lateral moraine. Evidence for this exists just northeast of the moraine's midway point. Across a ravine with a flowing tributary creeklet of Moraine Creek there rises a low granitic knoll, and its upper part is till-free. This knoll lies about ¼ mile west of the base of the western lateral moraine of site 7. The till on the knoll's slopes indicates that the ice surface of the glacier was about 50 feet higher than the moraine. Consequently, the Tahoe-age Roaring River glacier had overflowed its inner canyon to spill across the gentle slopes of the Moraine Meadows environs. Then, as the glacier began to wane, it left the conspicuous lateral moraine that exists today. Both of the lateral moraines of site 7 have weathered boulders, and so are Tahoe. They appear to be derived from overflow of a glacier from the Brewer Creek canyon, and their upper form indicates that they were deposited on bedrock ridges.

Whereas the east-side moraines examined appear to be quite correctly interpreted, the west-side moraines that I examined definitely were incorrectly interpreted. The lateral moraine stretching from site 4 to 5 is Tioga, not Tahoe, since it contains only fresh surface and subsurface boulders. Furthermore, it is a diachronous, late-Tioga moraine. The full-fledged Tioga glacier overflowed the bedrock ridge here, spilling westward across it and eradicated the Tahoe moraine that would have once existed. The mapped Tioga age moraine east of and below the sites 4-5 lateral moraine, is imaginary. The forest over which the trail descends is too open to hide the crest of a moraine. Not a vestige of it was seen. I suppose it was drawn to conform with the east-side pattern. Obviously, the standard method used to relative-date lateral moraines has not been entirely reliable here. What I suspect is that some of the lateral moraines in the Roaring River drainage are Tioga or late-Tioga, and that some of the Tahoe moraines actually are late-Tahoe. Additional glacial field work should readily prove if this was the case.

## Kaweah River drainage

Unlike all the previously discussed glaciated drainages to the north, this one lacks an inner canyon and the benches and gentle slopes often found above an inner canyon's rim. Since they lack steep sides, on which moraines are not preserved, and since they lack broad floors and recesses, in which glacial ice is stagnant and moraines are not constructed, glacial interpretation is straightforward, and the standard method appears to work, for the sides of the canyons are at optimal angles for moraine deposition and preservation. Consequently, the detailed glacial interpretation in the Marble Fork Kaweah River drainage by Wahrhaftig in 1984 probably is correct or nearly so. Given the results he obtained here, it is not surprising that he confidently used the standard method in his largely airphoto interpretation of Tioga and Tahoe moraines of Yosemite National Park. In addition, glacial mapping by Moore and Sisson of the Kaweah River drainage in the Triple Divide Peak 15' quadrangle, unlike that in the Roaring River drainage, appears to be more or less correct. Their mapped Tioga and Tahoe moraines of the Middle Fork Kaweah River drainage are reproduced in Map 39.

The Middle Fork is interesting because of its high drainage density. Nowhere else at mid-elevations in the Sierra Nevada is there such a superabundance of gullies (Photo 146). The gullied, unglaciated landscape depicted on Map 39—and in all of the Triple Divide Peak 15' quadrangle west of it—has developed entirely on granodiorite, which is the predominant bedrock of the Sierra Nevada. Apparently this drainage's landscape is different, even though it formed under the same climates and geomorphic processes that operate in other granitic drainages, simply because the bedrock was highly fractured. Therefore, glaciers were able to erode the fractured sides of the canyons, leaving them smoother, as the map shows. Note, however, that widening by glaciers has been minimal. Headward, the glaciated canyon floors are broader and the sides steeper, but not appreciably so. Whereas this change has been ascribed to glacial erosion, this view technically is not true, as discussed in the next two chapters.

## Kern River drainage

The almost universally accepted hypothesis that glaciers are very effective erosive agents for widening and/or deepening canyons was important for Matthes in Yosemite, but was absolutely vital for him in the upper parts of the Kern River drainage. Glacial deposits, if correctly interpreted, allow the reconstruction of the areal extent—

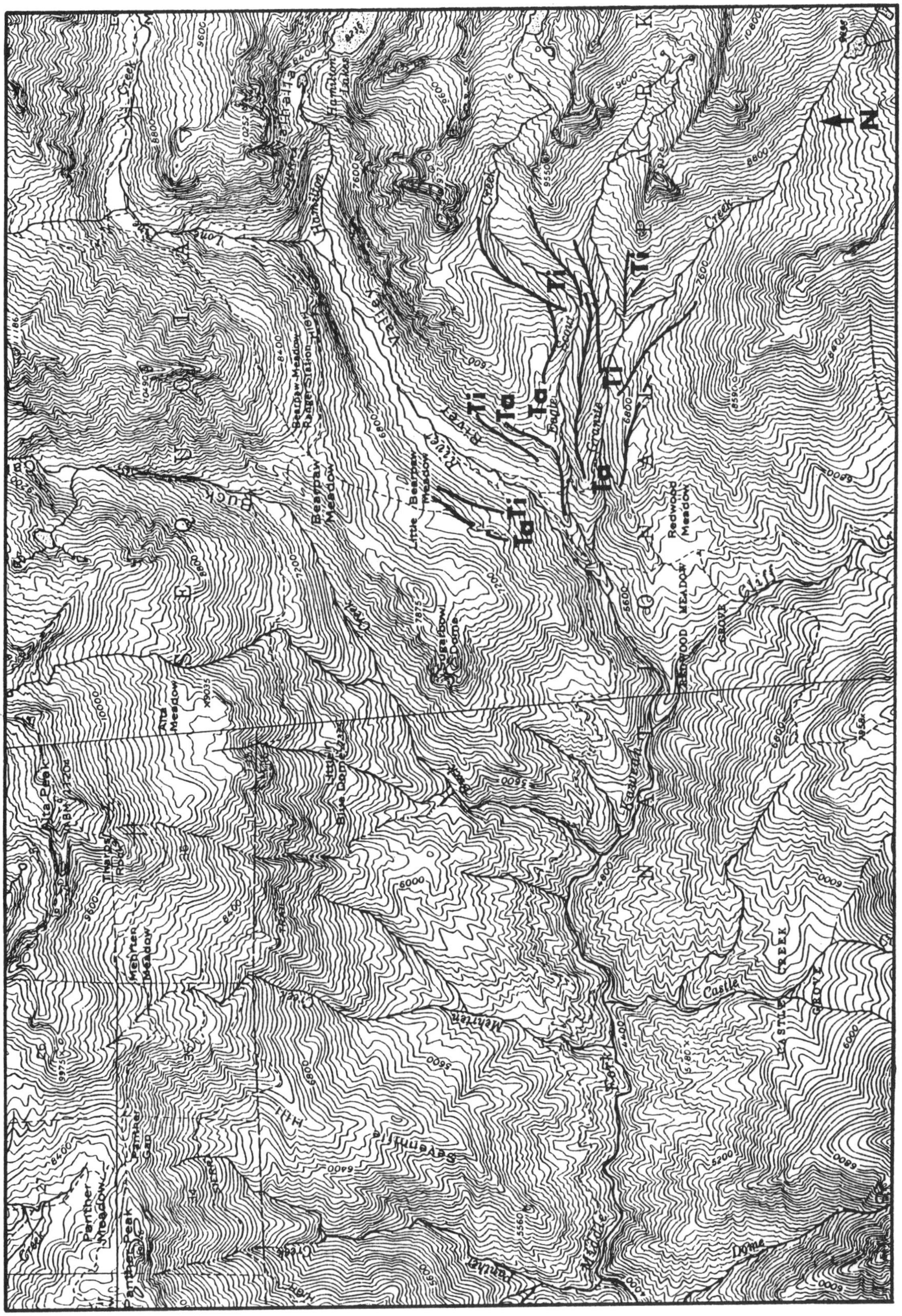


**Map 39. Tioga and Tahoe moraines of the Middle Fork Kaweah River drainage, Sequoia National Park. Scale 1:62,500.**

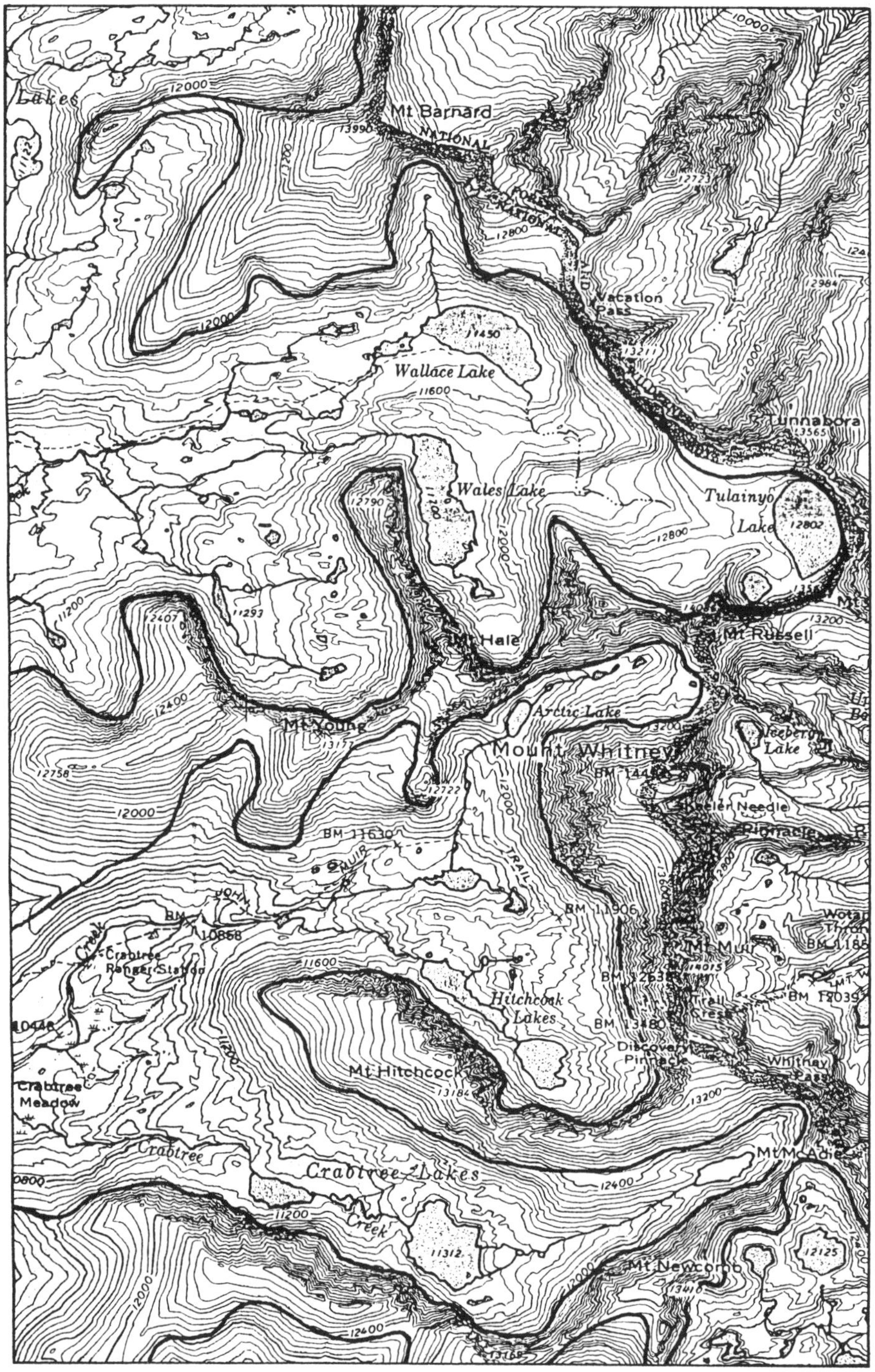


**Map 40. Matthes' Wisconsin glaciation in the upper northeastern Kern River drainage near Mt. Whitney, Sequoia National Park. Scale 1:62,500.**

and often the thickness—of glaciers. But how does one determine these measurements in the higher lands, where glacial deposits are almost entirely absent? Matthes, in papers published in 1937, 1938, and 1965, accomplished this by equating the *bases* of avalanche (snow) chutes (joint-controlled gullies on cliffs or steep slopes) to the ice surface of his Wisconsin glaciers. If glaciers effectively erode the side of a canyon, he reasoned, then they should plane away the lower parts of avalanche chutes, leaving a conspicuous trimline, or ice line, as Matthes called it. His "type locality" evidence was at the chutes of the highly gullied northern cliffs of the Mt. Hitchcock ridge, and to a lesser extent, those of the western cliffs of the Mt. Whitney ridge (Map 40). Elsewhere in the glaci-

**Photo 146. Glaciated Great Western Divide above unglaciated mid-elevation lands of the Middle Fork Kaweah River drainage, view east from Moro Rock.**

ated part of the drainage, which comprises the vast majority of the area, avalanche chutes are much more subdued.

While the erosion hypothesis is logical, as is the trimline hypothesis, the resulting glacial reconstruction is not. The bases of chutes correlate to change in gradient—where the base of the cliff or steep slope yields to gentler, canyon-bottom slopes. This change varies locally, so some chutes descend farther than nearby ones. If their bases actually were the result of truncation by glaciers, a row of them should have produced a relatively straight, nearly horizontal line. But as one of Matthes' photos shows (1965, Figure 49), variation does occur, larger gullies usually ending lower than smaller ones. Additionally, on the northeast face of Mt. Hitchcock (his Figure 50), all the chute bases are hidden by the upper parts of coalesced talus slopes. The elevations of the buried bases are unknown, and Matthes reconstructed his ice surface on the upper limit of talus, not on chute bases. If chute bases had been used, then the glaciers in this basin would have been on the order of 300 feet thick instead of about 500 feet thick. One has to question whether a glacier 300 feet thick—composed of an upper half of brittle ice and a lower half of flowing ice—could have advanced 6 miles to the Kern Canyon's trunk glacier.

The problems with Matthes' trimline hypothesis are most apparent in his "type locality" in the east-side cirque of Mt. Hitchcock, which is shown in Photo 147. In this photo the chutes below and in front of the summit descend only about halfway to the floor, while near the head of the cirque they descend much closer to it. A reconstruction derived from their bases would show the Wisconsin ice surface *increasing* rapidly from the bergschrund, then dropping precipitously as the glacier turned the corner to traverse along the base of the mountain's northeast cliffs. A final problem appears in the background of the photo, which was taken in 1971, before a series of drought years caused almost all of the Sierra's glacierets and permanent snowfields to disappear. The upper parts of these glacierets and permanent snowfields are *above* the bases of the chutes. Photo 148 presents a close-up view of this phenomenon, here a *seasonal* snowfield that nevertheless originates above the base of a chute below the southeast face of Mt. Whitney. If such an ephemeral feature can cover the base of a chute, what would a full-fledged Tioga or Tahoe glacier have done?

The problems are not unique to this basin, but exist throughout the range, especially near its crest, where voluminous talus has accumulated, largely due to numerous earthquakes generated by the fault zone at the east base of

Photo 147. **Mt. Hitchcock and its east-side cirque. View is southwest from the south shoulder of Mt. Whitney.**

the range. Photo 149 shows coalesced talus slopes below the Kearsarge Pinnacles, located in the Bubbs Creek headwaters (Mt. Pinchot 15' quadrangle), which are part of the South Fork Kings River drainage (the one immediately north of the Kern River drainage). In the background, avalanche chutes descend to varying extents, resulting in a reconstruction of a glacier of varying thickness, which would not have been the case.

Matthes' reconstructions at lower levels in the Kern River drainage, where glacial deposits exist, also have problems. The principal one is that in some basins his El Portal glaciers were significantly larger than his Wisconsin glaciers, but in other basins they are absent, implying that the Wisconsin glaciers were large enough to override the El Portal's deposits. Some of the best examples are in the Golden Trout Creek drainage (Kern Peak 15' quadrangle), which is covered in part on Map 41. In the Big Whitney Meadow area, the coalesced El Portal glacier was much larger than its three Wisconsin glaciers; in the Barigan Stringer canyon west of it, the glacier was slightly larger; and in the Johnson Creek canyon to its west, the El Portal glacier was presumably smaller, since only the Wisconsin glacier was mapped. To the south, on the northern slopes of the Toowa Range, crowned by Kern Peak, only Wisconsin glaciers existed, in Matthes' view.

A second problem is with relative sizes of glaciers. During historic times, glacierets and permanent snowfields have been larger in the higher cirques, especially in the north- and east-facing ones. However, the three examples just cited are much larger than two that formed just north of the divide above Rocky Basin Lakes, these two being southwest above the broad, relatively flat Siberian Outpost. The unnamed, deep, short canyon just east of Forgotten Canyon produced only a Wisconsin glacier, which advanced north a mere 0.4 mile, less than one eighth the length of the Rocky Basin Lakes glacier, which advanced southeast down the Barigan Stringer canyon. The glacier in the northeast-oriented canyon just west of the unnamed canyon fared better, advancing 1.1 miles almost to the southern edge of the Siberian Outpost plain, which, according to Matthes was never glaciated.

**Photo 148. Seasonal snowfield at the base of the southeast face of Mt. Whitney. Keeler Needle rises in the upper left.**

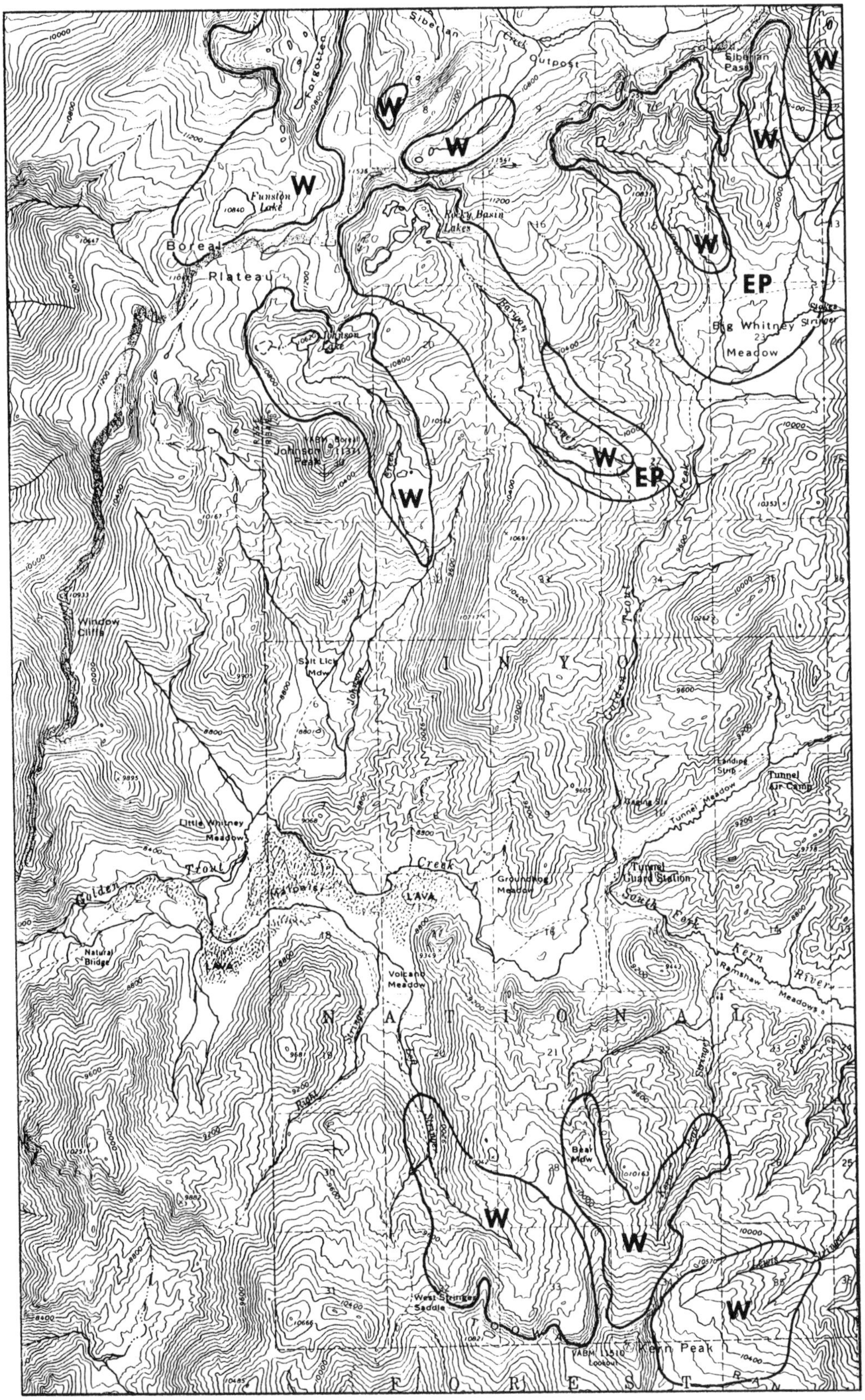


**Map 41. Matthes' Wisconsin (W) and El Portal (EP) glaciers in the Golden Trout Creek drainage. Scale about 1:90,000.**

This glacier was almost one third the length of the Rocky Basin Lakes glacier. The largest glacier north of the divide is the westernmost one. It originated in a shallow basin and advanced in part westward to just beyond Funston Lake and in part northward through a narrow gap to join the Forgotten Canyon glacier, giving it sufficient imput to allow it to descend all the way to the Rock Creek glacier.

Dozens of similar examples can be cited for Matthes' El Portal and Wisconsin reconstructions in the Kern River drainage, but his problems are matched by different reconstructions implied by glacial deposits mapped in the 1980s by Moore in the Mt. Whitney 15' quadrangle and by Moore and Sisson in the Kern Peak 15' quadrangle. Whereas in the Roaring River canyon, mentioned earlier, a pair of definable lateral moraines existed, in these two quadrangles, they essentially are absent. Not a single canyon has a crisp, linear ridge—readily observable on a topographic map—with another one just below it. Moore and Sisson have assigned presumed glacial deposits on canyon rims to Tahoe, and deposits on slopes beneath them to Tioga. Additionally, in the Kern Peak 15' quadrangle, their Wisconsin (Tahoe and Tioga) deposits indicate a much greater glaciation. Glaciers from the northern and southern slopes of the Golden Trout Creek drainage descended to meet on the broad, flat floor occupied today by the Malpais lava flow. These geologists substantiate this extent with the observation that glacial erratics are common on this pre-Tahoe lava flow. But as I have observed in similar flows in various parts of the Stanislaus River drainages, flows can entrain granitic boulders, which then weather out of them to resemble erratics. I reject their interpretation on the ground that their glaciers were largest where the rain shadow of the Great Western Divide was greatest—on the gentle upland slopes east of Kern Canyon. Their glaciers were smaller in the deeper canyons just east of and below the divide, where precipitation was greater, and their glaciers were few and meager on the west slopes of the divide, where precipitation was greatest. This is opposite the pattern to be expected, and indeed, to be seen elsewhere, such as the small rain-shadow glaciers of the White Mountains (east of Owens Valley), the larger glaciers of the eastern Sierra, and the much larger ones of the western Sierra.

Whereas one can argue with logic against an interpretation, arguing with evidence is better. Moore and Sisson have extended their glacial deposits to the southern part of the east edge of the Kern Peak 15' quadrangle, while du Bray and Moore, along the matching west edge of the

**Photo 149. Talus below Kearsarge Pinnacles, view southwest from Mt. Gould.**

adjacent Olancha 15' quadrangle, show no glacial deposits at all. Which geologic map is correct? To answer this, I backpacked to Movie Stringer, a creek originating west of Templeton Mountain, whose upper part should have abundant glacial deposits at the common border of the two quadrangles. There, I found no moraines, no till, no erratics; in short, no glacial evidence whatsoever. Moore and Sisson's mapped deposits, at least in this vicinity, are imaginary.

## Conclusion of Glacial Evidence

Chapters 13 through 20 have addressed the question, "How does one determine the length, thickness, and age of former glaciers?" These measurements are relatively easily and quite accurately determined for former glaciers in drainages east of the Sierran crest. But as I have shown for former glaciers in drainages west of it, the assumptions that work on eastern lands do not work on western ones. In particular, the overriding assumption—that inner lateral moraines are Tioga, outer ones are Tahoe, and deposits beyond them are pre-Tahoe—has wrought thoroughly incongruous results, except in the Kaweah River drainage, where slopes have been gentle enough to receive and preserve moraines of the Tahoe and Tioga maxima. Reconstructions based on this assumption have resulted in relative lengths (or ratios) of Tioga, Tahoe, and pre-Tahoe glaciers varying highly from one major drainage to the next as well as in many series of tributary canyons in a drainage.

These problems disappear when one relative-dates glacial deposits on the basis of the degree of weathering of their subsurface boulders. That this method works was proven when I dated Laurel Lake's lower sediments to determine the age of its end moraine. The age is pre-Tahoe, as determined by the U.S. Geological Survey's methods, but a radiocarbon date, lake-sediment thickness, and the freshness of adjacent subsurface boulders all indicate a Tioga age. By using as a criterion the degree of weathering of subsurface boulders, I have produced reconstructions of Tioga, Tahoe, and pre-Tahoe glaciers of the Merced River drainage that are similar to modern, large glaciers and that are in accord with glacial physics. My reconstruction of past Merced River glaciers, based on the degree of weathering of subsurface boulders, shows them gradually thinning with distance down-canyon from the equilibrium-line altitude, as do modern, large glaciers. Furthermore, my reconstruction recognizes that the velocity of thin glaciers is much less than that of thick glaciers.

In contrast, Matthes and Wahrhaftig, produced reconstructions in this drainage and elsewhere that do not resemble modern, large glaciers and that are discordant with glacial physics. Their mapped glacial evidence of the west slopes of the Sierra Nevada suggests that in the down-canyon direction from the equilibrium-line altitude, past Tioga and Tahoe glaciers: 1) locally made sudden, major drops; 2) locally thickened, causing a rise in their ice-surface elevations; 3) locally had ice-surface elevations on one side considerably higher than those on the opposite side; and 4) had stepped ice-surface profiles—glacier ice remaining quite uniform in thickness for a considerable length, then rapidly thinning, then resuming a quite uniform thickness for another considerable length, then rapidly thinning again. Furthermore, in discordance with glacial physics, past reconstructions had the underlying assumption that the velocity of thin glaciers equaled that of thick glaciers and that unbelievably high shear stress values were possible. Consequently, the conclusion is that almost every glaciated river drainage of the western Sierra Nevada, from the Yuba south to the Kern, needs to be restudied. The two exceptions are the Stanislaus, which I examined, and the Kaweah, which lacks problems.

Whereas Matthes and Birman did extensive field work to map moraines and till, most others following them relied heavily on aerial photographs. While this has saved time and money, it has produced detailed, believable maps that nevertheless are very inaccurate and that do a disservice to all who would use them. For example, in 1992 Moore and Nokleberg delineated glacial deposits on their geologic map of the Tehipite Dome 15' quadrangle that imply: 1) that the cirque-floors of former glaciers were far lower at this southern latitude than anyone has imagined (about 6000 feet); and 2) that some of the sequoia groves in this quadrangle were glaciated. Unsuspecting plant geographers then might use this information to devise a cause-effect relation between the glaciers and the groves.

The disadvantages of using subsurface boulders are that first, field work is required—and exhausting field work at that—and second, that such boulders are not always found, leaving the age of some deposits in doubt. Furthermore, I have argued repeatedly that even if one knows the relative age of a moraine, that moraine does not necessarily reflect the ice surface of the glacier that eventually left it. The great majority of lateral moraines that I have encountered have been diachronous, late-Tioga ones, each deposited over time as the glacier waned. They were deposited almost invariably on the rim of an inner canyon or on a bedrock ridge parallel or subparallel to the length of the canyon, and as such they say nothing about the actual elevation of the glacier's ice surface, which could have been at the moraine's elevation or up to even 1000+ feet above it. To find the actual elevation requires finding the glacier's highest deposits, which usually are scattered erratics resting on benches and gentle slopes. These are fairly good indicators of a glacier's thickness, but determining whether the erratic is Tioga, Tahoe, or pre-Tahoe often appears impossible. Many erratics need to be examined for degree of pitting. Highly pitted erratics may be pre-Tahoe, less pitted ones Tahoe, and unpitted ones Tioga. However, there are several pitfalls. Forest fires can cause the sides of an erratic to exfoliate, thus exposing a fresh surface, and so "youthful" boulders actually may be quite old. On the other hand, genuine erratics of exotic bedrock may intermingle with weathered, deeply pitted local bedrock that has become detached and rounded over time, and this assemblage could lead one to assign them a pre-Tahoe age, even though the age of the actual erratics may be Tioga.

Despite science making major advances in the twentieth century, the determination of the extent and age of former glaciers in the river drainages of the western Sierra Nevada now is more difficult than ever, and the results are less accurate. Future field mappers will replace the old, precise nonreality with a new, imprecise reality.

# 21

# The Case for Major Glacial Erosion

In the English-speaking world, modern geology began with Charles Lyell's three-volume treatise, *Principles of Geology*, published 1830-1833. In volumes I (p. 175-176) and III (p. 149-150), Lyell recognized that mass wasting occurs in glaciated ranges, especially after an earthquake, and that the rockfall debris lands on the surface of a glacier and is transported by it. He didn't mention the obvious: rockfall debris eventually may be deposited by the glacier in the form of moraines or till, or the debris may be carried downstream in glacial meltwater streams. In 1857 John Tyndall, a physicist and mountaineer, measured rates of flow of glaciers in the Alps. To do so along the side of a glacier, he had to dodge rockfall boulders precariously perched on ice above. But in the 1860s rockfall ceased in glaciated areas—at least in the textbooks. Apparently geology had become sufficiently robust that processes needed to be quantified. In other words, how much of a deep, glaciated canyon is the result of stream erosion, how much of glacial erosion, and how much of mass wasting? The first two processes lend themselves nicely to quantification, but mass wasting does not. It is highly erratic both in time and in volume, thereby defying quantification. Consequently, geologists discarded it!

If you look at an 1860s geology textbook, say by the esteemed Prof. James Dana of Yale, you will find that glaciers tremendously erode their canyons, but nowhere will you find any mention of rockfall. Modern textbooks are little better. There usually is a chapter on mass wasting, including rockfall, but then in the chapter on glaciers, there is no mention of mass wasting. Rockfall simply does not occur in glaciated mountain ranges. It does not matter that every year, mountaineers are killed by rockfall. Geoscientists apparently do not read the American Alpine Club's annual *Accidents in North American Mountaineering*. (I've been listed in it.) What you will find, perhaps without exception in any American book or article on glacial erosion, is that glaciers tremendously deepen and/or widen canyons. Obtain any American geology or geography textbook and turn to its glacier chapter. Odds are very high (I have yet to see an exception) that Yosemite Valley will be mentioned, and it will be used as an example of how glaciers severely erode a mountain canyon to produce a deep, steep-walled valley.

The evidence for this glacial-erosion paradigm is extremely theory-driven. Here are three examples: 1) glaciers *always* transform the cross-profile of a canyon from V-shaped to U-shaped—*but* the Sierra Nevada has some V-shaped glaciated canyons and some U-shaped *un*glaciated canyons; 2) glaciers *always* greatly deepen their preglacial canyons—*but* more often than not the floor of the glaciated part of a canyon hangs above the floor of its unglaciated part; and 3) glaciers *always* transform the longitudinal profile of a canyon's river from a nearly constant gradient to one with alternating gentle and steep stretches—*but* all rivers originating at the Sierran crest, glaciated and unglaciated alike, have alternating gentle and steep stretches. Quite obviously, reality defies theory.

So where is the proof for major glacial erosion? The "proofs" are what we will examine in this chapter, the word being in quotes because they are not proofs at all. Although apparently no one before me has made a major attempt to contest the proofs, a few have made minor waves, which broke harmlessly against the wall of glacial theory. Be aware that at least one glacial expert offered a caution against major glacial erosion. In his hefty 1971 treatise, *Glacial and Quaternary Geology,* Richard Foster Flint, in reviewing glacial erosion, concluded his section with:

> In summary, it is generally supposed that valley glaciers erode mountain terrain more rapidly than subaerial [stream] erosion would the same terrain. The supposition, though possibly true, is *not based on measurement.* (Italics mine.)

An example of a recent attempt to quantify glacial erosion was presented in 1994 by an aged glaciologist, Louis Lliboutry. He reviewed "the *old problem* of erosion by temperate glaciers." (Italics mine. After more than a century, major glacial erosion still had not been proved. Major erosion by glaciers has been difficult to prove, because non-events do not leave evidence of their non-existence.) Resolutely, Lliboutry demonstrated how one can use simple physical models and approximate mathematics to prove major erosion. For major erosion to occur, his methodology requires that grooving in the upper part of glaciers must have been "much more ubiquitous than reported." Hence his proof is valid only if the nearly infinite missing grooves actually had existed. In the Sierra's granitic rocks, remnants of Tioga-age glacial polish stand above adjacent rock by about ¼ inch, which represents the extent of post-glacial weathering. This extent is too small to have removed any traces grooves, which typically are ½ to 1+ inch deep. Therefore, virtually all of the grooves left at the time of deglaciation would have survived in some form. My observation, based on extensive hiking in glaciated Sierran canyons, is that grooves are uncommon. In this range the nearly infinite grooves never existed, and Lliboutry's methodology, properly applied, proves that glacial erosion has been minor. Not only is his analysis missing the required grooves, but his mathematical formulas require quantification of one probability and two parameters, which the author himself admits are totally unknown. Despite this, he assigns a value to each in order to achieve the desired result, that is, to validate major glacial erosion. But I do agree with his conclusion:

> ... in the absence of detailed pertinent field observations and ad hoc experiments, the models [his] remain *conjectural* and the required parameters [and probability] can only be assessed *crudely*. (Emphasis mine.)

Many others before him have attempted to demonstrate major erosion by alpine glaciers in areas of resistant bedrock. Not an issue is major erosion in nonresistant bedrock, such as incoherent sedimentary and volcanic rocks. Both occur in the northern part of the Sierra Nevada: the prevolcanic "Tertiary gravels" and the ensuing volcanic sediments. What follows is a short presentation of general arguments and field evidence in favor of major erosion by glaciers—and why they are faulty. Major glacial erosion has been inferred from the following: 1) U-shaped cross profiles, 2) deep glacial lakes, 3) longitudinal profiles, 4) stoss-and-lee topography, 5) glacial sediments, 6) field measurements, and 7) lab experiments and computer simulations.

## U-shaped cross profiles

Virtually every geology, geomorphology, and physical-geography textbook in North America proclaims that Yosemite Valley is a classic example of the ability of glaciers to radically transform a landscape. In Europe, Switzerland's Lauterbrunnen Valley is that subcontinent's classic example. However, the vast majority of glaciated canyons do not resemble either valley; these two are exceptional, and as such are poor examples of glaciated topography, which generally is much less dramatic. An essential assumption is that all glaciated canyons are initially V-shaped in cross profile, then are transformed by glaciers to U-shaped in cross profile. Without this assumption Matthes could not have made a quantitative determination of the amount of stream erosion and glacial erosion Yosemite Valley had experienced—exactly 50% for each. (I laboriously recalculated the areas of his cross profiles, plus three of my own, and also arrived at 50%. However, that assumed the valley's sediments were thin. Using thick sediments, as determined through Gutenberg and Buwalda's seismic work, the glacial erosion increased to 55% of the valley's total excavation.)

Although west-side canyons generally have a V-shaped cross profile along most or all of their lower, unglaciated part, there are exceptions. One striking exception is the totally unglaciated North Fork Tule River drainage, northeast of Porterville (Springville, Kaweah, and Mineral King 15' quadrangles). The floor of its upper part is about as broad and flat as that of any west-side glaciated canyon of similar size. Below it the next part defies explanation by conventional stream processes: a 3-mile-wide, gently sloping floor on knobby granitic bedrock that is locally buried under a veneer of sediments (Photo 150). Such a landscape is unlikely to form under modern climates, especially given the diminutive size of the North Fork Tule River, which above this wide floor has a drainage area of, at best, 31 square miles. Evidence elsewhere indicates that the North Fork Tule River canyon is very ancient. Lindgren in 1911 stated that at the time of deposition of Eocene-epoch gravels in the northern Sierra

**Photo 150. Aerial view up part of the lower North Fork Tule River drainage.**

Nevada, the range was about as high as at present. Furthermore, some of his mapped Eocene-epoch gravels lie close to the floor of the South Yuba River canyon, indicating that the relief back then was close to today's relief. One can infer that other granitic canyons, including the North Fork Tule River canyon, also were deep.

Support for this inference lies in the South Fork Kern River drainage. The upper South Fork flows through a broad, flat-floored valley now largely occupied by Templeton Mountain, a volcano dated at about 2.4 million years old (see topography and geology by du Bray and Moore, 1985). Lava flows descend to the edge of the South Fork's level floodplain, indicating that no measurable net incision has occurred in the last 2.4 million years. Hydrologic data by S. W. Anderson *et al.* (1993) show that the discharge for the North Fork Tule River above its wide-floor reach is about one fifth that of the South Fork Kern River above Templeton Mountain. Consequently, the North Fork Tule River should have been much less effective at incision than the South Fork Kern River. The canyons of each, like the South Yuba River canyon, must be very ancient, and exist today as relict tropical landscapes. (Evidence for this conclusion is presented in the next chapter.) Thus one cannot *a priori* assume that Yosemite Valley was V-shaped in cross profile at the onset of glaciation, especially since there is not a shred of evidence to suggest this. More likely it would have been broad and relatively flat floored, having achieved its general topography by Eocene time, as did the topography in the South Yuba River drainage.

Matthes had assumed that the valley's preglacial general cross profile was V-shaped, and then was transformed by major glacial erosion to U-shaped. But did major erosion occur? The two longest, thickest glacier systems in the Sierra Nevada were those of the San Joaquin and Tuolumne river drainages, producing pre-Tahoe glaciers that were about 60 miles long and up to about 4000 feet thick. Nevertheless, where the glaciers were at their thickest, and flow (and erosion) should have been at or near a maximum, the canyons were not noticeably widened. In Professional Paper 160 Matthes himself stated (p. 89) that the Grand Canyon of the Tuolumne River had barely been eroded by glaciers, but he did not question why glaciers there were so ineffective but in Yosemite Valley they were so effective. The impotence of glaciers in the San Joaquin River drainage is even more impressive, but because of the area's inaccessibility, few have realized how little the landscape has been modified. In its four main canyons, the North Fork, the Middle Fork, the lower part of the South Fork, and the lower part of Fish Creek, the glaciers were enormous, yet these four canyons are classic V-shaped in cross profile. Their bottoms are so narrow that it is highly impractical to construct a trail down any canyon—hence their inaccessibility. A final—and well known—example of a generally V-shaped, severely glaciated canyon is Tenaya Canyon in the vicinity of Clouds Rest.

The general explanation for merely minor glacial erosion in canyons such as these is that the bedrock is extremely resistant to glacial erosion. Indeed, Matthes way back in 1914 made a general statement about this:

> The main lessons learned last summer in this regard are that except in certain restricted localities, such as the Yosemite Valley proper, the rock character precluded extensive remodeling by the ice, and that as a consequence the landscape, although bearing the unmistakable stamp of ice work, still retains very largely the features given it by normal weather-and-stream erosion prior to the advent of ice.

Since Yosemite Valley proper is the premier example of massive bedrock, then to be consistent, Matthes should have concluded that remodeling of it by the ice would have been minimal. That he concluded the reverse—extreme glacial erosion in extremely resistant bedrock—was based not on field evidence but rather on the dictates of his mentor, Prof. William Morris Davis, who categorically stated that glaciers transformed the cross profiles of stream canyons from V-shaped to U-shaped. Matthes' reconstruction of Yosemite Valley's transformation by glaciers is purely hypothetical.

That Davisian analysis does not work is easily demonstrated by examining one of Matthes' cross profiles, his Figure 27, "Cross profile from El Capitan to the Cathedral Rocks," which shows an extreme transformation from V-shaped to U-shaped. This is presented in modified form as my Figure 32. In it I have added the bedrock floor—the solid curved line below the base of the cross profile—which I derived from Gutenberg *et al.*'s seismic work. This line shows that the sediments are much thicker than Matthes had conjectured (dashed line B-B below the valley's present floor, solid line C-C).

In brief, Davisian analysis required Matthes to first establish the preglacial longitudinal profile of the Merced River through the valley. He concluded that most of the hanging tributary streams had been hanging before glaciation, so their longitudinal profiles should not be extrapolated inward to the Merced River to establish points along its preglacial profile. Therefore, he *a priori* selected a hanging tributary from each end of the valley—the creeks of Bridalveil Fall and Royal Arch Cascade—and then extrapolated the profile of each from the valley's rim to its center in order to determine the elevation of the Merced River at these two points immediately before glaciation. A straight line connecting these two points then determined the profile of the preglacial Merced River through the valley. To construct a given cross profile Matthes drew an arbitrary V up from the preglacial river

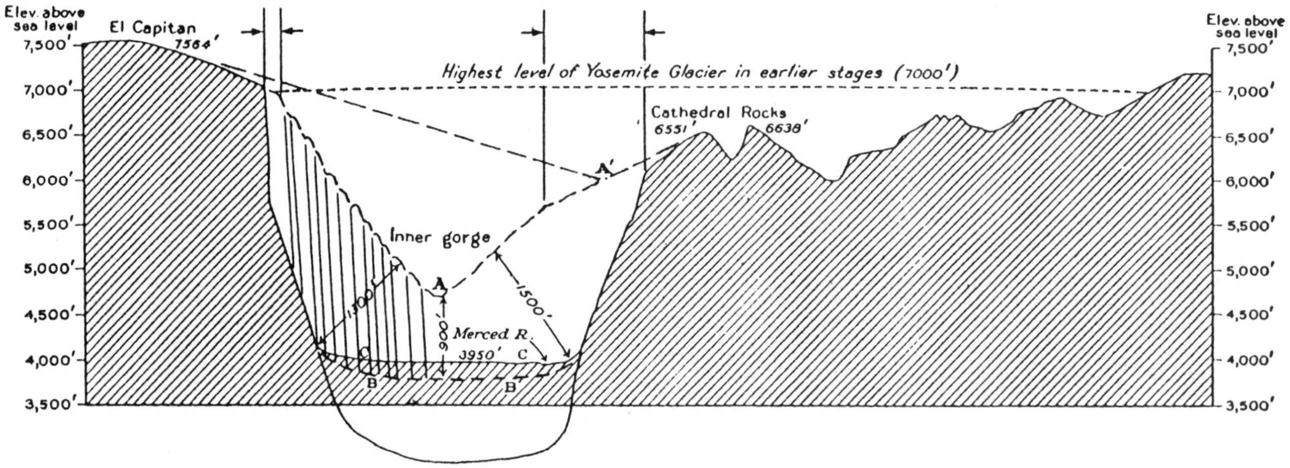


**Figure 32. Yosemite Valley cross profile from El Capitan to Cathedral Rocks. No vertical exaggeration. See text for explanation.**

and where possible extrapolated dipping surfaces inward from the valley's present rims. A point on the preglacial rim then was located where the rim extrapolation met the top of the V. Note in Figure 32 that Matthes determined only the south preglacial rim with this method. If he had used it to determine the north rim, as I have shown by the gently dipping dashed line from the summit of El Capitan, then construction of a plausible preglacial cross profile would have been impossible, for the preglacial Merced River would initially have been at A' instead of at A. Matthes circumvented this dilemma by positing a series of hypothetical, nearly vertical, parallel joint planes for preglacial El Capitan. These are shown in his Figure 15 (Yosemite Valley during the canyon stage, reproduced as my Figure 17) as parallel, exfoliating cliffs. Ensuing glaciers presumably then removed this highly jointed rock, leaving the El Capitan monolith whose vertical walls challenge climbers today. This was not a new idea. John Muir had proffered a similar view, although he had gone to greater extremes by stating that all of the canyon topography of the Sierra Nevada was due to glaciers quarrying away all of the jointed bedrock. Also, Matthes' reconstruction resulted in an inexplicable, asymmetrical widening of the rims of the preglacial inner gorge: the rim at El Capitan retreated very little, while that of the less steep, unjointed (and therefore very massive) Middle Cathedral Rock retreated much farther (the horizontal distances between the pairs of arrows on Figure 32).

Additionally, one should be aware that the topography implied in Matthes' cross profiles conflicts with his preglacial reconstruction. For example, in Figure 32, the inner preglacial slope of Middle Cathedral Rock is gentle, but in his reconstruction in his Figure 15 it is about twice as steep. Additional discrepancies exist in some of his other cross profiles. In particular in his Figure 15 the descending ridge down-valley from Half Dome was required by Davisian analysis, but there is no field evidence to support it. Furthermore, the seismic evidence argues against it, for it shows that glaciers cut deepest in the valley where this resistant ridge supposedly existed. Finally, Matthes was selective in choosing his cross profiles; none exist for the valley's recesses, such as Indian Canyon, Eagle Creek canyon, or the massive half bowl between the Cathedral Rocks and Sentinel Rock. In each, bedrock lies just below the talus, as is evident by the local outcrops one encounters along the dozens of stream beds that descend the valley's talus slopes. As on the Starr King bench or in the Little Prather Meadow side canyon, each time a glacier advanced through the valley, part of it spread laterally and buried at least the lower parts of these recesses. Consequently, the glaciers *preserved* their bedrock surfaces rather than eroded them, and thus today's surfaces are fairly accurate representations of the recesses' preglacial topography. This observation is particularly important for the current bedrock outcrops in the lower parts of the recesses, especially in the Eagle Creek drainage, for they demonstrate that the valley floor at the onset of glaciation was very close to its present elevation. Had glaciers severely deepened the valley, then the uneroded recesses all would have been left hanging. Instead, all their streams are accordant with the valley floor, despite being very ephemeral and lacking erosive power to incise down to an accordant level. In short, Matthes' major transformation of Yosemite Valley by glaciers is a theoretical construct contradicted by the field evidence.

## Deep glacial lakes

Large, deep glacial lakes suggest tremendous erosive power of glaciers. Two prime examples are Lake Chelan and former Lake Yosemite, both used to prove major canyon deepening by glaciers. Two unused examples of relatively large and deep (by general Sierran standards) bodies of water are Fallen Leaf Lake and Emerald Bay, both in the Lake Tahoe Basin. The moraine damming the bay is under such shallow water that during drought years, when Lake Tahoe's level drops, the bay almost becomes a lake. In past glacial times the lake's level was higher, and then Fallen Leaf Lake existed as a glacier-occupied bay, in essence, a fjord, as Emerald Bay today could be considered. Fjords generally are considered as examples of profound glacial erosion. However, that conclusion is based on the assumption that before glaciation fjords were much shallower. Clifford Embleton and Cuchlaine King (p. 300) present J. H. Winslow's alternative interpretation—that fjords represent glacially modified raised submarine canyons—and the two authors concluded:

> There is no doubt that both submarine canyons and fjords separately exist, and there is no *a priori* reason why glacially modified submarine canyons should not also exist. The difficulty lies in recognizing and distinguishing them in practice.

Researchers have avoided the difficulty by not attempting to assess a fjord's preglacial depth through field evidence. Theory is far easier. Hence fjords or fjord-like bodies such as Emerald Bay and Fallen Leaf Lake cannot be used to prove major glacial erosion. Indeed, these two argue against it. Each lies in a deep tributary canyon that emptied on the floor of the Lake Tahoe Basin before the lake existed. The lake began to form about 2¼ million years ago when lava erupted in the north, creating a volcanic dam and effectively impounding water south of it. Between 2¼ and 1¼ million years ago there were at least seven major lava flows, which at times dammed the lake as much as 800 feet above its current 6229-foot level. "Fjords" were created not through glaciation but to flooding by the rising lake.

## Lake Chelan

Lake Chelan, located in the North Cascades of Washington and having a length of 50 miles, has a surface elevation of 1100 feet and a maximum depth of 1486 feet, the floor descending to 386 feet below sea level. Superficially, this implies that the glaciers have eroded 1486 feet below the lake's outlet—a considerable feat. However, this area is in the western part of the earthquake-prone Great Basin extensional province, in which extensive fracturing of bedrock can occur. Did the massive glaciers erode down through massive bedrock, or did they erode down through pervasively fractured, readily quarried bedrock? To prove that glaciers readily erode through massive, resistant bedrock, you first have to demonstrate that the bedrock in question was indeed massive, not fractured. This has not been done.

## Lake Yosemite

Yosemite Valley presents a similar problem: did the glaciers erode down through solid bedrock or did they excavate subsurface, deeply weathered, granitic detritus (i.e., gruss)? As with Lake Chelan, the possibility of glaciers removing thick, incoherent material was never considered. This likely is because until I began my research, all have considered Sierran uplift and consequent landscape development to have been relatively recent—i.e., mostly late Cenozoic—hence there would have been insufficient time to produce deep weathering in fractured bedrock on the valley's floor. Furthermore, until now, all perceived that post-uplift stream incision has been occurring at a high rate; hence weathered bedrock on the valley's bottom would have been removed about as fast as it was forming.

However, the Sierran landscape is very old. Late Cretaceous uplift has been recognized at least since early in this century, when in 1911 Waldemar Lindgren (p. 37, 44) said that the range was tilted westward along an east-side fault. This postplutonic, late Cretaceous uplift date has been confirmed in a 1993 study, based on plate-tectonic terrane accretion, by Jason Saleeby *et al.*, who implied that uplift was greatest in the southern part of the range, close to where the terrane accretion had occurred (the forcing of a large piece of crust beneath the southernmost Sierra's crust). I suggest that progressively northward the uplift was less, and the northern Sierra Nevada may not have been affected at all, judging by the accumulation there of voluminous, largely Eocene-epoch river gravels (the Tertiary auriferous gravels of Lindgren). However, it had been assumed that the late Cretaceous range had been reduced to a low-elevation, low-relief landscape by mid-Cenozoic time, and therefore the high-elevation, high-relief landscape of the present range could have been explained only by major late Cenozoic uplift. But what Part IV of this book demonstrates is that the range has remained high since the late Cretaceous uplift and that most of the canyon cutting occurred before about 33 million years ago, when the moist and warm climate began to change toward the modern summer-dry climate. Since then, weathering and erosion have been relatively minor and late Cenozoic uplift, if it has occurred at all, has been too minor to measure accurately.

Whereas the range long has been known to have experienced major denudation through weathering and erosion, and the amount is known to have been greatest in the southern Sierra-Tehachapi area—15 to 18 vertical miles, according to Donald Ross—what has not been appreciated until recently are the rate and timing: enormous denudation during the late Cretaceous. In 1992 Linn *et al.*, who studied the Mesozoic sediments of the Great Valley, proposed that denudation rates of the Sierra Nevada's Mesozoic volcanic rocks were high, on the order of 650 to

1000 feet per million years, assuming that the relief of the Sierran volcanoes had been similar to that of the present Cascades, which they put at 6500 to 10,000 feet. This high rate of denudation is consistent with the high rates of sedimentation found in the Great Valley sequence.

How fast did denudation expose the granitic plutons underlying the Sierran volcanoes? Richard Fiske and Othmar Tobisch observed that in the vicinity of the Ritter Range, just beyond Yosemite National Park's southeast border, granitic rocks appeared in 100-million-year-old conglomerate, and since the oldest granitic rocks are the same age, exposing the top of the granitic rock was very fast, almost coincident with the formation of a volcanic caldera (a very large crater). Rapid rates occur today. Satoru Harayama claims to have found the youngest unroofed granitic rock on earth, the Takidani Granodiorite in the Japan Alps, which is less than 2 million years old. His article lists 14 other unroofed plutons around the world that are 5 million years old or less.

One might argue that these are exceptions to a much slower rate of denudation. However, the Chilliwack batholith of the northwest Cascades, Washington, is mostly 25 to 30 million years old, yet the relief of its granitic, unfaulted canyons is as much as 5800 feet, which rivals or exceeds that of the Sierra's deepest canyons. Paul Bateman, using dates determined by others at the USGS, stated in 1992 that the plutons in the Yosemite Valley area range from 114 and 86 million years old. Given these ages, the granitic rocks initially could have been exposed in Yosemite Valley before 80 million years ago. Since warm, wet climates lasted until about 33 million years ago, the valley could have been exposed to some 50 million years of tropical weathering, and if the Chilliwack batholith is an accurate indicator of canyon-incision rates, then after the Sierra's late Cretaceous uplift there would have been ample time for major rivers to create their deep canyons.

There is a modern analog of the Sierra's late Cretaceous extremely high rate of denudation: the D'Entrecasteau Islands just off the east tip of Papau New Guinea. These islands exist in a shallow sea atop continental crust. Some granodiorite intrusions exposed on the islands are as young as 2.1 million years, and the rapid rate of their exposure is due not so much to tropical weathering as to tectonics—the unroofing of active metamorphic-core complexes. The average reader need not know anything about such complexes, but southwestern geologists most certainly do. E. J. Hill *et al.* noted that these complexes were similar to those in North America's Basin and Range province. Moreover, in both areas extensional faulting was accompanied by voluminous magmatism, from which granitic rocks originate. Their observation accords with Bateman's suggestion that magmatism in the late Cretaceous Sierra Nevada was accompanied by extensional faulting. Thus a high rate of denudation can be attributed to several causes that would have been operating to some extent during the late Cretaceous: detachment faults that caused large-scale rifting apart of the earth's crust, tropical climates that produced intense weathering, locally high relief that promoted mass wasting, and large rivers that transported high amounts of easily erodible volcanic deposits. All would have contributed to producing a largely granitic Sierran landscape by late in the Cretaceous period.

As Gutenberg *et al.*'s seismic work revealed, the canyon west of Yosemite Valley is underlain by two massive, granitic sills: one at Pulpit Rock and one at the head of Merced Gorge (see my Figure 31). Both Muir and Matthes had implied that the bedrock in which the valley developed had been jointed, the jointing then dictating the valley's geomorphic evolution—this is the *only* major conclusion that everyone agrees on. The joints are old. Bateman (p. 50) stated that "probably they formed during the late Cretaceous as the Sierra Nevada was being uplifted."

One could add that the Sierra's prevolcanic and volcanic sediments are unjointed, despite the underlying granitic rocks being jointed. This shows not only that the jointing occurred before the sediments were deposited, but also that jointing did not occur during late-Cenozoic time, when uplift is believed to have occurred. If Bateman is correct about jointing being in response to uplift, then a lack of late Cenozoic jointing implies a lack of late Cenozoic uplift—as I so conclude in Part IV. Consequently, the two sills, which are merely unjointed bedrock, would be as old as the adjacent jointed bedrock. That glaciers have not been able to remove them implies that they are exceptionally resistant to erosion and that they have been around, relatively unchanged, for an exceedingly long time. As such, they have served as local base levels for the Merced River since the late Cretaceous uplift. (Sea level is the ultimate base level, below which a river cannot erode; resistant bedrock can also prevent downward stream erosion for a very long time.) Behind these sills, a weathering front would have extended deeply through the fractured rock, creating a deep basin, such as those found in granitic, tropical lands, as noted by Jean Tricart (p. 155-159). The depth to bedrock can be enormous. Julius Büdel (p. 24) asserted that extended weathering under tropical climates can penetrate to a depth of 2000 feet, which, coincidentally, is the maximum depth to bedrock through the sediments of Yosemite Valley. In short, given the tens of millions of years of tropical weathering, it would have been unlikely for the valley's jointed bedrock floor *not* to have experienced deep subsurface weathering. Massive glacial erosion through the valley's presumed, preglacial, unweathered, very resistant bedrock is an unnecessary, undocumented hypothesis.

Finally, we need to admit that Yosemite Valley is an exception. If glaciers eroded deeply, then every glaciated Sierran canyon should have a stretch of thick sediments like those in the valley. None have been found. At best, generally shallow lake basins exist. Often in a drainage several lakes lie one after the other. These are called paternoster lakes—after beads on a rosary. Tropical streams in granitic terrain interestingly can have a series of poorly graded, wide, marshy valley depressions that follow one another at different levels, "like beads on a string," to quote Tricart (p. 157). Streams of the tropical/paratropical Sierra Nevada range would have developed these myriad depressions. Much later, then, glaciers removed the loose materials that had accumulated in these generally shallow basins. This is hardly an example of major glacial erosion in resistant bedrock.

## Longitudinal profiles

Glaciers are believed to convert a stream's nearly constant profile into an irregular one composed of alternating low-gradient treads and high-gradient risers (rapids, cascades, and even falls). A corollary to this hypothesis is that because incision of the trunk canyon has been great, tributary canyons, not being eroded as deeply, are left hanging. Neither supposition is necessarily true, for in Montana's prime example of glaciation, Glacier National Park, most of the glaciated lengths of canyons have relatively smooth profiles not much different in gradient from unglaciated reaches, and most of the tributary canyons are accordant, not hanging. Only near the headwaters does the profile become noticeably irregular. Nevertheless, the view of longitudinal-profile conversion by glaciers has been widely published, and in the Sierra Nevada has been promulgated by Matthes and Huber. But as was just stated above, the Merced River had an irregular profile for tens of millions of years before the first glacier flowed down it, and even Matthes had stated that most of the tributaries were hanging before glaciation.

In the Sierra Nevada the major rivers have one of two longitudinal profiles: smooth or irregular (alternating treads and risers). The rivers with the relatively smooth profiles are North and Middle forks of the Feather River (Figure 33) and the North Yuba River (Figure 34). Except for the Feather River, whose North and Middle forks originate east of the range, the profiles begin at the headwaters divide and descend to the foothills, beyond any glaciation. These are derived from detailed measurements of rivers as drawn on 15' topographic maps. (The Sierran crest is quite obvious in the North Fork drainage, and the river crosses it near the Ben Lomond summit. However, it is not at all obvious in the Middle Fork drainage, and one can interpret its location to be near one of three landmarks: Dogwood Peak, Oddie Bar, or Crescent Hill.) Note on Figure 33 that east of the crest the profiles of the Middle Fork and the East Branch of the North Fork are low gradient, both reaches paralleling the east base of the range. In contrast the North Fork originates near the base of Lassen Peak and flows south across Neogene volcanic rocks to the Lake Almanor dam (where the figure's profile begins), then after descending a few miles downstream, its course turns southwest and flows through a metamorphic canyon.

Also worth noting is that the linear projection of the low-elevation parts of these profiles eastward to the crest. Sierran uplift determinations by Axelrod, Axelrod and Ting, Huber, Hudson, Lindgren, Matthes, and Unruh all were based on the fundamental assumption that Sierran rivers had possessed essentially constant, preglacial profiles up to the Sierran crest. The Feather River's North Fork does, but its Middle Fork does not. In the Yuba River drainage (Figure 34), the North Yuba River does have a fairly smooth profile, except for a steep-gradient reach in the foothills, where it crosses quartz diorite, and a steeper one between 6 and 12 miles, where it crosses Mesozoic metavolcanics along the east side of Sierra Buttes. Between miles 30 and 60 the gradient is quite constant, but if it is projected east to the crest, at Yuba Pass, the projected point lies about 3000 feet below it. Detailed mapping of moraines in the Sierra Buttes area by Mathieson demonstrates that the snouts of glaciers

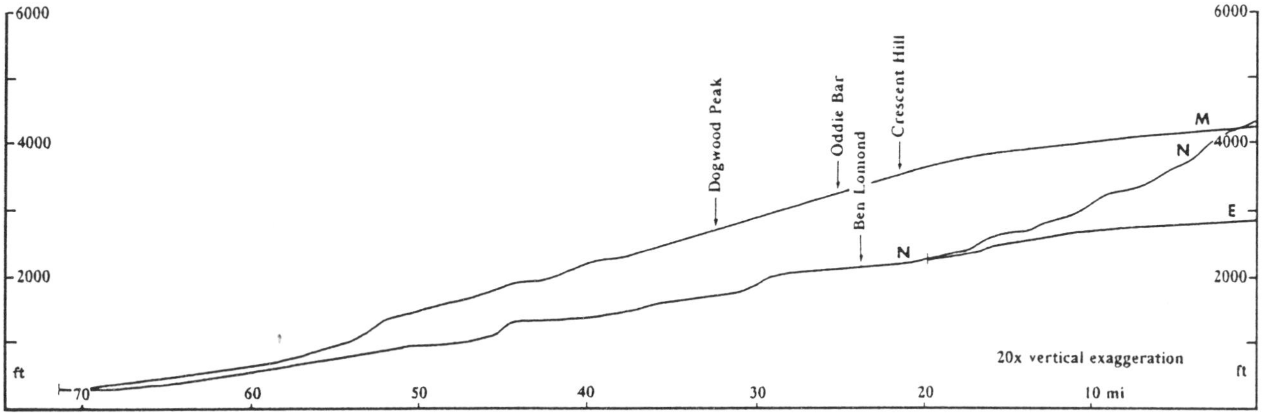


**Figure 33. Profiles of the Feather River's North Fork (N), its East Branch (E), and the Middle Fork (M).**

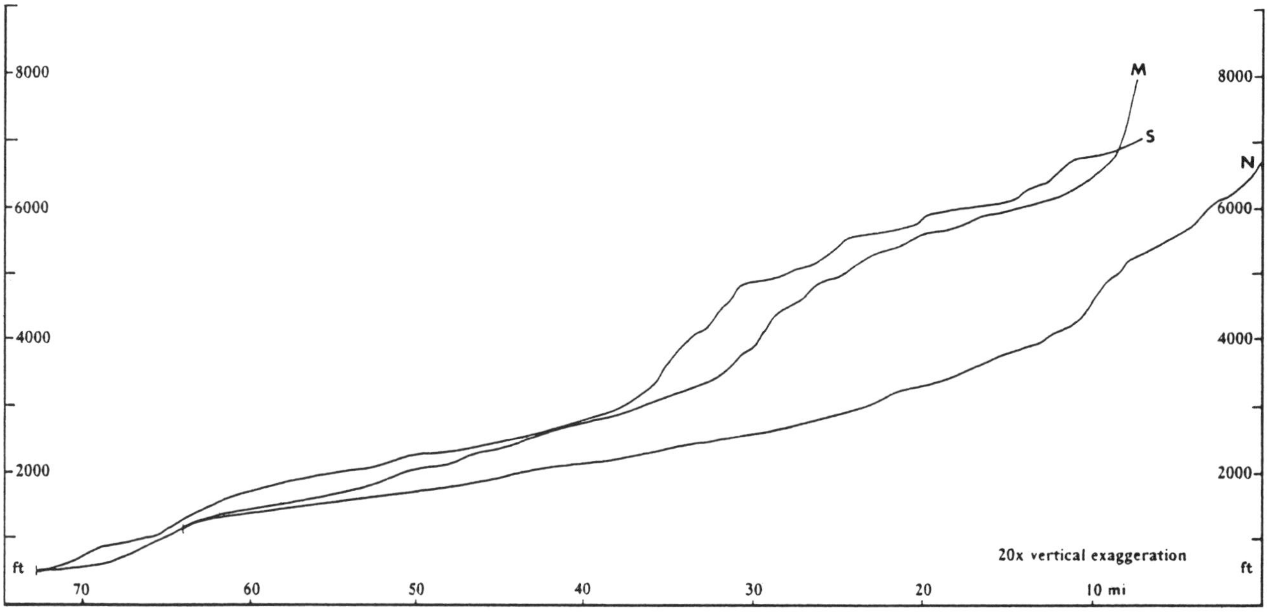


**Figure 34. Profiles of the North (N), Middle (M), and South (S) Yuba River.**

descended just barely to the North Yuba River, reaching it in the upper half of the steep-gradient reach along the east side of Sierra Buttes. Therefore, the irregular profile of the first 12 miles appears to be independent of glacial erosion.

If glaciers severely deepen their canyons through erosion, then the upper glaciated part of a Sierran river should not hang above the lower unglaciated part, but often they do. Both the Middle Yuba and South Yuba rivers have a major high-gradient reach about midway along their course, each incising through the Bowman Lake pluton. In both drainages glaciers descended to about the base of the high-gradient reach, yet they did not make any significant incision, for the glaciated canyon floors hang high above the unglaciated ones. In the Middle Yuba River drainage, the glaciers flowed mostly across metamorphic bedrock, while in the South Yuba River drainage, they flowed mostly across granitic bedrock—as is the case for all rivers south of them. South of the Yuba River are the American and Mokelumne rivers. However, the extent of glaciation in each drainage is poorly known, so the amount of glacial erosion cannot be analyzed.

South of the Mokelumne are the North, Clark, and Middle forks of the Stanislaus, whose profiles are shown in Figure 35. (The South Fork is relatively minor.) My field work, mentioned in the previous chapter, suggests that in the North Fork drainage the Tioga, Tahoe, and pre-Tahoe glaciers extended down to about 4000 feet, 3900 feet, and 3600 feet respectively. Consequently, the largest glacier advanced down-canyon only to about where the gradient dramatically increases. The never-glaciated, major granitic step is a product of stream erosion, not glacial. The profile then becomes less steep where the river meets metamorphic bedrock, and across it the North Fork flows a short distance to its junction with the Middle Fork. The extents of Tioga, Tahoe, and pre-Tahoe glaciers in the Middle Fork are poorly known. The Tioga and Tahoe glaciers probably advanced down to about 4000 feet, if not considerably farther, while the pre-Tahoe probably advanced down to between about 2500 feet and 1500 feet. Note that by examining the profile, one cannot accurately determine the boundary between stream and glacial erosion. The only conspicuous change in gradient in both the Middle and Clark forks is at about 6000 feet, well above the lower limit of any major glaciation. Below this, for the lower two-thirds to three-fourths of the total glaciated extent, the profile is indistinguishable from that of the unglaciated profile. One generalization that applies here—and to all Sierran rivers from the Yuba south to the Kern—is that the profile is low-gradient and essentially constant only below about 1500 or 2000 feet elevation, as one can judge from topographic maps that show these rivers.

The profiles of the Tuolumne River and its Middle and South forks, all across granitic bedrock, are shown in Figure 36. As was mentioned earlier, in the Tuolumne River canyon the Tioga glacier descended to about 2400 feet, the Tahoe presumably a bit lower, and the pre-Tahoe to about 2000 feet. However, the stretch of profile between about 2000 and 3000 feet, which contains these maximum extents of these glaciers, is a relatively smooth line. No accurate indication of maximum glacial advance is present. The Tuolumne contains the Sierra's grandest step, from the Tuolumne Meadows-Lyell Canyon tread down to the Hetch Hetchy Valley tread. Its existence is not correlated with the thickness of past glaciers but rather exists because neither stream nor glacial processes

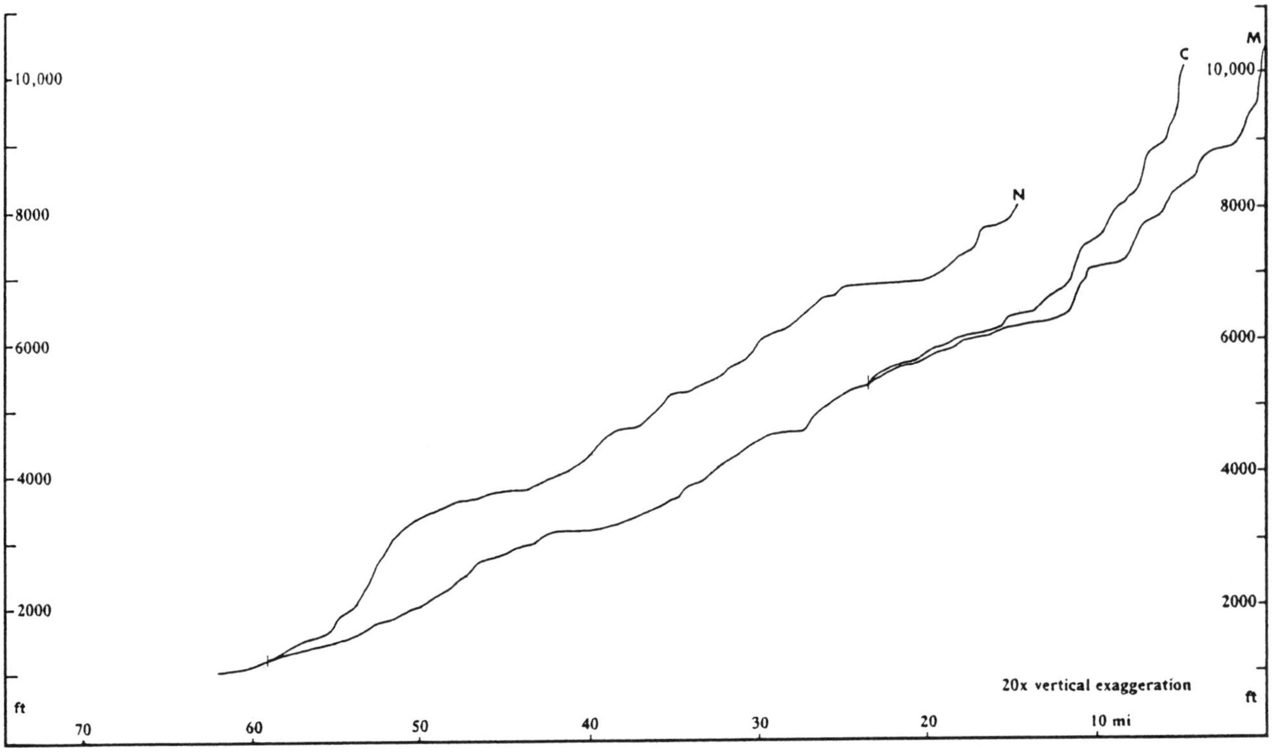


Figure 35. Profiles of the Stanislaus River's North (N), Clark (C), and Middle (M) Forks.

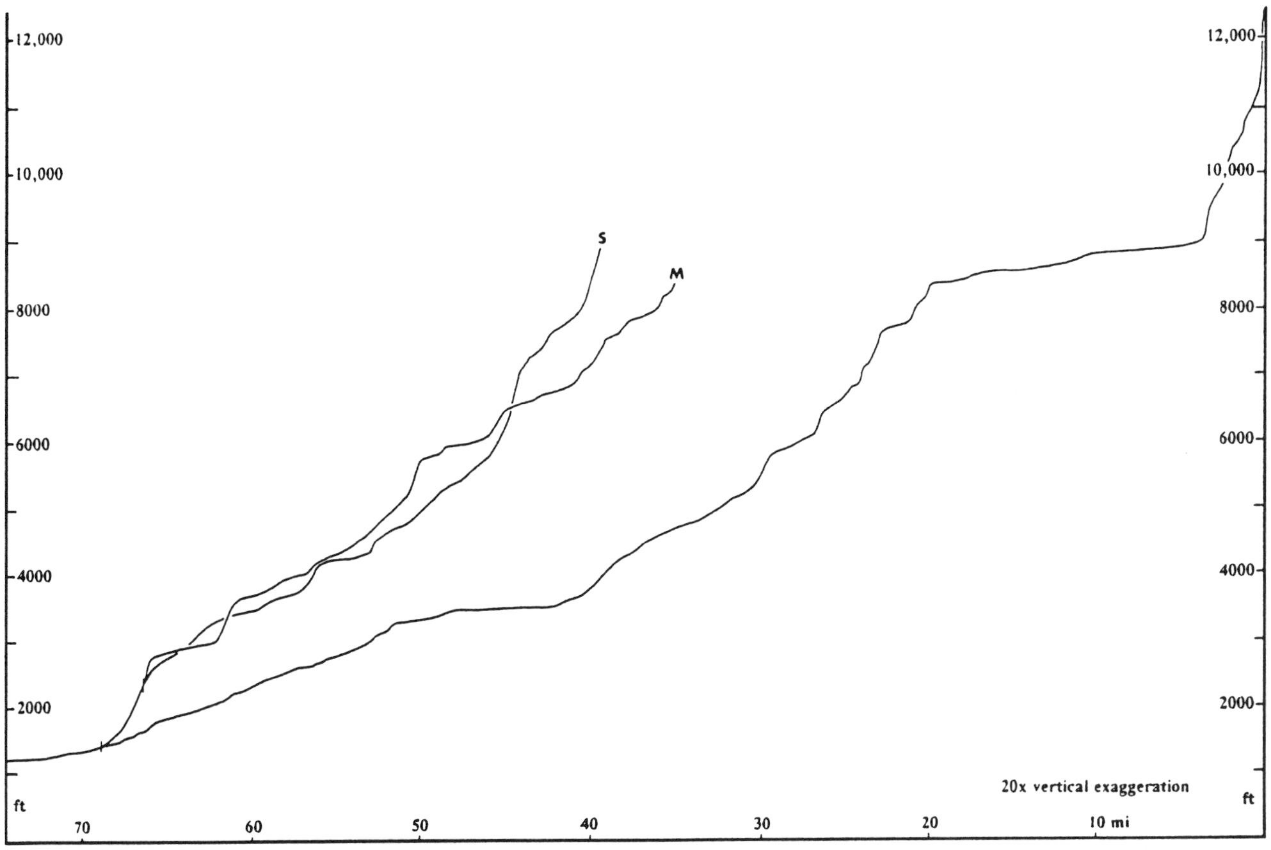


Figure 36. Profiles of the Tuolumne River and its Middle (M) and South (S) forks.

have been able to cut down through the resistant bedrock at the west end of Tuolumne Meadows, which has acted as a local base level for both river and glacier alike. In that area is a small volcanic plug, dated at 9.4 million years old, and known as "Little Devils Postpile." The base of this columnar-lava formation stands only a few feet above the Tuolumne River, indicating that only a very minimal amount of post-volcanic incision by both the river and its mammoth glaciers has occurred. From where the much smaller Middle Tuolumne and the South Fork Tuolumne join, the combined stream makes a major step to the Tuolumne River. Above that step the two forks have irregular profiles, the South Fork having a more "glacier like" profile, one with a relatively gentle middle gradient, a major step, and a "cirque" headwaters area. It, however, never was glaciated, while the upper part of the Middle Tuolumne experienced minor glaciation, but its profile offers no clue of this.

The profiles of the Merced River, its South Fork, and Tenaya Creek are shown in Figure 37. Of the three profiles, that of Tenaya Creek has the greatest step, one approaching in height that of the Tuolumne River. Like that one, its existence is not correlated with the thickness of past glaciers but rather with the areal pattern of resistant and nonresistant bedrock. The resistance of the bedrock was observed by Matthes, who concluded:

> Tenaya Canyon as a whole presents the most stupendous exhibit of massive granite in the entire Yosemite region, perhaps in the entire Sierra Nevada.... Not even the Grand Canyon of the Tuolumne nor the Kings River Canyon is hemmed in by walls so uniformly and continuously massive.

This also applies to the Merced River proper, which has three smaller, yet still major, risers, the lowest in Merced Gorge, the intermediate between Yosemite and Little Yosemite valleys, and the highest above Washburn Lake. Had glaciers been effective at eroding, then the Tuolumne glaciers repeatedly overflowing into the Tenaya Creek drainage should have eroded down through its headwaters divide. (This also should have occurred at other major overflow sites—but didn't—where the Mokelumne River glacier overflowed south into the North Fork Stanislaus River drainage, where it overflowed north into the East Carson River drainage, and where the East Carson River glacier overflowed west into the Clark Fork drainage.) The South Fork Merced River's glaciers descended to the vicinity of Wawona, which lies at an elevation similar to the floor of Yosemite Valley, 4000 feet. It is at about this elevation that the South Fork greatly diminishes its gradient. The relatively thin South Fork ice field left no significant imprint on the river's profile. Indeed, above Wawona the profile is a "classic" unglaciated one in that until its headwaters it maintains a fairly constant gradient. In contrast, in the unglaciated canyon lands from Wawona onward, the irregular profile down to the confluence of the Merced River is what one would expect for a glaciated canyon.

Those who have studied Sierran uplift have assumed that river profiles were transformed through glacial erosion from ones of fairly constant gradient to ones with treads and risers. However, the South Fork Kern River profile, shown in Figure 38, demonstrates that rivers across unglaciated, granitic topography also have profiles with treads and risers. At best, only its upper few miles were glaciated, as discussed in the previous chapter. The risers increase in magnitude downstream, the largest being the descent from Rockhouse Basin south down to where the river turns west. This entire reach is parallel to the Sierran crest and so should not have been steepened through westward tilting of the range (be it Mesozoic or Cenozoic); nevertheless, it descends just as much as any major, west-flowing river. The South Fork flows west to meet the Kern proper, which has flowed down a relatively smooth course, dictated in large part by the Kern Canyon Fault, not by glaciation or lack thereof. United, the Kern then makes a dramatic, cascading descent to the base of the range—a very major step perpendicular to the Sierran crest. Major risers occur regardless of orientation with respect to the crest. This is true not only in the Kern River drainage, but also in the others.

As suggested just above under the antiquity of the South Yuba River canyon, and as will be shown in the next chapter, the granitic landscape of the Sierra Nevada west of its crest has changed relatively little over tens of millions of years. It was created largely late in the Cretaceous and early in the Cenozoic in response to late Cretaceous uplift and to weathering and erosion under tropical climates. The South Fork Kern River and some of its tributaries have dramatic risers, and the same appears to be the case for other Sierran drainages, especially the Merced, with its finest stepped example, Yosemite Valley and environs. An example of a modern drainage developed under a tropical climate and eroded down through granitic bedrock is that of the Río Grande (central Colombia, equatorial South America), which has been cut entirely within the Antioquian batholith. None of the drainage has ever been glaciated. The profile of the Río Grande, from its headwaters east to its junction with a larger river, the Río Medellin, is shown in Figure 39. Its profile is hardly smooth and concave upward; rather, it is stepped very much like those of the Sierra's rivers that traverse granitic lands. The unglaciated Río Grande landscape has all the characteristics of a "classic" glaciated one, which are enumerated in the next chapter. The landscape, like that of the unglaciated South Fork Kern River landscape, lacks only the *real* glacial evidence: polish, striations, and till.

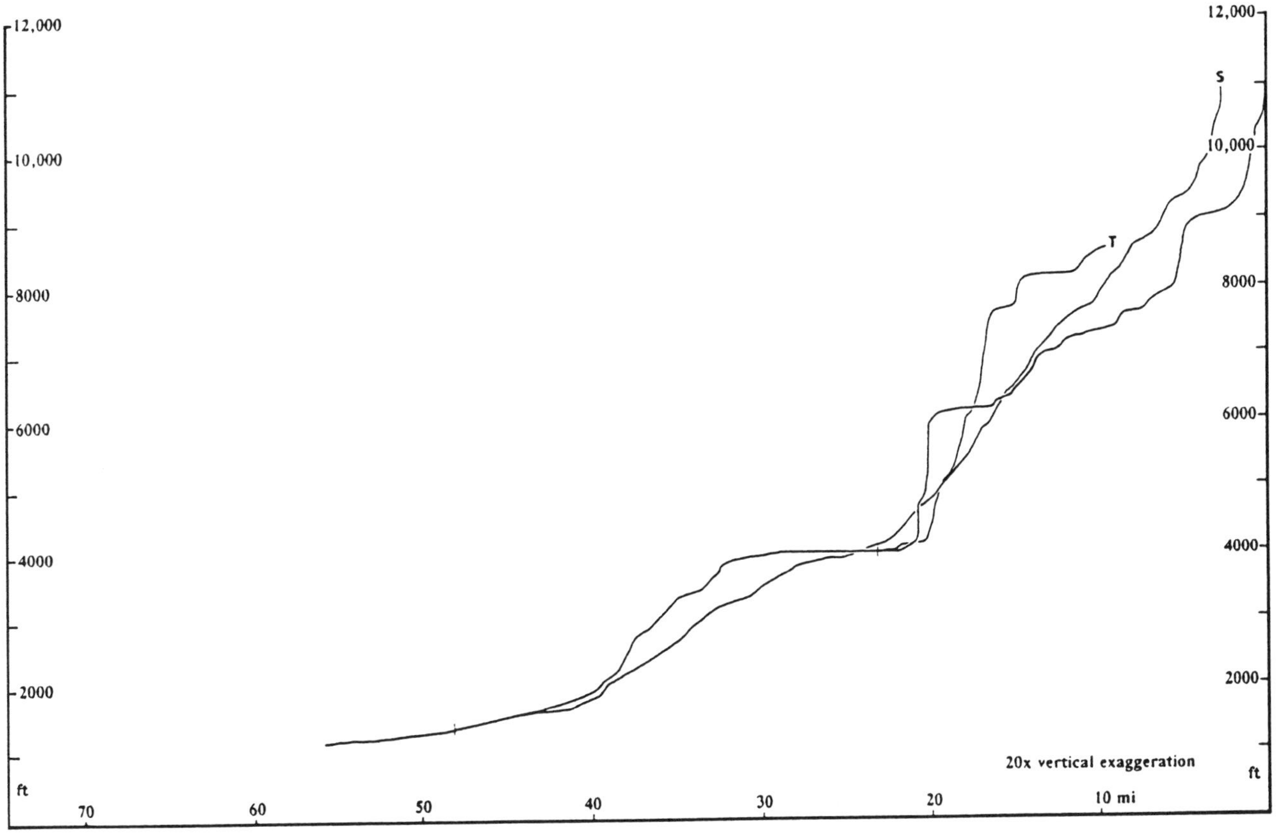


**Figure 37. Profiles of the Merced River, its South Fork (S), and Tenaya Creek (T).**

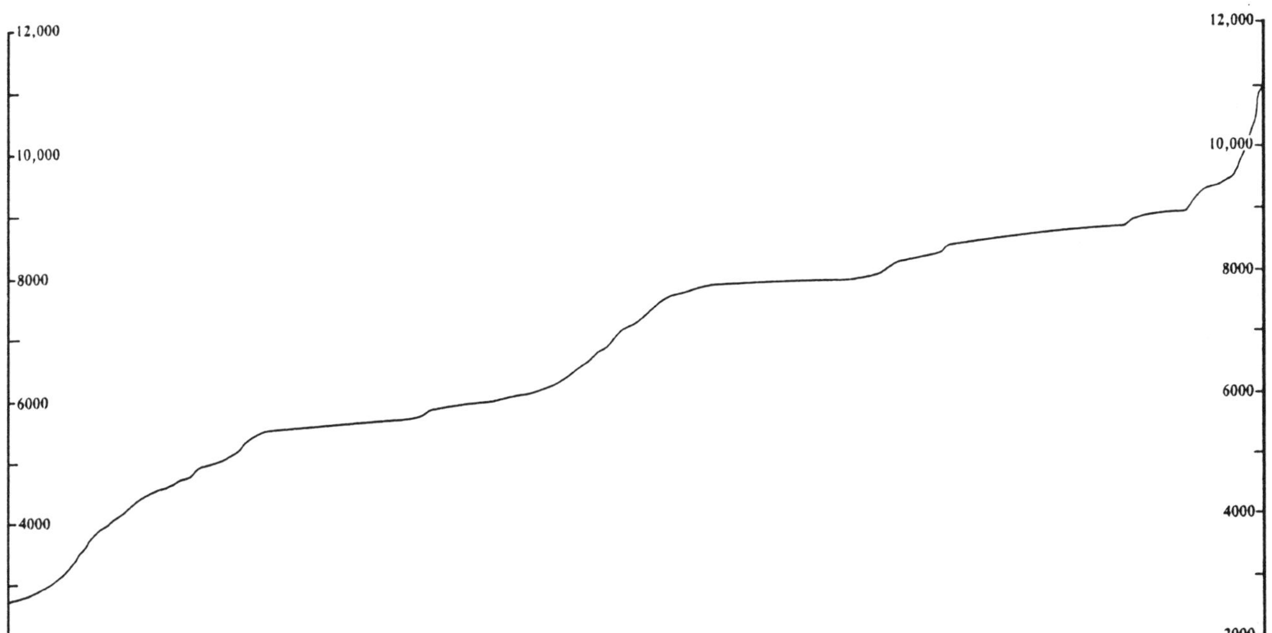


**Figure 38. Profile of the South Fork Kern River.**

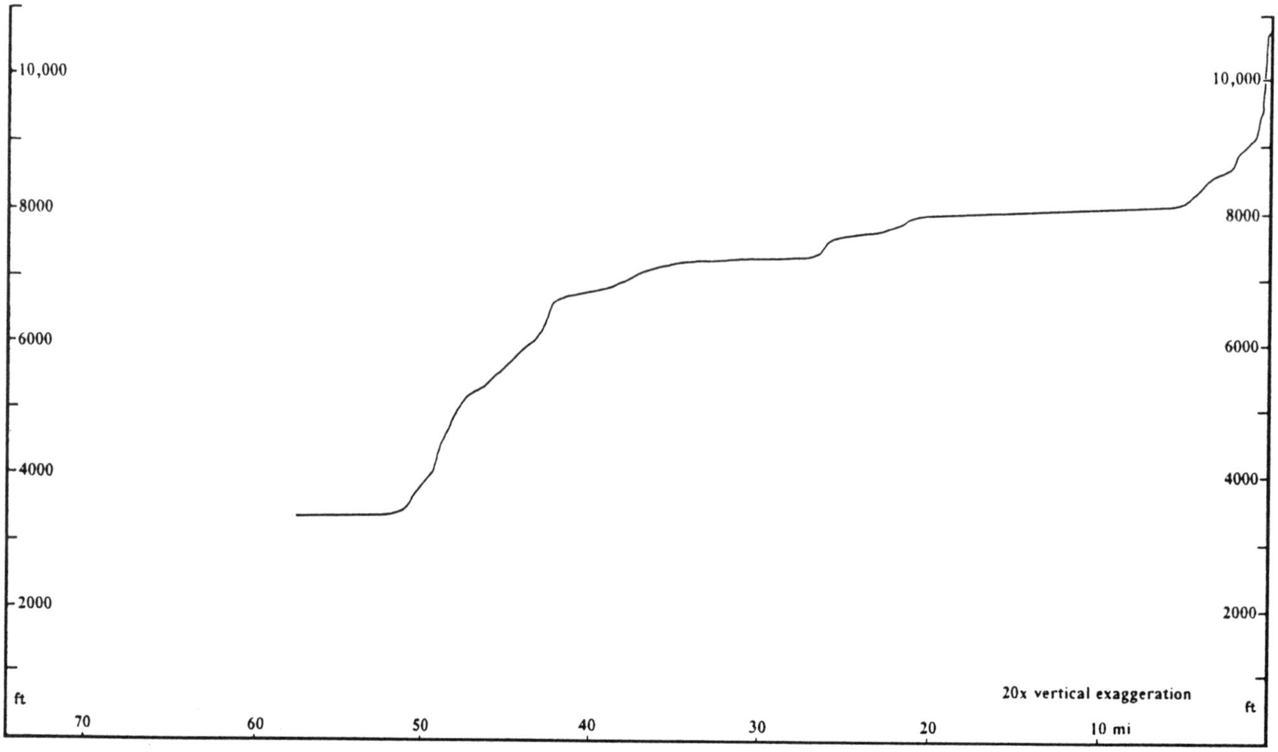


**Figure 39. Profile of the Río Grande, central Colombia (equatorial South America), from its headwaters east across the Antioquian batholith to its junction with a larger river, the Río Medellin.**

## Stoss-and-lee topography

Plucking (quarrying) and abrasion long have been considered primary mechanisms of glacial erosion, and these have been proposed to produce stoss-and-lee topography: an advancing glacier that flows across protruding bedrock abrades its stoss (up-canyon) side and plucks its lee (down-canyon) side. The asymmetrical forms often are called roches moutonnées, and such features can be minor or major. According to Robert Sharp and N. King Huber, an example of a major one is Lembert Dome, whose summit stands 850 feet above Tuolumne Meadows. If the dome originally had been symmetrical and a major part of its lee side had been plucked away by glaciers, then one could make a case for major glacial erosion—but only for the missing part of the dome. On the abraded part of the dome, erosion was minimal, as it was for the adjacent bedrock floor. (If erosion through abrasion had been effective, it would have planed the dome down to the level of the adjacent bedrock floor.) Such "domes" (most are not), however, are rare, so overall the glacial plucking in an entire glaciated drainage would be minimal.

In the Merced drainage there are two comparably large examples, Liberty Cap and Mt. Broderick (see Map 25 and Photo 151), which Matthes called "roches moutonnées of gigantic size." However, in his argument against the origin of glacial stairways through glacial plucking that caused up-canyon retreat of lee faces (p. 94-95), he noted that the riser (lee face) over which Nevada Fall leaps is essentially fixed and has barely retreated. The lee faces of Liberty Cap and Mt. Broderick are part of the resistant band of cliffs over which Nevada Fall leaps, and so by inference their lee faces also are not retreating. Nevertheless, in the same section Matthes provides an illustration of the formation of a roche moutonnée that closely resembles Lembert Dome, Liberty Cap, and Mt. Broderick. Just east of Lost Lake, this asymmetrical bedrock ridge differs from Liberty Cap only in scale, and both should have formed in the same manner in response to the same processes, through glacial plucking that caused up-canyon retreat of the lee face. If so, then the riser (cliff) over which Nevada Fall leaps *has retreated.* You can't have it both ways; either the Liberty Cap cliff and the Nevada Fall cliff have retreated through glacial plucking or they have not.

The formation of roches moutonnées has little to do with glaciers, which merely accentuate a preexisting, asymmetrical ridge. Countless examples of intermediate and large "roches moutonnées" exist across the granitic, *unglaciated* landscape west of the Sierran crest. As was mentioned earlier, many Sierran rivers and tributaries have stepped profiles, and paralleling them are asymmetric ridges, each with its summit closest to the steep lee side—just like the summits of Lembert Dome, Liberty Cap, and Mt. Broderick. These "roches moutonnées" are—like the streams they bound—relict, tropical features, having originating tens of millions of years ago. An indication of their longevity can be estimated from large erratics on the summit of Turtleback Dome which are believed to have been deposited about 800,000+ years ago, at the end of the Sherwin (El Portal) glaciation. The two large erratics on the dome rest on pedestals averaging about 4 inches high. If this average value is the amount of denudation to have taken place on gentle, broad ridges and summits in the last 800,000 years (in post-Sherwin time), then in one million years there would have been about 5 inches. The amount of denudation at this rate for the entire Cenozoic era (65 million years) amounts to only about 27 feet! Obviously, it must have been greater during warmer, wetter tropical times. Not coincidentally, these "roches moutonnées" exist today in unglaciated tropical-river drainages of granitic lands.

## Glacial sediments

Calculations of the volumes of till and outwash long have been used to determine the volume of material removed by glacial erosion. The assumption here is that till and outwash are derived solely from glacial erosion, even though Lyell (1833, p. 149-150), before the earliest glaciologists, had stated what mountain residents had long observed before him: 1) earthquakes occur in mountainous areas; 2) they produce prodigious amounts of rockfall; and 3) glaciers transport this rockfall. In the mid- and late-nineteenth century battle over the efficacy of glacial erosion, the pro and con arguments were couched solely in terms of glacial versus stream erosion. Mass wasting—the most obvious process observed by alpinists and mountain residents—was completely ignored. One can estimate the amount of glacial erosion by determining the total volume of glacial deposits and then subtracting the total volume of mass-wasting deposits. What remains is the volume derived through glacial erosion.

In most of the world's glaciated mountain ranges, the volume of mass-wasting deposits is very substantial. This is because glaciers tend to exist in high ranges, each of which became high through earthquake-generating uplift. Each earthquake generated enormous rockfall. During historic time, Alaska earthquakes of greater than magnitude 5.5 have occurred at the average rate of about one per month. The amount of rockfall generated even by a moderate earthquake or earthquake sequence can be significant. A must-get map is USGS Map I-1612, prepared by Edwin Harp *et al.* It shows that the May 25-27, 1980 Mammoth Lakes earthquake sequence (several 6+ quakes, up to magnitude 6.6) generated several thousand mass-wasting deposits in the east-central Sierra Nevada, hundreds of them large enough to be accurately drawn on the 1:62,500-scale topographic map. Future glaciers will transport this debris to leave significant moraines, which could be used as "proof" of major glacial erosion. Just how much debris can be produced in this earthquake

**Photo 151. Liberty Cap and Nevada Fall. Matthes argued that the Nevada Fall cliff has barely retreated while the adjacent Liberty Cap cliff has had major retreat. This is an impossibility.**

sequence is seen in the following example. In the Fourth Recess of the upper Mono Creek drainage (in the San Joaquin River drainage, Mt. Abbot 15' quadrangle), which has an area of about 1830 acres, about 15% of its area, roughly 270 acres, were covered by deposits (Map 42). If one assumes that the average thickness of the deposits was about 3 feet—a very conservative estimate based on my observation of deposits in the Fourth Recess and elsewhere—then the total volume would be about 560,000 cubic yards. This is approximately the volume of Yosemite Valley's "Medial Moraine," the valley's largest end moraine. The larger 1872 Owens Valley (Lone Pine) earthquake (about magnitude 7½) caused mass wasting over a far greater area, generating rockfalls even 120 miles from the epicenter. Earthquakes such as the Mammoth Lakes sequence have been more prevalent than most people realize. Carl Stover and Jerry Coffman's U.S. Geological Survey Professional Paper 1527, *Seismicity of the United States, 1568-1989 (Revised)* show that in or

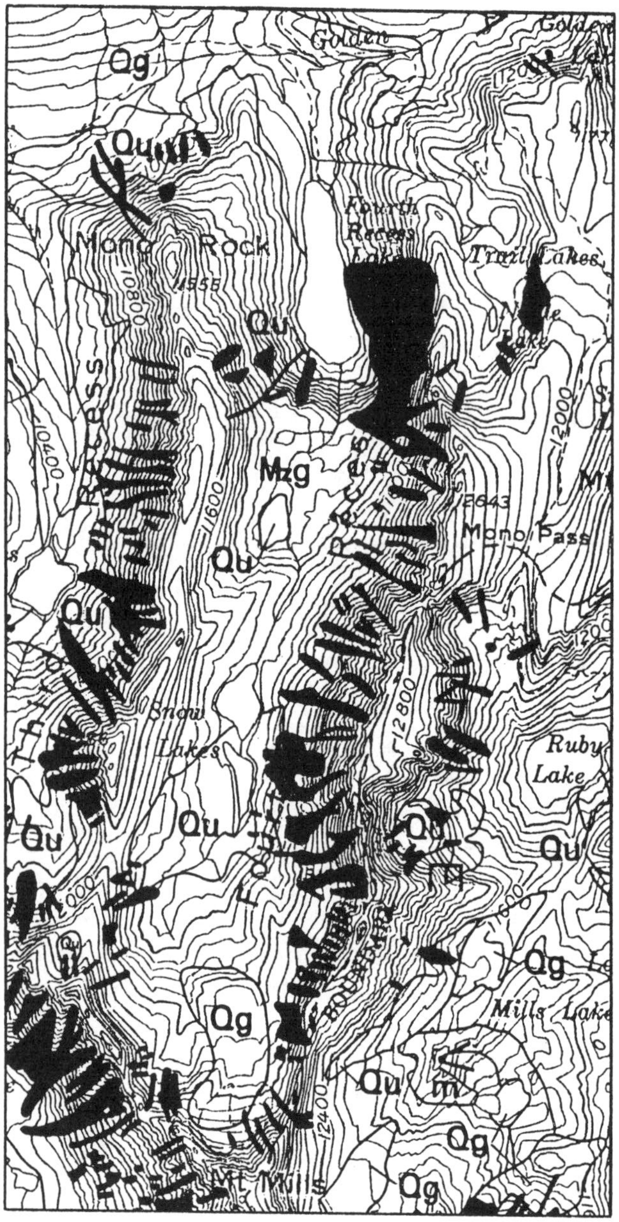


**Map 42. Rockfall in Fourth Recess, upper Mono Creek drainage. Qu = Quaternary surficial deposits, undivided; Qg = Quaternary glacial deposits; Mzg = Mesozoic granitic rocks; m = metamorphic rocks. Scale 1:40,000. 2x enlargement of part of USGS Map I-1612.**

near the Sierra Nevada a significant earthquake occurs about once every two years.

Some simple mathematical calculations can demonstrate the enormous role mass wasting likely has had on the widening of Sierran canyons. (East-side canyons are chosen, since west-side ones vary considerably in width, e.g., Middle Fork vs. South Fork San Joaquin River canyons.) The average width of a typical glaciated east-side canyon, rim to rim, is about 3000 feet. If (hypothetically) the canyons were about half their present width (i.e., 1500 feet) at the start of major glaciation about 2 million years ago, then what average rate of wall retreat (by any process) would be required to achieve the present 3000-foot width? The amount of retreat for each wall of the canyon would be 750 feet over 2 million years, which amounts to a rate of retreat for each wall of about ½ inch per century. In a matter of a few days the Mammoth Lakes earthquake sequence created myriad localized rockfalls that caused local retreat on the order of several feet. The total amount of rockfall scars in any given canyon was generally between 1 and 10% of the surface area of the walls.

If we assume the average cliff retreat was 4 feet, and if we use the lower value of rockfall area, 1%, then only one earthquake this size needs to occur every century (48 inches x 0.01% = 0.48 inch ≈ ½ inch). With the higher figure, only one earthquake this size needs to occur every millennium. Thus, earthquake-generated rockfall by itself can account for canyon widening; glacial erosion is not required (although glacial transport is useful). (In the absence of glaciation, such as in the southern Sierra's Dome Land Wilderness, the fallen rocks very slowly disintegrate, creating gruss, which then can be windblown or gravity-transported down to a stream.) Given that east-side faulting has been occurring for about 3 to 6 million years and that modern, major earthquakes appear to be occurring in any given part of the eastern Sierra Nevada at a rate of greater than one per century, the actual amount of rockfall should be even higher. The problem then is that too much rockfall seems likely to have occurred over the last few million years, not too little. And this model only addresses one kind of mass wasting: earthquake-generated rockfall. It does not address the substantial number of nonseismic (e.g., freeze-thaw generated) rockfalls. For Yosemite Valley, Gerald Wieczorek *et al.* concluded that where the causes of rockfalls were known, 47% of the rockfalls were due to nonseismic reasons.

Whereas the rate of rockfall production in crest lands seems unbelievably high, even Yosemite Valley, which lies about 20+ miles from any active fault, seems to have a high rate. In 1994 David Keefer determined the average rate of retreat of its walls to be about 14 inches per 1000 years, which extrapolated constantly over time would cause the valley to be filled to the rim in about 5 or 6 million years. Wieczorek *et al.* derived rockfall thickness by extrapolating Gutenberg *et al.'s* bedrock-floor laterally to beneath the talus. This produces a much greater volume than one that is based on field evidence. Furthermore, they assumed all talus had been removed by the Tioga glacier, which yields a higher production rate. My own estimates, based on extensive traversing of talus slopes over the years, is lower: the talus is not as thick as it superficially appears, and bedrock, as exposed in gullies, can lie just below the surface. I determined the average thickness of the valley's talus slopes to be about 10 feet, which represents an average backwasting of the valley's walls of about 3 feet during the last 14,000 years, or about 460 feet in the last 2 million years of glaciation. This rate, extrapolated constantly over time, would cause the valley to be filled to the rim in about 20 million years.

Keefer's rate and my rate are too high, since Yosemite Valley would have been in existence for 60+ million years if it developed a deep form early on as did the South Yuba River canyon, mentioned above. There are at least five reasons for these overly high rates. First, earthquakes along the east base of the Sierra Nevada have been occurring for only the last few million years, as lands to the east began to subside; before their commencement, rockfall would have been considerably less. Second, periglacial (near glacial) climates cause intense physical weathering and mass wasting, and before glaciation commenced, the Sierra's climates would have been milder and mass wasting less intense. Third, when a glacier advances down a canyon it can bury talus that collects in recesses such as Indian Canyon, Eagle Creek canyon, and perhaps in El Capitan Gully and the half-bowl east of the Cathedral Rocks. In Yosemite Valley, great volumes of talus occur in such recesses and would have been buried, not transported away, since glacial ice would have flowed into these recesses and would have been trapped there. Therefore, perhaps much or most of the valley's talus is pre-Tioga and pre-Tahoe, and the rate of *post*-Tioga production has been greatly overestimated. Fourth, the valley's rockfall appears to be in part a response to post-glacial pressure release at the bases of the cliffs. Without glaciation, rockfall production would be much lower, as in Dome Land Wilderness, where talus is minimal, despite steep cliffs and despite the area's lying closer to active faults. And fifth, when a glacier flows through a canyon, the part of its walls that is buried by ice is protected from mass wasting. Hence during the last 2 m.y., when glaciers were present more often than not, the rate of mass wasting would have been lower than it is during our current interglacial time, the Holocene.

Consequently, the amount of late Cenozoic widening of Sierran canyons due to mass wasting is lower than talus deposits would indicate, though nevertheless it has been substantial. In contrast, the amount of widening due to glacial erosion appears minimal, since mass wasting, by itself, can account for all of it. In conclusion the volume of till and outwash cannot be used as an indicator of severe erosion by alpine glaciers. This would be true even in the areas of continental glaciation, such as around Hudson Bay, where during interglacial times intense physical weathering (periglacial processes in particular) provides "fodder" for future ice sheets to remove. Before the 1980s, investigators concluded that major erosion due to glaciation has not occurred in areas of continental glaciation such as the Laurentide region. In contrast, Bell and Laine in 1985 concluded that an average of 400 feet of bedrock erosion by glaciation over the last 3 million years has occurred, this estimate being an order of magnitude (10x) higher than earlier estimates. Four hundred feet may sound impressive, but it is not. There have been about 48 glacial cycles in the last 2.5 million years. This amounts to only 8.3 feet of glacial erosion for each cycle. Interestingly, around the southern part of Hudson Bay, which has been ice free from about 9000 to 6000 years ago, the depth of active permafrost averages about 8 feet. So glaciers apparently remove just fractured bedrock. Plucking and abrasion of unfractured bedrock need not be invoked, since periglacial processes already provide the loose material for glaciers to transport.

## Field measurements

David Drewry (p. 83) has urged caution in the quantitative measurement of glacial erosion:

> It is questionable whether any attempt *should or could* be made in view of the uncertainties in measurements, the need to make wide, often *ill-considered generalizations,* the *pitfalls* in extrapolation and the marked *spatial and temporal variability* known to occur in nature.... [This ch. 6] highlights *serious shortcomings* in our knowledge of the glacial regime and methods of measurement and study. (Italics mine.)

Nevertheless, attempts have been made in the past and probably will continue to be made in the future. There are two types of measurements: 1) attach an object onto the bedrock floor and see how fast a glacier erodes it; and 2) drill a hole in the bedrock floor and see how fast a glacier erodes down to its base. Neither method is a legitimate proxy for glacial erosion. In the first case, the object protrudes into the path of the basal ice, in contrast to smooth bedrock lying unobtrusively under flowing ice. One large entrained boulder may be sufficient to remove the object, but may have a minimal effect on the bedrock. In the second case, the smooth surface of the bedrock is disrupted by the hole, which changes the rates of processes. For example, a microscopically thin film of water exists beneath some glaciers, and it would have no effect on smooth bedrock. However, it would accumulate in the drilled hole, could freeze, and then could lead to plucking. Finally, there is the matter of process rates. Exposed bedrock in periglacial environments can undergo intense physical weathering, and then this fractured, upper layer of bedrock is rapidly removed by an advancing glacier. The underlying bedrock then may be abraded smooth at a significant rate, but later erosion occurs at a much slower pace over the smoothed bedrock. Unfortunately, field measurements are most likely to measure erosion rates for a brief glaciation, not one lasting thousands of years. When such high rates are applied to the entire Quaternary period, the amount of erosion is enormous—and erroneous. Drewry's caveat should be well heeded.

## Lab experiments and computer simulations

The weakest demonstrations of significant glacial erosion come from lab experiments and computer simulations, the latter unfortunately having become increasingly popular. In an effort to give legitimacy to a branch of a soft science (such as geomorphology), its practitioners have turned increasingly to mathematics, often based on theoretical considerations or on quantitative analysis of grossly simplified, unrealistic field experiments. Says William Resetarits (*Science*, 1995, v. 269, p. 314) on ecology, "The push toward experimentation beginning in the 1960s was the result of 'physics envy.' We wanted to be a hard science." In the soft sciences, which have many positively and negatively interacting variables, many of their researchers have attempted to reduce the variables to a manageable few, and sometimes to just one. The results based on such simplistic research are, perhaps more often than not, utterly meaningless.

By assigning the appropriate assumptions and knowing the desired conclusion, one can set up a lab experiment or computer simulation to prove virtually any hypothesis. A 1992 example of this is a demonstration by Jonathan Harbor of how valleys such as Yosemite have their cross profiles transformed through glacial erosion from V-shaped to U-shaped (Figure 40). He ran a computer program for ice flow through a glacier cross profile with an erosion model, using Matthes' assumption that an initial valley cross profile is V-shaped. Harbor's simulation produced a deeply incised U-shaped cross profile that accords fairly well with Matthes' cross profile between El Capitan and the Cathedral Rocks. But as was discussed above, Matthes' method of analysis, although confirmed by Harbor, was faulty. Had Harbor used more computer time, his mighty glacier would continue to erode through the earth then beyond it out into space! Enough said.

As in the immediately preceding section, there is an appropriate caveat for this kind of research, one given by Julius Büdel (p. 17):

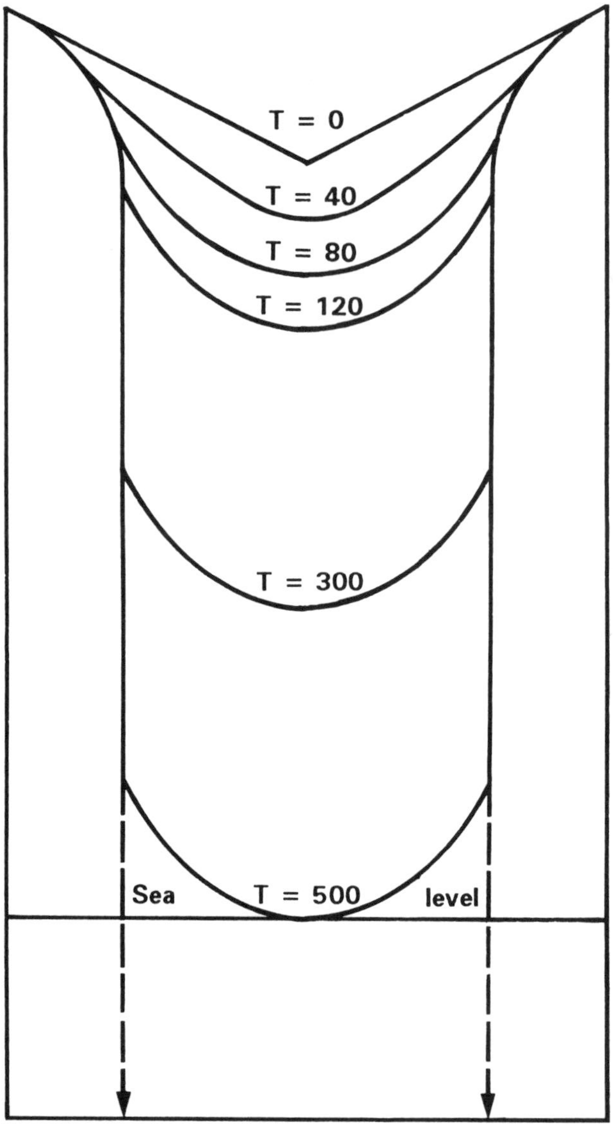


**Figure 40. Simulated canyon transformation from V-shaped to U-shaped (based on Harbor's Figure 5). The simulation of form development has erosion scaled to the local basal velocity squared. The figure shows the glacial-valley cross profile at different time steps (T). Harbor stopped at T = 300, but I have extrapolated his simulation farther. At T = 500 I show the glacier as having eroded down to sea level. Beyond T = 500 the glacier is eroding below sea level.**

> When individual analyses or mathematical models are applied without recognizing their place in the multi-leveled integrational system of geomorphology, no matter how rigorous or logical the attempt in itself, unnatural models and inappropriate simplifications may result.

# 22

# The Case against Major Glacial Erosion

In the previous chapter hopefully I convinced you that the evidence in favor of major glacial erosion is pretty weak. If not, then perhaps I can convince you with the evidence in this chapter. There are three lines of evidence which demonstrate that major glacial erosion does not occur in resistant bedrock. These are: 1) anomalous features in existing topography; 2) similarities of glaciated and unglaciated landscapes; and 3) exhumed unglaciated landscapes with "glaciated" features.

## Anomalous features in existing topography

N. King Huber (1987, p. 42-43), in discussing how the Yosemite National Park landscape was altered through glaciation, stated (as others have generally) that glaciers transformed its sinuous canyons with V-shaped cross profiles into straight ones with U-shaped cross profiles. The sinuous-to-straight transformation is believed to have occurred because glaciers eroded projecting spurs. The V-to-U transformation is believed to have occurred because glaciers filled the canyons to their rims, and so were able, unlike streams, to erode slopes and cliffs. Neither supposition is correct, for there are numerous exceptions to each. First, as was mentioned under the previous chapter's "U-shaped cross profiles," the major glaciated canyons of the Tuolumne and San Joaquin drainages, which contained the longest, thickest Sierran glaciers, have sections that retained their V-shaped cross profiles. Furthermore, both are sinuous, as is the glaciated part of the Merced River drainage (e.g., see Maps 22-31). Additionally, glaciers usually did not fill their canyons exactly to their rims. Especially in the lower parts of canyons and in southern ones (Kings, Kaweah, Kern) ice surfaces fell short of the rims. Nevertheless, none of them left trimlines that would have allowed one to determine the thickness and extent of the glacier. It is this lack of trimlines—that is, the lack of a distinct line (representing the glacier's ice surface) that separates unglaciated slopes above from glaciated ones below—which makes the reconstruction of some past glaciers so uncertain. Had measurable glacial erosion of canyon walls occurred—*even one single foot*—glacial reconstruction would be easy.

In the middle and upper parts of canyons, glaciers not only filled to the rims, but also spread laterally onto the uplands. This contradicts Huber's interpretation, that the glaciers should have eroded the canyon's walls, which would have retreated and would have usurped some area of the uplands. As was mentioned earlier, the existing topography is totally independent of a glacier's thickness or ice-surface gradient. Where is the "glacial fingerprint?" In the Sierra Nevada there is a profound lack of any tangible evidence for major glacial erosion.

Other evidence of glacial inadequacy are the major knolls and ridges that stood directly under the thickest parts of the glaciers. In the Merced River drainage, Bunnell Point and the spur immediately down-canyon from it project directly into the center of the canyon (see Map 23), forcing the river to flow around them. These monoliths were so resistant to erosion that glaciers not only failed to eliminate them, but also failed to transform the inner gorge beside them from V-shaped to U-shaped. Likewise, Liberty Cap and Mt. Broderick were not removed (see Map 25), and the joint-controlled gully between the two was not transformed from V-shaped to U-shaped. In like manner, glaciers failed to remove the Tuolumne's Kolana Rock (Hetch Hetchy area, Map 21) and its Tuolumne Meadows domes and roches moutonnées, and they failed to remove the San Joaquin's Junction Butte and Balloon Dome.

Perhaps the best example of Huber's straight, U-shaped, glaciated canyon is arrow-straight, steep-walled, flat-floored Kendrick Creek canyon, located in north-

**Map 43. Kendrick Creek canyon. Scale 1:62,500.**

western Yosemite National Park (Map 43; Pinecrest and Tower Peak 15' quadrangles). This canyon is one of many relatively straight canyons on the uplands north of the Grand Canyon of the Tuolumne River. However, glaciers on these uplands were only moderately thick and their ice-surface gradients were relatively low; consequently their velocities were relatively slow and glacial erosion would have been relatively minimal. Like Yosemite Valley, these uplands have retained most of their joint-controlled, *preglacial* topography. Notice on the map that as at the upper end of Yosemite Valley, at the upper end of the U-shaped part of Kendrick Creek canyon there are two major steps, the largest situated beneath Richardson Peak. From the bench beneath it, moderately steep slopes drop about 1300 feet to the broad floor of Kendrick Creek canyon.

## Similarities of glaciated and unglaciated landscapes

### Within the Sierra Nevada

If glaciers profoundly alter a landscape, then the difference between an unglaciated granitic landscape and a glaciated one should be readily apparent. Geologists have delineated the maximum extents of past glaciers by noting where a U-shaped stretch of canyon gives way to a V-shaped stretch. This does not work, as Maps 44 and 45 demonstrate. The maps' scales and contour intervals are the same (1:62,500 and 80 feet). Cultural features have been removed to hide the true identity of each, and both have been similarly oriented (rivers flow toward the top) to provide an easy comparison of the features. One map is part of the *glaciated* Tuolumne River drainage (Lake Eleanor 15' quadrangle), and one is part of the *unglaciated* North Fork Feather River drainage (Pulga 15' quadrangle). In the glaciated drainage, the Tioga glacier's thickness—even by Wahrhaftig's mapped evidence, which minimizes thickness—was between 1600 and 2000 feet thick—thicker than the Tioga glacier which flowed through Yosemite Valley. His Tahoe and pre-Tahoe glaciers were considerably thicker, on the order of 2400 to 3000+ feet. Such huge glaciers should transform a canyon's cross profile from V-shaped to U-shaped, that is, to one resembling Yosemite Valley. As you can see from these two maps, that did not happen. If you were to stand on the floor of either canyon, as I have, you would perceive both to be classic, unglaciated canyons. And in one of them, you'd be wrong.

Let's take another example. The V-shaped canyons can be glaciated or unglaciated—but what about U-shaped ones? Are some glaciated and others not? Check the following pair of maps, Maps 46 and 47. As above, the maps' scales and contour intervals are the same (1:62,500 and 80 feet). Elevations and cultural features have been

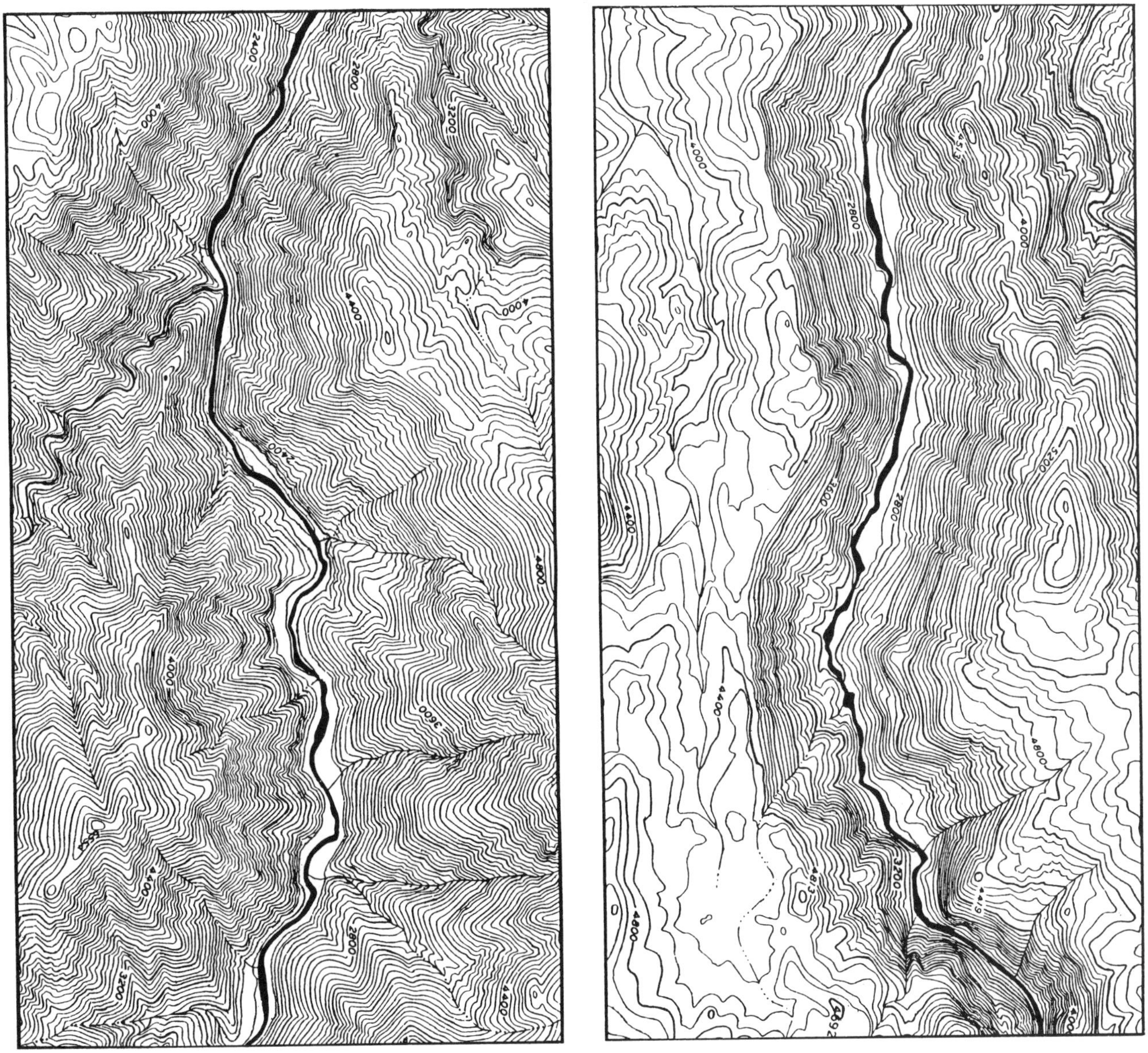

**Map 44 (left) and Map 45 (right). Comparison of two granitic, V-shaped canyons. One is glaciated, one is not. Scale 1:62,500.**

removed to hide the true identity of each, and both have been similarly oriented (rivers drain toward the top) to provide an easy comparison of the features. One map is part of the *glaciated* Merced River drainage (Tuolumne Meadows 15' quadrangle), and one is part of the *unglaciated* South Fork Kern River drainage (Lamont Peak 15' quadrangle), and is mostly within Dome Land Wilderness. Shallow lakes exist in the glaciated drainage, but these have been removed so as not to bias one's opinion about which drainage was glaciated. Both have the following "glacial" attributes (photos are shown for attributes of the unglaciated landscape):

1) hanging tributary canyons (as opposed to accordant tributary canyons) (Photo 152);
2) U-shaped cross profiles with broad floors and locally steep slopes that can be vertical to overhanging—both have at least one overhanging cliff (as opposed to V-shaped cross profiles that have narrow floors and moderate slopes) (Photo 153);
3) absence of projecting spur ridges, as if they were planed away by glaciers—the canyon walls are relatively smooth (as opposed to streams flowing around spur ridges) (Photo 153);
4) smooth, low, subordinate ridges between tributary canyons, as if glaciers had planed down once-higher ridges between them (as opposed to ridges of variable topography and elevations) (Photos 152, 154);
5) fins or finlike ridges, as if glaciers had later-

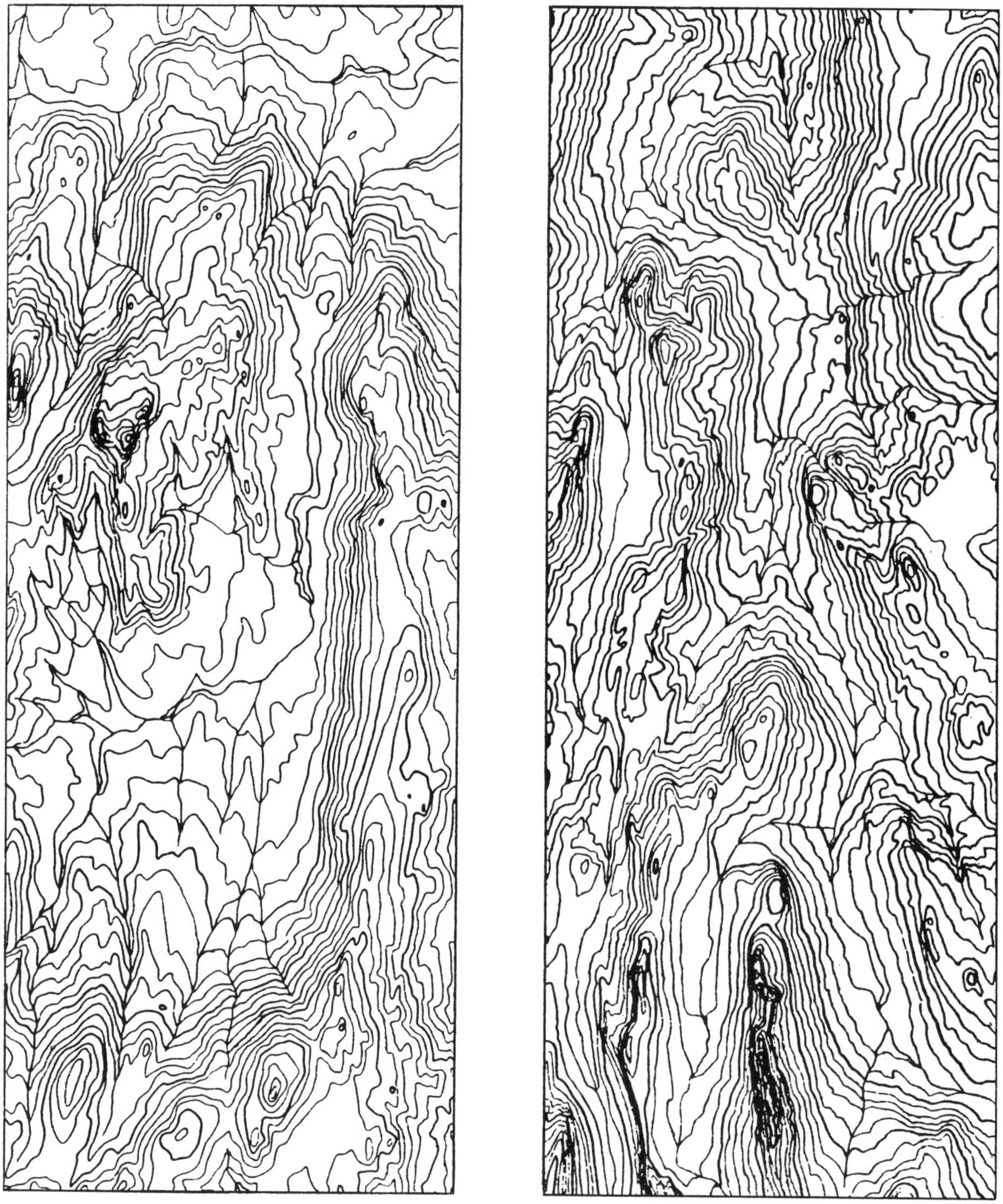

**Map 46 (left) and Map 47 (right). Comparison of two granitic landscapes. One is glaciated, one is not. Scale 1:62,500.**

ally planed back the sides of former domes and ridges (as opposed to broad summits or ridges) (Photos 153 and 155);

6) broad, cirque-like heads of canyons (as opposed to typical headwalls); and

7) irregular, stepped profiles of tributary streams (as opposed to smooth, relatively constant profiles).

Had the maps covered more area, two more attributes would have become evident:

8) irregular, stepped profiles of trunk streams (as opposed to smooth, relatively constant profiles) (Photo 156) ; and

9) roches moutonnées—asymmetrical linear

Photo 152. "Glacial" smooth-ridged lands west of Rockhouse Basin. View is southeast from north of Trout Creek. A very prominent "roche moutonnée" is on the right skyline; Stegosaurus Fin is to its left, and hanging tributaries are left of it (note cascading creek near center). On aerial photos the bedrock, forested ridges in foreground can appear as lateral moraines.

Photo 153. Tibbets Creek's flat-floored, hanging-tributary canyon. View is north from a small summit (in lower Section 15) atop its southeast ridge. Note slightly overhanging fin.

**Photo 154. Upper Tibbets Creek drainage. View is west-southwest from same spot in Photo 153. Smooth southeast ridge of Tibbets Creek canyon is in the foreground; smooth low, open-forested subordinate ridges between minor, subparallel canyons are in the middle ground. In distance are domes and domelike ridges, some resembling roches moutonnées.**

ridges with steep lee sides, aligned parallel to ice flow (as opposed to symmetrical ridges, some with random orientations) (Photos 152, 154, and 155).

Both of these features, along with the others, are identified on Map 48, which covers a larger area in the west-central part of Lamont Peak 15' quadrangle. Finally, there is a "glacial" attribute that cannot be discerned from the maps:

10) soils everywhere are thin and discontinuous, suggesting that glaciers had removed the preglacial soils (as opposed to thick, valley-bottom soils) (Photos 152-156).

On the two maps you cannot tell which landscape has been glaciated and which one has not because there are *no significant differences.* (Later in this text I identify the glaciated and unglaciated landscapes for Maps 44-47.) An interesting aside with regard to glacial studies in the southern Sierra Nevada is that both John Muir and François Matthes stopped their southward investigations where the glacial evidence ended. Had they pressed onward across mid-elevations of the Kern River drainage, they would have discovered that much of the unglaciated lands had an uncanny resemblance to the nearby glaciated lands. If erratics, till and moraines in the glaciated parts of the southern Sierra Nevada were lacking, then one could not determine the extent of glaciation, since there is no perceptible change from glaciated to unglaciated bedrock topography.

The lower, southern part of Rockhouse Basin—the vicinity around Rockhouse Meadow—is a fair analog for the preglacial Yosemite Valley, because it has a broad,

**Photo 155. Stegosaurus Fin. View is northwest from same spot in Photo 141. Low, curving, subordinate ridges in foreground resemble small-scale roches moutonnées.**

**Photo 156. Rockhouse Basin and South Fork Kern River. View is northeast from White Dome. Broad floor of basin is nearly flat and is covered with thin veneer of soil; river's gradient is 0.3%. Inner gorge in foreground is V-shaped (Tenaya Canyon analog); river's gradient is 5.5%.**

nearly flat floor, locally steep cliffs, and at its lower end yields to a V-shaped canyon with a high-gradient stream. However, Rockhouse Basin proper is a fair analog of what Matthes termed the "Half-Yosemite at Wawona," which is at intermediate elevations along the South Fork Merced River drainage. Both localities have high, relatively steep slopes on one side of a broad-floored basin and have subdued hills on the other side. Also, both have been incised largely in two adjacent plutons. In the Rockhouse Basin area the floor of the basin and the impressive topography west of it have been incised in Isabella granodiorite, whereas the subdued topography to the east has been incised mostly in Sacatar quartz diorite. Additionally, Rockhouse Meadow, the V-shaped South Fork canyon south of and below it, and lands east and west of this canyon have been incised in Isabella granodiorite. In the Wawona area the eastern floor of the basin and the impressive topography northeast and southeast of it have been incised in El Capitan granite, whereas the western floor and the subdued topography to the west have been incised mostly in Bass Lake tonalite.

Except for the "cirque" at the head of Long Valley, ones elsewhere on the map are poorly developed. A "classic cirque," that is, one which is nearly the shape of a half-bowl, requires highly fractured bedrock. Resistant bedrock may become highly fractured if it is repeatedly shaken by earthquakes. This is what I suspect has been occurring along the Sierran crest for about the last 3 to 4 million years, as the Owens Valley began to subside along the Sierran fault system. In the Sierra Nevada, earthquakes have been most prevalent east of the crest, so the range's eastern slopes have been most profoundly affected; west of the crest the amount of mass wasting decreases exponentially, as USGS Map I-1612 demonstrates. (In my approximate 6000 miles of hiking in the range, I repeatedly have noted this pattern.) It is not a coincidence that rock glaciers also are found just beneath the crest of the High Sierra, but that they diminish westward. Compare the amount and size of rock glaciers on both sides of the Sierran crest in the Mt. Whitney area (Mt. Whitney 15' quadrangle, Moore, 1981) with those to the west on both sides of the equally glaciated Great Western Divide (Mt. Whitney 15' quadrangle, Moore, 1981; Triple Divide Peak 15' quadrangle, Moore and Sisson, 1987). Also in cirque formation, sheeting (unloading) and, increasingly at the higher elevations, freeze and thaw may play roles. Neither is significant in the "cirques" of Dome Land Wilderness, an area that overall lacks significant mass wasting and consequent talus slopes. In my opinion, the amount of talus is the key diagnostic feature separating glaciated from unglaciated granitic mountain landscapes. Glaciated ones have abundant talus,

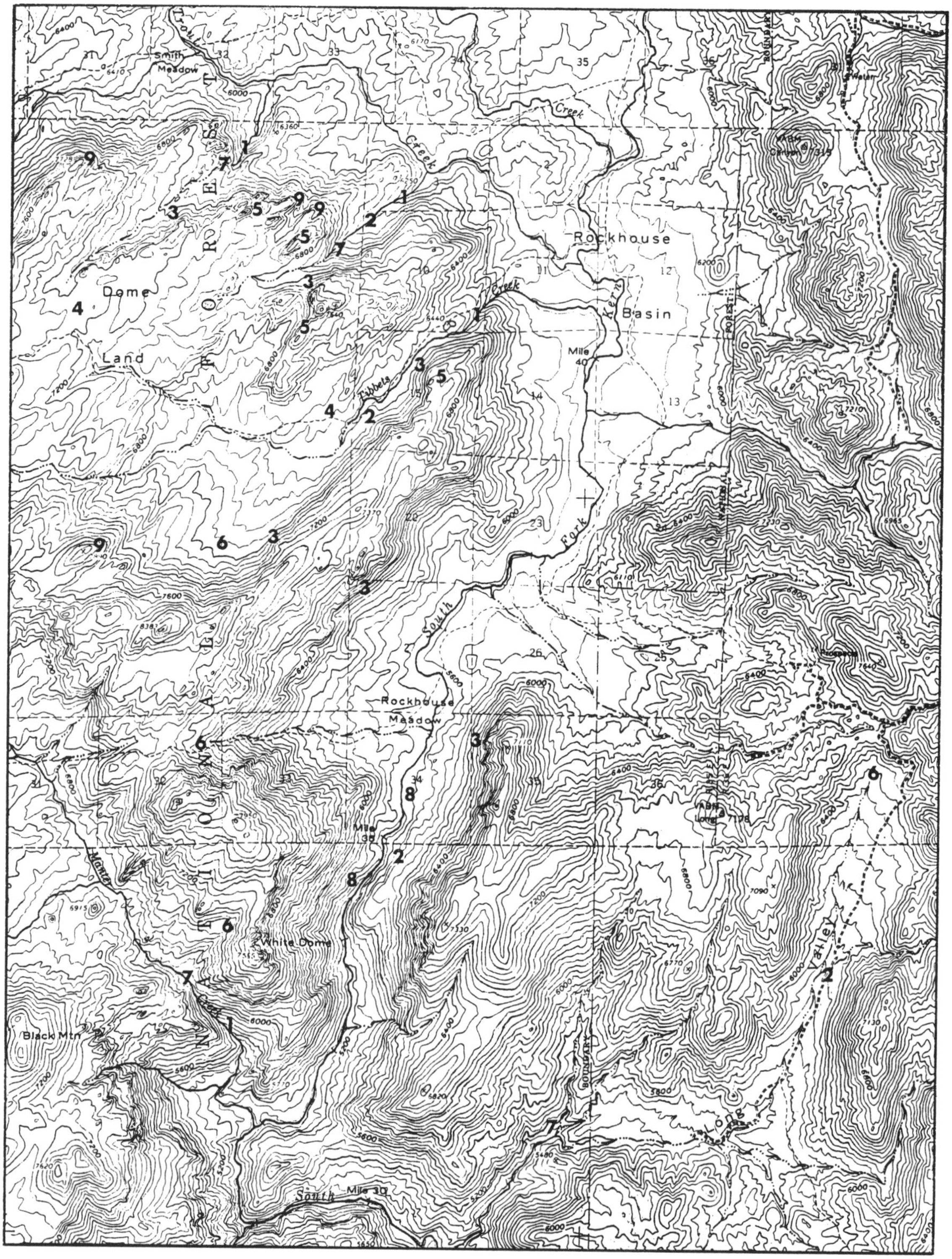


**Map 48. Rockhouse Basin and vicinity. Scale about 1:80,000. Numbers are for these "glacial" attributes: 1, hanging tributary canyons; 2, U-shaped cross profiles; 3, absence of projecting spur ridges; 4, relatively smooth, low, subordinate ridges between tributary canyons; 5, fins; 6, broad cirque-like heads of canyons; 7, irregular, stepped profiles of tributary streams (as opposed to smooth, relatively constant profiles); 8, irregular, stepped profiles of trunk streams; and 9, roches moutonnées.**

and hence have talus slopes, while unglaciated ones have sparse talus and rarely have talus slopes. They are absent not only in the unglaciated Sierra Nevada, but elsewhere, such as at the very bouldery hills along Interstate 15 between Escondido and Temecula, mentioned earlier under Chapter 12's McGee till. One would expect that in this earthquake-prone country countless thousands of boulders on steep slopes would have tumbled down to the bases of hills, forming massive talus slopes, but this has not happened.

However, true cirques definitely are much better developed than the false cirques of unglaciated, granitic lands. They have been atttributed to severe plucking of bedrock by glaciers. But a bergschrund separates the head of the glacier from the cirque's wall, making plucking impossible. This gap between glacier and headwall was observed in Yosemite National Park by Willard D. Johnson, mentioned earlier, who in 1904 descended the bergschrund at the head of the Lyell glacier, Yosemite's largest. There he noted that some of the rockfall was becoming attached to the glacier. Cirques appear to be more the result of enhanced physical weathering than glacial erosion.

Whereas Sierran earthquakes cause extensive mass wasting near the crest, they have a preservation effect on canyon bottoms. These become buried with rockfall, preventing streams from eroding their bedrock channel. South of Rockhouse Meadow, the South Fork Kern River enters a high-gradient reach not unlike the Merced River's reach through Merced Gorge, below Yosemite Valley. In each reach the channel is littered with large blocks, preventing the flowing water from scouring the channel. And in Merced Gorge the accumulation of blocks is very thick, which also prevents local mass wasting along the canyon bottom.

Lava flows and volcanoes in the South Fork Kern River drainage have been dated, and they allow one to measure the amount of denudation of the underlying and/or adjacent granitic bedrock. Templeton Mountain is a large volcano dated at about 2.4 million years old, and constructed on a flat-floored valley, and it forced the South Fork Kern River to detour around its base. Over the last 2.4 million years how much incision has occurred below the base? Its lava flows descend to the edge of the South Fork's level floodplain, indicating that the elevation of the floodplain has remained unchanged over the last 2.4 million years (Photo 157). The bed of the adjacent South Fork typically lies 3 feet or less below the floodplain, but because streams always lie below their floodplains, even this minimal depth cannot be ascribed to incision. Contrary to our logical expectations for stream erosion, no measurable incision has occurred. Southward the contacts would seem to indicate about 200 feet of incision has occurred in a minor gorge. However, the rhyolite contacts on the west slopes of this gorge demonstrate that the gorge had already existed essentially as it does today back when the rhyolite partly buried its west slopes.

About 7½ miles south of Templeton Mountain is a second, equal-size, equal-age volcano, Monache Mountain. It too was constructed on a flat-floored valley, although there is no evidence that it forced the river to detour around its base. As at Templeton Mountain, the river here, flowing through Monache Meadows, has made virtually no incision over the last 2.4 million years. Along the northeast base of the mountain at a site about 150 yards northwest of a cow camp, a lava flow descends to the edge of Monache Meadow, indicating that locally this part of the meadow has not decreased in elevation over the last 2.4 million years (Photo 158). Because the

**Photo 157. East base of Templeton Mountain's satellite volcano. A weathered lava flow descends to the edge of the level floodplain. View is from the mouth of a granitic gorge, through which the South Fork Kern River makes a high-gradient traverse. The gorge predates Templeton Mountain, for intact remnants of its flows lie on the gorge's slopes.**

**Photo 158. Monache Mountain, view southwest from South Fork Kern River. The cow camp's cabin is along the volcano's northeast base, and about 450 feet to the right (northeast), a weathered lava flow descends to the edge of the level floodplain.**

meadow, while very flat is not absolutely flat, the South Fork Kern River's floodplain, which lies about ½ mile to the east, may be a foot or two lower, and so one might be able to infer that much denudation.

One could argue that due to the dry climate of this part of the southern Sierra Nevada, which lies in the rain shadow of the Coast Ranges and the Sierra Nevada's Great Western Divide, this area is not a good one in which to measure rates of Sierran incision. However, the Coast Ranges rain shadow is a relatively recent one, due to major uplift in the last 615,000 years. This date is based on evidence presented by Robert Anderson for marine terraces in the Santa Cruz Mountains, and more importantly on evidence presented by George Davis and Tyler Coplen for sediments in the San Joaquin Valley.

The South Fork drainage differs from the Kern River drainage in that it lacks the latter's spectacular canyon, which has been cut largely along the linear Kern Canyon Fault. One could say that the straight, deep canyon exists because of the fault. Indeed, back in 1903 Andrew Lawson had proposed that the canyon's floor was a narrow graben, much as Whitney had proposed a similar structure for Yosemite Valley back in the 1860s, both deep, steep-walled canyons resulting through subsidence. François Matthes in 1937 and Robert Webb in 1946 rejected this view of the Kern Canyon, both stating that it developed through erosional processes along the fault. Webb gave evidence against Lawson's graben, while Matthes merely stated that although the Kern Canyon Fault possibly is of great antiquity, the canyon nevertheless is young—solely the result of glacial erosion. If one accepts Matthes' deep-canyon origin, which is essentially the same one he used for Yosemite Valley, then one can infer that the reason the South Fork Kern River lacks a similar canyon is that it is essentially unglaciated. Had glaciation occurred, Matthes would have argued, a deep canyon would have developed.

One could attempt to attribute the absence of a deep canyon in the upper part of the South Fork Kern River drainage to the ineffective, minor discharge of the South Fork, which averages about 14% the volume of the Kern River. But that is not enough. Webb acknowledged a problem when he stated (p. 367):

> The South Fork has a most anomalous course. Through 50 of its 70 miles, it flows in valleys far too large to have been cut by it with its present volume. The stream apparently inherited pre-existing valleys and integrated them by subsequent erosion. It staggers except in segments between valleys where it plunges rapidly through gorges, only to meander across the next graded reach.

And later, under "Problems for future study:"

> Two important topographic features of the Kern Basin have been neglected in this treatment: (1) origin of South Fork Valley [the valley of the lower, west-flowing part of the South Fork Kern River], and (2) origin of the extensive pattern of upland meadows so prominent in the southern Sierra [in the upper South Fork drainage]. It seems best to omit these features, since data on their origin are meager. At the present writing, the origin of South Fork Valley is still exceedingly obscure.

Webb had analyzed the landscape according to Matthes (i.e., according to William Morris Davis), and so, not surprisingly, ran into the same problems Matthes had encountered in his 1937 Mt. Whitney work. Of Webb's first statement, one could attempt to explain the broad, flat-floored valleys by concluding they were incised by a higher-discharge South Fork that would have existed before the Coast Ranges' rain shadow. But rivers of the central and southern Sierra Nevada also had higher discharges and yet did not develop this topography. As Webb noted, the topography cannot be explained in terms of currently operating geomorphic processes. Therefore, is best explained in terms of past processes, namely, tropical ones, mentioned earlier. Indeed, the broad upland valleys, the stepped course of the river, and the broad South Fork Valley (of Webb's second statement) all have parallels in the Río Grande drainage of central Colombia.

The antiquity of the landscape can be gauged because in 1982 Joel Bergquist *et al.* produced a geologic map of Dome Land Wilderness, which included detailed mapping of remnants from nine basalt flows that then were dated. I inspected seven of these to verify the accuracy of the mapping, which was necessary in order to estimate the amount of postvolcanic denudation of the granitic bedrock. Map 49 and Photo 159 show the topography in the Black Mountain area as well as the 12.2-million-year-old Black Mountain basalt flow and, to the south-southeast, the 11.8-million-year-old Peak 7620 basalt flow. That each is a separate entity and is not a part of a former, larger flow is very obvious in the field; color and textural differences are readily apparent. Actually, the Peak 7620 flow is composed of two remnants, as shown on Map 49. The extent of each remnant is limited by the size of the relatively level summit area. I found no evidence of flow remnants on the northern slopes, so either the flow was very restricted or else what accumulated on those slopes has been eroded away.

For Black Mountain, which I investigated more thoroughly, the basalt flow, presumably originated from the highest (southwest) part of the split-level summit area, and then flowed across the summit area as well as down the slopes. The map by Bergquist *et al.* indicates two flow remnants, one on the west summit and one on the lower

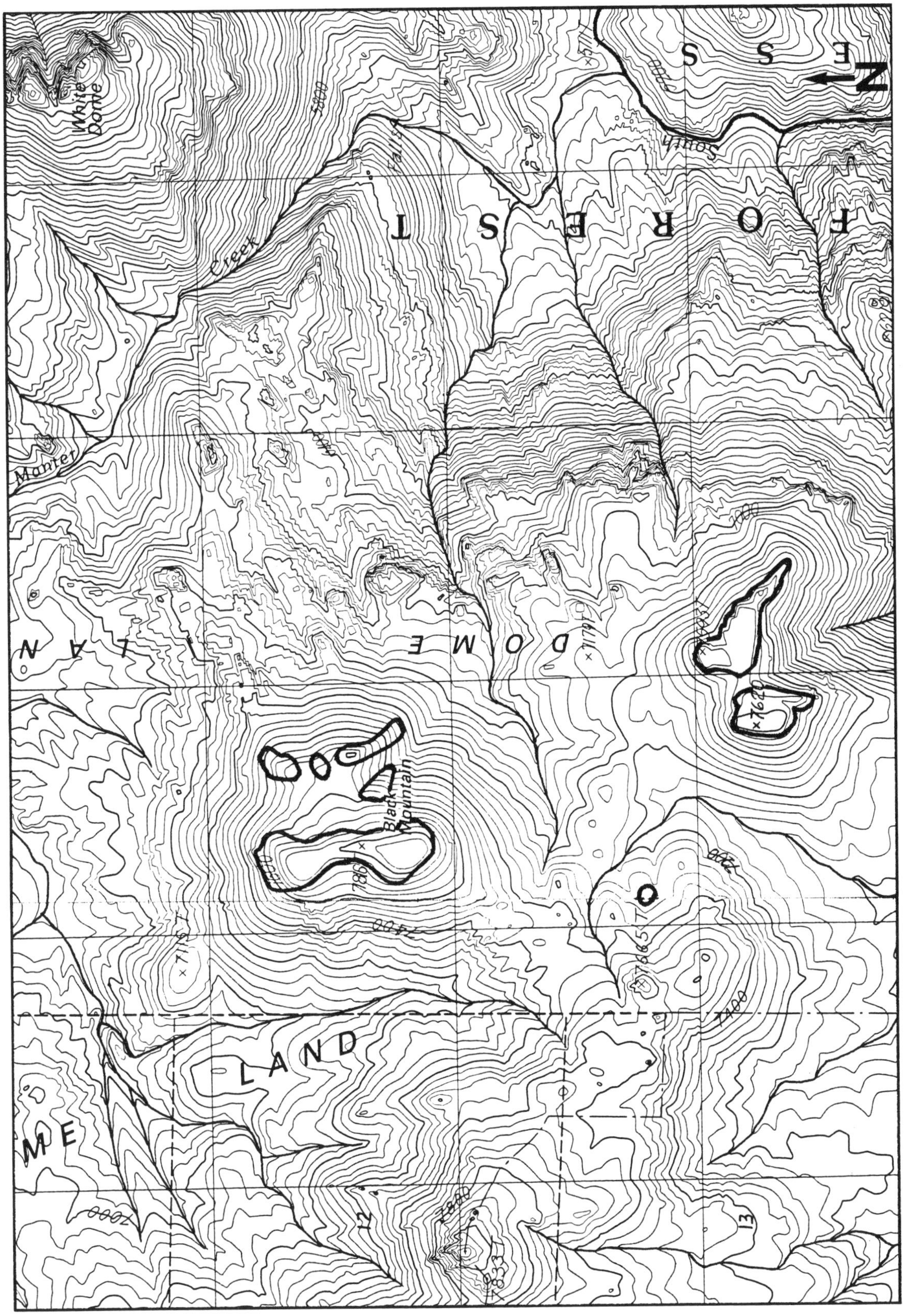


**Map 49. Remnants of basalt flows in the Black Mountain vicinity, Dome Land Wilderness (Cannell Peak and White Dome 7.5' quadrangles). Scale 1,24,000.**

**Photo 159. Black Mountain (center) and Peak 7620 (far left). Serrated Church Dome stands on skyline above Black Mountain. View is west-southwest from White Dome.**

east summit, separated by a relatively broad, flat saddle. In contrast to the flow on Peak 7620, the flow on Black Mountain is preserved on some slopes. Bergquist *et al.* depicted flow remnants on the upper slopes immediately below the north end of the west summit, more extensive remnants on slopes below the north end of the lower east summit, and a similar, extensive amount on a ridge below the south end of the lower east summit (both down to about 7360 feet). But field evidence indicates a slightly greater extent. The south-end ridge flow descends to 7250 feet.

Furthermore, to the west, immediately beyond a conspicuous gully that heads at the saddle, what appears at first to be a large talus slope that descends to about 7400 feet more likely is a flow remnant. No intact flow remnant was seen, merely blocks of lava. However, talus slopes have a random assortment of boulders, and this one does not. There are local lenses of large boulders in a matrix of smaller boulders. This pattern could not have been derived through rockfall from the summit area, but rather must be the result of breakage in place of the underlying lava, the patterns of the talus conforming to the fracture pattern of the underlying lava. In very extensive mountaineering for some 30+ years in the Sierra Nevada, which included myriad traverses, descents, and ascents of talus slopes, I have never encountered such a patterned talus slope anywhere in this range. Additionally, the broad, flat saddle has a veneer of basalt, the contact of the basalt with the underlying granitic bedrock showing an absence of soil development, as is the case also for contacts at Peak 7620. Also, a thin remnant of the flow descends about 100 feet down the shallow gully on the east side of the east summit. It is quite broken, and appears to be mostly talus, like the talus slope just mentioned. However, patches of solidified lava in contact with the granitic bedrock do occur, confirming that it is a flow, weathering in place.

Finally, I discovered one flow outlier, now essentially a lag deposit, in a small gully about midway between Black Mountain and Peak 7620 (south of the former, west of the latter). I suspect that it is a remnant from the former, a course of lava descending southwest to the broad southwest-trending saddle and then spilling both north and south from it. Had this lag deposit been from the Peak 7620 basalt, that flow would have had to cross the relatively deep gully at the northwest base of the peak and pond up to a *minimum* thickness of 200 feet to reach the lag-deposit site. Had this occurred, the flow would have buried the northwest-trending granitic ridge that connects the base of Peak 7620 to the base of Black Mountain. Due to an absence of stream erosion on this ridge, the flow would have been well preserved, but no trace exists.

The basalt-flow remnants on the slopes of Black Mountain and in the small gully are important because they show that this peak had essentially the same granitic topography 12.2 million years ago as it does today. Over this span perhaps a few feet of granitic bedrock—but not much more—have been weathered away from the slopes. This rate is consistent with the amount of post-Sherwin weathering seen on ridges and gentle slopes above Yosemite Valley. As mentioned earlier in "Stoss-and-lee topography" the post-Sherwin denudation rate atop Turtleback Dome is about 5 inches per million years. This rate applied to Black Mountain and Peak 7620 amounts to about 5 feet for the flat summit areas and perhaps several times more for the slopes. The five other flows I visited in the area also show no detectable denudation of granitic bedrock below the base of any lava flow. Whereas the absolute rate of denudation may be unknown, it appears to be a very slow rate.

A slow rate has been independently derived for the Alabama Hills, which lie in Owens Valley between Lone Pine and Mt. Whitney. Sandwiched between the active Independence and Lone Pine faults, these low, granitic hills have become intensely fractured and should be experiencing a denudation rate of one or two orders of magnitude (10x to 100x) faster than the much less fractured granitic bedrock of Dome Land Wilderness. In 1995 Paul Bierman, using cosmogenic-isotope dating, determined that the Alabama Hills' bedrock was being denuded at a rate of only 6½ feet per million years. Thus postglacial denudation rates for Yosemite Valley are in agreement with postvolcanic denudation rates in Dome Land Wilderness, which in turn are compatible with higher denudation rates in an actively faulted area. Bierman also noted that in the wet tropics the denudation rates appear to be on the order of tens of meters (perhaps 100-200 feet) per million years. This too would have been the case for the granitic Sierran landscape under tropical and paratropical climates that lasted until about 33 million years ago.

A final point to be made is the uncertain age of Kern Canyon, which Matthes had suggested was young, although he suggested that the Kern Canyon Fault could be very old. The age of the fault has been unknown, being poorly limited between the age of the canyon's 80-million-year-old granitic bedrock and a 3.5-million-year-old lava flow. However, as I have just shown for the topography and streams of the South Fork Kern River drainage (like that of the ancient South Yuba River drainage, which hasn't changed appreciably in 50 million years), very little change has occurred over millions of years. If the fault had been active only in the last few million years, then there would have been evidence of fault-induced drainage rearrangement, which there is not. The concordant arrangement of tributary canyons with respect to the Kern River argues in favor of the fault forming early in the history of that granitic landscape, perhaps even at the same time as the late Cretaceous uplift that Saleeby *et al.* have put at 75 to 65 million years ago. Recently, Jason Saleeby and Mihai Ducea made a similar observation, identifying the fault as a "latest Cretaceous dextral ductile shear zone."

All of the foregoing indicates that this part of the Sierra Nevada—as well as the rest of the granitic part of the Sierra Nevada (that is, most of it)—is largely a relict tropical landscape, since it appears to have changed little during last 33 million years of post-tropical climate. Like other workers in the tropics, Avijit Gupta showed that tropical mountain valleys that have developed in resistant bedrock can be broadly U-shaped in cross section. This cross-sectional shape is found in some of the Sierra Nevada's canyons, both glaciated *and* unglaciated. Additionally, as Jean Tricart noted, tropical valleys can also be steep-sided, even having vertical walls, as does Yosemite Valley. In short, glacial erosion is not required to explain the complex topography of the Sierra Nevada. And oh, by the way, in the two sets of maps the unglaciated lands are in the left map, and the glaciated lands in the right map.

### Outside the Sierra Nevada

"Glacial" features appear in unglaciated parts of California as well as in arid and semiarid lands outside the state. For example, "cirques" also occur in the San Jacinto Mountains of southern California, a very prominent one existing on the north side of Tahquitz Peak (Palm Springs 15' quadrangle). This small, fault-bounded range has classic, relict, stepped topography that rivals any found in the Sierra Nevada.

Marginal transformation by glaciers is not restricted to granitic landscapes. A comparison of semiarid Zion National Park, Utah, with Glacier National Park, Montana, is one case in point. Both parks contain somewhat horizontal strata of varying thickness and rock type, the thicker, more-resistant strata forming prominent, nearly vertical cliffs. Glacier has all the characteristic landforms a famous, glaciated landscape should have (Photo 160)—and so does Zion (except for lake basins). South of Angels Landing, Zion Canyon is steep-walled and relatively broad-floored (Photo 161). Farther south, near the park's south entrance, the floor is very broad (Map 50). This area's tributary canyons—Birch Creek (Court of the Patriarchs), Oak Creek, and Pine Creek—are equally impressive U-shaped canyons. Hanging valleys abound, as do irregular, stepped stream profiles. Furthermore, the heads of many tributaries end in cirque-like half bowls. Three of these surround the West Temple, an analog of a glacial horn. The development of the topography in this

Photo 160. St. Mary Lake, Glacier National Park.

Photo 161. Zion Canyon, viewed from Angels Landing.

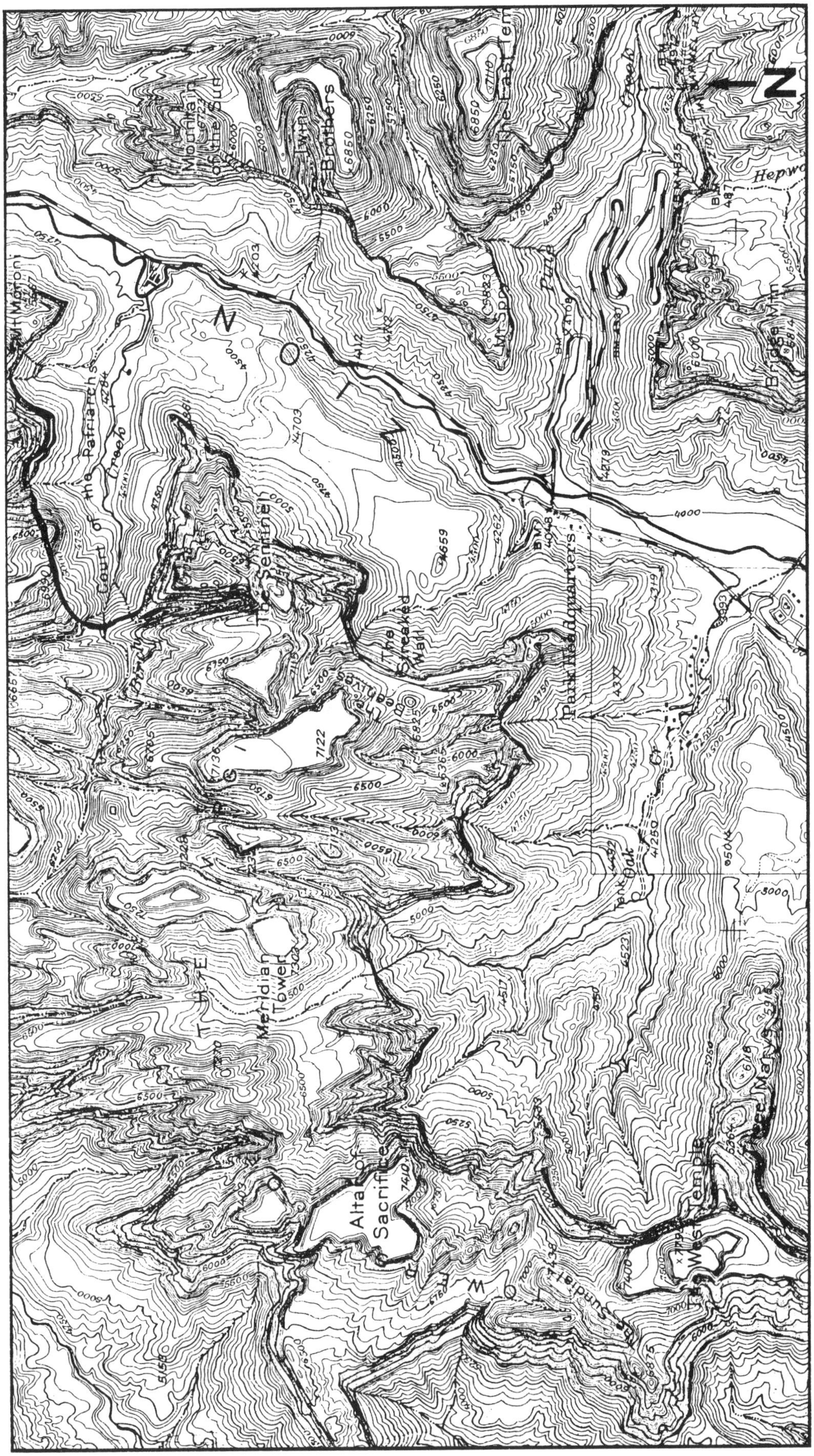


**Map 50. Southern Zion Canyon and its tributary canyons. Scale 1:31,680 (2 inches = 1 mile).**

area is well known and uncontroversial: it is largely the result of mass wasting, the Virgin River transporting the debris. Glacier National Park has abundant evidence of mass wasting in the form of myriad talus slopes, yet its topography invariably is declared to be largely the result of glaciation. For that park, no one has attempted to determine how much denudation is due to mass wasting and how much to glaciation. Because the difference between the landscapes of the two parks is minor, the impact of glaciation probably is minor. Likewise, in other glaciated landscapes around the world, the volume of mass wasting should be ascertained before ascribing the total volume of moraines, till, and outwash to glacial erosion.

## Exhumed unglaciated landscapes with "glaciated" features

The burial of the northern Sierra Nevada under Cenozoic prevolcanic sediments and under more-voluminous volcanic sediments, lahars, and lava flows is well documented, especially by Waldemar Lindgren in 1911, Victor Allen in 1929, A. M. Piper *et al.* in 1939, Burton Slemmons in 1953 and 1966, Paul Bateman and Clyde Wahrhaftig in 1966, and Cordell Durrell in 1966 and 1987. In this area the Stanislaus River drainage has been used to make perhaps the most convincing case for major late Cenozoic uplift, which then caused major canyon incision first by streams and later by glaciers. However, as the evidence actually shows here—and in all the northern Sierra Nevada drainages, for that matter—a range with high elevations and relief comparable to those of today already had existed at the initiation of the earliest deposits during the Paleocene and Eocene epochs. Lindgren stressed this (p. 37), but was ignored. If the range already had been high at the start of late Cenozoic time (start of the Miocene epoch), there was no need to infer late Cenozoic uplift; and if the canyons already had been deep, there was no need to infer late Cenozoic incision by streams and glaciers. Hence, erosion by glaciers would have been minimal. Everyone chose to ignore Lindgren, who easily had performed the most comprehensive Sierran field work in his day, and instead all believed in a range reduced to low elevation and low relief by late Cenozoic time. Why they ignored the evidence of the rocks is a "Riddle in the Rock" to be answered in the final chapter.

The Stanislaus River drainage contains the 9.2-million-year-old Table Mountain latite flow, which coursed along what was presumably the bedrock floor of the Stanislaus River's main canyon. In the 1890s Leslie Ransome mapped most of the flow, and he was the first to suggest (p. 69-70) that it could be used to estimate Sierran uplift. He observed that the longitudinal profile of the base of the flow is "more nearly graded" than those of the modern North and Middle forks, so the canyon down which it flowed must have had a lower gradient and must have been at a lower elevation than these forks. Because this trans-Sierran latite flow was not faulted or warped, he concluded (as others would) that this part of the range had been tilted westward as a simple block. Consequently at any point along the flow, the difference in elevation between the assumed initial elevation of the 9.2-million-year-old river canyon and the elevation of the base of the flow today would represent the amount of uplift at that point over the last 9.2 million years. Some uplift is believed to have occurred before the latite flow originated, so the amount of uplift determined by the flow represents only part of the total uplift.

Similar reasoning has been applied by N. King Huber to volcanic deposits in the San Joaquin and Tuolumne river drainages in order to quantify timing and magnitude of late Cenozoic Sierran uplift. In such studies, the view has been that over time the flow down a canyon bottom would have evolved to become a ridge. This transformation would have occurred as first the walls of the canyon were eroded away by the river, here the Stanislaus, which would have been displaced laterally to the side of the latite flow, and then with major uplift the river would have incised thousands of feet through granitic bedrock to create today's deep, steep-walled canyons adjacent to the latite flow. Under this interpretation, major Sierran canyons such as the Middle Fork Stanislaus are quite young.

However, mapping of the area by Ransome, Slemmons, and Huber confirm what I have repeatedly observed in the field: the Table Mountain latite flow far more often than not lies on volcanic deposits, not on granitic or metamorphic bedrock. This means that the flow, rather than advancing down-canyon along the actual bedrock floor, was advancing across deposits of *unknown* thickness, occasionally lapping onto bedrock. Ransome had mapped as far east as the Dardanelles, a dramatic butte composed of volcanic flows and sediments. This butte is viewed readily from Highway 108's Donnell Vista, although more appreciably from the air (Photo 162). Just west of and below the Dardanelles, Ransome found an older deposit—massive rhyolite—which he mapped as lying on a granitic bench about 1000 feet below the base of the Table Mountain latite flow. Obviously the bench had been in existence when the rhyolite had buried it. Much later, in 1983, and on a more accurate topographic base map (Dardanelles Cone 15' quadrangle), Huber mapped the same granitic-bench rhyolite deposit, and his mapping indicates that it lay 1200 feet below the base of the flow (see Figure 41).

The age of the deposit (like others in the quadrangle) is considered by Huber to be similar to that of a dated rhyolite deposit in the drainage, and they are considered to be part of the Valley Springs Formation. In 1964 G. Brent Dalrymple determined two ages for it, 26.1 and

**Photo 162. The Dardanelles, aerial view northwest. Note the light-colored rhyolite outcrop at the base of the Dardanelles. Ransome's deposit is hidden in the forested, shadowy area behind the lower, southwest (left) end of the Dardanelles. Beyond it lies the upper North Fork Stanislaus River drainage.**

23.3 million years, and he put more confidence in the younger date, which makes the formation earliest Miocene. Therefore, when the latite flow was advancing down-canyon, it was doing so across sediments with a wide range in age that were at least 1200 feet thick, if not much thicker, since below the bench the inner gorge of the Middle Fork may have been almost as deep as it is today. Thus the base of the latite flow cannot be a proxy for the former base of a granitic canyon, and it cannot be used to determine uplift. This point, unfortunately, was not realized by Ransome, and others repeated his oversight, including Slemmons *et al.* for the Stanislaus River drainage, and Huber for the Tuolumne River and San Joaquin River drainages.

Another point not realized is that the present stream pattern could not have been derived in the manner espoused. A flow down a canyon could have displaced the Stanislaus River laterally, and then it would have eroded downward along that side of the flow. However, the other side of the flow, lacking a stream, would not have been eroded, so it and the adjacent canyon slopes should have been preserved. (The flow's original width and sinuosity

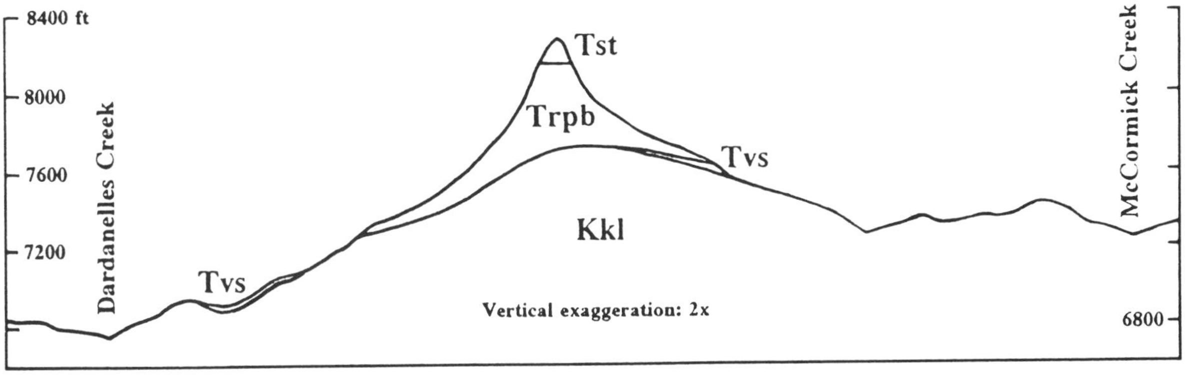


**Figure 41. Cross section through the Dardanelles along the county-line boundary. Hidden contacts based upon evidence both up- and down-canyon from the Dardanelles. Tsd = Dardanelles Formation; Tst = Table Mountain Latite; Trpb = Relief Peak Formation—basalt; Tvs = Valley Springs Formation; Kkl = Granodiorite of Kinney Lakes.**

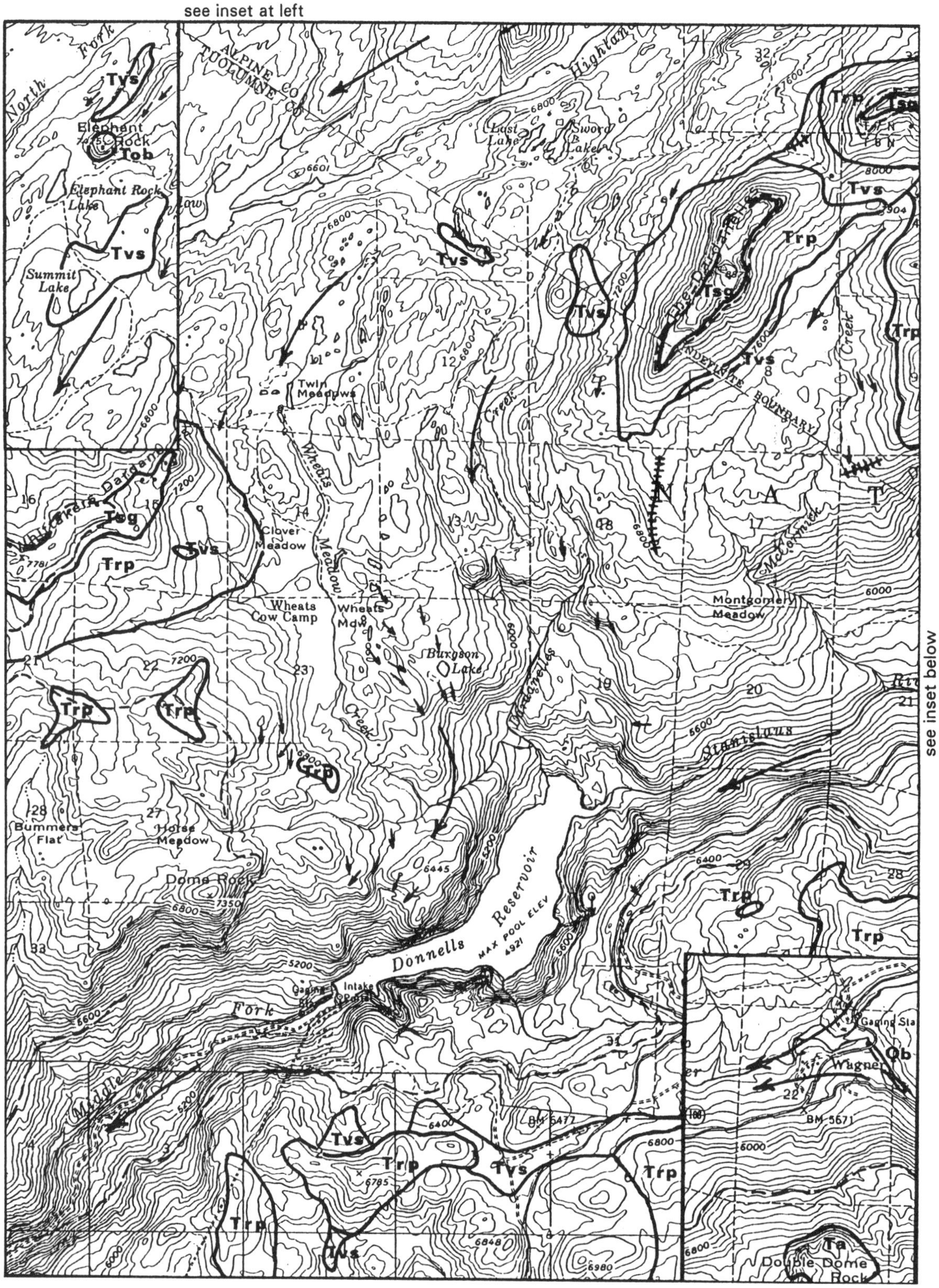


Map 51. Volcanic deposits and Tioga-glacier evidence near the Dardanelles, Stanislaus River drainage. Geologic units (after Huber, 1983), youngest to oldest, are: Qb = late-Quaternary olivine basalt; Tob = Tertiary olivine basalt; Ta = Hornblende andesite; Tsg = Stanislaus Group, undivided (includes 9.2-million-year-old Table Mountain latite flow); Trp = Relief Peak Formation; Tvs = Valley Springs Formation. Bedrock is Cretaceous granodiorite. Glacial evidence: long arrows = general ice-flow direction; short arrows = striations; dashed lines = upper limit of glacier, based on highest erratics (dots); crossed, solid lines = recessional moraines. Scale 1:62,500.

**Photo 163. Aerial view northwest across the upper North Fork Stanislaus River drainage. Union Reservoir and Utica Reservoir behind it are in the center of the photo, among the barren, granitic benchlands. Union has flooded a smaller, previously existing post-Tioga lake. In the lower right foreground is Summit Lake, one of several in the area. (Post-glacial ponds are far more abundant, about 100 in number.) In contrast to other lakes and reservoirs, Summit Lake has slightly murky water, due to its lying totally within a rhyolite deposit.**

are unknown.) Over time the Stanislaus River should have incised a canyon essentially parallel to the edge of the flow, but that also did not happen. The down-canyon course of the flow is first between the Clark and Middle forks, "leaping" across the Clark Fork, migrating eventually over to and "leaping" across the North Fork, migrating back and "leaping" across it and eventually flowing over to the Middle Fork, then finally "leaping" southeast across it.

How does one explain the modern drainage pattern, which is so discordant with the latite flow? As in the South Yuba River drainage, the Stanislaus River drainage developed by *early* Cenozoic time, then its canyons buried up to and over their rims mostly during middle Cenozoic time (as documented, for example, by Bateman and Wahrhaftig), and finally were exhumed from their volcanic cover during late Cenozoic time through stream and glacial erosion. One rhyolite deposit, mapped by Huber, lies on upper south slopes between Donnells Reservoir's dam and a summit, 6785 on the map, about 1.0 mile to the south (Map 51). This deposit extends down to about 6070 feet elevation before pinching out. It demonstrates two points. First, the lowest elevation is well below the north-side bench, indicating that the inner canyon already existed in some fashion and was not initially incised after the latite flow. And second, the topography of this thin deposit grades imperceptibly to exhumed, adjacent granitic slopes, which shows that granitic slopes have changed very little since they were first buried by the formation some 23+ million years ago. The great majority of the Sierra's volcanic sediments, like its prevolcanic sediments, are readily erodible, and late Cenozoic rivers, after having removed much of these sediments, would have reached bedrock and then would have become entrenched in their former canyons, which then would have been exhumed.

Geologic maps of the Sierra Nevada clearly show a complex pattern of deposition and erosion of volcanic sediments and flows in the last 33 million years, that is, since the end of the Eocene epoch and the essentially contemporaneous initiation of volcanism in the Sierra Nevada. Major volcanic eruptions, like those in the Cascade Range today, were few and far between, with long periods of time during the Sierran river systems eroded much of the relatively loose volcanic sediments. However, when the streams reached granitic bedrock, incision then occurred at a nearly imperceptibly slow rate, as discussed later, under Part IV.

This lack of major incision is seen particularly well in the upper North Fork Stanislaus River drainage, which includes its major tributary, Highland Creek. At first, this corrugated benchlands topography appears to be the result of major glaciation: most of the volcanic deposits have been removed, and signs of glaciation—polish, striations, erratics—are ubiquitous. Furthermore, as is typical of Sierran glaciated lands, much of the area is devoid of soil (Photo 163). My field mapping throughout this area indicates that in a transect perpendicular to the flow of glacial ice, one extending from Lake Alpine southeast to the Dardanelles, the ice surface of former glaciers was at about 8500 feet. The undulating topography has an average elevation of about 7000 feet, and so the average thickness of the glacial ice was about 1500 feet. (This is about the thickness of Tahoe and Tioga glaciers in the eastern part of Yosemite Valley.) Did a thick ice sheet severely erode this North Fork landscape?

The answer is seen by examining the rhyolite remnants of the Valley Springs Formation, which lie on granitic topography. These remnants obviously have existed since the time of the formation's deposition. Of most interest is the remnant between Elephant Rock and the North Fork Stanislaus River (mapped by Huber and verified by me in the field), for it shows that the river has incised only about 30 feet below the remnant, despite some 23 million years of elapsed time. Had glaciers been even moderately effective, then the North Fork Stanislaus River at a minimum (and the rest of the granitic topography in general) should have been deeply incised below the remnant. These remnants exist on flats as well as on slopes, indicating that both types of topography existed back then. Furthermore, since none of the adjacent granitic rocks have been even shallowly eroded below the adjacent volcanic rock, then likely all of the equally resistant granitic surfaces also have not been significantly eroded. The evidence shows that the prevolcanic granitic topography at the start of the Miocene epoch was essentially the same as it is today; net denudation of granitic bedrock has been absolutely minimal—a few tens of feet, averaging about *a foot or so per million years*.

Note also that Huber mapped the mid-Miocene Relief Peak Formation as lying in direct contact with the granitic rock. This formation, with a date of 12.1 million years (by Morton *et al.*) for strata on the major ridge in the southeast corner of Silver Peak 15' quadrangle and located about 4½ miles northwest of the North Fork rhyolite remnant, is an assemblage of andesite and basalt flows, lahars, and sediments. For this formation to lie in contact with granitic rock, streams first must have eroded entirely through all of the Valley Springs Formation in at least several locales. As with the rhyolite-granite contacts, the andesite-granite contacts show no sign of any significant erosion of granitic bedrock by streams or by glaciers.

Supporting evidence lies just to the north, in the upper Mokelumne River drainage, where the same pattern occurs: volcanic deposits of varying ages in contact with one another and with the granitic bedrock (Markleeville 15' quadrangle). For example, along the Sierran crest about 1.7 miles northwest of Ebbetts Pass there is an outcrop of rhyolite, equivalent in composition and in age to part of the Valley Springs Formation, and dated at 21.2 million years (Map 52). It lies atop a broad, granitic bench, and in turn is overlain by the Raymond Peak and Silver Peak Andesites, believed by Armin *et al.* to be no older than 7 million years. This younger formation also overlies the Relief Peak Formation (about 12 million years, as just mentioned) and granitic bedrock. These contacts indicate a history of: 1) deposition of the Valley Springs Formation followed by its partial eradication to expose granitic bedrock; 2) deposition of the Relief Peak Formation followed by its partial eradication to expose the Valley Springs Formation and granitic bedrock; and 3) deposition of the Raymond Peak and Silver Peak Andesites, followed by their partial eradication to expose the Relief Peak Formation, the Valley Springs Formation, and granitic bedrock. If stream and glacial processes had been effective at eroding granitic bedrock, then where it is exposed, it should lie considerably lower than where it has been protected by a volcanic cap. However, in the field it is impossible to detect any measurable denudation of the exposed granitic bedrock, based on adjacent buried, granitic bedrock. Again I emphasize that net denudation of granitic bedrock has been absolutely minimal.

With the previous evidence in mind, one can reconstruct the history of the North Fork Stanislaus River's broad-basin lands. The basin was already more or less in existence by the end of the Oligocene epoch, with a relief similar to that today, about 1600 feet, judging by the elevation difference in the Lake Alpine area between the highest exposed granitic ridge and the lowest volcanic deposit. The basin then was buried by the Valley Springs Formation, but was stripped of much of it before mid-Miocene andesitic volcanism began. Such volcanism buried all or virtually all of the landscape, though by 9.2 million years ago, major drainages had been incised into these deposits, and the Table Mountain latite flow descended one of them. Since then, streams have eroded much of the loose volcanic sediments, and in the last 2 million years glaciers have aided them to an unknown extent.

Perhaps the basin was mostly freed of its volcanic cover by the onset of glaciation, but evidence is lacking in the basin. However, a clue exists at Tryon Meadow, which lies about 5 miles east of Pacific Valley (southeast corner of Map 52), where there is a small rhyolite intrusion dated by Armin *et al.* at 6.3 million years. On the west edge of this meadow this intrusion is in direct contact with the Relief Peak Formation as well as with granitic bedrock, these contacts demonstrating that by 6.3 million years ago at least some of the Relief Peak Formation had been totally removed to expose granite. Nothing more than three ephemeral, snowmelt creeklets are at work in this vicinity today, and back at 6.3 million years ago, the

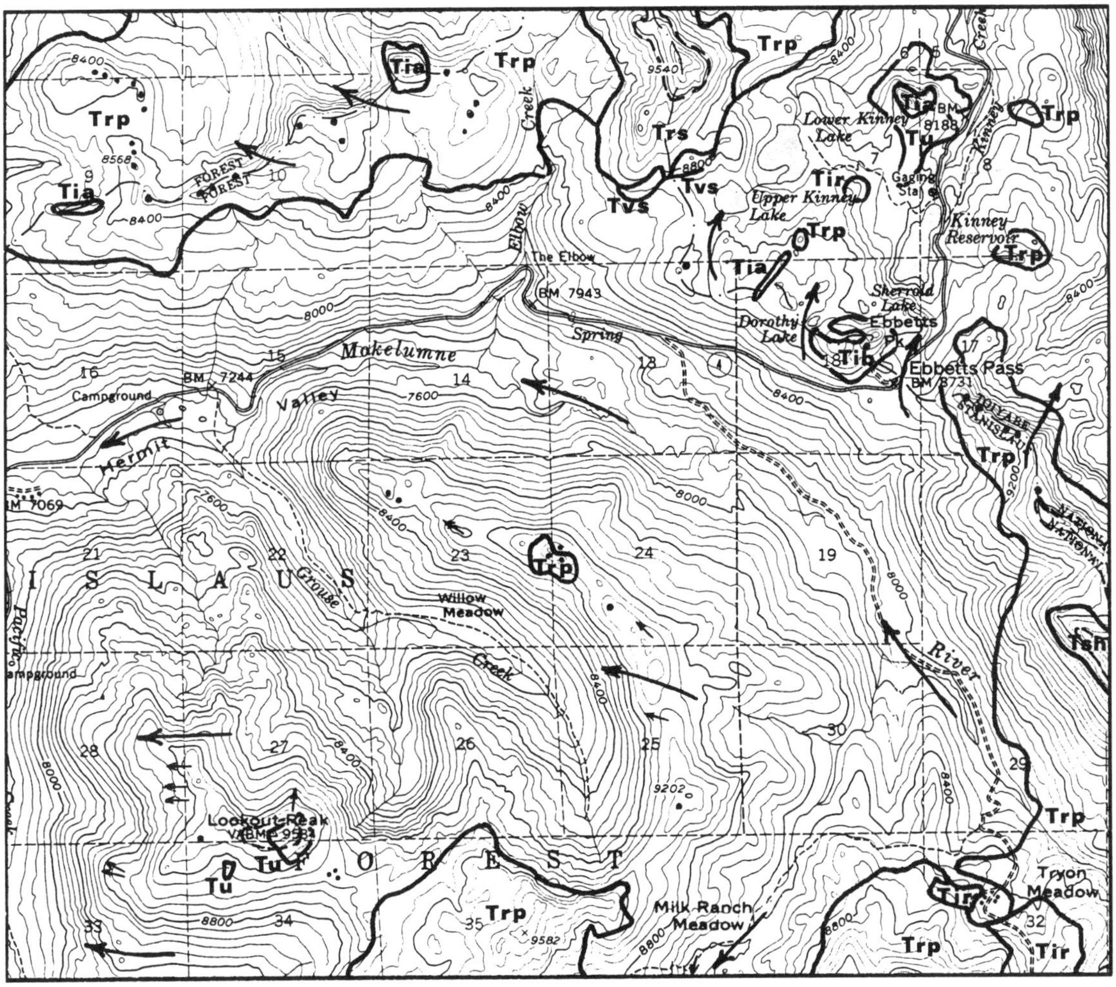


**Map 52. Volcanic deposits and Tioga-glacier evidence near Ebbetts Pass. Geologic units (after Armin and others, 1984), youngest to oldest, are: Tsh = Silicic volcanic rocks of Silver and Highland Peaks (4 million years old); Tir = Intrusive rhyolite (6.3 million years old); Tu = undivided, intermediate-age lava and deposits; Tib = Intrusive basalt; Tia = Intrusive andesite; Trs = Raymond Peak and Silver Peak andesites; Trp = Relief Peak Formation; Tvs = Valley Springs Formation (21.2 million years old). Bedrock is Cretaceous granodiorite. Glacial evidence: long arrows = general ice-flow direction; short arrows = striations; dashed lines = upper limit of glacier, based on highest erratics (dots). Scale 1:62,500.**

erosional processes may not have been much more active. Thus in the North Fork's broad basin, its many minor tributaries may have accomplished as much erosion by that time, exposing patches of granitic bedrock, and by the onset of major glaciation some 4+ million years later, much if not most of the granitic bedrock may have been exposed. So at best, glaciers removed whatever loose volcanic deposits the streams missed; they did little to the granitic bedrock. Glaciers on benchlands performed minimal erosion, but what about major canyons, where glaciers presumably were very effective, because they were supposedly thicker and their ice-surface gradients were supposedly steeper?

Just above the confluence of the Stanislaus River's Clark and Middle forks, the Clark Fork Road bridges the Middle Fork across its small inner gorge (Map 51, bottom right inset). According to Matthes, inner gorges such as these were carved by their rivers in post-glacial time, that is, in about the last 13,000 years. He specifically mentions (1930, p. 101) the inner gorge of Tenaya Canyon. That inner gorge, cut in massive granite, is up to 300 feet deep, which implies dramatic erosion (by the *seasonal* Tenaya Creek), "in striking contrast to the almost negligible depth to which the relatively powerful Merced River has cut the massive rock." Matthes, like others, has assumed that both streams and glaciers performed major erosion in late Cenozoic time, so the post-glacial river erosion of the inner gorge seems likely, while the lack of measurable erosion by the Merced River seems very unlikely. Evidence against a post-glacial origin is found at the

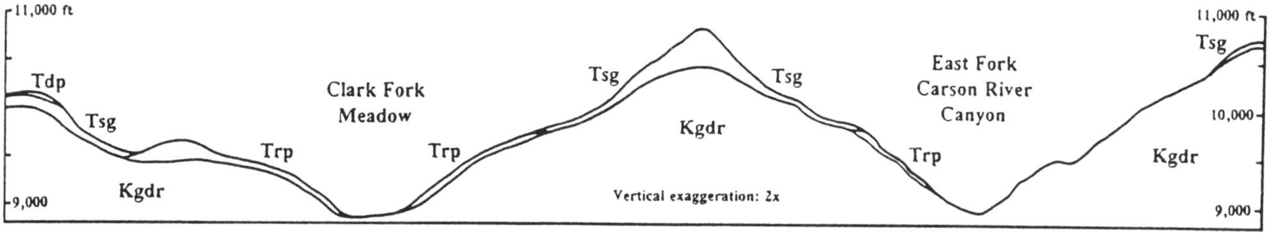


**Figure 42. Cross section through the upper Clark Fork Stanislaus River Canyon and the upper East Fork Carson River Canyon. See Map 53 for location. Tdp = Disaster Peak Formation; Tsg = Stanislaus Group, undivided; Trp = Relief Peak Formation; Kgdr = Cretaceous granodiorite, undivided (Giusso, 1981).**

Middle Fork Stanislaus River's inner gorge. A remnant of a columnar-lava flow dated as 150,000 years old exists along the inner gorge's south side. If the lava is that old, then it was glaciated by a Tahoe glacier and a Tioga glacier—both about 1300 feet thick—and perhaps by three smaller glaciers that may have existed between these two (oxygen-isotope stages 3, 4, and 5a-5d, as mentioned under "Glacial Chronology—Tenaya"). This columnar lava extends all the way down to the river, so measurable net incision into granitic bedrock by the river and its glaciers is essentially zero. This granitic, inner gorge is much older, perhaps originating about 2 million years ago when the first glacier might have advanced down this canyon and beneath it roared a swift, mighty, boulder-choked, subglacial river. If you drive to this bridge, you ought to drive a bit farther to the Clark Fork bridge. It spans an inner gorge that is more spectacular than the one by the Middle Fork bridge. An even more spectacular one is higher up, by mile 55.03 along Highway 108. This spot is about ½ mile above the Pigeon Flat Picnic Area, at the Columns of the Giants—a huge mass of lava that was extruded at that site, some of it flowing down to the Middle Fork bridge. Were it not for a dated lava-flow remnant at the Middle Fork bridge, we would have no evidence for the negligible rate of stream and/or glacial erosion of the Sierra's inner gorges that lie within its canyons.

To the northeast, in the upper parts of the Stanislaus River's Clark Fork and Middle Fork canyons, the relief is impressive, on the order of 4000 feet, and it appears to argue for major glacial downcutting. However, as in other issues, close inspection of the volcanic-granitic contacts reveals the opposite. (See Map 53, which is a slight simplification of a geologic map of the Sonora Pass 15' quadrangle by Giusso. Earlier, Slemmons also had mapped this quadrangle, and the volcanic deposits on both maps are in general agreement.) Near the center of Map 53 lies Clark Fork Meadow, and two thin, lower remnants of the Relief Peak Formation descend granitic slopes to it. To the south, on a canyon floor just west of Brown Bear Pass (Tower Peak 15' quadrangle), Morton *et al.* dated the base of this formation at 19.2 to 20.0 million years. This site is in the headwaters of the Middle Fork, and because the base of the Relief Peak Formation at Clark Fork Meadow occupies a similar topographic setting, it too is presumed by Giusso to have a similar date. Because this formation descends to the edge of the meadow, it indicates that approximately the same amount of relief existed in this canyon 20 million years ago as exists today: for both times the relief of granitic bedrock from the saddle at the east base of Stanislaus Peak south down to the granitic bedrock about the meadow amounts to about 1600 feet. Here the former glaciers, about 1200 feet thick based on field evidence, failed to excavate the meadow area any deeper than the lower part of the Relief Peak Formation.

More important is a small remnant northwest of Clark Fork Meadow that extends down a steep wall of a hanging tributary canyon to just about 50 feet above the Clark Fork (this spot is near "Mile 15"). Again, it indicates very little net incision in 20 million years. More significantly, however, it indicates that the tributary canyon was hanging by at least 20 million years ago, that is, long before this landscape was glaciated. The inner gorge of the Clark Fork already existed much as it does today. Additionally, the profile of the Clark Fork had to be stepped, as the following analysis shows. About 0.4 mile northwest of this small remnant another one descends almost as close to the river, and the two together indicate that this reach of the Miocene Clark Fork had a low gradient comparable to that of today, about 2°. The

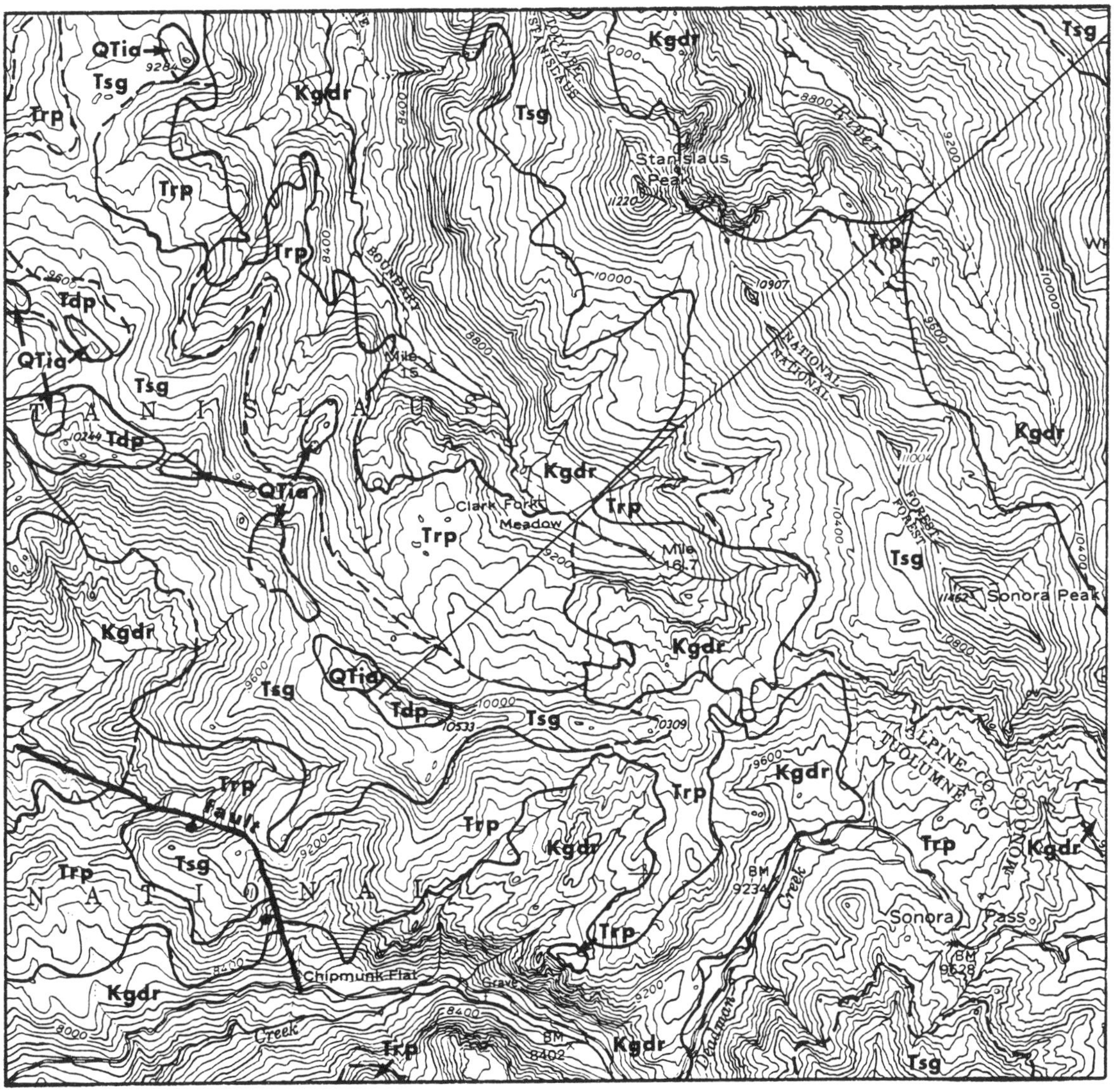


Map 53. Neogene volcanic deposits in the headwaters areas of the Clark Fork and Middle Fork Stanislaus River and the East Fork Carson River, near Sonora Pass. Geologic units, youngest to oldest: QTia = intrusive andesite; Tdp = Disaster Peak Formation; Tsg = Stanislaus Group, undivided; Trp = Relief Peak Formation; Kgdr = Cretaceous granodiorite, undivided. Contacts dashed where approximate. Scale 1:62,500. The straight line southwest from the upper right corner of the map is for a cross section shown as Figure 42.

remnants at Clark Fork Meadow indicate that the reach there was about as flat as it is today, about 0°. Between the two lower, small remnants and those at Clark Fork Meadow, the modern Clark Fork cascades for about 0.3 mile along an average gradient of 14°. The Miocene Clark Fork must have had a similar cascade. Finally, the prevolcanic topography of the Clark Fork's upper canyon was complex; it did not maintain a simple V-shaped cross section along its length but rather varied in shape just as it does today. Note additionally that the west-side remnants reaching down toward the canyon's floor below Clark Fork Meadow indicate that the prevolcanic canyon was about as wide as it is today (again, see Figure 42). The floor is relatively flat, but the canyon was not significantly broadened by glaciers; rather, it was relatively flat before volcanism occurred. All this prevolcanic topography, presumably etched in the granitic bedrock during warmer, wetter climes, resembles the topography of central Colombia's Río Grande drainage (Map 54), and the river's longitudinal profile resembles the Río Grande's (Chapter 21's Figure 39). Compare the topography of this area with that of Rockhouse Basin and vicinity (Map 48). The topography of each unglaciated, granitic area resembles that of the Sierra's glaciated, granitic lands. All have cross sections ranging from V-shaped to U-shaped, floors being narrow where the stream gradient is high and broad where it is low. Also, all have tributaries ranging from accordant to hanging. Such features in the Sierra Nevada have been attributed to glacial erosion, but clearly in the Clark Fork drainage, they existed long before glaciers appeared.

The Clark Fork's upper canyon is not the sole example of an exhumed relict tropical landscape. Note the East Fork Carson River's upper canyon, located in Map 53 between Stanislaus Peak and White Mountain. In this canyon a finger of the Relief Peak Formation descends to within about 150 feet of the canyon's floor, indicating that this mostly granite-walled canyon too has not changed much over 20 million years. This raises an important point. Lands east of the Sierran crest, particularly in Owens Valley and the Mono Basin, are believed to have downfaulted to their current levels in the last few million years, but no faults have been found in the East Fork Carson River canyon that indicate any substantial subsidence. Thus the Sierran crest in this vicinity—from Sonora Pass north at least to Ebbetts Pass, where the Valley Springs Formation straddles a broad bench on the crest—the Carson River drainage predates major faulting by millions of years.

Along the southern part of Map 53 is Deadman Creek, the Middle Fork Stanislaus River tributary that originates at Sonora Pass. A finger of the Relief Peak Formation descends south over 1000 feet to within about 200 feet above Chipmunk Flat, the granitic floor of the creek's canyon, which lies at about 8000 feet elevation. Hence, this heavily glaciated canyon was *at least* five-sixths as deep some 20 million years ago as it is today (*at least* 1000 feet relief then versus 1200 feet now). Massive glaciers not only were ineffective at incising through the granitic bedrock in this tributary drainage, they also were ineffective at removing a granitic monolith (flat summit on south border of map) and the opposite granitic ridge that cause a constriction in the canyon.

Besides the Stanislaus and Mokelumne river drainages, the Yuba, San Joaquin, and Kings river drainages also have volcanic flows and deposits which show that their preglacial topography was very similar to their modern, postglacial topography. In Chapter 21's "Longitudinal profiles," I mentioned that if glaciers severely deepen their canyons through erosion, then the upper glaciated part of a Sierran river should not hang above the lower unglaciated part. But often they do, as is well demonstrated by the longitudinal profiles of Middle Yuba and South Yuba rivers.

The South Yuba River is particularly interesting in that its headwaters are truncated at the east end of broad, flat-floored Summit Valley, that is, at Donner Pass (Norden 7.5' quadrangle). Between the pass and Norden, the north side of the valley floor is bounded by Beacon Hill, which is composed of a sequence of volcanic deposits. The lowest and oldest is 33.2-million-year-old rhyolite, and about ¼ mile west on old Highway 40 from Donner Pass, the base of the rhyolite is in contact with granitic bedrock that comprises the floor of upper Summit Valley. What the rhyolite and volcanic deposits on the valley's south margin demonstrate is that Summit Valley already was broad and flat-floored long before glaciation commenced, indeed, by 33 million years ago, when the warm, moist climate was beginning to evolve toward the modern summer-dry climate. Polish and striations about this area, particularly on granitic rocks west of Donner Peak, indicate that Summit Valley lay under about 650 to 1000 feet of Tioga ice, whose principal source area was Mt. Lincoln, about 2 miles to the south. Despite this thickness, this glacier and previous ones have, at best, eroded only a few feet to a few tens of feet below the base of the rhyolite. This rhyolite occurs locally in glaciated lands to the west. One important site is about 6 miles west of Donner Pass, near Interstate 80's Kingvale exit. There, a road switchbacks southward up toward some railroad tracks, and along it rhyolite is exposed as low as only 200 feet above the river. The rhyolite probably occurs lower down, but without additional roadcuts (or trenching), its lowest limit is unknown. Glaciers may have removed all of the rhyolite along the floor, but still, with persistence some, I suspect, could be found. Anyway, we can safely conclude that over some 33 million years that in this vicinity the South Yuba River canyon has been deepened by no more than 200 feet. Furthermore, because the rhyolite continues up the slope to benchlands holding

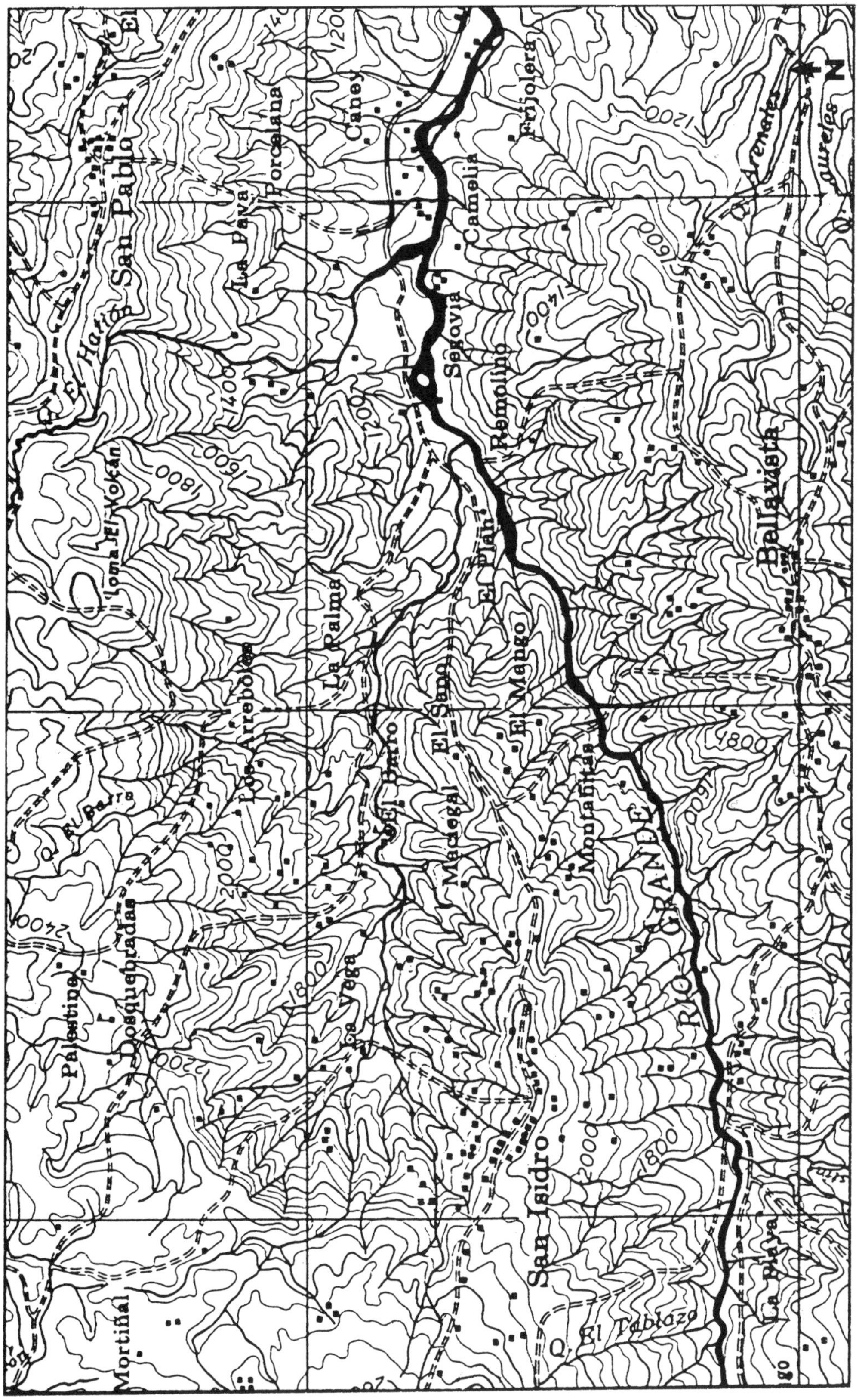

Map 54. Topographic map of the lower part of the Río Grande drainage, Antioquian batholith, central Colombia. Contours, 50 m interval. Each grid is 5 kilometers (3.1 miles) on a side.

Kidd Lake and the Cascade Lakes (Chapter 3, Photo 5), the canyon in this vicinity also is about as wide as it was some 33 million years ago. Despite glaciation having occurred over at least the last 2 million years, glaciers have not removed all of the rhyolite that buried the canyon's floor and walls.

South of Yosemite, the San Joaquin River and Kings River drainages were not buried under voluminous volcanic deposits, but they nevertheless experienced some volcanism. Additional evidence of minimal glacial erosion lies to the south, in the San Joaquin River drainage. This drainage has several dozen remnants of lava flows presumed to be about 3.5 million years old, based on radiometric dates on some of them. One flow remnant—Devils Table, in the South Fork San Joaquin River drainage—is particularly interesting (Chapter 20, Map 36). At Devils Table, no measurable incision into the adjacent granitic bedrock has occurred, despite the flow remnant lying directly in the path of the immense glaciers, about 1500 to 2000 feet thick, that flowed down this drainage. At spots along the northeast, southwest, and southeast bases of Devils Table, granitic ridges adjacent to the lowest parts of the flow actually rise 2 to 4+ feet above it (Photo 164). This indicates that the flow buried an undulating, granitic topography, and that despite perhaps dozens of glaciations, these ridges have not been worn down below the base of the flow.

One could argue that those dozens of glaciations were required just to remove thick, resistant lava flows to expose the underlying granitic bedrock, and that future glaciations then will carve deeply into this bedrock. But the lava flows never were extensive; they occurred locally in various sites. So when the first glacier advanced down the San Joaquin River drainage, the landscape, as today, was mostly granitic. Be aware that lava locally was erupted at various elevations, from canyon bottoms to mountain tops, and all types of topography in between. Some of these lavas were severely glaciated, some moderately, some slightly, some not at all. All show the same amount of post-volcanic denudation: essentially zero.

Fewer flows occur in the Kings River drainage, but as in the San Joaquin, they occur on unglaciated land as well as on glaciated land that was buried under ice varying in thickness from thin (cirques and adjacent low ridges) to thick (canyon bottoms). In the glaciated areas of the North, Middle, and South forks, these flows appear on the following USGS Geologic Quadrangles: Blackcap Mountain (Bateman, 1965), Marion Peak (Moore, 1978), and Tehipite Dome (Moore and Nokleberg, 1992). The most important volcanic remnant is the largely intrusive, Tertiary volcanic complex of Windy Peak (Map 55). This has been dated at 4.3 million years by Luedke and Smith, and it extends continuously from summit of the peak down to the broad, flat, floor of the glaciated Middle Fork canyon, a relief of about 2860 feet, and it also occurs on

**Photo 164. Exposed base of lava flow, northwest side of Devils Table. This lies 3 feet below the granitic rock in the background.**

the opposite side of the canyon down to the floor. These volcanic remnants demonstrate that at the time they buried the local topography, which was before glaciation commenced, this canyon had already achieved its present width and depth. No major widening or deepening by glacial erosion had occurred.

Thus it appears that for more than a century glacial geomorphologists have been operating under a logical though unproved and incorrect paradigm. As the presented evidence demonstrates, glacial erosion in the Sierra Nevada, and probably elsewhere, has been negligible. The major glaciated ranges of the world are high because they have experienced rapid uplift that has far outpaced the forces of erosion. That uplift was generated by myriad earthquakes, which also generated tremendous amounts of mass wasting. This mass wasting is particularly important during interglacial time, when glaciers are smaller. As Map USGS I-1612 (mentioned in the previous chapter) shows, mass wasting that occurs in the Sierra Nevada today can occur at any point along a canyon's walls, and

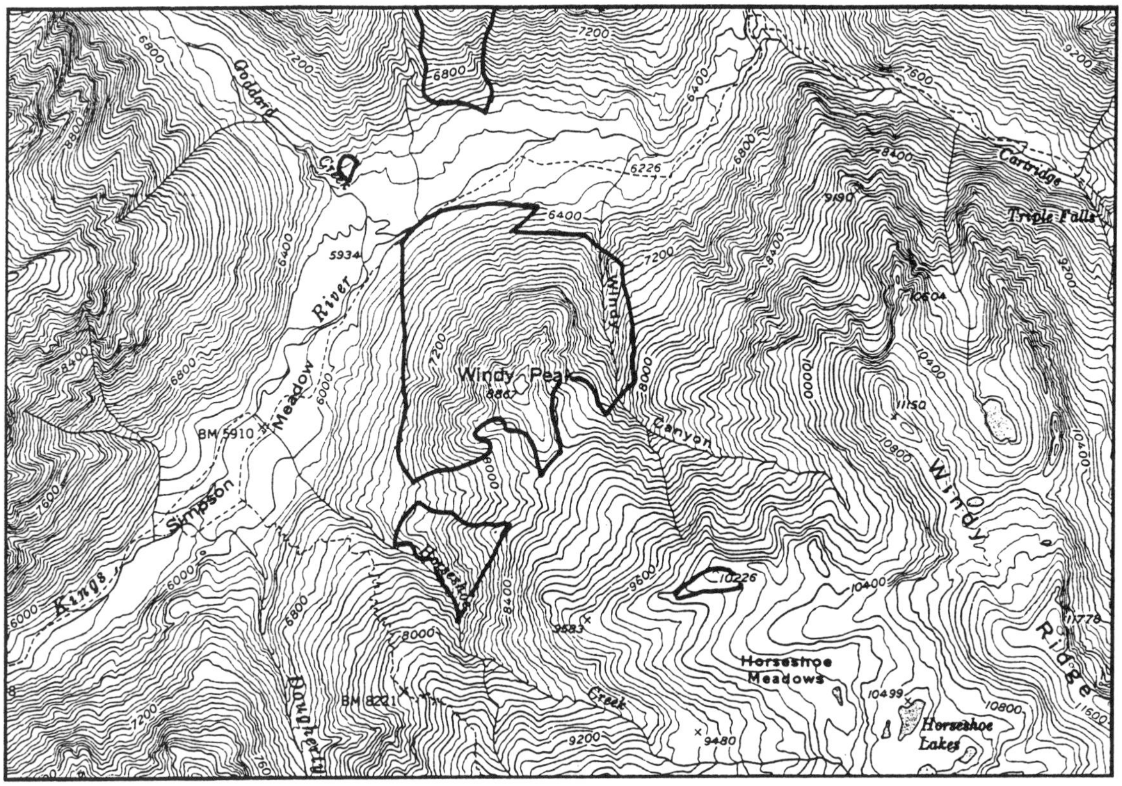


**Map 55. Remnants of 4.3 million-years-old Windy Peak dacite in Middle Fork Kings River canyon. Bedrock is granitic and metavolcanic rocks. Scale 1:62,500.**

this causes them to maintain their angle of steepness as they retreat. During maximum glaciation, most of the slopes of a canyon are buried under ice and are protected from mass wasting. However, once deglaciation has occurred, pressure is released from the lower slopes, which then undergo a spurt of mass wasting. This seems to be what is occurring in Yosemite Valley, where one can trace sequences of exfoliation that began with a lowermost slab that broke off, leaving a roof. That, in turn, destabilized the immediate bedrock, which then broke off, leaving a higher roof, and so on up the nearly vertical cliffs. These two processes of mass wasting, one seismic-induced and the other glacier-induced, may be pervasive in the High Sierra and can operate in concert to cause parallel retreat of canyon walls, making canyon floors broad. Hence U-shaped cross profiles do not need glacial *erosion* to explain the shape, although a minor amount of broadening may be due to such erosion. Glacial erosion seems most apparent at the uppermost parts of a canyon, but as I have argued, mass wasting even there is more important, due to proximity to earthquakes generated by faults along the east base of the range, and due to enhanced freeze and thaw that occurs near the Sierran crest. Increasingly farther down-canyon, one has an increasingly more difficult task of finding *real*, factual and quantitative evidence of glacial erosion, which is why the maximum extents of glaciers in most of the western canyons are poorly known.

I am slowly figuring out that if you have opinions you want to express, and they are not part of the general consensus, there is a price to be paid.

—Patricia Moehlman

# Part IV

# Sierran Uplift

# 23

# Previous Methods of Evaluating Uplift

## Introduction

How do you create a mountain range? The late Cretaceous Sierra Nevada was lifted at its southern end by the continental lithosphere of a large terrane being thrust under what is now the southern part of the range. The Alps, the Himalaya, and the Alaskan ranges exist for the same reason: convergence of two continental-crust lithospheric plates. This has *not* happened in the Sierra Nevada during middle or late Cenozoic time, and perhaps not even during early Cenozoic time. There are no proven plate-tectonic motions or processes that document late Cenozoic Sierran uplift, although a number have been proposed. These are theory-driven. For over a century Sierran earth scientists have asserted that major late Cenozoic uplift has occurred, and some looked for mechanisms to explain it. For example, in USGS Professional Paper 1197, N. King Huber proposed 11,320 feet of total Cenozoic uplift, almost all of it in late Cenozoic time. Geologists and geophysicists have taken his assumptions as factual and his results as precise, and have constructed plate-tectonic models to conform to both Huber's timing and his amount of uplift.

This Part IV presents evidence against major late Cenozoic uplift. Possibly some minor uplift has occurred, although it is impossible to quantify even a minor amount with certainty. A principal argument for little or no late Cenozoic uplift, as has been shown in Part III, is that the Sierra's canyons—contrary to popular notions—already were about as deep in mid-Cenozoic time, if not earlier—as they are today. Conventional wisdom dictates that the canyons had been shallow and that they had deepened only due to major late Cenozoic uplift, which had caused rivers and ensuing glaciers to incise deeply. However, in the South Yuba River drainage there are remnants of sediments that indicate the lower part of its river canyon (at the confluence of Spring Creek, elevation 1930 feet) was very close to its current depth when the sediments were deposited during the Eocene epoch, and that at best relatively minor post-Eocene river incision has occurred. Therefore, the river canyon, when measured from its bedrock floor to the highest adjoining bedrock ridges, has had almost the same relief for some 50+ million years. Volcanism existed from the Oligocene through the Pliocene, so for half of the Cenozoic era the bedrock topography of the northern half of the range has been buried under sediments. Hence, it is not surprising that only negligible incision by the South Yuba River has occurred. What is surprising, as detailed in this section, is that minor post-Eocene incision also occurred in drainages never buried by sediments.

The current paradigm—a low-elevation, low-relief range before late Cenozoic uplift—does not work. As will be shown in the next chapter, if one takes various estimates of uplift and then depresses canyon floors by these uplift amounts, not only do much of their lengths have mid-Cenozoic elevations *below* sea level, but also some of the benches on their rims were below sea level. But there is no evidence (such as marine sediments or pillow basalts) to indicate that the pre-uplift topography had existed as submarine canyons, nor has any been proposed. Additionally, the unfaulted, lower *eastern* slopes of the Sierra Nevada would be depressed to impossibly low levels.

On the western slopes there are four major problems with the current paradigm. First, the west-side canyons appear to be incising at highly variable rates that bear no relation to the size of their rivers or to the extent of their former glaciers. Second, moraine-age correlation both within individual canyons as well as between river drainages has been impossible (addressed under Part III's "Glacial Evidence"). Third, the present distributions of conifers such as giant sequoia, foxtail pine, lodgepole pine, and Shasta red fir in the Sierra Nevada and elsewhere in California cannot be explained if some of California's ranges—and the adjacent Great Basin—were low until late Cenozoic time. Fourth, the rate of evolution to create alpine flowering plants, based on the perceived, relatively recent development of the Sierra's alpine climates, is anomalously rapid. These problems disappear if, as was mentioned earlier, the range has been high since Eocene time, if not since the terminal Cretaceous.

The Sierra Nevada is not an island. Its uplift history is tied to that of eastern lands. This was recognized back in the 1880s, but stated much later, in 1966, by Mark Christensen:

> The sequence of uplift of the range followed by faulting on the eastern escarpment indicates that parts of the Basin [and] Range province rose along with the Sierra. The origin of the modern Sierra, therefore, can be fully comprehended only in the context of the origin of the Basin [and] Range structures.

Under the current paradigm, both the Sierra Nevada and the Basin and Range province were low until late Cenozoic time. But detailed, 1994 work in west-central Nevada by Jack Wolfe and Howard Schorn suggests that mountain elevations about 16 million years ago were about 10,000 to 13,000 feet, and that by 12 million years ago extensional faulting had caused the lands to collapse to about their present altitudes. If the origin of the modern Sierra Nevada can be fully comprehended *only* in the context of the origin of the Basin and Range structures, then it too must have been high before extensional faulting. Logically, it is still at its 16-million-year height. Under this interpretation, the third and fourth problems of the paradigm disappear.

At least nine methods have been used to identify, date, and/or quantify late Cenozoic uplift. These are presented in an order starting with the least known and least cited. My critiques of methods 8 and 9, respectively espoused by Matthes and Huber, are more detailed because each appears to have substantial field evidence to support it. Additionally, both have been widely cited as evidence for major late Cenozoic uplift. Whereas Matthes proposed three pulses of increasingly greater uplift, Huber proposed a more uniform uplift whose rate became increasingly greater with time. Although the two methods are different, they yield essentially the same timing and magnitude of the uplift; hence it is not surprising that Huber accepted Matthes' uplift *results*, and by doing so, promulgated his questionable method.

## 1) Warps and faults within the Sierran block

Frank Hudson wrote four papers in which he first suggested a method of uplift, developed it in detail, and then refined his timing and estimates of uplift. His seminal 1948 paper discussed the Donner Pass zone of deformation, an area that would be the basis for further study. This short paper, based on a brief stay in the field, concluded that most of the crest-area faults were reverse faults, which raised the Sierra Nevada above the eastern lands. This position stands in contrast to the longstanding widely held position that the faults are normal. He also noted another zone of deformation west of Dutch Flat, which is about midway between the pass and the west base of the range, and he suggested that other faults might exist. A number of faults do exist, and these are shown on a map by George Saucedo and David Wagner. However, these faults are very old, preceding late Cretaceous uplift, and they became inactive more than 100 million years earlier than Hudson had suspected. Nevertheless, these faults later would be important in his uplift determinations.

Hudson's 1951 paper was a longer discourse based on more intensive studies in the same area and just northwest of it. These studies discovered folded granite with folded andesitic pyroclastic rocks resting conformably on it. He concluded that after recurrent periods of reverse faulting and folding, normal faulting became important in the Quaternary. As in his previous paper, his position on folded Sierran crest rocks stands in contrast to the long prevailing one that granitic and volcanic rocks near the Sierran crest have been faulted. A better explanation is twofold. First, folding granite can form when upwelling, flowing magma spreads laterally before finally solidifying to form a pluton. The flow pattern, which can be folded, will be preserved, and it has nothing to do with folding solid granite, which did not occur here during the entire Cenozoic era. Second, folded andesite can originate when a flow slowly solidifies on a slope, as is seen elsewhere in the range. For example, G. Brent Dalrymple in 1964 discussed and illustrated variably dipping volcanic flows on granitic slopes in the upper San Joaquin River drainage, and Garniss Curtis and Howard Wilshire, respectively in 1951 and 1956, described variably dipping volcanic flows in the Ebbetts Pass area.

Having identified two zones of deformation, Hudson went looking for more, and in his 1955 paper found *indications* of five zones based on gradient calculations, although three could not be confirmed (and the other two have problems). As in his two earlier papers, he departed from conventional wisdom—that uplift of the Sierra Nevada had occurred as simple tilting of a rigid block—and stated that it had occurred along these five zones of deformation.

His uplift analysis required a primary assumption: the middle Eocene Yuba River had a fairly smooth profile. He divided the length of this river into many reaches and determined the average gradient of each. Next he assumed that there were series of three consecutive channel segments that originally had the same gradient but do not today, and that the original gradient could be calculated from differences in today's gradients. With laborious calculations he solved for gradients for each of some 19 triplets of consecutive channel segments. The results more often than not were unfavorable, for only eight yielded realistic results, as he so admitted (p. 850):

> By assuming the equality of the original gradients of the three reaches of a triplet, a set of three simultaneous equations, containing three unknowns, was produced. These can always be solved, but often produce absurd results, such as negative gradients, or absurdly high gradients or amounts of tilt, which had to be discarded. The remainder were regarded as proper solutions.

Despite discarding half the length of the middle Eocene Yuba River profile, he defended his method as valid. Like most others listed in this chapter, he was selective with his evidence, tossing out contrary evidence. He concluded that uplift at Donner Pass since middle Eocene time has been only 1986 feet—a value considerably lower than Lindgren's for the same area. But as I discussed in Chapter 21, Sierran river profiles have long been irregular and stepped—knowledge readily available to him. By ignoring *real* profiles, Hudson *had* to discard most of triplets.

His 1960 paper was a criticism of Matthes' and Axelrod's uplift methods. (His criticism of the latter is mentioned just below in the third uplift method.) Matthes had concluded that his broad-valley stage evolved through the Miocene epoch until late Miocene uplift initiated the mountain-valley stage. Hudson, in contrast, concluded that the broad-valley stage was early Pleistocene, for it was correlated with the eastern San Joaquin Valley's China Hat pediment. In 1981 Denis Marchand and Alan Allwardt concluded that the China Hat surface is not a pediment, but rather is a remnant of a late Laguna gravely fan deposit. It is risky enough to extrapolate a pediment uniformly eastward, as Hudson did, but it is fatal to extrapolate a remnant of an alluvial fan. Hudson concluded that the uplift was only about 3900 feet, not the 9000 feet determined by Matthes. That Hudson's age of the broad-valley stage is clearly wrong is proved by the many dated lava flows that buried ancient topography (e.g., 33-million-year-old rhyolite deposits in the South Yuba River drainage). Given that Hudson was at odds with virtually everyone in regard to his field observations and to his low estimates of uplift—as well as at odds with evidence in the form of dated, inner-canyon lava flows—it is not surprising that his work never was taken seriously.

## 2) Diminishing size of the White Mountains' late Pleistocene glaciers

In 1987 Deborah Elliott-Fisk observed that over time the late Pleistocene glaciers of the White Mountains east of the Sierra Nevada diminished in size. Presumably this was due to significant Quaternary uplift of the Sierra Nevada that would have caused an intensifying rain shadow. However, as was mentioned in the previous chapter, George Davis and Tyler Coplen demonstrated that major uplift in the South Coast Ranges of central California commenced just after the Sherwin glaciation. The growing rain shadow cast by these ranges explains not only the diminishing White Mountains glaciers, but also the diminishing central Sierran glaciers. Had major late-Quaternary Sierran uplift occurred, then the central Sierran glaciers should have increased with time, not diminished, as lands rose into colder climates.

## 3) Changes in the composition of Nevada's paleofloras

Daniel Axelrod in 1957, 1959, 1966, and 1980, and with William Ting in 1960, claimed to have seen changes in the composition of Nevada's Neogene paleofloras (plants of the Miocene and Pliocene epochs) that imply the creation of a rain shadow, presumably due to Sierran uplift. Mark Christensen, Herbert Meyer, and Jack Wolfe have addressed serious problems with Axelrod's data, methodology, and paleoaltitude-reconstruction assumptions. Additionally, Howard Schorn, a very meticulous paleobotanist, reevaluated Axelrod's fossil-plant identifications and found that most of the fossils were misidentified. When they are properly identified, they are found to be plants of former highlands, not lowlands, for Nevada. My examination of Axelrod's Mt. Reba paleoflora site (Photo 165), north of Highway 4's Lake Alpine, confirmed that his mapped faults and erosion surfaces do not exist. This conclusion accords with mapping in the area by Richard Armin *et al.* of the USGS. Gravity maps also fail to show his major proposed Desolation Wilderness-Mt. Reba fault. Finally, his methods, as presented in a

**Photo 165. The Mt. Reba paleoflora site, at about 8600 feet elevation. The actual site is in a low road cut immediately past the metal post.**

1986 paper, dictate that the Mokelumne River canyon (below Mt. Reba) was incised in the last 4 to 5 million years. However, as was mentioned in the previous chapter, volcanic flows in various Sierran drainages confirm that the canyons have deepened very little over the last 20 to 50+ million years. Axelrod unfortunately established himself as the authority of giant-sequoia migration, a subject that needs to be reassessed, as I briefly discuss in this book's Epilogue.

## 4) Measured, ongoing uplift in the Donner Pass area

John Bennett *et al.* claimed that 1953, 1969, and 1976 leveling surveys along the Southern Pacific Railroad tracks from Roseville northeast to Donner Pass and down to Reno indicate about an inch or so of uplift at the pass since the 1947 survey. However, the accuracy of measuring changes in elevation to the nearest fraction of an inch in days before GPS precision is questionable, and their data show some short-term subsidence as well as uplift. Taken at face value, their results support Hudson's conclusions, which have been refuted. Even if an inch or so of uplift actually occurred, it could be due to glacial isostatic rebound, which is common to all areas that have experienced major deglaciation. In the vicinity of the pass the Tioga ice sheet was on the order of 1000 feet thick, and over thousands of years the range would have sank several hundred feet with this added weight. After deglaciation it would rebound to its preglacial position. This takes time, and it should still be occurring today, though at a diminished rate. Since Tioga ice was thick as far south as the Mt. Whitney environs, isostatic rebound should be expected along this entire length of Sierran crest. Unfortunately, no one has addressed this issue, despite it being real. Measurable isostatic rebound still is occurring in Canada and New England, and it has been known since the late 1800s. Additionally, we have long known that where former large, glacier-fed lakes have greatly decreased in volume, such as the demise of Lake Bonneville to create today's (not so) Great Salt Lake, the basin's lands have isostatically rebounded. Conversely, the creation of Lake Mead led to isostatic subsidence.

## 5) Normal faulting along the east base of the Sierra Nevada

A fault scarp at least 100 miles long was created in the March 26, 1872, Owens Valley earthquake. It caused a horizontal displacement up to 23 feet and a vertical displacement that averaged about 3 feet. From the latter figure, one could conclude that uplift is ongoing, as Paul Bierman *et al.* did in 1991, like many others before them. Unfortunately no accurate surveys existed back in 1872 from which post-quake surveys could be made to determine actual elevation change. Was there 3 feet of Sierran uplift, 3 feet of Owens Valley subsidence, or a combination of mountain uplift and valley subsidence? Some valleys in the Basin and Range province have bedrock floors lying well below sea level, which is possible only through subsidence. The greater the crustal extension, the greater the subsidence, which is why Death Valley and the lower Colorado River zone have such low elevations, as Craig Jones observed in 1987.

Bierman *et al.* stated that the maximum thickness of sediments in Owens Valley were around Independence and Lone Pine is about 6900 feet. Since these towns lie at about 3700 feet elevation, the bedrock floor beneath them is at about 3200 feet below sea level. In the Long Valley caldera northwest of Owens Valley and east of the San Joaquin River drainage, the bedrock floor is up to 2000 feet below sea level. As elsewhere in the province, the only way that the bedrock floors in these two areas could have gotten below sea level is through subsidence. Sierran uplift, on the other hand, is purely hypothetical, or is based on faulty evidence, such as the following. In 1981 Nelson stated that uplift of mountains in the province (and presumably in the Sierra Nevada) is *certain,* because the: "paleobotanical evidence shows clearly that the region has experienced absolute uplift." This evidence is merely Axelrod's work, mentioned above, which has been discredited by reputable botanists, paleontologists, and geophysicists. Hence, it is not real evidence at all.

## 6) Tilted Cenozoic strata in the eastern Great Valley

The floor of the Great Valley lies atop sediments derived primarily from the Sierra Nevada but also from the Coast Ranges. These sediments are thin at the base of the Sierra Nevada, but thicken westward, and generally reach their greatest thickness fairly close to the base of the Coast Ranges. Just as the weight of thick Ice Age glaciers would have caused the High Sierra to isostatically subside by up to several hundred feet, so too the weight of thousands of feet of sediments—locally up to about 30,000 feet thick—on the bedrock floor of the Great Valley would have caused it to isostatically subside by a great amount. The valley's bedrock floor would have subsided the most where the sediments are the thickest. Hence the floor would subside very little near the base of the Sierra Nevada, where the sediments are thin, but would subside increasingly westward as the sediments became progressively thicker. Consequently, the strata dip westward. The lower, older strata dip the most. No Sierran uplift is required to explain this dip pattern.

Nevertheless, the dipping Cenozoic strata and angular unconformities in the eastern Great Valley have been attributed to late Cenozoic westward tilting of the Sierra Nevada. The sedimentary history of the Great Valley is complex, for at times sediments were deposited in a shallow sea and at times on land. Furthermore, an inland sea varied over extent, depth, and time before finally disappearing around the end of the Pliocene epoch. Alan Bartow in 1991 documented episodes of regression of this inland sea, and each time the sea regressed, some strata where shallow water formerly lay would have been exposed. These dipping strata would have been partly eroded. Then atop them sediments were later deposited, but because the surface of the partly eroded strata was not quite the same as the dip of the strata, there was a slight angular discordance. The gap in the strata is known as an unconformity, and the slight angular discordance makes it an angular unconformity (Figure 43). Sierran uplift is not needed to explain either the changing dip or the regional unconformities.

However, an *increased* rate of sedimentation and a change in dip imply uplift, which should cause increased erosion and transportation of sediments, and this is what Jeff Unruh suggested. But this increase, found in the northern Sierra, appears to be no more than the removal of voluminous deposition of late-Cenozoic volcanic sediments and lahars. This removal would have led to an increased amount of isostatic uplift of the range and to an increased amount of isostatic subsidence of strata in the Sacramento Valley. We can test Unruh's proposed uplift. Presumably the greatest amount of uplift occurred in the vicinity of Mt. Whitney, for the crest there stands about twice as high as it does in the Feather River drainage, where Unruh worked. In the eastern part of the San Joaquin Valley west of Mt. Whitney, the sediments should be coarse, due to the high gradient of the High Sierra's rivers; they should be voluminous, due to the high rate of erosion in this rugged landscape; and they should dip the most, due to the greatest amount of isostatic subsidence. But in actuality, the sediments are fine, not coarse; are thin, not thick; and dip slightly, not greatly—exactly the opposite of what uplift should dictate.

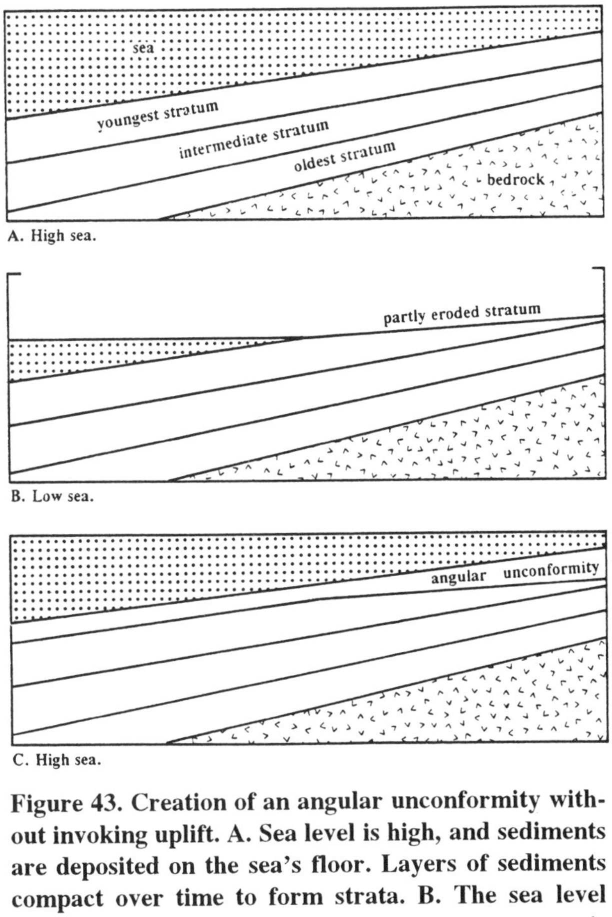


**Figure 43. Creation of an angular unconformity without invoking uplift. A. Sea level is high, and sediments are deposited on the sea's floor. Layers of sediments compact over time to form strata. B. The sea level drops, and the upper part of the youngest stratum is partly eroded. C. The sea level rises, and sediments are deposited atop the youngest stratum. An angular unconformity separates the base of these sediments from the eroded top of the underlying stratum.**

The Great Valley's basement floor in contact with the base of the lowest strata exhibits complex topography, which one can verify by reviewing cross sections and maps of the San Joaquin Valley by Alan Bartow and of the Sacramento Valley by David Harwood and Edward Helley. Since late in the nineteenth century geologists have known that the bedrock beneath the Great Valley is merely the continuation of the Sierra Nevada—the western part that is buried under strata. Therefore, had the Sierra Nevada been tilted as a block, the bedrock floor beneath the Great Valley also would have been similarly tilted, and it should be a rather uniform tilt, not a tilt with complex bedrock topography.

Likewise, the stratigraphic record is complex. Different east-west cross sections produce different histories, and you can find evidence to support your own model on the timing and magnitude of uplift, even when none has occurred. For example, Unruh's 1991 uplift history, based on the northeastern Sacramento Valley's strata, conflicts with Bartow's 1991 uplift history, based on the eastern San Joaquin Valley's strata. Unruh concluded that uplift for the entire range began about 5 million years ago, contradicting Thomas Crough and George Thompson's 1977 plate-tectonic/geophysical uplift model. Bartow, on the other hand, concluded that uplift began in the southern Sierra Nevada about 8 million years ago and migrated northward, in concert with the northward-migrating Mendocino triple junction, supporting their model. Their theory-driven model, by the way, has its problems, for its mechanism has been applied to South America's Altiplano uplift, but George Zandt *et al.*, 1994, concluded that the mechanism fails to explain uplift.

As should become apparent in this chapter, all of the Sierran-uplift models I evaluate have been theory driven. In short, we geoscientists "know" that Sierran uplift has occurred over the last few million years, so let's find some "evidence" to prove this "fact." Unruh's manuscript was critiqued by N. King Huber, and Daniel Axelrod's discredited Mt. Reba evidence figured heavily in Unruh's arguments. And Bartow uncritically accepted Huber's timing and magnitude of Sierran uplift as fact.

## 7) The gradient of a reach perpendicular to the crest minus the gradient of a reach parallel to it

Writing in 1900 and then in 1911, Waldemar Lindgren logically assumed, as others before and after him have, that the pre-uplift gradient of any Sierran river was essentially uniform throughout its length. After uplift, a reach of it perpendicular to the crest would be steepened the most, while a reach of it parallel to the crest would be unchanged. The difference in the two gradients would reflect the amount of uplift. For example, if the gradient of a reach parallel to the crest was 20 feet per mile, while a reach perpendicular to it was 90 feet per mile, then the difference would be 70 feet per mile. And if the range was 60 miles wide, then the uplift at the crest would be the gradient difference times the width, or 4200 feet of uplift. However, as I have discussed under Chapter 21's "Longitudinal profiles," Sierran rivers possessed irregular, stepped profiles before late Cenozoic time. Consequently, Lindgren's basic assumption is invalid.

In the Yuba River drainage of the northern Sierra Nevada this method produces a moderate—and credible—uplift. However, it is also produces unexplainable local subsidence not found anywhere else in the range and never substantiated by later field work. Additionally, under Lindgren's method, reaches having the same orientation should have similar gradients. More specifically, a reach with a constant bearing should have a constant gradient. But unglaciated reaches through granitic canyons have highly variable gradients. For example, in the North Fork Stanislaus River drainage my field work suggests that in the North Fork drainage the pre-Tahoe glaciers extended down to about 3600 feet, an elevation in accord with 1898 mapping by Turner and Ransome. From that elevation down 400 feet to the 3200-foot contour line (on Blue Mountain and Columbia 15' quadrangles), the gradient averages 140 feet per mile. Lower down, between the 2800-foot and 2400-foot contour lines, a similar descent, the gradient averages 507 feet per mile, which is a value 3.6 times greater, despite similar rock type and similar orientation. Quite obviously, the gradient of a reach is not related to its orientation with respect to the crest.

Finally, Lindgren's method does not produce similar results in other river drainages. Progressively farther south, uplift should be greater, since the Sierran crest increases in height, but I found the opposite occurs (drainages listed from north to south):

> Yuba River, about 4400 feet, or moderate uplift;
> Tuolumne River, about 3080 feet, or relatively minor uplift;
> Merced River, about *minus* 7980 feet, or *major subsidence.*

Because the fault-free headwaters of the Tuolumne and Merced rivers are adjacent, their uplift histories should be virtually identical, not irreconcilable.

## 8) Hanging tributary canyons record cycles of uplift and erosion

François Matthes in his 1930 Professional Paper 160 assumed that before any uplift every Sierran river had a smooth longitudinal profile and all the tributaries of each were accordant, not hanging. Following his mentor, Harvard's William Morris Davis, he envisioned uplift as occurring in short-lived major pulses, and he proposed three of them for the Sierra Nevada, each one greater than the previous one. After the first pulse in the early Eocene the Sierran rivers were rejuvenated, each river incising a shallow canyon that then widened over tens of millions of years to form the broad-valley stage of Sierran topography. After the second pulse late in the Miocene the rivers again were rejuvenated, and they incised deeper canyons into the first stage and then widened over a few million years to form the mountain-valley stage. After the third pulse late in the Pliocene the rivers incised even deeper canyons into the second stage and created the canyon stage. A more detailed, illustrated account (Figures 15-17) was presented in Chapter 9. Matthes (p. 28-29) gave specific dates for the three episodes of uplift responsible for the generation of the three stages of canyon cutting: nearly 60 million years ago (early Eocene), 12 million years ago (second half of Miocene), and 1 million years ago (Pliocene/Quaternary boundary). With a modern time scale, these three episodes become 65, 14, and 2 million years ago respectively. Hence although the durations of the Miocene and Pliocene epochs of the two time scales differ significantly, the three episodes of uplift are in relatively close agreement.

One should be aware that Matthes proposed an additional period of uplift. He proposed it began about 40 million years ago, during a period of intermittent volcanic outbursts. (These were rhyolitic eruptions, which now are known to have occurred mostly between 33 and 20 million years ago.) Interestingly, he did not propose a stage of canyon cutting. This issue is further complicated by a statement (p. 23) to the effect that after these gradual uplifts there was a prolonged interval (late Oligocene-early Miocene, about 20 million years ago by his time scale) that was marked by minor warpings of the earth's crust, up and down. Like the gradual uplifts, these warpings were not addressed. Matthes stated (1930, p. 27):

> The history of the successive uplifts that gave the Sierra block its present height and tilted atti-

> tude is as yet imperfectly known. Only the major events stand out clearly. The reason will be apparent when the nature of the evidence is learned; *it is mostly indirect or circumstantial.* (Italics mine.)

Despite this uncertainty, he reconstructed the evolution of the Yosemite landscape *in detail.*

Matthes' method of uplift estimates was based on hanging tributary streams of the Merced River. Most of the tributary streams below the glaciated part of the drainage were not hanging, but Matthes identified three that he claimed were. Then for a given reach that had a fairly constant gradient, he correlated it with a specific uplift-erosion stage: low gradient, broad-valley stage; moderate gradient, mountain-valley stage; and high gradient, canyon stage. The low-gradient reaches of his chosen tributaries occur near their headwaters, the high-gradient ones near the Merced River, and the moderate-gradient ones in between. (In reality, many creeks have a reach/stage missing.) For each tributary he then projected the longitudinal profile of each reach to the center of the Merced River canyon. Elevation differences there between the projections indicate the amount of uplift between any two projections/stages. Because he used three hanging tributary streams, he could connect their points to establish longitudinal gradients for the Merced River.

There are a number of reasons why Matthes' method of uplift estimates, which was of paramount importance for interpreting the evolution of Yosemite Valley, does not work. The first is that like Lindgren and others he assumed a smooth, essentially constant, pre-uplift gradient for Sierran rivers. However, he went farther, hypothesizing that after each pulse of uplift the Merced River achieved a smooth profile before the next pulse. This was a totally theoretical approach, one in direct conflict with the field evidence for stepped profiles during the Cenozoic era.

### Correlation with Other Researchers and Areas

Matthes (p. 43) summarized Lindgren's uplift work in order to form a correlation of his rivers' gradients with those of the Merced River, and he arrived at a value of 60 feet per mile for the mountain-valley stage, a stage that he believed dated back to the Pliocene epoch, formed after a late-Miocene uplift. This correlation seemed plausible back then because Lindgren had concluded in 1911 (p. 64) that the Tertiary river gravels were Miocene (despite his p. 37 statement that the *oldest* gravels were probably Eocene). In more recent time Warren Yeend dated the earliest gravels as possibly deposited during the Paleocene epoch, but most of the gravels were deposited during the following epoch, the Eocene, and some in the Oligocene epoch. Thus Matthes' Merced River gradient of 60 feet per mile should have been mostly for Eocene time, not Pliocene, that is, for about 50 million years ago, not 5 million years ago. But according to Matthes, the Eocene Sierra Nevada was low, and his high mountain-valley gradient suggested otherwise.

Matthes also correlated the Merced River and San Joaquin River drainages, because bedrock in the former may have created a problem. Back in 1924 in a typed, double-spaced, 2½-page manuscript he stated that the key to his method of analysis:

> is found in the lower Merced Canyon which, although unglaciated, has a set of imperfect hanging valleys. This set, it was found by a careful reconstruction of the former profile of the Merced to which these valleys were graded, corresponds to the middle set of the Yosemite region. Unfortunately, a complete comparison between the Merced Canyon and the Yosemite region is not possible, owing to the fact that the metamorphic rocks through which the former is laid are much less resistant to erosion than the massive granite of the Yosemite region and are now so thoroughly dissected that no hanging valleys of the upper set remain preserved on them. Again, the comparison is unsatisfactory because the boundary between the granite and the metamorphic rocks happens to coincide with the extreme limit reached in the canyon by the Pleistocene ice. For these reasons it was deemed desirable to institute a comparison with the San Joaquin Canyon, which has passed through the same morphologic history but is carved in essentially massive granite through its length.
>
> Inspection of the San Joaquin Canyon in 1921 and 1923 has revealed the existence [sic] throughout its unglaciated lower part and as far up in its glaciated middle part as the Pleistocene gorge cutting has progressed, of two sets of hanging side valleys that correspong [sic] in elevation to the two higher sets found in the Yosemite region. These features are in fact splendidly developed to within a few miles of the foothills, and the analysis made in the Yosemite region accordingly stands fully confirmed.

But according to Paul Bateman and Clyde Wahrhaftig, Matthes' Miocene and Pliocene profiles of the San Joaquin River (1960, p. 41-44 and Plate 2) are in error, for they are not parallel to the ancient river profile as defined by the river canyon's basaltic flows. These indicate that the canyon was deeper than Matthes had supposed. Additionally, as I demonstrated back in Chapter 22's description of the granitic lands in the Rockhouse Basin area, tributaries have been hanging for a much longer period of time than Matthes imagined. Furthermore, they hang regardless of orientation to the Sierran crest (refer back to Map 48).

Finally, while Matthes proposed two post-Miocene erosion surfaces (mountain valley and canyon) in the Yosemite National Park area, Slemmons in his dissertation used Matthes' methods and arrived at five post-Miocene erosion surfaces. Slemmons' area was just to the north, in the Sonora Pass area. Is it possible that one area had three more uplifts than an adjacent area? More likely, the method of analysis is invalid.

### Three Crucial Points along the Lower Merced River

Had modern-day maps existed in Matthes' day, he never would have derived the profiles he did. Maps of those days were too poor to derive accurate stream profiles. Matthes in fact complained in his diary about the park map's inaccuracies. August 9, 1913: "Unable to locate myself on Marshall's map." August 16, 1913: "Made numerous corrections to Marshall's map [in the upper Merced River drainage between Little Yosemite Valley and Merced Lake]." Such errors in topography were bound to cause problems.

For example, the Merced River's mountain-valley profile below Crane Creek is about 17½ miles long, yet this paleoriver's smooth profile is controlled by *only three points*. This is far too few to establish that the river had a smooth profile. Do today's unglaciated rivers have smooth profiles? One can compare Matthes' profile with that of a similar length of today's unglaciated South Fork Merced River between its junction with the Merced River and the lowest glacial deposits in the Wawona area. Three similar points drawn on the 1:125,000 topographic map used by Matthes (his Plate 2) would indicate that today's South Fork does indeed have a smooth profile. However, examination of the more recent, more detailed Kinsley, El Portal, Buckingham Mountain, and Wawona 7.5' quadrangles demonstrates that the South Fork's profile is not smooth (refer back to Figure 37). An examination of unglaciated reaches of the Sierra Nevada's major rivers (e.g., the South Fork Kern River in Figure 38) yields the same conclusion. Therefore, since these rivers do not have smooth profiles today, one cannot assume that they had smooth profiles in the past. As was shown earlier, they did not.

Furthermore, the three points Matthes chose to fit the mountain-valley profile of the Merced River are of questionable utility. The creeklet of his "hanging vale east of Big Grizzly" (his Plate 27) is not a legitimate point because the vale's stream does not plunge south to the Merced River. There is an error on the 1:125,000 topographic map he used. The real course of this vernal creeklet is correctly shown on the Kinsley 7.5' quadrangle. This creeklet first flows south almost to the brink of the Merced River gorge, then turns abruptly west and skirts the north slope of a small knoll (x 3188) along its descent west to Miller Gulch. Today we can drive along Highway 140 and verify that the hanging vale does not exist (Photo 166). That highway opened back in 1926, but Matthes' diary shows that he never took it before completing Professional Paper 160, hence he never verified the existence—or nonexistence—of the hanging vale.

**Photo 166. Matthes' "hanging vale east of Big Grizzly." The early topographic map used by Matthes showed a stream descending from the shallowly concave slope in the sunlit area in the middle of the photo. However, all traces of a stream, including riparian vegetation, are absent.**

The remaining two points downstream from the El Portal (Crane Creek) area are extrapolations of the *upper* profiles of Saxon Gulch (Matthes' Figure 5) and Feliciana Creek (Matthes' Figure 6). In selecting these two creeks, he considered the varying rock types of the area's nearly vertical metamorphic strata. Therefore, streams lacking hanging valleys could be eliminated since they might be incising along weak strata. Matthes had considered that perhaps the hanging valleys were supported by bodies of extremely resistant rock, but he stated that for his selections the reverse is actually true: Feliciana Creek and Saxon Gulch are hanging not because of bedrock resistance but in spite of the unresistant nature of their floors. Both creeks more or less follow contacts, the former between two Triassic formations that are largely phyllite, the latter between Jurassic greenstone on the east and Triassic phyllite, according to Paul Bateman and Konrad Krauskopf. Whether the streams' floors are unresistant is debatable, since Matthes presented no detailed analysis of the actual rock outcrops along their floors.

His mountain-valley extrapolations for these two creeks are significantly different than ones derived from contours on the Feliciana Mountain 7.5' quadrangle. This is shown on Figure 44, which has Matthes' tributary profiles and

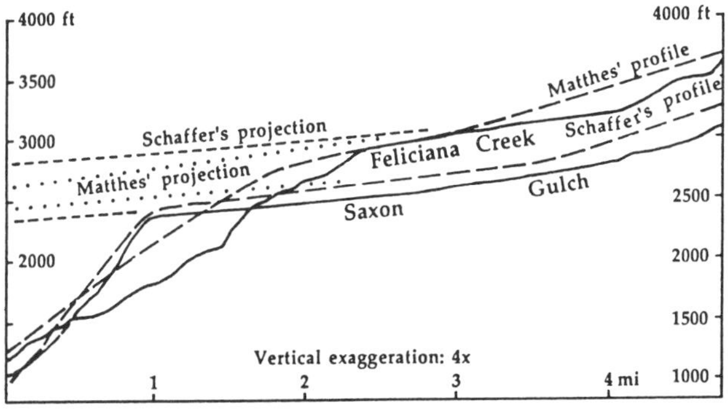


**Figure 44. Comparison of mountain-valley and present profiles of Feliciana Creek and Saxon Gulch, using old and new maps.**

extrapolations and mine ("Schaffer's"), derived from the quadrangle. This 1947 1:24,000 map is substantially more accurate than the error-prone map Matthes had to use. On the newer, large-scale map, the Feliciana Creek mountain-valley profile reaches the Merced River at 2850 feet, the Saxon Gulch profile at 2375 feet. This 475-foot elevation difference produces a river gradient of 102 feet per mile. Matthes' values are, respectively, 2650 feet, 2450 feet, and 43 feet per mile. At Indian Creek, in central Yosemite Valley (see Map 29 for location), Matthes determined a gradient of 126 feet per mile, a 2.93 increase over his Merced River mountain-valley gradient at Saxon Gulch. If one applies a similar increase to the more accurate value of 102 feet per mile, then the river's gradient at Indian Creek is 299 feet per mile. The mountain-valley elevation of the Indian Creek-Merced River confluence then changes from Matthes' value of 6350 feet to one of 11,600 feet. This is about 4000 feet higher than the valley's rim—an impossibly high value. In 1960 Frank Hudson provided similar criticisms. For example, he used Matthes' own data to show that one reach of the Merced River, when restored to its gradient during the mountain-valley stage, would have flowed *uphill.*

Matthes reconstructed the Merced River profiles for the broad-valley and canyon stages solely in Yosemite Valley. Only his mountain-valley stage extends the entire length through the valley and down to Saxon Gulch. Evidently there was no evidence by which he could reconstruct the two other stages, and therefore evidence of three uplifts and three periods of valley deepening along the lower, unglaciated part of the Merced River is nonexistent.

### Three Stages in Yosemite Valley

To reconstruct the broad-valley, mountain-valley, and canyon stages in Yosemite Valley, Matthes applied the same method he used for the lower Merced River. As I have just shown, Matthes' method does not achieve the desired results in the river's lower, unglaciated reaches, so his mountain-valley profile cannot be extrapolated up-canyon through Yosemite Valley. If the tributaries indeed had responded to three episodes of uplift, they should have responded to them with a short, steep-gradient reach of tributary closest to the valley, followed by a short, moderate-gradient reach, and then by a long, low-gradient reach. The ideal profile is for Bridalveil Creek (his Figure 10), which contains all three stages with proper lengths and gradients. But most of the Merced River's tributaries are missing one or two stages, as shown in Table 8. Furthermore, there is no order to the absences: the bedrock is almost entirely granite and granodiorite, not alternating layers of resistant and unresistant metamorphic bedrock. (Fireplace Creek flows across about 0.4 mile of diorite and Bridalveil Creek across about 0.5 mile, but the profile of neither appears to be affected—see the valley's geologic map by Frank Calkins *et al.*, 1985.)

Also disconcerting is Matthes' omission of other tributaries, particularly Ribbon Creek. Ribbon Creek gives rise to the valley's tallest waterfall, 1612-foot Ribbon Fall. This creek's omission at first seems to be a mystery, since except for its headwaters, it appears to display all three

**Table 8. Representation of Matthes' stages in Yosemite Valley's tributaries.**

| Tributary Stage: | Broad-Valley | Mountain-Valley | Canyon |
|---|---|---|---|
| Tamarack/Cascade cks | present | present | present |
| Fireplace Creek | **absent** | present | **absent** |
| Meadow Brook | present | **absent** | **absent** |
| Bridalveil Creek | present | present | present |
| Sentinel Creek | present | **absent** | **absent** |
| Yosemite Creek | present | **absent** | **absent** |
| Indian Creek | **absent** | present | **absent** |
| Royal Arch Creek | **absent** | **absent** | present |
| Illilouette Creek | **absent** | present | **absent** |

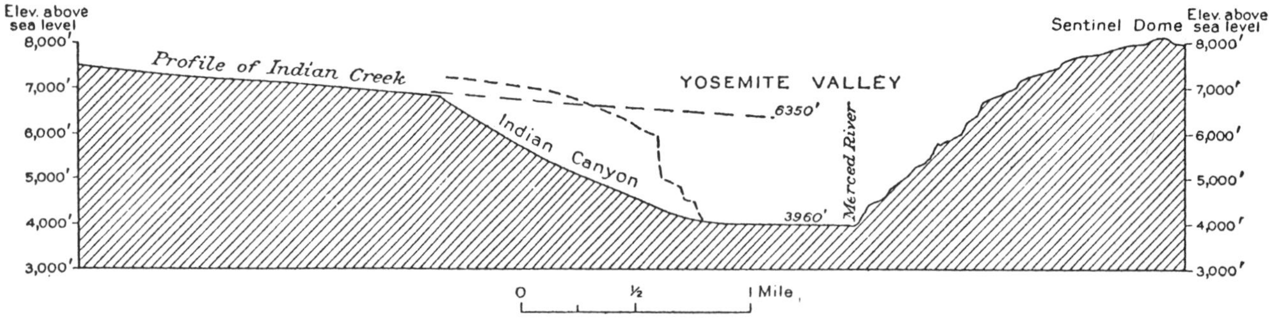


**Figure 45. Longitudinal profile of Indian Canyon Creek. The profile of the upper part is designated as the mountain-valley stage. The broad-valley stage is completely absent.**

stages with the proper lengths and gradients. However, when one reconstructs its broad-valley profile, which would have joined the Merced River opposite where Bridalveil Creek, one can see perhaps why Matthes omitted it. On his Plate 27 Bridalveil Creek's confluence with the Merced River is at 6700 feet. Ribbon Creek's confluence, in contrast, is at about 7250 feet, or about 550 feet higher. This is too large a discrepancy to reason away, and perhaps explains Ribbon Creek's omission. Matthes (p. 38) acknowledged the creek, saying that it is "closely accordant" with Bridalveil Creek—but this is not so. At the Bridalveil-Merced confluence, the difference in elevations between the broad-valley and canyon stages is 900 feet. If he had used Ribbon Creek for the broad-valley-stage control point, then the difference at the confluence would have been 1450 feet, a discordant value about 60% greater. One can only conclude that tributary profiles are not accurate indicators of uplifts.

Finally, the evolution of the Indian Canyon Creek drainage offers a specific example that demonstrates the failure of Matthes' analysis to adequately explain the origin of a side canyon in Yosemite Valley. In his Figure 8 (reproduced here as Figure 45), Matthes showed the profile of Indian Canyon Creek (Indian Creek in his work) and the extrapolation of it to the center of Yosemite Valley (elev. 6350 feet). This creek profile has been designated without explanation as the mountain-valley stage. It enigmatically lacks a broad-valley stage in the upper part of its profile, and so does its equally long, similar-profile tributary, Lehamite Creek. Into this creek profile has been incised Indian Canyon—but when? Matthes stated (p. 37) that the canyons of Indian Canyon and Lehamite creeks,

> though repeatedly buried under ice to a depth of fully 600 feet, have suffered but slight excavation, for the ice came into them through the passes to the north and northeast and spread out as a partly inert mass that was held back by the powerful current of the passing Yosemite Glacier, much as backwater is held stagnant in a tributary channel by a swollen river.

This situation is similar to the one in the Little Prather Meadow area, mentioned in Chapter 14. However, in his statement Matthes was referring only to the low-gradient upper part, which was buried *only* by pre-Tahoe (his Glacier Point and El Portal) glaciers. The much later Tahoe and Tioga (his Wisconsin) glaciers filled only the lower half of Indian Canyon, flowing laterally into it to bury and preserve it from erosion. If glaciers had been erosive, then the lower half of the canyon should be conspicuously different from the upper half, since it would have experienced more glacial erosion. Instead, the topography and creek gradient throughout this canyon are so similar that it is impossible to identify even a shred of glacial erosion or to establish any former ice surface. Since no glacial erosion occurred, today's Indian Canyon has changed very little since the onset of glaciation. This ineffectiveness leads one to conclude that the only time under Matthes' scheme that Indian Canyon could have been created was during the canyon stage, which is believed to have existed briefly late in the Pliocene epoch. If so, then the preglacial valley floor would have been at about the same elevation as it is today, about 4000 feet, since it has been unchanged by glaciers. However, his Plate 27, which shows the longitudinal profiles for the three stages of the Merced River through the valley, places the canyon-stage elevation at Indian Canyon at about 5200 feet—a full 1200 feet higher. The existence of Indian Canyon absolutely cannot be explained under Matthes' method of uplift determination.

## 9) Longitudinal profiles of "canyon bottom" flows and deposits

A number of geologists concluded that major late Cenozoic uplift occurred, basing their conclusion on the difference in gradients between modern rivers and their Miocene-age lava flows (or other Cenozoic deposits). Each flow presumably had advanced down the bedrock floor of a former river canyon, and the base of each flow represents the longitudinal profile of the river that the flow had buried. Surprisingly, river gravels at the bases of these flows generally are absent, whereas they should be ubiquitous. These geologists assumed a low, nearly constant gradient and a low elevation for the river at the time of burial. These geologists include Leslie Ransome in 1898, whose Stanislaus River drainage uplift was discussed earlier under Chapter 22 and was disproved by field evidence he himself (as well as Huber) had mapped and described. This drainage's uplift plus the Mokelumne's (i.e., from about Sonora Pass to Carson Pass) was determined by David Slemmons *et al.* in 1979, presumably using the same procedure, but since they did not identify their method, it cannot be addressed. They determined uplift for the Valley Springs Formation, the Relief Peak Formation, the Stanislaus Latite Group—Ransome's latite flow—and a 4-5 million-year-old erosional surface. Cordell Durrell in 1987 used the Lovejoy basalt flows for the Feather River Drainage.

However, this method came most into prominence when used by N. King Huber in 1981 in his Professional Paper 1187 on uplift in the San Joaquin River drainage. His later, shorter article in 1990 on uplift in the Tuolumne River drainage followed the same reasoning, but is less cited. I have not found a single post-1981 Sierran uplift article that does not tailor its plate-tectonic model or geophysical model to conform to the timing and magnitude of uplift as proposed in Huber's professional paper. Therefore, all the models are suspect if his uplift studies are invalid.

The arguments Ransome, Durrell, and Huber used are essentially the same, and they are approximately as follows. Figure 46 presents six stages for the Sierra Nevada. 1) Profile of a pre-volcanic river from its source to the edge of the range. The river's profile across the granitic terrain is stepped, and the only smooth profile is for the reach below about 1500 or 2000 feet elevation. 2) Over millions of years the river and its canyon are buried by volcanic deposits. Periodically the river incises deeply, only to be buried again. In these incoherent materials the river generally maintains or approaches a nearly constant profile. 3) During a major hiatus in volcanic activity, the river incises below its former highest level (long-dash line). 4) A major lava flow then descends the river's nearly constant profile. 5) Volcanism soon ceases, and the pre-volcanic profile gradually becomes exhumed. In the lower part of the profile, where the river canyon is low and wide, parts of the lava flow, which there has spread laterally, are preserved on the sides of the exhumed river canyon. 6) In the narrow middle and upper reaches of the river canyon, mass wasting along with stream and glacial erosion have removed all traces of the lava flow. Glaciation has occurred in the upper third of the drainage and has excavated two shallow basins. Elsewhere along the glaciated section, glaciers have not significantly eroded below the pre-volcanic profile (long-dash line just above the glaciated profile). Two remnants of the lava flow remain, both off to the side of the present river. The lower remnant is on pre-lava-flow sediments (e.g., Table Mountain of the Stanislaus River drainage); the upper one is on a granitic bench (e.g., Kennedy Table of the San Joaquin River drainage). From these two lava-flow remnants one can extrapolate their bases headward to the crest (long-dash line). One can also assume that at the time of the origin of the lava flow, the river had a low gradient, and so its profile can be extrapolated headward to the crest (short-dash line). The amount of uplift at the crest, then, is the elevation difference between these two extrapolated, slightly concave-headward profiles. *The derived uplift in this example is major, despite no uplift having taken place!* Actual uplift can vary from the total amount just derived to none at all—what my next chapter demonstrates.

### San Joaquin River drainage

For the San Joaquin River drainage Huber projected the gradients of the Great Valley's Eocene Ione Formation and the 10-million-year-old Kennedy Table trachyandesite flow uniformly eastward to the crest. He also used the lowest parts of local Pliocene lavas in the upper part of the drainage to reconstruct a Pliocene profile. He used the same method for the Tuolumne River drainage, discussed next. Because volcanic flows in that drainage merge lower down with those in the Stanislaus River drainage (mapped in part by Huber in 1983), that drainage also can be evaluated by his method (i.e., Ransome's).

There are at least 10 objections to Huber's most-cited 1981 study. First, as was mentioned under method 6, Great Valley sediments locally vary, and an uplift history derived by evidence in one locale is contradicted by that derived in another. Consequently, the dip of the Ione Formation in this locale cannot be used with any degree of certainty. Second, also mentioned earlier, Sierran rivers have had variable gradients across granitic bedrock, so the gradient of the trachyandesite flow cannot be projected constantly eastward. Third, while the flow is locally preserved on a granitic bench, elsewhere it rests on 11+ million-year-old alluvial (stream-transported) pumice (Chapter 10's Photo 29). Forty years earlier, Gordon MacDonald named this the Friant Formation, but it also has been called pre-volcanic alluvial deposits (Bateman and Busacca, 1982), alluvial deposits (Bateman *et al.*, 1982), and nonmarine sedimentary rocks (Bartow, 1985). Significantly, in the eastern Great Valley this alluvial pumice occurs only near the San Joaquin River, not near rivers to the north or south of it. Obviously, the lava flow locally lay atop the alluvial pumice, whose original thickness is unknown. If the base of that pumice still existed in the bedrock canyon, it would serve as a proxy and would indicate far less uplift, since the floor of the canyon 11+ million years ago definitely was below the base of the lava flow, perhaps by only 300 feet, or perhaps by well over 1000 feet. If the pumice originally had been 1400 feet thick in the vicinity of the Kennedy Table flow, then

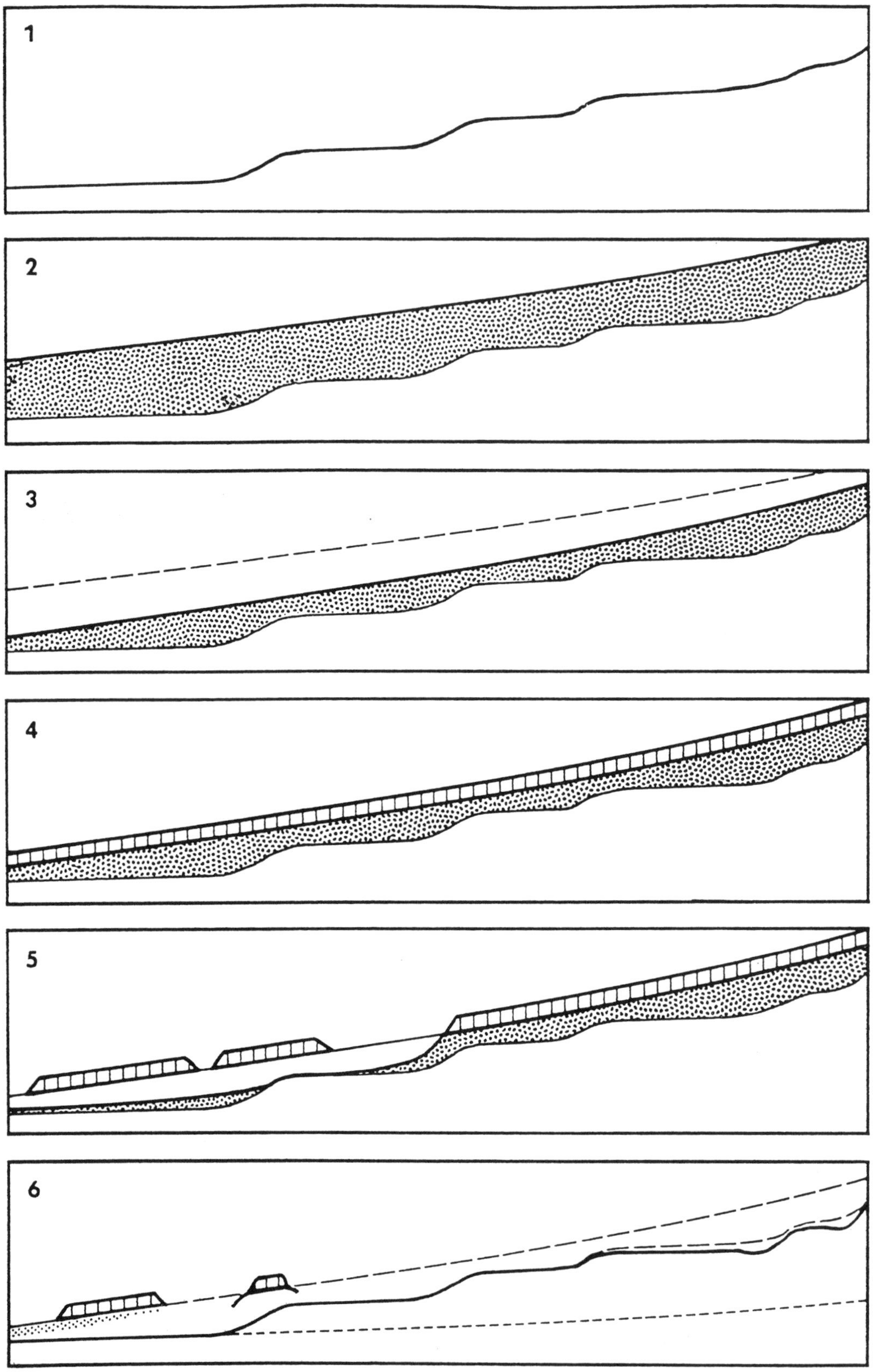


**Figure 46. Sequence of burial and exhumation of a Sierran drainage. Vertical exaggeration: 6x.**

Huber's 11+ million-year river profile would indicate absolutely no uplift. Huber went to some detail to describe this alluvial pumice, but surprisingly, when he determined uplift as gauged by the base of the flow, he totally ignored the underlying alluvial pumice.

Fourth, had post-flow uplift occurred, the river, through headward erosion of the loose underlying pumice, would have entrenched straight upslope across the flow, not along a course of giant meanders through *resistant* bedrock. Durrell in 1987 concluded that similar giant meanders in the Feather River drainage (which has a similar "table mountain" flow) must have been incised into the bedrock before the late Cenozoic. Huber does not explain how uplift can transform a river's course from relatively straight to excessively meandering—through extremely resistant bedrock, at that. Meanders the size of the lower San Joaquin's occur along the lower Mississippi River, where the gradient is nearly zero and the river is flowing across sediments, not bedrock.

Fifth, if the base of the flow really had solidified on the floor of the San Joaquin River canyon some ten million years ago, we can use it to reconstruct that canyon's width. The base of Kennedy Table, on the northwest side of the canyon, and the base of Squaw Leap, on the southeast side, are at an elevation of about 2200 feet. By extending a line of this elevation northwest from Kennedy Table and southeast from Squaw Leap, we find that the flat-floored canyon here was over 14 miles wide, being bounded on the northwest by the slopes of Ward Mountain and on the southeast by the slopes of Black Mountain. That is quite a wide floodplain across resistant bedrock. I am unaware of any similar size floodplain in a mountain canyon. The world's largest river, the Amazon, has a larger floodplain, but that is near the Atlantic Ocean, very far from any mountain canyon of the Andes. Conventional geomorphic analysis cannot explain the origin of this excessively broad San Joaquin River floodplain. Actually, its explanation is simple. The "floodplain" is merely a benchlands topography that is the result of former tropical processes operating over tens of millions of years. The meanders were incised at some time during the Mesozoic era when the range was relatively low and the sea was high, and so this part of the river was very close to sea level. By the time the alluvial pumice was deposited about 11+ million years ago, the San Joaquin River had incised down very close to its present level. The trachyandesite flow then buried the pumice and adjacent bedrock. The vast majority of the loose, alluvial pumice was eroded, exhuming the former bedrock landscape.

Sixth, the current bases of local upper-drainage lavas, which spilled into the inner canyon of the Middle Fork San Joaquin River, cannot be used to gauge the canyon's pre-glacial floor because glaciers would have removed the lower parts of these lavas. Just how far the local flows descended into the canyon cannot be determined, and that amount should not *arbitrarily* be set at the bases of today's glaciated lava flows.

Seventh, to justify a nearly constant gradient, Huber—like Matthes—assumed a lengthy, powerful river originating east of the crest. It would explain how the 11+ Ma pumice, derived from the northeastern part of the Mono Basin (and straddling the California-Nevada border), got into the San Joaquin drainage. However, in the catastrophic Long Valley eruption of 730,000 years ago, voluminous tephra (materials forcibly expelled in the eruption) accumulated as Bishop tuff in the drainage and down to the Great Valley despite absence of a trans-crest river. In the upper Middle Fork drainage near the confluence of Fish Creek are remnants of Bishop tuff on some higher slopes of the river gorge, and they range in elevation from 6900 feet to 8700 feet, as mapped in 1965 by N. King Huber and Dean Rinehart. In 1967 they suggested an original *minimum* thickness there of 1000 feet. The 11+ Ma eruption would have produced the same results. Field relations at Huber's crest-area site, a V-shaped volcanic remnant below and south of Deadman Pass (and just above the road to Devils Postpile), argue against the river. He interpreted the remnant as a volcanic infill of a V-shaped trans-crest canyon.

The most convincing of nine lines of evidence against it is that at the bottom of the V, river gravels should lie sandwiched between the basal flow and the bedrock, but are absent (Photo 167). Where former rivers existed in the Sierra Nevada, they left unmistakable evidence, such as the large, rounded boulders in Highway 108's cuts through conglomerate not far above Strawberry (Photo 168). At the bottom of the Deadman Pass V I could not find a single river boulder or cobble. Nor did Huber and Rinehart, or Roy Bailey later, propose or identify any. The V-shaped crest canyon is purely imaginary.

The other lines of evidence against a trans-Sierran river include the following. Second, the base and lower north slope are in contact with bedrock, which should show evidence of pre-burial weathering, but I found none. Third, there is no V-shaped remnant on the *east* side of the crest. Had there been a valley filled with lava, it would be continuous throughout the V-shaped crest canyon. Fourth, the east side has outcrops of granitic and metamorphic bedrock (see in particular the detailed map in Bailey *et al.*, 1990), and these constrain the bed of the supposed river, which would have flowed across this bedrock, to an elevation of about 9020+ feet on the east side, versus 8300 feet on the west side. A drop in elevation of 720+ feet over a distance of 1.2 miles is a high-gradient reach—a major step—and if the trans-Sierran river had existed, this step would undermine one of Huber's essential assumptions, a river of nearly constant, low gradient. Fifth, the angle at which the proposed V-shaped valley joins the existing main canyon is incorrect

**Photo 167. Vesicular basalt at base of hypothetical V-shaped crest canyon. Hidden immediately under the basalt is bedrock. No hint of river boulders or cobbles exists. Peter Bernard provides scale.**

for a former main drainage. It may be suitable for a former tributary, but not for the trunk canyon. Sixth, the topography is wrong. For example, compare this site with the one where the South Yuba River plunges northwest through its major "stream capture" gorge. There should be indentations in the adjacent bedrock to indicate the walls of a major gorge, but in the field I could not find a vestige of a gorge. Seventh, the present *upper* Middle Fork canyon maintains its width down-canyon past the beheading site. However, if the present upper Middle Fork was only a tributary to the major, eastern main fork, then from the confluence downward the canyon should be noticeably wider, reflecting the erosive work of a much more powerful river. Eighth, the east-side topography of the bedrock both north and south of this site indicates that prevolcanic streams flowed eastward into the Mono Basin. And ninth, Bailey's cross sections of the Mono Basin and the Long Valley caldera clearly show a drainage divide north of, at, and south of Deadman Pass.

The eighth major objection to Huber's 1981 study is that where uplift presumably was the greatest, in the Mt. Whitney area, Great Valley sediments to the west (atop the Sierra Nevada-Great Valley composite block) should dip steepest (as mentioned above under uplift method 6). Huber estimated about 11,300 feet of total uplift at the San Joaquin's 11,500-foot crest, so the Mt. Whitney area, being about 20% higher, should have experienced that much more. His uplift history indicates that about 650+ feet of uplift occurred there since 615,000 years ago, when Lake Corcoran (San Joaquin Valley, west of Mt. Whitney) began to drain and lake deposits were replaced by stream deposits (see Davis and Coplen). From the Mt. Whitney area the range is about 52 miles wide, so 615,000 years of uplift should have increased average gradients by 12.5+ feet per mile for the range and, presumably, the entire composite block. However, between Visalia (331 feet elevation), near the western edge of the range, and Guernsey (218 feet elevation), 21 miles to the west-southwest (along a line perpendicular to the Sierran crest), the gradient of the surface is only 5.4 feet per mile. Furthermore, this low gradient is more likely due to the construction of a broad alluvial fan extending outward from the range than to uplift of a flat surface. Had uplift occurred, the river should have incised into this area's sediments, which occurred farther north. There, massive Tertiary volcanism led to the burial of the northern Sierra Nevada topography under sediments and lahars. These were readily eroded and deposited in the Great Valley, but

**Photo 168. River conglomerate in a Highway 108 cut ¾ mile west above Strawberry.**

as the source-area sediments diminished, so too did deposition rate, and Quaternary rivers incised through the older sediments. Invoking uplift to account for that incision is unnecessary.

Ninth, Huber applied his results to Yosemite National Park, stating (1987, p. 28) that "total uplift in the vicinity of Mount Dana during late Cenozoic time to the present is estimated at about 11,000 ft." Therefore, before late Cenozoic time (the last 20 million years, according to Huber) the elevation at the crest would have been about 11,000 feet lower. Whereas geologists have assumed that major erosion greatly deepened Sierran canyons, none have suggested this for broad, nearly flat surfaces, such as the Dana Plateau, above Tioga Pass, which has been recognized as old and barely denuded. So, it would have been close to sea level. Since the range has been tilted westward as a block, then lands increasingly farther west would have been raised increasingly less by late Cenozoic uplift. A transect of the range, from crest to base and passing through Yosemite Valley, is about 65 miles wide, and if the crest was depressed 11,000 feet to its pre-uplift level, then depression at Half Dome would have been 7800 feet, and at the head of Merced Gorge, 6400 feet. This would result in putting some of Yosemite Valley's most resistant topography below sea level back about 20 million years ago. The summit of Liberty Cap would have been about 700 feet below; the summit of Mt. Broderick, 1100 feet below; the brink of Royal Arches, 2000 feet below; the summit of Middle Cathedral Rock, 400 feet below; and the summit of Turtleback Dome, 1100 feet below. Therefore, Huber's uplift estimate requires massive denudation of these features over the last 20 million years. For example, if you assume that Mt. Dana initially was about 4000 feet high (as Matthes had—see his p. 44), and if it was raised by 11,000 feet, then it should be about 15,000 feet high, instead of 13,000, so 2000 feet of denudation must have occurred. However, in the last 20 million years Yosemite Valley's resistant masses should have been denuded by about 10 to 20 feet, based on pedestal heights of ancient erratics, and on minimal amounts of denudation in river drainages that were evaluated in Chapters 21 and 22. Because these masses have been so resistant over time, they serve to constrain the maximum amount of Sierran uplift, as will be shown in the next chapter.

Tenth, as I mentioned under the first method to determine late Cenozoic uplift, Denis Marchand and Alan Allwardt concluded that the China Hat surface is not a pediment, but rather a remnant of a late Laguna gravely fan deposit. It is risky enough to extrapolate a pediment uniformly eastward, as Hudson did, but it is fatal to extrapolate a remnant of an alluvial fan. Surprisingly, Huber, also in 1981, recognized the China Hat surface for what it was, but nevertheless used it to determine post-late Pliocene uplift. This new date carries with it the inference that tremendous erosion has occurred during the Pleistocene.

Finally, Huber's method, like Matthes' and Lindgren's, produces disparate results, as is shown in four consecutive drainages, listed north to south:

Stanislaus: His method indicates *moderate subsidence* between about 23 and 9 million years ago followed by major uplift. This uplift history is based on dated volcanic rocks in the Dardanelles Cone quadrangle, which he mapped.

Tuolumne: no change in the river profile between about 30 to 10 million years ago, then major incision between 10 million years ago and the onset of glaciation. Glacial erosion was negligible.

Merced: Huber changed Matthes' three-stage uplift pattern to the San Joaquin's pattern. He accepted Matthes' conclusion that glaciers performed half of the erosion in Yosemite Valley, but this contradicts the Tuolumne's negligible glacial erosion, despite glaciers there being much larger.

San Joaquin: this drainage experienced an exponentially increasing rate of uplift from at least 25 million years ago, and the canyon increasingly deepened in response to it.

### Tuolumne River drainage

As was mentioned immediately above, the uplift and glacial histories for this drainage conflict with those of the Merced and San Joaquin river drainages. This indicates serious problems for Huber's uplift method, so it doesn't warrant additional evaluation. It is as problem-prone as the San Joaquin, but I will raise only two points. The first is about Huber's proposed giant, *singular* 10-million-year-old lahar, whose principal remnants supposedly are the volcanic sediments of Rancheria Mountain, standing just north of the Grand Canyon of the Tuolumne River (Hetch Hetchy Reservoir 15' quadrangle; geologic map by Ronald Kistler, 1973). However, Huber's Figure 3 photograph (one taken by Henry Turner, 1902) clearly shows a major contact—as judged by a continuous horizontal layer of vegetation—about half-way up the deposit and another less-continuous contact closer to the summit. Therefore, there are at least three units, and excavation may reveal many more. As elsewhere, these could represent a long period of deposition and erosion. The lahars lie atop a basal stratum of gravels, which both Huber and Turner interpreted as those of the ancestral Tuolumne River.

Turner's photograph moreover shows the sediments as occupying a prominent, broad, V-shaped notch through which the ancestral Tuolumne River presumably flowed west before being buried by the sediments. Turner notes that these are mostly volcanic pebbles, but that "there are numerous pebbles of slate and metamorphic lavas such as make up the mass of Mt. Dana, and one pebble was found of epidotiferous sandstone, precisely like the rocks of the summit of Dana." Given that Mt. Dana's rocks are quite unique and limited in scope, the Tuolumne River at one time must have flowed from the Mt. Dana area to the gravels. What is to be questioned is: Did the ancestral Tuolumne River flow through the V-shaped notch, or did it merely flow alongside the lower, granitic slopes of Rancheria Mountain? In either case, river sediments would have been deposited. The size of the sediments provides evidence for a solution to this question: *they are pebbles*. The climate 10 million years ago was quite modern, and so there is no basis for proposing that intense chemical weathering comminuted large clasts to pebbles. Myriad *un*glaciated California streams today, including ones in the Sierra Nevada, possess abundant boulders in their beds. These are absent here. On the other hand, if the Tuolumne River meandered across a broad floodplain that was constructed atop lahars, it would leave finer sediments in pockets of slow water. The pebbles are located in a shallow recess of granitic Piute Creek canyon, a locality where slow water would be expected. As with the hypothesized V-shaped canyon below Deadman Pass, this one appears hypothetical for similar reasons: first, it lacks a conspicuous V-shaped notch on the opposite side of Rancheria Mountain; second, the lowest volcanic deposits on that side are some 660 feet lower, indicating an overly steep (i.e., stepped) Miocene-river gradient that contradicts his constant-gradient assumption; third, gravels are absent on that side; and fourth, granitic outcrops exist between the highest and lowest volcanic deposits on that side, proving that a bedrock ridge, not a V-shaped canyon, existed.

The second point is that as Ransome in his 1898 USGS Bulletin 89 ignored the earlier, underlying pre-volcanic gravels in the Stanislaus River drainage, so did Huber do the same for gravels along the Highway 120 corridor between Groveland and the park. Along the rim in the vicinity not far northwest of the ranger station at Buck Meadows there are quartz gravels of the middle Cenozoic Tuolumne River that contained a rhyolite boulder dated in 1964 at 28.5 million years by Brent Dalrymple. Huber said that the gravels at the base of the lahar define the channel of the Tuolumne River when it was buried by the lahar 10 million years ago. But this is impossible, since the older gravels dated by Dalrymple, where they occur with the volcanic deposits, underlie them, such as at the site shown in Photo 169. Obviously, the Tuolumne River some 28.5 million years ago had to be lower than the base

**Photo 169. Quarry of prevolcanic gravels north of Highway 120. These lie north of and below a volcanic ridge. Quarry is in the northwest quarter of Section 28, T. 1 S., R. 17 E. (Jawbone Ridge 7.5' quadrangle). Peter Bernard provides scale.**

of the lahar to deposit those older gravels beneath it. The original thickness of these older gravels is unknown, since they have been mostly eroded away, so we don't know if the depth of the Tuolumne River canyon in the vicinity of Rancheria Mountain was about 1400 feet, as Huber concluded, or if it was closer to today's depth of 4300 feet. Because the Miocene Stanislaus and Tuolumne rivers and their volcanic and prevolcanic sediments merged close to the base of the range, they shared the same sedimentary—and presumably, uplift—history. Therefore, since canyon-bottom evidence in the upper Stanislaus River drainage indicates that it was a high-relief, high-elevation landscape at least 20 million years ago (refer back to Chapter 22's Map 53 and Figure 42), the same should apply to the upper Tuolumne River drainage.

### Stanislaus River drainage

Long before Huber, Leslie Ransome used the Table Mountain latite flow in the Stanislaus River drainage to time and quantify the amount of Sierran uplift. Both assumed that before late Cenozoic uplift every Sierran river had an essentially uniform profile throughout its length and that each had a low gradient. Then about 9.2 million years ago, lavas making up the Table Mountain latite flow advanced along gravels atop the bedrock floor of the Stanislaus River canyon and then solidified. With uplift, the fable goes, the rivers incised deeply leaving the old river gravels and overlying lava flows high above the bottom of today's canyons. But as I stated back in Chapter 22, both Ransome and Huber correctly mapped a thick sequence of volcanic sediments below the base of the lava. (See that chapter's Figure 41: Cross section through the Dardanelles along the county-line boundary.) The lava had advanced across volcanic sediments, not along the bedrock floor of a canyon, as explained for this chapter's Figure 46: Sequence of burial and exhumation of a Sierran drainage.

With the aid of Figure 47 we can now analyze Sierran uplift, based on the sequence of volcanic sediments in this drainage and on the stepped topography that exists in the granitic lands of this range. The left side shows the evolution of stepped topography as Matthes envisioned it, and it is required for Huber's and Ransome's uplift studies to be valid. Erosion following minor uplift creates the broad-valley surface, BV. More uplift causes river

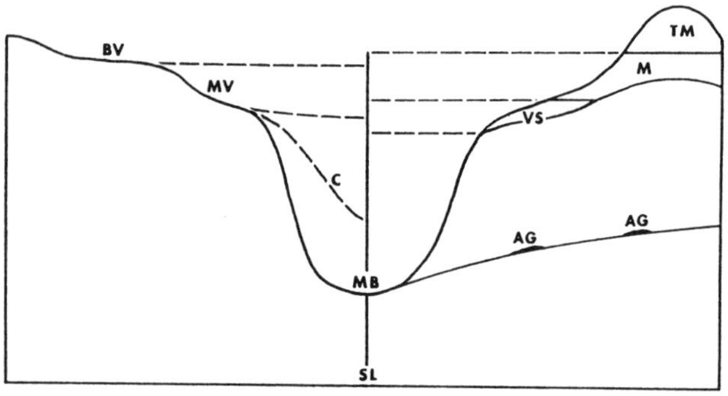


**Figure 47. Idealized cross section of a typical west-side Sierran canyon. BV = broad-valley surface; MV = mountain-valley surface; C = canyon surface; MB = modern base; SL = sea level; TM = Table Mountain latite flow; M = Mehrten Formation; VS = Valley Springs Formation; AG = auriferous gravels. See text for explanation.**

incision into this surface, and erosion creates a moderately sloped mountain-valley surface, MV. Later uplift causes river incision into this second surface and creates a steep sloped canyon surface, C. These three uplift-erosion cycles are shown for Yosemite Valley by Matthes in his Professional Paper 160. There, as here, glaciers then are believed to have widened the inner gorge and eroded it down to its modern base, MB. Matthes then projected the three erosion surfaces (dashed lines) to the center. The vertical distance between the modern base and any projected surface represents the amount of uplift that has occurred since that surface was formed. The broad-valley surface would have been raised the most, the mountain-valley surface less, and the floor of the canyon surface the least.

If sediments or flows had been left on these surfaces, the oldest ones, such as the mostly Eocene-age auriferous gravels, would have been left only on the broad-valley surface, since the mountain-valley surface developed later. Intermediate-age ones, such as the San Joaquin's Table Mountain flow, would have been left on the mountain-valley surface, and youngest ones on the bottom of the canyon surface. Then the base of each group of sediments, projected to the center, would produce the same uplift results that Matthes suggested: the oldest sediments would have been raised the most, and the intermediate ones less. The youngest ones generally would not have been preserved, due to glaciation in the upper elevations and to mass wasting on overly steep slopes in the lower ones.

But typical Sierran stratigraphy, shown on the right side of Figure 47, indicates *a reverse order in age*. As in accord with Nicholaus Steno's universally accepted principle of superposition, formulated back in 1669, the lowest sediments are the oldest and the highest are the youngest. The 20 to 30 million-year-old Valley Springs Formation (VM), is overlain by the 10 to 20 million-year-old Mehrten Formation (M), which in turn is capped by the 9.2 million-year-old Table Mountain latite flow (TM). Projection of the base of each layer to the center yields the impossible result that the oldest layer has experienced the least amount of uplift, the youngest the most. This is the situation in the Stanislaus River drainage—and the situation is similar elsewhere. The lowest layer is the 30 to 60 million-year-old Auriferous (gold-bearing) Gravels (AG), and remnants of it exist on the floor of shallow Spring Creek canyon, a tributary of the South Yuba River canyon. Because Spring Creek has not incised below these remnants, along those remnants it has the same profile that it had when it was buried about 50 million years ago. And because the modern tributary is nearly accordant with the modern river, the river has experienced little or no net incision, and by implication, *no uplift*. Therefore, Ransome's and Huber's method, applied to *all* the sediments, actually argues against late Cenozoic uplift. Ransome and Huber were able to perceive uplift only because they *ignored* the sediments underlying the prominent lava flows.

### Feather River drainage

The late Cenozoic uplift, as analyzed in 1987 by Cordell Durrell, is a moot point, because like Ransome and Huber, he too ignored the older sediments underlying the prominent Lovejoy basalt flows of this drainage. Unlike Huber, who claimed that the San Joaquin's large meanders incised into resistant bedrock were created in response to uplift, Durrell recognized that the giant meander along the lower part of the North Fork Feather River (Map 56, lower left) must have been created before uplift. He recognized that if the trans-Sierran North and Middle forks had established their courses in response to uplift, then these two forks (and others like them, such as the San Joaquin River) would have flowed directly down the western slope of the range, which they did not. Note on the map that the upper part of the giant meander flows east toward the crest, which is certainly not the course a river would take in response to westward tilting.

I question two of Durrell's points: that there was major late Cenozoic uplift, and that today's North, Middle, and South forks meandered across what he calls the Old Erosion Surface, the gentle uplands between the river canyons. Durrell's uplift determination is highly uncertain, varying from 1500 to 5000 feet, depending on evidence found at different localities within and east of the Feather River drainage. He believes that not only has the Sierra Nevada been uplifted but also the western Great Basin—the latter, however, experiencing less uplift than the Sierran crest. Unfortunately, Durrell used Axelrod's numerous paleoaltitude estimates (uplift method 3) of the Sierra Nevada and the Great Basin. Axelrod's plant misidentifications indicated a very low range and a low

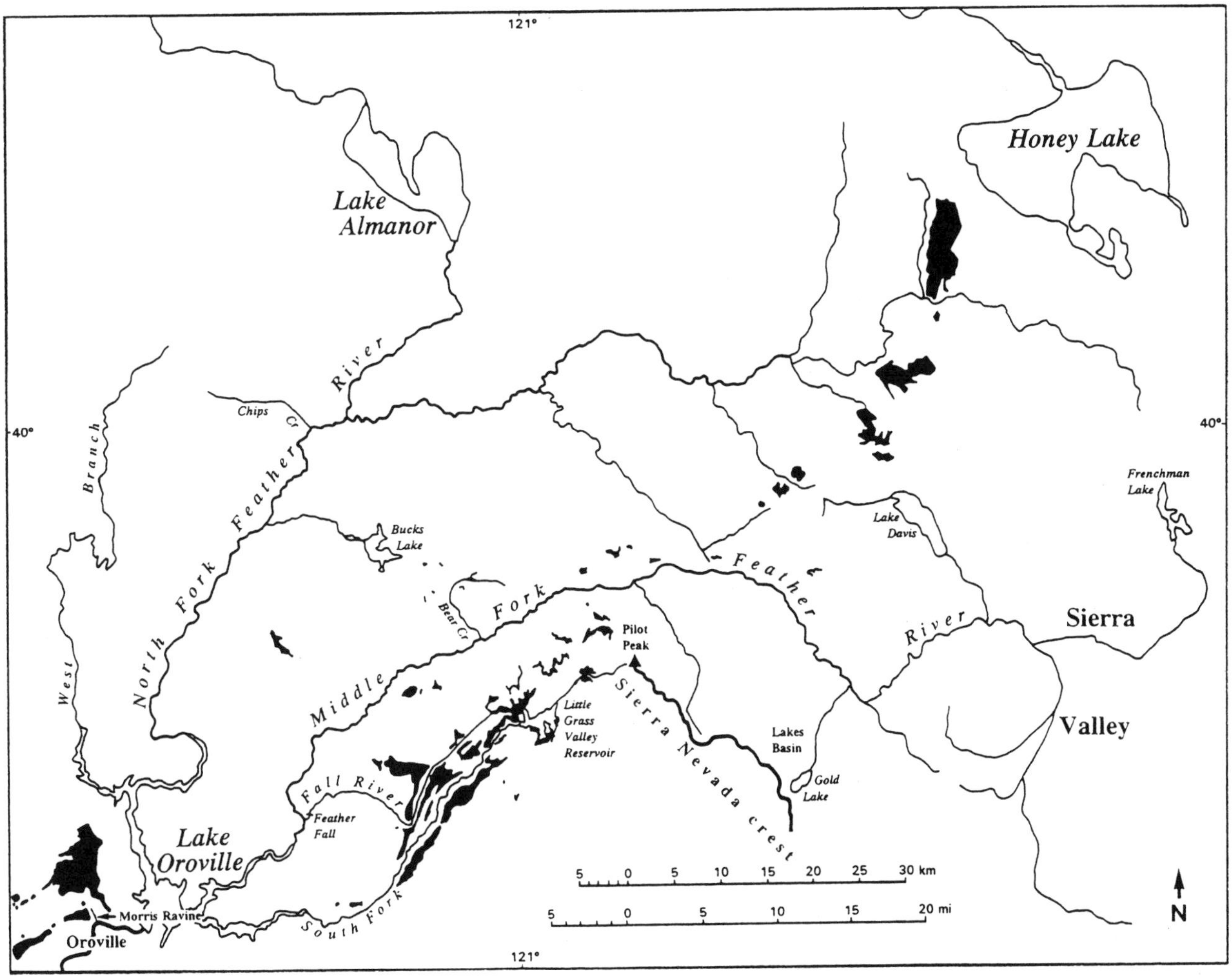


**Map 56. Feather River drainage, including remnants of Lovejoy basalt.**

Great Basin until Late Cenozoic uplift. Axelrod, in turn, used Durrell's geology to substantiate his uplift views. This circular reasoning from two mutually reinforcing professors at U.C., Davis, spread into the broader question of Sierran uplift relative to geophysics and plate tectonics, with even Warren Hamilton in 1989 citing Axelrod's now discredited paleoaltitudes as proof that the Great Basin lands have been low until relatively recent geologic times.

The second point is that today's North, Middle, and South forks flowed across the Old Erosion Surface (gentle uplands between the river canyons) instead of through deep, ancient canyons. If indeed the range had been low before late Cenozoic time, as Durrell and Axelrod would have us believe, then the gentle uplands of the middle and lower slopes west of the crest were barely above sea level. Consequently, it would be logical to assume that the forks of the Feather River would have flowed across the Old Erosion Surface. However, back in 1911 Lindgren proved that the South Yuba River and the South Fork American River both had deep canyons early in the Cenozoic, so one must question why the Feather River's prevolcanic canyons were shallow. Today the Feather River's North and Middle fork canyons have impressively greater relief than do today's South Yuba and South Fork American river canyons. How did they get so deep so fast, especially since Feather River crest uplift was less than crest uplift to the south? One could argue that because the Feather River's North and Middle forks have more discharge, they were capable of greater erosion. However, the South Fork, which originates immediately west of the Sierran crest (on the lower slopes of Pilot Peak, Quincy 15' quadrangle), has a short, narrow drainage area and a small discharge to match, and yet the relief of its lower canyon matches those of the lower canyons of the much larger North, Middle, and South Yuba rivers. Given that, as discussed earlier, the rivers from the Stanislaus north through the Yuba have demonstrably deep prevolcanic canyons that were incised far below the broad uplands, one can conclude that the prevolcanic Feather River's North, Middle, and South forks likewise incised deeply below the Old Erosion Surface. There is no *a priori* reason—nor any field evidence—to assume they had a unique history.

The undeniable evidence for existence of prevolcanic gorges lies with the history of the Lovejoy basalt, whose age, according to Edward Helley and David Harwood, lies between 22.2 and 23.8 million years. At least nine flows, ranging in thickness of about 10 to 50 feet, originated near the present Honey Lake and flowed southwest more or less directly downslope before any faulting had occurred. The path the basalt flows took was mostly down the South Fork and Middle Fork drainages (Map 56). In the former, where the flows are best preserved, the course was a generally narrow one for the Sierra Nevada, about 4 miles or less in width except for one westward diversion, which today is a ridge, Lava Top. The Middle Fork drainage may have had a similar breadth of flows, but today only isolated remnants exist, so the former areal extent is unknown. Did the flows of the Lovejoy basalt descend along shallow river canyons etched across the Old Erosion Surface, or did they descend through deeper, essentially modern canyons? As in earlier analyses the contacts between various flows and sediments with the granitic and metamorphic bedrock provide the answer. Detailed 1973, 1976, and 1981 maps by Anna Hietanen show that the Lovejoy basalt and underlying sediments of the Eocene-epoch Ione Formation buried today's deep canyons, which must have existed, like the South Yuba River canyon, at least as far back as the Eocene epoch. Had the North Fork and Middle Fork canyons formed later on, then none of the basalt or the Ione Formation would occur on slopes within today's deep canyons. These would occur only on the Old Erosion Surface, which is not the case.

In addition to these nine methods of determining late Cenozoic uplift, there are three more which I've devised. All have a common theme: major uplift and/or massive erosion is perceived in other mountain ranges, so the Sierra Nevada should behave similarly.

## 10) Major late Cenozoic uplift and erosion in the Teton Range

The glacially eroded Teton Range, which is believed to have been uplifted during the late Cenozoic, is a close analog to the Sierra Nevada, so the latter must have a similar history. But the former's recent geologic map by Love *et al.* clearly shows that in the deep canyons of South Boone Creek and Conant Creek, in northwestern Grand Teton National Park, Eocene and Paleocene deposits extend from high ridges down to the canyon floors. In other words, these ancient deposits buried deep canyons. The relief back then was essentially the same as today's, despite glaciation. Furthermore, remnants of the Pliocene Huckleberry Ridge Tuff on pre-Mesozoic bedrock near the crest and on a slope of Owl Canyon demonstrate that crest cirques and the canyon had developed their "glacial topography" before major glaciation occurred. Therefore, the Teton Range's history must be like the Sierra Nevada: late Cretaceous-Paleogene (i.e., Laramide) uplift and geomorphic processes operating in warmer, moister climates, followed much later by late Cenozoic down-faulting of the (Jackson Lake) lands along the east base of the range.

## 11) Major late Cenozoic erosion in the Santa Lucia Range

As I've stated several times earlier, the last verifiable major Sierran uplift occurred during late in the Cretaceous period. This early uplift implies that the rate of denudation during most of the Cenozoic has been very slow, given the relief and elevation of today's range. However, some of California's largely granitic ranges appear to be eroding very fast. A prime example is the Santa Lucia Range, along central California's coast. Today the relief of its major canyons is about 3000 to 4000 feet, comparable to that of the Sierra's major canyons. But since this range has been raised above sea level only during the last 5 million years, its major canyons seem to have an average rate of incision of about 600 to 800 feet per million years. This is far greater than what I propose for the Sierra Nevada.

However, that range, like the Sierra Nevada, developed most of its topography late in the Cretaceous period. It then subsided and was buried under marine sediments that accumulated from the Paleocene through the Miocene epochs. These sediments buried the range's pre-Paleocene topography. (Karen Loomis and James Ingle, Jr. presented a similar history for other parts of the Coast Ranges). In response to uplift, most of the sediments were eroded, thereby exposing the earlier topography, although some sediments remain by which we can judge previous relief. The highest peak in the range is Junipero Serra Peak, at 5910 feet, and north of it Victor Seiders *et al.* mapped Paleocene sediments lying near the bottom of Santa Lucia Creek canyon. The actual base of these sediments has been eroded, but nevertheless, their lowest exposure today is at about 2270 feet, indicating a *minimum* relief before Paleocene deposition of about 3640 feet—similar to that of major Sierran canyons. (The relief would have been more since the original base of the Paleocene sediments is unknown, and the summit of Junipero Serra Peak has undergone some erosion.) Therefore, the major canyons of the Santa Lucia Range had been created prior to uplift, and they had an incision rate very similar to that of the Sierra Nevada.

## 12) Perceived late Cenozoic uplift in the Peninsular Ranges

Just as geologic evidence has been used to confirm late Cenozoic Sierran uplift, so have plate-tectonic and/or geophysical models, especially the often cited ones by Edward Hay in 1976 and by Thomas Crough and George Thompson in 1977. I too proposed a plate-tectonic model, in 1987, which was criticized by Huber—rightly so, but entirely for the wrong reasons. Consequently, I proposed a testable hypothesis, and using field evidence was able to disprove my model, so in 1989 I retracted it. Unfortunately, others have not proposed testable hypotheses for their models.

None address the Peninsular Ranges, which along with the Sierra Nevada were once part of a much larger range that ran along the western coast of North America. The Santa Lucia Range also was a part of that extensive range until it was dislodged westward, sinking below sea level as the lithosphere (the earth's crust and its associated, underlying upper mantle) was extended. Before the breakup, all three shared a common *late Cretaceous* uplift history. In the northern part of the Peninsular Ranges is a fault-block range, the San Jacinto Mountains, which has an ancient, unglaciated highland much like those of the unglaciated southern Sierra Nevada. Like the Sierra Nevada, this block is believed to have experienced major uplift in the last few million (perhaps some minor uplift has occurred at its north end, where the range abuts against the Transverse Ranges). Also as with the Sierra Nevada, geoscientists have ignored the indisputable fact that the lowlands surrounding it, which have bedrock well below sea level, like the Owens Valley have *dropped.* Subsidence is proved; uplift is not proved.

In 1997 the team of Wolf, Farley, and Silver could find no evidence of major uplift of the San Jacinto Mountains during the last 30 or so million years. When did it occur? Peter George and Roy Dokka identified two major uplifts, one between 98 and 85 million years ago and the other between 80 and 74 million years ago, the latter dates being essentially contemporary with uplift dates in the southern Sierra Nevada. Moreover, the authors suggest that widespread unconformities *throughout Baja California* (the southern part of the Peninsular Ranges) indicate that the entire Peninsular Ranges batholith was uplifted at this latter time. This would include 10,154-foot Picacho del Diablo, in Parque Nacional Sierra San Pedro Martir. In late Cenozoic time the western edge of Baja California became a passive continental margin, according to Jon Spencer and William Normark, and passive continental margins, unlike active ones, lack the compressional forces to bring uplift. Furthermore, as D. S. Gorsline and L. S.-Y. Teng observed (as had 1890s geologists), a former part of the Peninsular Ranges, the California Continental Borderland, has subsided mostly below sea level (southern California and northern Baja "fell into the sea"). The Santa Lucia Range had done this much earlier, by the Paleocene epoch, before compressional forces began to resurrect it during the Pliocene epoch. If anything, the lands of southern California's Peninsular Ranges should be undergoing subsidence, since the lithosphere of which they are composed is experiencing more extensional forces than compressional forces. Like the Peninsular Ranges, the Sierra Nevada is not in a compressional environment. Indeed, its faulted southern end, the Tehachapi Mountains, appears to be going down, not up. There is no reason to assume, nor any plate-tectonic motions to justify, major uplift in either the Sierra Nevada or the Peninsular Ranges. (The motion between the Pacific plate and the North American plate along the San Andreas Fault System is mostly lateral slip, but is also partly compressional, and the compressional forces created coastal ranges such as the Santa Lucia.) Those who use geophysical models to demonstrate late Cenozoic Sierran uplift must also address contemporary uplift in the Peninsular Ranges. Both ranges were high late in the Cretaceous period. There is no evidence of later major erosion or major subsidence for either range, so there is no reason to assume either was reduced to low elevations. Late Cenozoic uplift for each is an unnecessary hypothesis.

## Reflections on late Cenozoic Sierran uplift

The older we get, the more we should appreciate history. In doing research for this book, I read almost every book and article that had been written on Sierran uplift and glaciation. Apparently the younger scientists who are currently doing this kind of research in the Sierra Nevada, have not, for in the 1980s and '90s they are resurrecting the same issues that were raised in the 1880s and '90s. What has changed in a century? To find out, one can read Leslie Ransome's 1896 lengthy paper, "The Great Valley of California: a criticism of the theory of isostasy," which discusses views by Joseph Le Conte, George Becker, and others, and also presents a good picture of the state of 1890s geological knowledge in California. In the 1890s, wells had been drilled through the Great Valley's sediments down to at least 3000 feet, so people knew that the valley's sediments were thick. Also back then, most of the Sierra Nevada had been mapped at least on a rudimentary level, so the general field evidence we know today was known back then. What happened in the passing 100 years are several developments. First, in the early 1900s geologists finally were able to accurately date rocks that had radioactive isotopes. Today, hundreds of dates exist for granitic and volcanic rocks in the Sierra Nevada. Second, in the 1960s plate tectonics ("continental drift") came of age, and since then has become very useful in showing how landscapes rise and fall. Back in the 1890s, by the way, hints of plate tectonics had been suggested with regard to Sierran uplift. Finally, seismology, which existed back then, has made great progress, allowing us to reconstruct with fair confidence the earth's lithosphere (crust and upper mantle), the asthenosphere and lower mantle, and the inner and outer cores.

But despite great advancements in the geosciences, none of them have helped with regard to Sierran uplift, for it, like glacial erosion, has always been tied to faulty assumptions and/or to nonexistent or misinterpreted field evidence. A timely example of geoscientists' inability to quantify Sierran uplift is from the Southern Sierra Nevada Continental Dynamics Working Group, mentioned at the start of Chapter 20. This group included some of the country's foremost geologists from some of the country's foremost schools (Cal Tech, Stanford, Princeton, Duke, etc.) and from the United States Geological Survey. The geoscientists set off explosions across the San Joaquin Valley, across the southern Sierra Nevada, and across the western part of the Great Basin, recording the seismic waves reflected off various layers in the lithosphere. With results in hand, Brian Wernicke in the January 12, 1996 issue of *Science* (and a nationwide press release of the same date carried in many newspapers), suggested that in the late Cenozoic the Sierran crest had *subsided* by about 13,000 feet during the last 20 million years. Remember that N. King Huber had suggested about 11,000 feet of *uplift* for that period. Not everyone in the group agreed. Two other researchers, Moritz Fliedner and Stanley Ruppert, on the same team and using the same data, produced a model in the April 1996 issue of *Geology* that conformed with major late Cenozoic uplift according to N. King Huber. To reach such diametrically opposed propositions of major uplift and major subsidence, the two models require *a mere 5%* difference in their estimates of varying densities and temperatures along a 50-mile descent through the Sierran lithosphere (Sierran crust and underlying mantle). And both models ridiculously assume invariant densities and temperatures for the 20 million years, despite major plate-tectonic changes. Talk about unreality! A century later, geoscientists still are trying to prove late Cenozoic uplift—by increasingly sophisticated methods—while ignoring the abundant, tangible, verifiable field evidence that argues against it. "The Blind Men and the Elephant" are alive and well in the field of alpine geomorphology.

# 24

# Quantifying Late Cenozoic Uplift

## Longitudinal profiles of canyon-bottom flows and deposits

Ransome's and Huber's method can be used to quantify uplift—when applied *properly*. The problem with their conclusions is that they did not follow the two essential assumptions they themselves had proposed. The first is that past rivers had nearly constant gradients. The second is that the base of each geologist's chosen lava flow solidified along the bottom of a river canyon. Notice that the subheading just above is almost the same as in the previous chapter's method 9, where I enclosed *canyon bottom* with quotes. They mean that the flows really were *not* along canyon bottoms, geologists' claims to the contrary. As I have mentioned earlier, all Sierran rivers, glaciated or unglaciated, now have—and have had in the past—irregular, or stepped, profiles. However, as I have also mentioned, their lower reaches, below about 1500 to 2000 feet elevation, today have profiles with nearly constant gradients (with minor irregularities), and probably have had similar profiles in the past. Therefore, to use Ransome's and Huber's method, we have to restrict our analysis to a reach of river at or below 2000 feet elevation that has a lava flow—or suitable sediments—which would have rested on the river canyon's floor.

No such reach of river exists, but one stream comes close enough for analysis: Spring Creek, a tributary of the South Yuba River. This creek and its river were discussed with regard the previous chapter's Figure 47. In it, for the sake of simplicity I smoothed out the profile of Spring Creek. The creek's profile is more accurately portrayed in Figure 48. Notice that the vertical exaggeration is four, so the "steep" reach on the left part of the figure actually is only one-fourth as steep, that is, a 5° gradient. Along the creek's low-gradient reach, its bed is very young, since before hydraulic mining began in the 1850s, the creek along this reach flowed across the auriferous gravels, not across the bedrock exposed by hydraulic mining. These gravels mostly are of Eocene age, that is, they were deposited roughly 50 million years ago. Consequently the part of Spring Creek's bed above the "steep" reach, which has just been exposed in historic time, has the same stream profile as when this creek was buried about 50 million years ago.

But what about the "steep" reach? There were no historic auriferous gravels along it. Perhaps that reach, in response to Sierran uplift, has begun rapid incision to match that of the South Yuba River. Let's assume this was so. Then before uplift, the South Yuba River at its junction with Spring Creek would have been higher. Like Matthes, we can project the low-gradient reach of Spring Creek over to the river, as I have done with a dashed line. This indicates about 400 feet of river incision has occurred, which implies 400 feet of uplift over the last 50 million years at this locality. The Sierra Nevada is about 57 miles wide in this drainage, and the Spring Creek-South Fork confluence is located about 25 miles up from the base of the range, or about 44% of the distance from

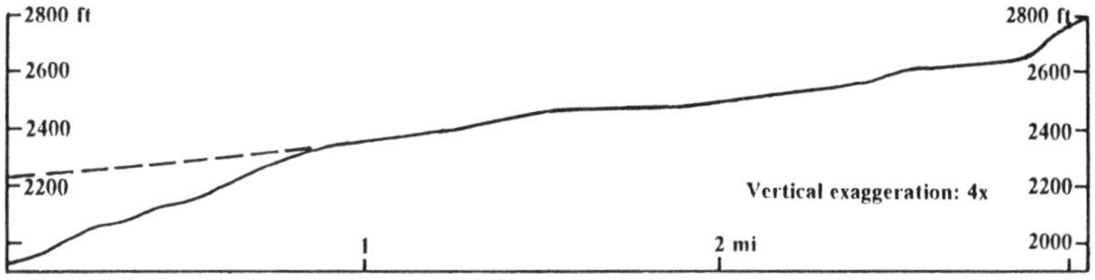


**Figure 48. Profile of Spring Creek, a South Yuba River tributary.**

the base to the crest. To determine uplift at the crest (100% of the distance), we merely divide 400 feet by 44% to arrive at a figure of 909 feet, which I'll round down to 900 feet. (One should not suggest that uplift over millions of years can be measured to the nearest foot.)

What does this say about uplift in the San Joaquin River drainage? The average elevation of the *granitic* crest in the Donner Pass vicinity, at the headwaters of the South Yuba River, is about 8000 feet (volcanic deposits on the crest locally rise significantly higher). In contrast, the average elevation of the *granitic* crest in the Deadman Pass vicinity, at the headwaters of the San Joaquin River, is about 10,000 feet (volcanic deposits on the crest also locally rise significantly higher), or about 25% more. Therefore, crest uplift in this drainage should be about 25% more, or about 1125 feet. This is essentially one tenth of the 11,320 feet of uplift N. King Huber has proposed for the last 50 million years.

However, the profile of Spring Creek above the "steep" reach is not perfectly smooth along the low-gradient reach, which is across resistant metamorphic bedrock (see Warren Yeend's USGS Professional Paper 772 for a good geologic map). It is faintly stepped. At the right side of Figure 46 there is part of a major step, this occurring across resistant granitic bedrock. But the lowest fourth of the "steep" reach also is across resistant granitic bedrock. Therefore, the reach across this granitic bedrock, for the sake of consistency, also should be a step. And because there is no significant change in the creek's profile along the entire "steep" reach, the metamorphic bedrock comprising three-fourths of the reach was likely just as resistant as the granitic bedrock. Consequently, the entire reach appears to be one large step. If this is so, then instead of 400 feet of South Yuba River incision over the last 50 million years, there has been none. Remember that for almost all those 50 million years the South Yuba River has flowed across sediments, not bedrock, so the river has been eroding along bedrock only a geologically short time. The sediments not only buried and preserved the early Cenozoic Spring Creek, they buried and preserved the South Yuba River canyon, which back then would have looked—save for the vegetation—very much as it does today (Photo 170).

No local incision means no local uplift. And no measurable uplift at Donner Pass. And no measurable uplift at Deadman Pass. Thus Ransome's and Huber's method, when applied *properly,* indicates little or no Sierran uplift. But there are two additional methods, which I have devised, that lead to the same conclusion. The first is based on the undisputed fact that during the Cenozoic era the Sierra Nevada has always been above sea level. No one has ever found Cenozoic-age marine sediments in the range, nor has anyone ever found Cenozoic-age pillow-basalt flows, which form under water. Certain granitic summits, ridges, and benches are extremely resistant to

**Photo 170. South Yuba River canyon at the lower end of North Canyon. The river's elevation here is about 2050 feet, and the view from a bedrock bench is southeast up through a section of resistant metamorphic rock.**

weathering and erosion and have lost only a few feet of elevation over millions of years, and they can be used to gauge uplift. For example, if one of these benches today stands at 3000 feet, and originally it formed at sea level (an absurd proposition) then the *maximum* uplift there was 3000 feet. With resistant granitic summits, ridges, and benches we can determine a maximum amount of late Cenozoic uplift. (The actual uplift can be much less than the maximum amount.)

The second method is based on the undisputed fact that streams flow downslope, not upslope. As I will demonstrate, where small (and even moderate) Sierran streams flow across very resistant granite, they are essentially powerless to incise through it, even over a period of several million years. Consequently, their gradients are essentially the same as they were several million years ago, assuming no uplift. But what if uplift has occurred over that period? Then their gradients have steepened. Therefore I have made the assumption that the lowest-gradient reach of a stream—one close to zero—had a zero gradient before uplift (another rather absurd proposition). Like Lindgren, we can use the change in gradient to measure uplift. For example, if a reach of stream today has a gradient of 20 feet per mile, and before uplift it was zero feet per mile, that reach obviously has been steepened by 20 feet per mile. And if the Sierra Nevada in that stream's vicinity is 60 miles wide (which is a good average figure for the range), then uplift at the crest would have been 1200 feet.

## Resistant granitic summits, ridges, and benches

N. King Huber proposed that over the last 50 million years about 11,320 feet of Sierran-crest uplift occurred in the upper San Joaquin River drainage and in the river drainage north of it, the Tuolumne. (The Merced River drainage, sandwiched between the two, doesn't extend quite far enough east to reach the crest.) His data suggest that very little uplift occurred between 50 and 33 million years ago, only about 300 feet. Therefore, in the last 33 million years—that is, after the demise tropical climates and their high denudation rates—about 11,000 feet of uplift occurred. However, some parts of the crest are below 11,000 feet in elevation, for example, Mammoth Pass, located between Devils Postpile National Monument and the town of Mammoth Lakes. It stands at about 9400 feet elevation, and if no erosion occurred over the last 33 million years, then before uplift it was 1600 feet below sea level. So 33 million years ago the San Joaquin River must have ascended 1600 feet as it flowed west to the shoreline at the base of the range. Actually, if would have ascended about 2200 feet, since the oceans were about 600 feet higher. (This is based on a history of past sea level elevations constructed especially by Bilal Haq and Peter Vail. Although this research initially was met with a healthy dose of skepticism, others performing later research have reached similar conclusions. Their fluctuating sea-level curve is reproduced on Plate 2 of Alan Bartow's Professional Paper 1501.) Like others before and after him, Huber got around this absurdity by proposing that major erosion occurred over the last 33 million years. The range had been denuded by thousands of feet, so initially the range at the crest was above sea level by several thousand feet. But no one before me made an effort to see if thousands of feet of denudation had occurred.

In previous chapters I have suggested two methods by which we can measure denudation rates. The first is by measuring the heights of pedestals beneath ancient erratics. I estimated the rate of denudation of the resistant granitic bedrock surrounding two pedestals to be about 5 inches per million years, based on an estimate of 4 inches over the last 800,000 years since the Sherwin glaciation ended. This low rate is quite amazing, especially when you realize that for most of the last 800,000 years the climate more often than not was glacial, not interglacial, as our climate today is, and weathering processes should have been greater than today's. Still, even this may be too much, since so many ancient erratics lack pedestals, indicating no weathering whatsoever. Furthermore, some of the erratics could have been left as early as 2 million years ago, so their pedestals would have had up to 2½ times longer to form. The weathering rate, using the 2-million-year date, then would be about 2 inches per million years.

The second method is by measuring the amount of denudation of granitic bedrock either adjacent to a dated volcanic unit or beneath it. At Black Mountain in the southern Sierra's Dome Land Wilderness the amount of denudation of granitic bedrock beside its 12.2-million-year-old lava flow is too small to accurately measure, suggesting equally low denudation rates. And in the North Fork Stanislaus River drainage, which had Tahoe and Tioga glaciers that rivaled in thickness those through Yosemite Valley, glacial erosion of resistant granitic bedrock floors was minimal. Since the Valley Springs Formation was deposited in the Sierra Nevada over a period roughly from 30 to 20 million years ago, in the North Fork only a few tens of feet of granitic bedrock have been removed, an average of about a foot or so per million years. Since Huber mapped this area, he could have made this assessment himself. Then again, since he mapped the Devils Postpile area, he could have measured the negligible amount of glacial erosion and denudation at glaciated Mammoth Pass by using a remnant of datable volcanic rock on it.

But there is a third, recently developed method: measure the amount of cosmic-ray-generated isotopes that occur just beneath the surface of unvegetated bedrock. In 1995 Paul Bierman measured these extremely rare isotopes, and determined a denudation rate of about 6.5 feet per million years for resistant granitic rocks in the Alabama Hills, in the Owens Valley just west of Lone Pine. This rate, while much lower than conventional wisdom would dictate, nevertheless is considerably higher than my denudation rates. But there is a logical explanation for this discrepancy. These hills have been repeatedly shaken by earthquakes for about the last 3 to 4 million years, since the Owens Valley began to subside along the Sierran fault system. Experiencing tens of thousands of earthquakes over this time, it is no wonder that the hills are so highly fractured. What is a wonder, though, is that the denudation rate isn't considerably higher (Bierman chose the most resistant granitic bedrock). Note that others using this method have concluded that in wet tropical climates, such as the one the granitic Sierra Nevada experienced for tens of millions of years, the denudation rate can be as high as about 300 feet per million years. This is an important fact to remember when we reconstruct how Yosemite Valley evolved late in the Cretaceous and early in the Cenozoic. The dramatic changes in the Sierran landscape have occurred not in the last 33 million years, when the climates have been close to modern, but before then, when they were warm and wet.

Consequently, we can't use today's denudation rates before 33 million years ago. Fortunately, we don't have to. Huber suggests that uplift began about 50 million years ago, but if on log-log paper you plot his four uplift/time coordinates (0 feet of uplift at 50 million years ago, 4265 feet by 10 m.y.a., 8200 feet by 3 m.y.a., and 11,320 feet by today), you get a fairly straight line, and from it you can determine that between 50 and 33 million years ago only about 300 feet of uplift occurred. Consequently, we can use 11,000 feet, rather than 11,320 feet, as the amount of uplift over the last 33 million years.

Along the Sierran crest near the headwaters of the Middle Fork San Joaquin, Huber and Matthes each proposed a site crossed by a hypothetical trans-Sierran river: Huber, near the canyon floor below Deadman Pass; Matthes, at Mammoth Pass. Midway between the two lies the crest pass of Minaret Summit, which provides a glorious northwest view of its namesake, the Minarets, and which also is the start of a winding road down to Devils Postpile National Monument. Just south of this summit lies a broad, nearly level exposure of resistant, unglaciated granodiorite bedrock. Today it is at about 9200 feet elevation, but if we use Bierman's rate of denudation, then about 200 feet has occurred over the last 33 million years. I will use this rate, even though it is probably too high for two reasons. First, for about the first 30 of those 33 million years there were no nearby active faults, so rates of rock fracture and mass wasting would have been less. And second, for about the same 30 of those 33 million years the climate was milder than it has been over the last 3 million years, so weathering rates would have been less. Restoring 200 feet to the granodiorite bedrock gives an elevation of 9400 feet. However, back at 33 million years ago, sea level was about 600 feet above today's level. Therefore, the actual restored elevation of the non-denuded crest would have been 8800 feet above sea level, if no uplift or subsidence has occurred over the last 33 million years. Conventional wisdom would have put the crest back then at about 3000 feet above sea level. If this were the case, then the most uplift that could have occurred would be 5800 feet. (Math: 3000 feet initial elevation + 5800 feet uplift + 600 feet sea level drop - 200 feet denudation = today's 9200-foot elevation.) This is about half of the uplift Huber proposed. Note that Huber proposed 11,000 feet of uplift over this time span, which would have put this local part of the Sierran crest initially at 1600 feet below sea level. (Math: today's 9200-foot elevation + 200 feet of restored denudation - 11,000 feet uplift = -1600 foot initial elevation.)

Let's take another example in this river drainage, Chawanakee Flats, a broad, barren, resistant granitic benchland, that averages about 3200 feet in elevation. Like Yosemite Valley, it is located far away from any active faults, so denudation should be much less than at the crest. And like the broad, barren summits above the valley, it should be experiencing a similar denudation rate, about 5 inches per million years, which amounts to about 14 feet over 33 million years—an insignificant amount. Back at 33 million years ago, when the sea level was about 600 feet higher, this benchland would have stood about 2600 feet above sea level. Because it is located almost exactly halfway between the range's base and its crest, it should have experienced half the uplift. If we make the untenable assumption that it initially was at sea level back then (the river having a zero gradient), then 2600 feet of uplift has occurred here and twice that, or 5200 feet, has occurred at the crest.

But directly across the San Joaquin River is a broad ridge that descends to about 2600 feet elevation to the brink of the river's inner gorge. That would have been about 2000 feet above sea level back at 33 million years ago. This implies that only 4000 feet of uplift has occurred at the crest—*if* the brink had been at sea level, and *if* the river back then had had a zero gradient. Neither assumption is logical or tenable. Huber used an arbitrary, very unrealistic, pre-uplift gradient: one meter per kilometer, which equals about 5.3 feet per mile. This may be a good gradient for the river near the base of the range, but certainly not for the entire length of river across the range. Nevertheless, if such a gradient existed, then at the site of the brink about 30 miles above the base of the range, the elevation of the river would have been about 160 feet. The brink then would have been about 1840 feet above the river, and the uplift of the crest would be twice that amount, or 3680 feet, which is about 33% of Huber's 11,000 feet of uplift. But *if* the inner gorge had existed in some fashion back then, and *if* the river had possessed a realistic gradient, the amount of crest uplift would have been significantly less, perhaps not even 2000 feet.

There are other ancient, resistant, granitic benches and broad ridges west of the Sierran crest that one can use to reach similar results. In these western lands, this method limits late Cenozoic uplift to about 4000 feet if you force the data; to about half that if you do not. This amount is a *maximum* figure; there could have been no uplift whatsoever. And this is what an analysis of eastern slopes suggests. While west-side rivers flow down to about sea level, east-side streams from about Deadman Pass southward flow into the Owens River, which in turn flows south through the Owens Valley. As François Matthes recognized back in 1937, this valley is very old. He had concluded that the precursor of Mt. Whitney, which he thought was in existence 35 to 40 million years ago, was hill-like, and so it must have had elevations comparable to modern hills. Its summit elevation, he determined through Davisian landform analysis, was about 2000 feet, and it stood about 1500 feet above its surrounding lowlands. Therefore, under his interpretation, Owens Valley back then was at low elevation and had low relief. Also, to reach its current elevation near 14,500 feet, Whitney Hill had to be raised about 12,500 feet—much more, if its summit has been significantly denuded.

I suspect that the Owens Valley is much older than Matthes thought, originating either during late Cretaceous uplift about 75 to 65 million years ago—or earlier. As I mentioned under Chapter 21's "Lake Yosemite," extensional faulting occurs along with voluminous magmatism, which in the eastern Sierra occurred before 80 million years ago. And if Ronald Kistler's 1993 controversial fault reconstruction is correct, then Owens Valley—and other east-side valleys north of it—should have begun forming along a major fault system that existed before 117 million years ago. This lengthy fault system may have been a precursor of the current fault system that runs along the east base of today's Sierra Nevada. By the time the modern system came into existence 3 to 4 million years ago, the Owens River, with the aid of tens of millions of years of tropical and semitropical weathering, had excavated a very deep, broad valley, essentially the one we see today. This is quite a different picture than the generally accepted one painted by Matthes. By using Alan Bierman's rate of denudation in the Alabama Hills, 6.5 feet per million years, we can answer two questions. First, what was the relief of Owens Valley before the modern episode of downfaulting? And second, how much uplift (or subsidence) has occurred along the Sierran crest?

Conventional wisdom dictates that Owens Valley acquired most of its relief in response to the last 3 to 4 million years of faulting, which caused the valley's bedrock floor to sink below sea level, and caused Mt. Whitney to rise thousands of feet to its present elevation. However, resistant, barely eroded benches near the bases of unfaulted ridges that descend east from the Sierran crest can be used to determine the approximate elevation of the valley's prefaulted floor. Such benches occur at elevations as low as 4200 feet just north of Cottonwood Creek's alluvial fan (Olancha 15' quadrangle; geologic map by Edward du Bray and James Moore). In 33 million years the benches near the bases of the unfaulted ridges should have experienced a maximum denudation of about 200 feet. The figure actually would have been lower because until about 10 to 15 million years ago there would not have been any nearby faults, and so the denudation should have been comparable to that on the west side of the range, about 10 feet. From about 10 to 15 million years ago, faults developed east of Owens Valley which should have shaken the bedrock about the valley, so the denudation rate would have been greater, perhaps 10 to 30 feet. Only over the last 3 to 4 million years has nearby faulting occurred, during which time, if Bierman is correct, about 20 to 26 feet of denudation should have occurred. Therefore, in the last 33 million years a realistic denudation estimate is about 50 feet, not 200.

With regard to relief of Owens Valley, which is about 10,800 feet between Mt. Whitney and Lone Pine, the difference between 50 and 200 feet is inconsequential. During the last 33 million years the Sierran crest lands would have experienced seismic activity comparable to that along its east base, and so would have become quite fractured. (As Sierran mountaineers know, the eastern escarpment of the range is intensely fractured—and rockfall can be a serious hazard.) Moreover, even by conventional wisdom the crest of the range over the last 2+ million years would have been sufficiently high to have experienced intense physical weathering, such as by the freezing and thawing of ice, and so it would have experienced a greater denudation rate than down along the east base. So while resistant benches near the east base of the range may have undergone 50 to 200 feet of denudation in the last 33 million years, the crest lands high above them would have experienced the same amount, if not slightly more. In sum, the amount of relief back then would have been virtually identical to what it is today (Photo 171).

In the vicinity of 14,495-foot Mt. Whitney, the relief from the bases of the granitic ridges up to the peak's summit is about 8000 feet. But the Owens Valley floor is about 2800 feet lower than the bases of the ridges, so the total relief is about 10,800 feet, as was just mentioned. How low can we depress the Sierran crest in this vicinity and still have a plausible pre-uplift reconstruction? By Lone Pine the Owens River is at about 3650 feet elevation, and from there it parallels the range south about 70 miles to Indian Wells Valley, at about 2300 feet elevation, in which it dries up. During glacial and deglacial times it turned east to descend more or less along the northern edge of the Mojave Desert ultimately to Death Valley, whose floor today is below sea level. It is a known fact that a stream cannot flow below sea level unless that land has sunk below sea level through extensional faulting. So the question becomes what were the elevations of the prefaulted lands in the Owens Valley and east along the northern edge of the Mojave Desert? South of Mt. Whitney the ridges descending east from the Sierran crest can also be used to define pre-uplift relief, and then we can address faulting.

Today Olancha Peak, about 24 miles south-southeast of Mt. Whitney, stands at 12,123 feet elevation, and ridges near it descend east and have resistant benches at about 4600 feet elevation. If these descending ridges are extrapolated eastward to the center of the valley, its prefaulted, bedrock-floor elevation could have been lower than that of adjacent Owens Lake, which today is at about 3600 feet. Farther south, the Sierran crest both north and south of Walker Pass has resistant, unfaulted ridges that descend east almost to the western edge of Indian Wells Valley, which is also at about 3600 feet elevation (Inyokern, Little Lake, and Onyx 15' quadrangles; geologic map by Diggles *et al.*, 1987).

The bedrock floors of both the Owens and Indian Wells valleys have subsided in response to extensional tectonics. At what elevation these floors were before faulting commenced is unknown. If they had been at about 3600 feet, as they are today, then no uplift has occurred, and the pre-faulting relief was essentially the same as it is today. If they had been at sea level, then about 3600 feet of uplift has occurred. But *unfaulted* closed basins do not have

**Photo 171. Aerial photo of the east escarpment below Mt. Whitney. The bedrock base is just below the bottom of the photo and the total bedrock relief at this site is about 8000 feet. It would have been similar before uplift.**

bedrock floors at sea level. Death Valley, the Salton Sea trough, and many other structural depressions have bedrock floors well below sea level, but only because of extensional tectonics. Before faulting, the southern Owens Valley likely had its bedrock floor well above sea level, and the bases of east-descending Sierran ridges lay at similar elevations.

Roy Dokka and Timothy Ross in 1995 presented a synthesis of the evolution of southern California, based in part on their detailed work in the Mojave Desert. In this study they documented the timing of the collapse of the desert lands, this occurring between 24 and 16 million years ago, considerably before downfaulting began in the Owens Valley. Unfortunately, they did not say how much collapse has occurred. Before faulting, the southern Owens Valley, like the northern Mojave Desert and Indian Wells Valley, would have had its bedrock floor above sea level, but was it at about 3000 feet, 4000 feet, 5000 feet? The first two elevations imply minor uplift or subsidence, but if the last figure is correct, it implies a measurable amount of subsidence of the Sierran crest in the vicinity of Mt. Whitney and also along the crest south of it. The bases of east-descending ridges would have subsided about 1000 to 2000 feet to reach their current elevations, and likewise the Sierran crest, at the top of these ridges, would have subsided by the same amount.

If you've ever studied a map of the Sierra Nevada, you've probably noticed that its southern part—essentially, the Tehachapi Mountains—curves increasingly westward. This was not always so. Dokka and Ross place its rotation between 21 and 16 million years ago, the rotation occurring in concert with extensional faulting of adjacent lands. Rotation of the range was greater increasingly southward, and the height of the Sierran crest decreases increasingly southward, suggesting a cause-and-effect relation. Additionally, lands increasingly southward are increasingly faulted, and subsidence could have occurred along these faults. In the faulted lands south of Highway 178 and Walker Pass, one cannot accurately evaluate subsidence. However, in the mostly fault-free lands north of the highway, one can. This is because the South Fork Kern River, which flows parallel to the nearby crest, has several reaches that have a gradient close to zero, as is discussed in the next section. If we attempt to raise the southern Sierra to a higher, prefaulted elevation, we quickly change the river's gradient from low, to zero, and then to negative—flowing uphill. The low reaches limit Sierran-crest elevation change to minor subsidence at best. Likewise, the high prefaulted relief of the Owens and Indian Wells valleys, coupled with formerly higher elevations in the Mojave Desert, limit Sierran-crest change to minor uplift at best. So the best interpretation over the last 33 million years is that the crest elevations opposite the Kern River drainage in the southern Sierra Nevada have remained virtually unchanged.

Farther north, a similar history unfolds. In the central Sierra Nevada the Walker and Carson rivers today flow from the eastern Sierra Nevada into closed basins in west-central Nevada. The East Fork Carson River is particularly interesting. As was mentioned earlier under Chapter 22's "Exhumed unglaciated landscapes with 'glaciated' features," this river's upper canyon has a finger of the Relief Peak Formation that descends to within about 150 feet of the canyon's 8960-foot floor, which indicates that this part of the mostly granite-walled canyon had essentially the same depth and width 20 million years ago as it does today. This unfaulted, heavily glaciated canyon has Miocene-age volcanic remnants all the way down to its mouth at 6500 feet elevation in Silver King Valley (see David John *et al.* for geology), where again the depth and width are essentially unchanged. The river then turns northwest and flows down to about 5580 feet elevation in the vicinity of Markleeville. This reach of river is across unfaulted volcanic deposits (see Richard Armin *et al.* for geology), so the prevolcanic, granitic canyon floor beneath them must lie lower than the current elevation. Downstream, today's river quickly encounters faulted terrain (see John Stewart *et al.* for geology), so determination of the river's prefaulted base level is uncertain. Today it is in the Carson Sink, at 3870 feet. Given that the prefaulted river must have flowed below Markleeville, it quite certainly descended to an elevation hundreds of feet below its Markleeville elevation. Like the extended lands of the Mojave Desert, the extended lands of the Great Basin had a higher prefaulted elevation, and the vicinity of Carson Sink could have been significantly higher, perhaps even at 5000 feet. Therefore, one cannot say whether minor late Cenozoic uplift or subsidence has occurred in the vicinity of the upper drainages of the Carson and Walker rivers. The best interpretation is that measurable uplift or subsidence has not occurred. An implication here is that prefaulted west-central Nevada must have had high relief like its eastern-Sierra counterpart. Furthermore, its ranges must have had high elevations, since in the absence of faulting there could not have been a high Sierra Nevada attached to a low Great Basin. Jack Wolfe and Howard Schorn have reviewed fossil floras in west-central Nevada and have concluded that this was the case.

In contrast to the central and southern Sierra Nevada, the northern Sierra Nevada may have experienced some uplift, albeit minor. The evidence is in the glacial history of the Sierra Buttes area, mapped and described in detail by Scott Mathieson. Whereas to the south the pre-Tahoe (presumably Sherwin) glaciers were the largest, no pre-Tahoe evidence exists here. Therefore, the Tahoe glaciers must have been large enough to overrun the evidence left by any of their predecessors. The implication is that minor uplift occurred between the Sherwin and Tahoe glaciations, roughly between 800,000 and 200,000 years ago, which resulted in larger Tahoe glaciers. However, Tioga glaciers, being the youngest, should be the largest, as some uplift should have occurred in the intervening 100,000+ years between the Tahoe and Tioga glaciations. Instead, like those in the central and southern Sierra Nevada, they are slightly smaller than the Tahoe glaciers. In those parts of the range, this slight diminution could be explained by an increasing rain shadow cast by the rising South Coast Ranges of central California, although a more likely explanation is that, as elsewhere in North America, late Wisconsin glaciers were slightly smaller than their predecessors, indicating a large-scale climatic cause, not a regional-uplift one.

Because late-Quaternary uplift in the North Coast Ranges, north of the San Francisco Bay Area, has been minimal, one cannot use an increased rain shadow to explain diminution of the northern Sierra Nevada's Tioga glaciers. Precipitation reaching the northern Sierra should have been about the same in both the Tioga and Tahoe glaciations. Hence it appears that sufficient Sierran uplift occurred to cause the Tahoe glaciers to be larger than their predecessors (especially the 800-900,000-year-old Sherwin), yet insufficient to cause any measurable increase in the Tioga glaciers. Not much uplift would be required; 500 feet or less would suffice. I base this estimate on elevation/glacier-length ratios in the Klamath Mountains and the Sierra Nevada, which show that a minor increase in elevation can cause a substantial increase in a glacier's size. It is also based on the fact that warming by less than 2°F over the last century has caused a major retreat (and, in general, extinction) of the Sierra's nineteenth-century cirque glaciers. Only minor late-Quaternary uplift is required to have caused the Tahoe glaciers to be larger than their predecessors.

## Low-gradient reaches of streams

In the South Fork Kern River drainage of the southern Sierra Nevada there are reaches with a nearly zero gradient. Had even a minimal subsidence occurred, some of these reaches would have developed negative gradients; that is, they would have flowed uphill. Perhaps the best reach to evaluate subsidence is the South Fork from the 7880-foot contour to the 7840-foot contour (Monache Mountain 7.5' topographic map). Along this reach, parallel to the Sierran crest, the meandering river drops 40 feet in a straight-line distance of 17,000 feet, yielding an average gradient of about 12.5 feet per mile. Thus if the average gradient of the crest had subsided by this amount, then before subsidence the river's gradient would have been (very unlikely) zero (in order to have reached its current gradient). If it had subsided by a slightly greater amount, then the initial gradient would have been slightly uphill. But given that all rivers have gradients greater than zero, the actual amount of subsidence along the Sierran crest would have to have been less than 12.5 feet per mile—in essence, minimal to nonexistent.

We can look at this analysis in another way. In the vicinity of 14,495-foot Mt. Whitney, the Sierran crest has an average elevation of about 14,000 feet. To the south of it Walker Pass is about 64 miles away. Walker Pass, on the Sierran crest, lies at about 5245 feet elevation, and in its vicinity are broad, resistant ridges and knolls between 5000 and 6000 feet elevation. Could this area have subsided some 8000 to 9000 feet to reach these elevations? We can restore Walker Pass to its prefaulted elevation by multiplying the 64-mile distance times the 12.5 feet-per-mile gradient, arriving at a *maximum* permissible subsidence of 800 feet. This raises Walker Pass from about 5250 feet to 6050 feet, about 8000 feet short of the crest elevations near Mt. Whitney. Significant subsidence can be ruled out.

Conventional wisdom (as expressed in the first nine arguments for late Cenozoic uplift) suggests that rivers have greatly incised their canyons over the last several million years. With this in mind, one could argue that the preceding analysis is invalid, since much incision in the South Fork Kern River drainage would have occurred over this time span, altering the river's gradient, so today's river profile cannot be used over time. Fortunately, the rate of incision by the South Fork is well constrained by two large volcanoes, Templeton Mountain and Monache Mountain, as was mentioned under Chapter 22's "Similarities of glaciated and unglaciated landscapes—Within the Sierra Nevada," which showed no measurable erosion beside each volcano over the last 2.4 million years. It thus appears that under modern summer-dry climates, which have existed for about the last 15 million years, this river's incision has been only several feet. Consequently we can use today's river profile as a proxy for ancient river profiles, since so little change has occurred. One could argue that during glacial times, the relatively larger rivers of the Sierra Nevada severely eroded their bedrock floors. However, during these times, the higher discharges allowed the rivers to transport great amounts of sediment, which accumulated in the rivers and thus protected their beds from erosion. Remnants of such outwash exist in the Merced River today.

Note that this method of analysis used to constrain subsidence along the crest also can be used to constrain uplift. In the South Fork Kern River drainage some low-gradient streams flow westward away from the crest. If significant westward tilting had occurred with uplift, then these streams, restored to their pre-uplift gradient, initially would have been flowing uphill. What follows is a number of examples of low-gradient, westward-flowing streams from the southern and central Sierra Nevada that constrain the amount of Cenozoic uplift. Note that all have a discharge considerably smaller than that of the South Fork, and so their erosional powers would have been much less. For each stream the length of reach in question has been corrected for divergence from the perpendicular to the Sierran crest. Also, meanders are ignored. These corrections result in a steeper gradient than the actual one along each reach. For each example the maximum uplift at the crest should be accurate to within about 100 feet. The streams are listed from south to north.

1) Mulkey Creek through Mulkey Meadows.

Location of reach: between 9360 and 9320 feet; 36° 24' north latitude, 118° 11' west longitude; Cirque Peak 7.5' map.
Gradient: 40-foot drop over 6740-foot length = 31.3 feet per mile.
Width of range: 50 miles.
Maximum uplift at crest: 1565 feet.

Notes: The broad headwaters area of Mulkey Creek is at about 9700 feet on the Sierran crest, and as such is comparable to the elevations of the Deadman Pass-Minaret Summit stretch (10,000-9175 feet) along the Sierran crest above the San Joaquin River. For Deadman Pass, Huber had estimated total uplift of about 11,320 feet in 50 million years. The Mulkey Creek reach constrains uplift to only about 13.8% of this value, assuming that the reach originally had a zero gradient, which is unlikely. With a non-zero gradient, the actual amount of maximum uplift would have been less than 1565 feet. One could argue that the lands between the Kern Canyon Fault and the Sierran crest were lifted as an essentially untilted block, and hence the Mulkey Creek reach cannot be used to constrain uplift in the northern half of the range. Actually, however, a transect through broad basins from the crest west to the Kern River canyon shows a nearly constant drop of about 100 feet per mile, which definitely is not level. Furthermore, Donald Ross in 1986 concluded that motion along the Kern Canyon Fault is right-lateral, not normal, and so vertical block uplift is not supported by the fault pattern.

Ross also concluded that initial movement and deformation along the fault began about 90 to 80 million years ago, and this date accords fairly well with ones by Peter George and Roy Dokka, who obtained fission-track ages of Sierran plutonic rocks and concluded that major uplift occurred between about 80 and 74 million years ago. As was mentioned earlier, about the same time Saleeby *et al.* determined a date for major Sierran uplift at 75 to 65 million years ago. Also note that the eleventh method for concluding late Cenozoic uplift implies late Cretaceous dislodgment of what is now the Santa Lucia Range from the original cordillera. Since its bedrock canyons received Paleocene sediments, extensional tectonics to produce subsidence likely occurred during the first part of the Paleocene.

Clarence Hall, Jr., suggests this range was initially thrust westward from the Mojave and eastern Peninsular Ranges provinces during early Paleocene time, which he puts at about 65 to 62 million years ago. However, he does suggest a second pulse about 57 to 55 million years ago, and additionally, Ross mentions possible "but doubtful" deformation along the Kern Canyon fault at about 55 to 50 million years ago. After that, there do not seem to have been any *demonstrable* tectonics that could have raised the Sierra Nevada.

2) South Fork Kaweah River across the Hockett bench.

Location of reach: between 8520 and 8480 feet; 36° 21' north latitude, 118° 39' west longitude; Moses Mountain 7.5' map.
Gradient: 40-foot drop over 5000-foot length = 42.2 feet per mile.
Width of range: 52 miles.
Maximum uplift at crest: 2194 feet.

Notes: This reach is in lands west of the Kern Canyon Fault. Part of this reach is very level, and if only it were used, the actual amount of maximum uplift would be less. However, this broad bench has been glaciated, so the level part could be due to infill of a basin with glacial sediments. Still, at each end of this reach lie granitic outcrops, which serve as points from which the average gradient can be calculated. Note that glaciation has not made any significant change in the pre-glacial topography of this bench. If the shallow tarns and the till did not exist, then it would be impossible to conclude that the bench had been glaciated: it is no different than unglaciated benches elsewhere in the southern Sierra Nevada.

Clyde Wahrhaftig (1965), Mark Christensen (1966), N. King Huber (1981) have assumed that only the northern half of the range was tilted westward, while the southern half was raised more or less as an untilted block, and therefore the method of uplift constraint derived from streams in the southern Sierra Nevada is not valid. Actually, there is no field evidence for this bimodal uplift. Had it occurred, then stresses should have developed in the zone between the two areas, that is, in the San Joaquin river drainage. The stresses should be expressed in the form of faults, joint patterns, and gravity anomalies, none of which exist. Even the classic stepped topography in the western part of the range, described and analyzed by Wahrhaftig, extends uninterrupted across this zone. South of the zone there is an increase in the amount of faulting, particularly near the mouth of the Kern River canyon. However, the faulting is almost entirely in the southeastern part of the San Joaquin Valley. Had the range been lifted as a block, then there should be a major fault system along the west edge of the range, and this west edge should have developed a major escarpment. Neither fault system nor escarpment exists. The hypothesis of a westward-tilted Sierra Nevada block that was then uplifted at its southern end during the late Cretaceous obviates the unsupported bimodal-uplift hypothesis. Furthermore, it provides a framework to explain the asymmetric drainage patterns in the range: from the Tuolumne River to the Kings River, the south part of each drainage's headwaters is larger than the north part, implying southward growth at the expense of the north part. This pattern cannot be explained by the bimodal-uplift hypothesis.

3) Fine Gold Creek northwest of Kennedy Table.

Location of reach: between 1220 and 1280 ft; 37° 11' north latitude, 119° 38' west longitude; Millerton Lake 15' map.
Gradient: 60-foot drop over 13,020-foot length = 24.3 feet per mile.
Width of range: 62 miles.
Maximum uplift at crest: 1507 feet.

Note: This stream drainage never was glaciated. The stream is a tributary of the San Joaquin River. Because this tributary lies north of the river, there is no doubt that at this site—and at northern ones below—the Sierra Nevada was tilted westward as a block.

4) Middle Tuolumne River northwest beyond the Long Gulch "stream piracy" site.

Location of reach: between 6800 and 6720 feet; 37° 52' north latitude, 119° 43' west longitude; Tamarack Flat 7.5' map.
Gradient: 40-foot drop over 3900-foot length = 54.2 feet per mile.
Width of range: 60 miles.
Maximum uplift at crest: 3252 feet.

Notes: While stream piracy near this site looks very certain, the pattern of the river could be due merely to joint control. An aggressive stream that captures the headwaters of another should have a steeper gradient. However, the upper reach of seasonal Long Gulch creek is about 50% steeper than the capturing reach of the Middle Tuolumne River, suggesting that stream piracy has not occurred.

Small Tahoe- and Tioga-age glaciers occupied the upper parts of this drainage far above this reach of the Middle Tuolumne River, and they should have had no impact of the reach's gradient.

5) South Fork Tuolumne River through Harden Flat.

Location of reach: between 3520 and 3440 feet; 37° 48' north latitude, 119° 56' west longitude; Ascension Mountain 7.5' map.
Gradient: 40-foot drop over 3600-foot length = 58.7 feet per mile.
Width of range: 60 miles.
Maximum uplift at crest: 3522 feet.

Note: This stream drainage never was glaciated.

6) Middle Tuolumne River above Columbia Camp.

Location of reach: between 3760 and 3720 feet; 37° 51' north latitude, 119° 56' west longitude; Ascension Mountain 7.5' map.
Gradient: 40-foot drop over 4400-foot length = 48.0 feet per mile.
Width of range: 60 miles.
Maximum uplift at crest: 2880 feet.

Note: Upstream this drainage received minor overflow of nearly stagnant ice from the large glaciers that occupied the Tuolumne River canyon just to the north. This ice may have advanced to this reach.

7) Miguel Creek through Miguel Meadow.

Location of reach: between 5040 and 5000 feet; 37° 58' north latitude, 119° 50' west longitude; Lake Eleanor 7.5' map.
Gradient: 40-foot drop over 6600-foot length = 32.0 feet per mile.
Width of range: 60 miles.
Maximum uplift at crest: 1920 feet.

Notes: This reach was buried under as much as 1000 feet of glacial ice, yet glaciers did little to modify the landscape, as attested in this general part of the Tuolumne River drainage. Even Huber concluded that "the Tuolumne River through Hetch Hetchy Valley was close to its present profile before glaciation," and the glaciers flowing through that valley were three times thicker than those flowing across the Miguel Creek drainage. Although there is some till along this reach, bedrock outcrops at each end of it show that a till-free reach across bedrock would have the same high and low points and the same gradient. The maximum amount of uplift is 17% of that suggested by Huber for the Sierran crest above the San Joaquin River.

The most reliable of these seven examples is Fine Gold Creek northwest of Kennedy Table. First, like most it is north of the San Joaquin River, and so is definitely within the realm of uncontested westward tilting. Second, like some its entire drainage was never glaciated, and so modification of the creek's profile through glacial erosion is not an issue. Third, the length of this reach is far longer than that of the six others, and so it should have the most precise gradient determination, which would give the most precise value for maximum possible uplift at the Sierran crest—a mere 1507 feet at the crest. And finally, the stream lies in the San Joaquin River drainage, so its uplift value is the one best to compare against N. King Huber's same-drainage crest uplift of 11,320 feet. The reach of stream indicates that crest uplift was only about 13.3% of Huber's figure. Actual uplift most likely was less, since it is very doubtful that the stream's long reach had a zero gradient before uplift. Furthermore, absolutely no uplift may have occurred, since the actual values determined by this method lie between zero and those stated.

The evidence from the three methods presented in this chapter to quantify Sierran uplift all point to the same conclusion: little or no uplift over at least the last 30 million years. Additionally, methods two and three indicate little or no subsidence. It appears that back in 1911 Waldemar Lindgren, the most-experienced Sierran field geologist, was correct when he stated that the late Cretaceous Sierra Nevada had essentially the same elevations and relief that it has today. It is unfortunate that this statement, which was contrary to everyone's (false) perceptions, was universally ignored. Had it been accepted, Yosemite geomorphology in particular and Sierran geomorphology in general would have made great strides, which would have changed the course of worldwide alpine-geomorphic thought.

# Part V

# The Big Picture

# 25

# Birth of a Granitic Yosemite Valley

## Introduction

Parts III and IV hopefully have demonstrated five major points. First, Tioga and Tahoe glaciers on the west slopes of the Sierra Nevada in every instance had relatively the same proportions; that is, every Tioga glacier was slightly smaller than its Tahoe predecessor. Second, these Tioga and Tahoe glaciers overall were larger than previous investigators had believed. Third, the enormous Tioga, Tahoe, and pre-Tahoe glaciers that flowed down deep, major canyons were quite ineffective at eroding them through either abrasion or plucking. Fourth, these canyons, therefore, were not the result of glacial erosion, but rather they must have evolved over tens of millions of years, achieving modern or nearly modern depths by about 20 to 30 million years ago, if not earlier. And fifth, the last major, verifiable uplift of the range was late in the Cretaceous period, not late in the Cenozoic era. With this background information, we can begin to reconstruct the geomorphic evolution of Yosemite Valley. But to do this, we need to answer three major questions. First, when did the present drainage of the Merced River originate? Second, was the drainage pattern dictated by joints in the granitic bedrock, or was it inherited from streams incising through overlying metamorphic or volcanic rocks? And third, how fast did the valley deepen and widen? If these can be answered, then the valley's geomorphic history can be reconstructed, as well as the geomorphic history of the Sierra Nevada.

## Plutonism, Volcanism, and Uplift

It is obvious that the maximum age of a granitic landscape can be no older than the age of its plutons, and that it must be somewhat younger, since millions of years could be required for erosive processes to strip away the overlying rock that the plutons had intruded. Some of the oldest plutons are found in the southern Sierra Nevada near Walker Pass, these dated by Ronald Kistler and Donald Ross as having solidified within the crust about 240 million years ago. Farther north, in the vicinity of Yosemite National Park, they date back to about 210 million years ago. Geologists have long assumed that volcanism accompanied plutonism, so while plutons were solidifying at depth, volcanoes were erupting at the surface. For example, Matthes stated in his 1914 "Studying the Yosemite Problem": "great masses of molten rock pressed up from beneath, invaded the surface rocks and, possibly finding vent in orifices here and there, issued forth in the form of lava flows." Much later, in 1966, Paul Bateman and Clyde Wahrhaftig made a similar statement: "Some magma would perhaps break through to the surface in volcanic eruptions, but much of it would crystallize at depth as plutons." Despite these statements no one considered studying whether stream drainages developed in a volcanic range would imprint their patterns into underlying metamorphic and granitic rocks.

In the 1960s, this volcanism-plutonism hypothesis gained greater support in the light of the new paradigm of plate tectonics. (The development of this field is documented in a large collection of articles assembled by Allan Cox in 1973.) Later, Warren Hamilton in 1988 stated that granitic plutons underlie large calderas and stratovolcanoes (such as today's Mts. Rainier and Shasta), and furthermore, that many plutons are capped largely by their own volcanic ejecta (materials forcibly expelled). Such a cap would be rapidly eroded, geologically speaking. He noted that the largest pluton yet mapped in the Cretaceous Sierra Nevada batholith, the Mt. Whitney pluton, is about the same size (15 by 50 miles) as the largest known young caldera in an active magmatic arc, the late Pleistocene Lake Toba caldera of Indonesia's island of Sumatra. Paul Bateman (1992, p. 95) was more cautious in his assessment:

> correlations between intrusive suites and metavolcanic sequences [in the Sierra Nevada] are dubious at best. Furthermore, except for a few small felsic intrusions such as the Johnson Granite Porphyry and the granite of Hogan Mountain, none of the plutonic rocks show evidence of an eruptive phase. However, in regions less deeply eroded than the Sierra Nevada, such as the San Juan Mountains of Colorado ... or the southern Andes of central Chile ..., volcanism and plutonism have been shown to be closely related. In these areas, magmatism began with andesitic and basaltic eruptions from small centers and was followed by the outpouring of voluminous and widespread but easily erodible ash-flow sheets. The ash-flow sheets were erupted from calderas that were underlain by silicic-magma chambers comparable in size to large Sierran plutons. Wes Hildreth (written commun., 1986) suggested that the small masses of diorite and gabbro so widely distributed in the Sierra Nevada, especially in the margins of the batholith, were feeders for early mafic volcanism.
>
> Isotopic age data indicate that magmatism in the central Sierra Nevada was more or less continuous from about 210 Ma (Late Triassic) to about 85 Ma (early Late Cretaceous), but that the distribution of ages suggests that the peaks of volcanic activity alternated with peaks of plutonic activity.
>
> ... [There were both] uplift and deep erosion that accompanied and followed emplacement of the Cretaceous plutonic suites. The easily erodible ash-flow sheets of Cretaceous age were completely removed, whereas the steeply tilted [and metamorphosed] Jurassic volcanic rocks were protected. The presence of volcanic detritus in the Late Jurassic and Cretaceous Great Valley sequence to the west and of voluminous ash falls in Jurassic and Cretaceous strata to the east (in the direction of the prevailing winds) is evidence of rampant volcanism during the timespan of plutonism.

Thus Bateman and Hamilton are in general agreement. Note that Bateman mentions, but does not quantify, uplift. It has been assumed, for example by Richard Schweickert and David Howell that during the Cretaceous the Sierra Nevada was part of a North American volcanic arc that resembled today's Andean volcanic arc, complete with stratovolcanoes.

Elevation is another matter: did the Mesozoic Sierra Nevada resemble today's relatively low Cascades, the more imposing Andes, or neither? In his geologic guide to Yosemite National Park, N. King Huber suggests the former, but presents no supporting evidence. In contrast, Paul Bateman in his 1992 Professional Paper 1483 provides evidence that during the Early Cretaceous epoch the land near today's crest was close to sea level, but later was uplifted. If the Lake Toba caldera is a good analog for the former topography of the Mt. Whitney Intrusive Suite (Hamilton's Mt. Whitney pluton, which solidified about 86 to 83 million years ago), then elevations would have been about 5000 to 6500 feet. Because the Tuolumne Intrusive Suite closely resembles this suite in age and composition, its landscape too should have had similar elevations. The Lake Toba caldera has a range of elevations similar to those of Owens Valley, Long Valley, and the Mono Lake basin, which border the eastern side of the central Sierra Nevada. Bateman suggested that this line of down-faulted basins may be an analog for the late Cretaceous Sierra Nevada, the White-Inyo Mountains, and the down-faulted lands between them.

The Sierra Nevada well could have been at about 5000 to 6500 feet before its late Cretaceous uplift. Topographic maps show that the average elevation of today's crest lands in the drainages of the North and Middle forks of the Feather River is about 6200 feet. As was mentioned in the previous chapter, possible late Cenozoic uplift in the northern part of the range appears to be about 500 feet or less. Therefore, before that period of time, the range would not have been much lower. However, because some denudation would have occurred between the late Cretaceous uplift and the possible late Cenozoic one, this amount has to be added to today's elevations. This amount, like the possible late Cenozoic uplift, would be minimal, because the crest lands are very resistant benches, and would have been denuded very slowly. Therefore, before late Cretaceous uplift, the elevations of such lands in the drainages of the North and Middle forks of the Feather River would be comparable to today's, which are within the 5000-to-6500-foot range of elevations.

Jason Saleeby *et al.* in 1993 placed the late Cretaceous uplift from about 75 until 65 million years ago. They attribute this postplutonic uplift to terrane accretion near the southern end of the Sierra Nevada, so that uplift there was greatest. Today the crest of the range increases in height southward to Mt. Whitney, from where it diminishes. This decrease in crest elevations in the southern part of the range may be due to late Cretaceous subsidence caused by extensional faulting. As was mentioned in the previous chapter, today's nearly zero gradient along certain reaches of the South Fork Kern River limits Cenozoic subsidence to a negligible amount. Subsidence had to be after the late Cretaceous uplift but before the major climate change of

33 million years ago, after which weathering and erosional processes greatly diminished and the landscape underwent negligible change. I tend to favor an early Cenozoic date of about 65 to 55 million years ago, subsidence occurring in response to extensional faulting as proposed by Clarence Hall, Jr., mentioned earlier. However, south of the Kern River drainage, in the faulted Tehachapi Mountains, additional subsidence could have occurred later, between 21 and 16 million years ago, as these lands were rotated west through extensional faulting.

If the Mt. Whitney area had been at about 5000 to 6500 feet in elevation before the late Cretaceous uplift, then during it that area experienced about 8000 to 9500 feet of uplift. The Feather River crest lands, just discussed, experienced negligible uplift. Crest lands between these two areas experienced intermediate amounts, the amount of uplift diminishing northward from Mt. Whitney. Hence the crest of the range achieved its asymmetric, northward-decreasing elevation that exists today. As was mentioned earlier in Chapter 21's "Deep glacial lakes," the Sierra's plutons were exposed quite rapidly, even within a few million years of solidification. And since the plutons in the Yosemite Valley area range from about 114 to 86 million years in age, then granitic rocks initially should have been exposed there before 80 million years ago, that is, before the late Cretaceous uplift. Most of the Sierra's plutons also would have been exposed, since most are of this age or older. It is only near the crest, where the plutons are on the order of 86 to 80 million years old, that granitic rock may not have been exposed to any significant degree. Consequently, at the onset of uplift the Sierra's rivers already would have been entrenched in granitic canyons at middle elevations and entrenched in granitic and/or metamorphic canyons at lower elevations, as they are today. Opportunity for drainage change would have existed perhaps only in untrenched lands near the crest.

The late Cretaceous uplift, being greatest in the southern part of the range, would have increased the gradient of north-flowing tributaries, which flowed approximately parallel to the crest, and would have decreased the gradient of south-flowing ones, which also flowed approximately parallel to it. Hence, the steeper, more erosive, north-flowing tributaries should have expanded their headwaters southward at the expense of the slackened, south-flowing ones, resulting, over millions of years, in a drainage asymmetry. This is what exists today from about the American River south to the San Joaquin River (Chapter 3's Map 4). Perhaps the Kings River began to extend its headwaters southward, but the Kern River prevailed. Working in its favor were the Kern Canyon Fault, which facilitated erosion, and accompanying land subsidence, which steepened the river's gradient. Consequently the Kern River's headwaters today are north of the range's highest point, Mt. Whitney. Conventional wisdom may suggest headward migration during the late Cenozoic. However, the partly exhumed headwaters area of the Clark Fork Stanislaus River conclusively demonstrates that the divide between it and the Middle Fork has not changed measurably during the last 20 million years, so there is no reason to expect the Kern River's headwaters to have migrated substantially northward during the same period of time.

## Major pre-Cenozoic Denudation

In his Professional Paper 1381, Donald Ross concluded that the southernmost Sierra Nevada has experienced some 15 to 18 miles of denudation. However, as was stated in the previous section, that area contains some extremely old plutons. Therefore, up to 165 million years of plutonism occurred before late Cretaceous uplift, which allowed ample time for tropical weathering and erosion to account for the vast majority of the denudation. Denudation was much less in the central Sierra Nevada. There, Paul Bateman in his Professional Paper 1483 concluded that only about 2 to 4 miles of denudation occurred after the batholith in the central Sierra Nevada was emplaced. Much of the denudation could have occurred along with or shortly after emplacement of plutons, as was mentioned above. Geologists have known for some time that by the end of the Cretaceous period, the Sierra Nevada was largely granitic, and it would have been during that time that stepped topography first began to develop. About this, Clyde Wahrhaftig in 1965, p. 1179) stated:

> A possibility is that stepped topography has characterized the Sierra Nevada since the first exposure of granitic rocks beneath their metamorphic cover. Büdel (1957) and King (1953) have shown that plains and plateaus characterize the topography of crystalline terranes in tropical climates such as that which may have prevailed in the Sierra Nevada during Late Cretaceous and early Tertiary time.

However, he rejected this possibility in part due to an analysis of the origin of the Big Sandy Bluffs, which are a segment of the highest step front between the San Joaquin and the Kings rivers (Photo 172). His analysis indicated that the major steps of the western granitic Sierra Nevada formed largely during late Cenozoic time, not early. However, that analysis was based on much of the denudation there being post-volcanic, which was not the case. Like N. King Huber later on, he acknowledged the existence of alluvial pumice underlying the 10-million-year-old Kennedy Table trachyandesite flow, then dismissed it in his analysis, thereby greatly underestimating both the depth and age of the San Joaquin River's

**Photo 172. Big Sandy Bluffs, rising 1600 feet above Big Sandy Valley.**

canyon. (Nevertheless, for this faulty research, he received in 1967 the Geological Society of America's Kirk Bryan Award—the highest acclaim that can be given to a North American geomorphologist—which says something of the society's ability, or rather inability, to judge research.)

Wahrhaftig also recognized that in the northern half of the Sierra Nevada "the bedrock is predominantly metamorphic, and granitic rocks occur as isolated plutons" while in the southern half "the predominant bedrock is granitic, and metamorphic rocks are in narrow septa and roof pendants." In the Sierran foothills this change is quite abrupt, occurring in the Fresno River's drainage, which is situated between the Merced River's and San Joaquin River's drainages, and its headwaters lie only 60% of the distance across the range toward the crest. If indeed the range's stepped topography had developed largely in response to a perceived late Cenozoic uplift, then this quite abrupt boundary would not exist, since this uplift presumably was a *westward* tilting of approximately equal amounts at the heads of the Tuolumne, Merced, and San Joaquin river drainages. Thus from the Tuolumne River to the San Joaquin River, there should be only a minor reduction in width and continuity of the foothills metamorphic belt, which is not the case. Finally, the northernmost major stepped topography occurs in the Fresno River's drainage, and I suggest that its appearance and the metamorphic rocks' disappearance are linked. The greater uplift in the southern Sierra Nevada was responsible for the weathering and erosion of most of the metamorphic bedrock and for the development of the stepped topography in the underlying, exposed granitic bedrock. Had the northern Sierra Nevada experienced similar uplift, a similar landscape would have developed.

## Importance of a Tropical Landscape

Through the Cretaceous period and until early in the Oligocene epoch at about 33 million years ago, the Sierra Nevada landscape was subjected to geomorphic processes operating in tropical or nearly tropical warm, wet climates. Evidence for such climates was known quite well even before Matthes' Professional Paper 160 appeared in 1930. Waldemar Lindgren in 1911 had discussed Tertiary fossil plants, particularly those of the auriferous gravels. Those gravels had yielded a significant number of fossil plants before the turn of the century, such as those identified by Leo Lesquereux in 1878, although the age assigned to the plants and gravels by Lindgren and by others before him was Miocene. In 1929, Victor Allen, having studied the Ione Formation in quite some detail, agreed with earlier investigators that the Ione Formation correlated with the auriferous gravels. Just after Lindgren had published his professional paper, Eocene marine fossils were discovered in the Ione Formation by R. E. Dickerson, thereby making the auriferous gravels, through correlation, also Eocene, as they are so recognized today. Noting the fossil-plant assemblage of the auriferous gravels, Lindgren stated (p. 56) that their [Eocene] climate "was like that of the southern temperate zone of the Atlantic coast region to-day. Species like those of laurel, maple, beech, fig, magnolia, walnut, and oak predominated and the climate was assuredly characterized by heavy rainfall." Allen noted evidence of that warm, humid climate as expressed in the residual quartz sands and gravels, the clays, and the seams of lignite found in the Ione Formation. Furthermore, he noted (p. 383) that "At Jones Butte, the Ione clays rest unconformably on argillaceous laterite derived *in situ* from the older rocks of the Sierra Nevada." After describing the laterite (presumably pre-Eocene), he discussed laterites in low-latitude lands, particularly in India. Additionally, he mentioned (p. 389) that "the writings of Whitney, Le Conte, and other observers record deeply weathered surfaces below the auriferous gravels of the Sierra Nevada. The association of the white residual clays with laterites offers the possibility that the voluminous literature on laterites that has accumulated in recent years, may give some suggestion of the [climatic] conditions of pre-Ione time."

How does a mountainous granitic landscape evolve under tropical or nearly tropical warm, wet climates? In 1993 Avijit Gupta, reviewing the past, present and possible future directions of geomorphic studies of the humid tropics, considered that the evaluation of a landscape in terms of rock types and geological history may be more important than in terms of a model such as proposed by Julius Büdel, in which climates dictate landforms. Wahrhaftig, in his just mentioned paper on stepped topography, considered Büdel's theory, but overall rejected it in favor of differential weathering. He did, however, make an exception, stating that the eastward extension of the San Joaquin Valley across beveled granitic bedrock from the Fresno River southeast to the White River—a lengthy strip of land along the base of the central and southern Sierra Nevada—developed in a manner suggested by Büdel. It is not coincidental that the northern limit of this beveling is also the northern limit of the granitic Sierra Nevada's foothills. North of the river the foothills are metamorphic, and stepped topography is nonexistent.

An excellent place to observe this beveling is the Four Corners junction (Highways 41/145 intersection) north of Fresno (Raymond 15' quadrangle; geologic map by Paul Bateman *et al.*, 1982). This vicinity essentially is a peneplain—a nearly flat bedrock surface—atop Jurassic or Cretaceous gabbro and minor hills of Cretaceous tonalite (Photo 173). One of these, Little Table Mountain, about 200 feet high, is flat-topped and is mantled with the Eocene-epoch Ione Formation (Photo 174). Between the solidification of the tonalite pluton at about 114 million years ago and the deposition of the Ione Formation at about 50 million years ago, the top of the pluton had to be exhumed and beveled. Beveling would have been possible because the world's oceans were considerably higher, and this vicinity would have been at sea level.

**Photo 173. Four Corners area: gabbro plain and tonalite hills. View is northwest from intersection.**

**Photo 174. Four Corners area: Little Table Mountain. View is east from intersection.**

A middle part of the North Fork Tule River drainage, mentioned in Chapter 21's "U-shaped cross profiles," is a 3-mile-wide, gently sloping floor on knobby granitic bedrock that is locally buried under a veneer of sediments. At first it does not seem to have been beveled, because this surface ranges from about 1600 to 3600 feet in elevation, so it seems to have been above sea level. However, before late Cretaceous uplift, which at the crest could have amounted to about 8000 to 9500 feet, the lower lands to the west could have been around sea level, and therefore could have been beveled. This middle part of the North Fork Tule River drainage cannot be explained in terms of late Cenozoic uplift, processes, and climates. Rather, it is an ancient feature explained in terms of late Cretaceous uplift, processes, and climates.

## Origin of the Merced River Drainage

Of importance to the initial development of Yosemite Valley is the matter of when the Merced River's drainage first became established. Since there likely had been voluminous eruptions from stratovolcanoes and calderas, as alluded to earlier, then the Sierra's major canyons likely would have been totally inundated, as was the case for the northern Sierra Nevada during the latter half of the Cenozoic era. Consequently, new rivers would have been initiated, and their courses would have grown headward through the volcanic strata. About 80 million years ago or shortly thereafter, both plutonism and volcanism at the crest ceased. Hence it was about at this time that canyons ceased being buried and having their streams across volcanic sediments buried or displaced. Thus the ancestral Merced River, if it had not already been in existence, would have come into being at about this time, during the late Cretaceous. Does the present Merced River drainage, particularly in the vicinity of Yosemite Valley, owe its origin to inheritance of a drainage incised in a late Cretaceous volcanic landscape, or does it owe its origin to the inheritance of a drainage developed in an earlier, metamorphic landscape? Or, was neither that important, the valley instead developing in response to the joint pattern of the granitic rocks?

For as long as the Mesozoic Sierra Nevada existed, there would have been rivers, although not necessarily the ones that exist today. River canyons could have been incised into the metamorphic bedrock, but if these were buried under lava flows and volcanic sediments sufficient to completely inundate the canyons, the river courses could have changed appreciably. As was implied above by Wes Hildreth, there could have been a significant amount of early mafic volcanism, judging by the widespread small masses of diorite and gabbro. Note, however, that the easternmost mafic intrusive rock in the present Merced River drainage is diorite, located at Yosemite Valley's Rockslides, just west of El Capitan, as well as located southeast across the valley and up the Bridalveil Creek drainage. Because this rock does not occur eastward, one might then infer that mafic volcanism did not exist in the upper part of the Mesozoic Merced River drainage.

Because mafic volcanism typically produces extensive basaltic lava flows, the lower Merced River canyon could have been buried and the river could have taken another course. In contrast, the upper part of the canyon may have been exposed to only andesitic and rhyolitic volcanism, as was mentioned earlier. If Bateman's "rampant volcanism during the timespan of plutonism" was real, then the upper part of the drainage, as well as the lower part, may have been buried and the river's course changed. However, the andesitic and rhyolitic deposits, unlike basaltic lava, would have been readily eroded, as was the case later on for the northern Sierra's Cenozoic volcanic deposits. Therefore the pre-volcanic, Cretaceous-period, *upper* Merced River canyon may have been exhumed more or less unchanged from its pre-volcanic form, as was the case in the upper Middle Fork and Clark Fork Stanislaus River drainages (and upper North Fork Mokelumne, American, Yuba rivers, etc.) after Cenozoic volcanism there ceased. Thus the underlying granitic bedrock of the upper Merced River drainage might have had superimposed on it the drainage pattern of the pre-existing canyon incised in the metamorphic bedrock.

If the lower drainage, which included western Yosemite Valley, had been buried under resistant basaltic lava, then that part of the drainage may not have been exhumed. However, today's *western* Yosemite Valley is quite straight and shows no evidence of a channel deflection around a former mass of lava. Although the late Cretaceous Sierra Nevada was blanketed by volcanic deposits, these do not seem to have affected the course of the Merced River. Perhaps in the lower Merced River drainage the underlying granitic bedrock also might have had superimposed on it the drainage pattern of the pre-existing canyon incised in metamorphic bedrock. A somewhat similar hypothesis was put forth in 1930 by Matthes (p. 30) for the origin of the modern Merced River's drainage pattern:

> A direct consequence of the upwarping and tilting of the Sierra region in early Tertiary time was the rearrangement of its drainage system. Before the uplifts began, the master streams, with few exceptions, flowed in northwesterly or southeasterly directions, the trend of the valleys being determined in general by the parallel ridges of the ancestral mountains of Cretaceous time. But when the land acquired a definite slant to the southwest, this old drainage system was replaced by a new system whose master streams flowed prevailingly in southwesterly directions. The change doubtless took place very gradually

> and must have required considerable time, for it was effected mainly by the progressive headward growth of certain streams that already flowed in southwesterly directions and by their capture of other streams that were less favored by the tilting.

Matthes had assumed that the ancestral mountains were solely metamorphic and that they had the same structure that gave rise to the trellis drainage pattern that exists in the current metamorphic foothills of the Sierra Nevada, particularly from about the latitude of Yosemite National Park northward. With regard to the ancestral master streams flowing parallel to what is now the Sierran crest, if the above had been the case, then these streams would *not* have deposited their sediments to the west in what is now the San Joaquin Valley. It should be noted that the foothills' trellis drainage pattern has master streams that drain westward, and so to be consistent in his analogy, Matthes should have proposed the same for the ancestral master streams. The late Cretaceous strata of the Great Valley sequence indicate that this was the case. Alan Bartow in 1991 observed that in the east part of the San Joaquin Valley these sediments are deltaic to shallow marine, but westward give way to deep-sea fan deposits. Additionally, Raymond Ingersoll in two earlier articles determined that the San Joaquin Valley's late Cretaceous sediments show west-directed paleocurrents, and he concluded that the Sierra Nevada range was the only source of sediment. Therefore, Matthes' hypothesis of widespread stream capture and resulting drainage rearrangement during the Cenozoic lacks any supporting field evidence. His hypothesis finds support only in the writings of William Morris Davis on the Appalachian's drainage pattern in Pennsylvania.

In the quote above, Matthes asserted that the trend of the valleys has been determined in general by the parallel ridges of the ancestral mountains of Cretaceous time, which were aligned more or less parallel to the ridges of the modern Sierra Nevada's metamorphic foothills. Superficially this assertion seems to have some merit, for the remnants of metamorphic belts do tend to be aligned more or less parallel to the range's belt of metamorphic foothills. Furthermore, many of them have streams more or less parallel to the ridges of these remnants, which is an orientation somewhat parallel to the Sierra Nevada's crest. One might then assume that such streams would continue to incise through the underlying granitic bedrock, giving it an inherited drainage. The upper parts of some drainages indeed are skewed southward, as was mentioned above and shown on Map 4, and their rivers there are nearly parallel to the crest. A conspicuous example is the upper, southernmost stream of the Tuolumne River, the Lyell Fork. However, in Yosemite National Park most of the river's streams are perpendicular to the crest, not parallel to it. These cannot be explained by Matthes' hypothesis.

Still, there is the possibility that some streams may be inherited from metamorphic-rock drainages, while others may have developed in response to joint patterns in granitic bedrock. If the Lyell Fork was inherited from an ancestral river incised into metamorphic rock, then perhaps the same applies to the Merced River through Yosemite Valley. Visitors to the valley are seldom aware that metamorphic rocks exist in it or on the slopes above its rim, since the outcrops are minor, and, as shown in Map 57, are for the most part off the beaten track (data from a geologic map by Calkins *et al.*). The presence of these remnants of metamorphic rock, especially within Indian Canyon, indicate that the pre-erosion metamorphic rock was in contact with the granitic rock close to the surface of today's topography. These contacts demonstrate that at least locally about them very little granitic rock has been weathered and eroded, supporting my very low Cenozoic denudation rates. These remnants occur in a north-south zone, but the valley cuts right through them. Additionally, the valley does not seem to show any preference for a northwest orientation as it should, if Matthes and Davis were correct. The valley has a sinuous form to it (Maps 29 and 30), its trend from its eastern end being, respectively, west, southwest, west, and southwest. Nowhere is it oriented northwest (parallel to the metamorphic foothills). Furthermore, the major cliffs in the valley, which from early on have been recognized by John Muir and others as joint controlled, not unsurprisingly parallel the valley's trends, and they are perpendicular to any trellis drainage pattern that may have existed.

One can make a quantitative analysis of the influence of northwest-trending belts of metamorphic rocks on the underlying granitic bedrock. N. King Huber and Dean Rinehart's geologic map of the Devils Postpile 15' quadrangle shows lineaments observed on aerial photographs, which they called "probably joints or faults with small displacements." Lineaments occur in both metamorphic and granitic rocks. Furthermore, some continue uninterrupted from one rock type into the next. Map 58 is a generalized geologic map of a northern part of this quadrangle, which shows the upper part of the Middle Fork San Joaquin River. This reach flows southeast parallel to the strike of the metamorphic strata, indicating that the rock structure dictates the stream course. Is that the case, or is it coincidence? Table E in the Appendix provides a basis for analyzing the major patterns of the quadrangle's lineaments, and its data are analyzed here.

What Table E clearly shows is that very few lineaments in the Devils Postpile 15' quadrangle are parallel to either the strike of the metamorphic strata or to the bearing of the Sierran crest. The 36 groups of lineaments (of 5°) are arranged into 6 larger groups (of 30°), and the larger

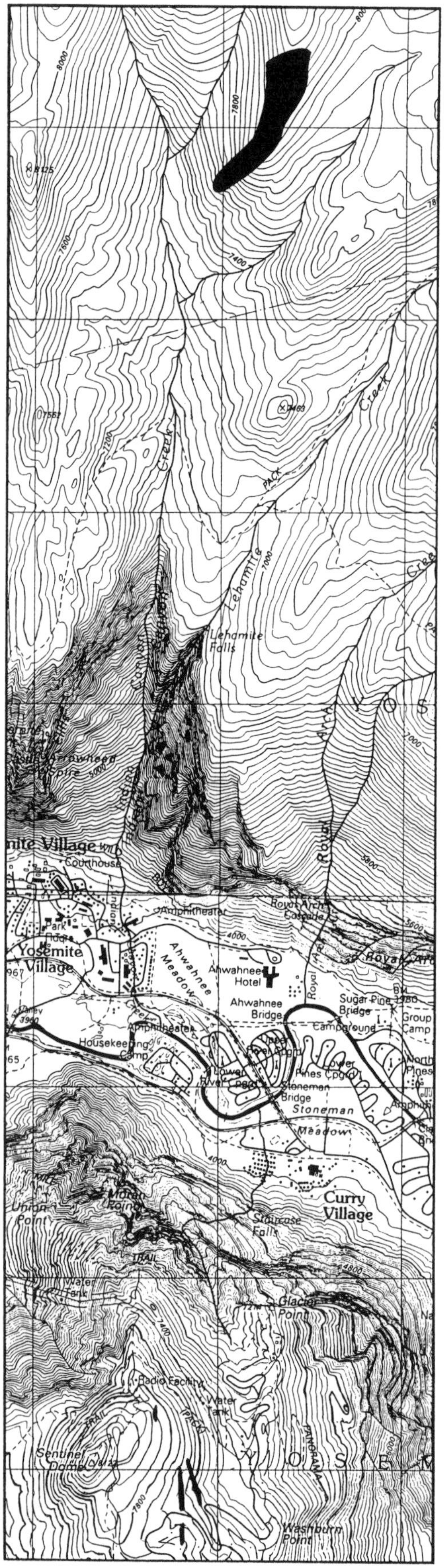


**Map 57. Metamorphic-rock outcrops in central Yosemite Valley. Scale 1:36,000.**

group that contains both the *strike* of the metamorphic strata and the *bearing* of the Sierra's crest (120-149°) has the *fewest* lineaments, not the most. Rather, lineaments cluster in a bimodal distribution pattern for both the metamorphic and the granitic rocks. For metamorphic rocks these clusters are at about 25-49° and 85-99°. For granitic rocks they are at about 175-179/0-34° and 65-99°. For the category of both metamorphic and granitic rocks there is only one cluster, at about 10-24°. Thus the clusters fall mostly within the northeast quadrant (0-90°). Had lineaments been dictated by the strike of metamorphic strata, they should have fallen mostly within the southeast quadrant. Thus the pattern of lineaments in the Devils Postpile 15' quadrangle argues against the generation of granitic landscapes in response to drainage patterns incised into formerly overlying metamorphic rocks. Consequently the modern Merced River drainage, particularly in the vicinity of Yosemite Valley, likely owes its origin and development to a response to the joint pattern of the granitic rocks.

Joints have been recognized as being very old. Bateman and Wahrhaftig in 1966 observed that in the northern Sierra Nevada deep weathering beneath the Eocene epoch's auriferous gravels extends downward along joints into unaltered rock, indicating that the joints there are at least as old as Eocene. Therefore, by the time the Merced River of the Late Cretaceous epoch began incising into the uppermost granitic rock, which I have suggested was by about 80 million years ago, the joints likely were already in existence. Furthermore, as was shown in the preceding analysis, joints extend from granitic rock into metamorphic rock, so the river, incising through the metamorphic rock just above the contact, should have been modifying its course in response to a joint pattern that was generated from below, in the granitic bedrock of the earth's brittle upper crust.

## The early, granitic Yosemite Valley

Yosemite Valley's adjacent domes and broad ridges rise above gentle upland surfaces that have been recognized, as have others in the range, as quite old. Dated lava flows, mentioned in this book, support an ancient age for upland surfaces. Perhaps the domes are older, since they are the highest features, and they *logically* should have been the first of the valley's granitic features exposed through stripping away of the overlying metamorphic rock. Matthes in 1930 (p. 115) observed that "the domes of the Yosemite region have been a long time in the making; they are among the oldest features of its landscape." He was aware of 700-foot-high Stone Mountain, a giant granitic dome in warm, moist Georgia, and noting its bald summit and sides protruding above a dense forest, apparently he concluded that the domes of Yosemite would have been likewise vegetation-free during warmer, moister times. Jean Tricart, in his monograph, *The*

**Map 58. Lineaments in northern Devils Postpile 15' quadrangle. Scale 1:62,500.**

*Landforms of the Humid Tropics, Forests and Savannas*, repeatedly mentioned the presence of monolithic domes in the humid tropics, and stated that such domes are tens of millions of years old. They are still standing despite intense tropical weathering processes.

From this observation one may conclude that weathering of the monolithic domes about Yosemite Valley (and elsewhere in the Sierra Nevada) was minimal under the tropical and nearly tropical climates of the Late Cretaceous through Eocene epochs. At the current weathering rate, broad, resistant summits such as El Capitan and Half Dome are being denuded only several inches per million years. This rate would apply for the last 15 million years of summer-dry climates, resulting in about 5 or so feet of denudation. Somewhat similar climates before them, from 33 to 15 million years ago, perhaps had more summer rain and warmer winters, but the weathering processes and rates on barren granitic bedrock should not have been that different. So during this time there was perhaps 10 or so feet of denudation. During the 50 or so million years before then, the granitic summits would have experienced warm, wet climates, which promote chemical weathering, so denudation rates should have been higher. How much higher is unknown, but it is very unlikely that the valley's domes would have protruded thousands of feet above today's remnants of metamorphic rock. Consequently, the amount of denudation of the summit of El Capitan and of the summits of granitic domes such as Half Dome and Mt. Starr King appears to have been relatively minimal, perhaps on the order of 100 feet between 80 and 33 million years ago. François Matthes, in his reconstructions both in Yosemite Valley and around Mt. Whitney, showed virtually no denudation. Therefore, we may be fairly correct to assume that today's resistant upland surfaces are fair representations of former metamorphic-granitic contacts, which would have existed not far above them.

Because stepped topography originated by or during the late Cretaceous uplift, other, resistant features such as Turtleback Dome and the Foresta bench west of it may have developed roughly at the same time as the domes, despite being much lower. This would not have been possible under conventional wisdom, which dictated low elevation and low relief until late Cenozoic time, so that these features would have been well below sea level. However, with late Cretaceous uplift, the simultaneous exposure of resistant bedrock at varying elevations would have been possible—if the Merced River had incised a fairly deep canyon through the Yosemite Valley area. Today's tropical landscapes indicate that this could be the case.

Under such climates the landscape (with the exception of the monolithic domes) generally would have been covered with a lush forest and with deeper soils than the ones cloaking the glaciated and unglaciated Sierran lands today. Consequently the jointed bedrock would have been very susceptible to the climates' weathering processes, and so differential weathering between jointed and unjointed granitic bedrock would have begun early on. Jean Tricart (p. 39 and 41) discussed this for granitic bedrock in tropical climates:

> A weathering front thus progresses downward [beneath the surface] from the large and permeable joints and the base of the regolith. All the rocks are affected but at unequal rates which depend on the petrochemical composition and the density of the joints. A high density, of course, multiplies the surfaces that are being attacked.
>
> ....
>
> This [subsurface] front is either irregular or jagged in rocks with large fractures, revealing deep pockets in highly fractured zones, *cryptopinnacles* in more widely fractured areas, and spheroidal weathering in places where joints are regularly spaced at about one metre (3 ft); or it is regular and continuous with a rapid transition from the unweathered rock to a rather evolved regolith in areas of massive rocks (that is, a non-jointed intrusive mass or the top of a thick massive bed without joints).
>
> An alternating wet-dry tropical climate, where chemical weathering is still intense although discontinuous in time, favors, as we have seen, irregular weathering fronts. But even in a wet tropical climate the difference between the two types of fronts shows up clearly. Similar rocks, whether acidic or basic, rich or poor in quartz, may reveal one or the other type within a short distance. But the irregular fronts progress more rapidly than the regular fronts, the rocks disintegrate more deeply, making them apt to slide or prone to stream erosion. Areas of regular weathering fronts therefore tend to lag behind and thus to produce residual reliefs. Many inselbergs of the wet-dry tropics and nearly all the monolithic domes of the wet tropics have this origin. (Italics and parenthetical comments his.)

Thus for Yosemite Valley and the Sierra Nevada, differential weathering would have wrought a granitic landscape that even initially was complex, and was not a widespread, essentially featureless peneplain. Over time the landscape became increasingly more differentiated as major Sierran streams incised deeply below local, relatively joint-free areas. For what the valley may have looked like, Tricart (p. 157) provides some insight:

> In granito-gneissic regions of intense weathering, the relief of valley bottoms is most irregular. Valleys are poorly graded, with marshy widenings alternating with narrowings frequently caused by rockbars [bedrock outcrops in stream channels] with or without rapids or cascades. Such widenings take the shape of depressions, almost without gradient, often 1 or 2 km (0.6 to 1.2 miles) in diameter on small streams. Their bottom, flooded or swampy in the rainiest season, is filled with grey kaolinitic clay mixed with siliceous precolloids and a little organic matter. They are the products of chemical weathering of the surrounding hills. Such deposits are normally several meters thick. Below there is the rotten rock, which may reach a great depth. Many depressions, often the most typical, are located at valley heads, as the sandstone amphitheatres of Sautter. Frequently depressions follow one another at different levels (like beads on a string) separated by minor rockbars even without a marked narrowing of the valley. Narrow passes occur only when sills produce important obstructions, with cascades or foaming rapids. In such cases the valley constrictions are directly dominated by monolithic domes.

Very important in the development of Yosemite Valley is the Pulpit Rock bedrock sill, which would have been a very major rockbar. This sill must have been—and still is—incredibly resistant, for it exists today despite repeated major glaciation. The largest glacier was about 2500 feet thick at Pulpit Rock (see Chapter 19's Figure 31 for location), and from it that glacier advanced down-canyon about 8¾ miles, if Matthes' reconstruction of the largest glacier is correct. (It is at least approximately correct.) Additionally, I have shown that the Tioga and Tahoe glaciers also must have advanced well beyond Pulpit Rock. Today the sill lies just beneath the bed of the Merced River at an elevation of about 3600 feet. Assuming that the joint-free bedrock that comprises this sill was unroofed with the rest of the local pluton by about 80 million years ago, and that its initial elevation was at the metamorphic/granitic contact, which may have been near the summit of El Capitan at about 7500 feet, then the Merced River has been incising through it over the last 80 million years at an average rate of only about 50 feet per million years. However, it may be more likely that the contact was considerably lower, judging by contacts within the valley at Indian Canyon, which are 5050 to 6050 feet in elevation, and by the contacts in the lands between Crane Flat and the Merced Grove, west of Yosemite Valley, which are about 4900 to 6200 feet in elevation. These elevations have an average value of about 5550 feet, 1950 feet above today's elevation. If we use this value for the initial elevation of the sill, then the average rate of incision through the sill over the last 80 million years was about 25 feet per million years.

The reason this second value may be more correct is that the first value has the implicit assumption that the metamorphic/granitic contact in and about Yosemite Valley was essentially a horizontal plane. However, the minor contacts of Indian Canyon clearly show that the metamorphic rock dipped some 1300 feet from a small metamorphic-rock exposure near the head of the canyon to those deep within the valley, demonstrating that the contact was quite irregular, as is also the case with the exposed contacts in the lands between Crane Flat and the Merced Grove. That the Merced River took the course it did may have been because about 80 million years ago it was incising through metamorphic rock that was fractured by the upward continuation of joints in the underlying granitic bedrock, as was mentioned earlier. The late Cretaceous uplift of 75 to 65 million years ago should have increased the river's rate of incision, and by the end of this uplift, if not earlier, incision should have reached the top of the Pulpit Rock sill. If the Merced River was at about the suggested elevation of 5550 feet, then Yosemite Valley already would have had over half of its present relief. The summit of El Capitan would have been about 2000 feet above the west end of the valley, while those of Sentinel Dome and Half Dome would have been, respectively, about 2500 feet and 3200 feet above the east end—assuming that the east end was about 100 feet higher than the west end. Matthes and others after him have assumed that the Merced River would have had a fairly high gradient through Yosemite Valley before glaciation. But as Tricart and other tropical geomorphologists have observed, behind a resistant sill a tropical river has a very low gradient—one across very deeply weathered bedrock.

Today one can observe in the Sierra Nevada, especially in crest lands, that talus slopes derived from metamorphic rock weather more rapidly than do those derived from granitic rock. Although the weathering there is physical, chemical weathering would have the same pattern, since the minerals of most metamorphic rock are more readily weathered than are those of granitic rock. Therefore, where the ancestral Merced River encountered a metamorphic/granitic contact, it should have more effectively eroded the metamorphic rock, leaving the resistant granitic rock to stand above the incising river, initiating the stepped topography. When this late Cretaceous river finally incised through the lowest metamorphic rock to reach the underlying granitic rock, the river was firmly entrenched, and there was no way for it to meander around the top of the Pulpit Rock sill. Had it been able to do so, there would have been no significant local base level, and Yosemite Valley would not have evolved into

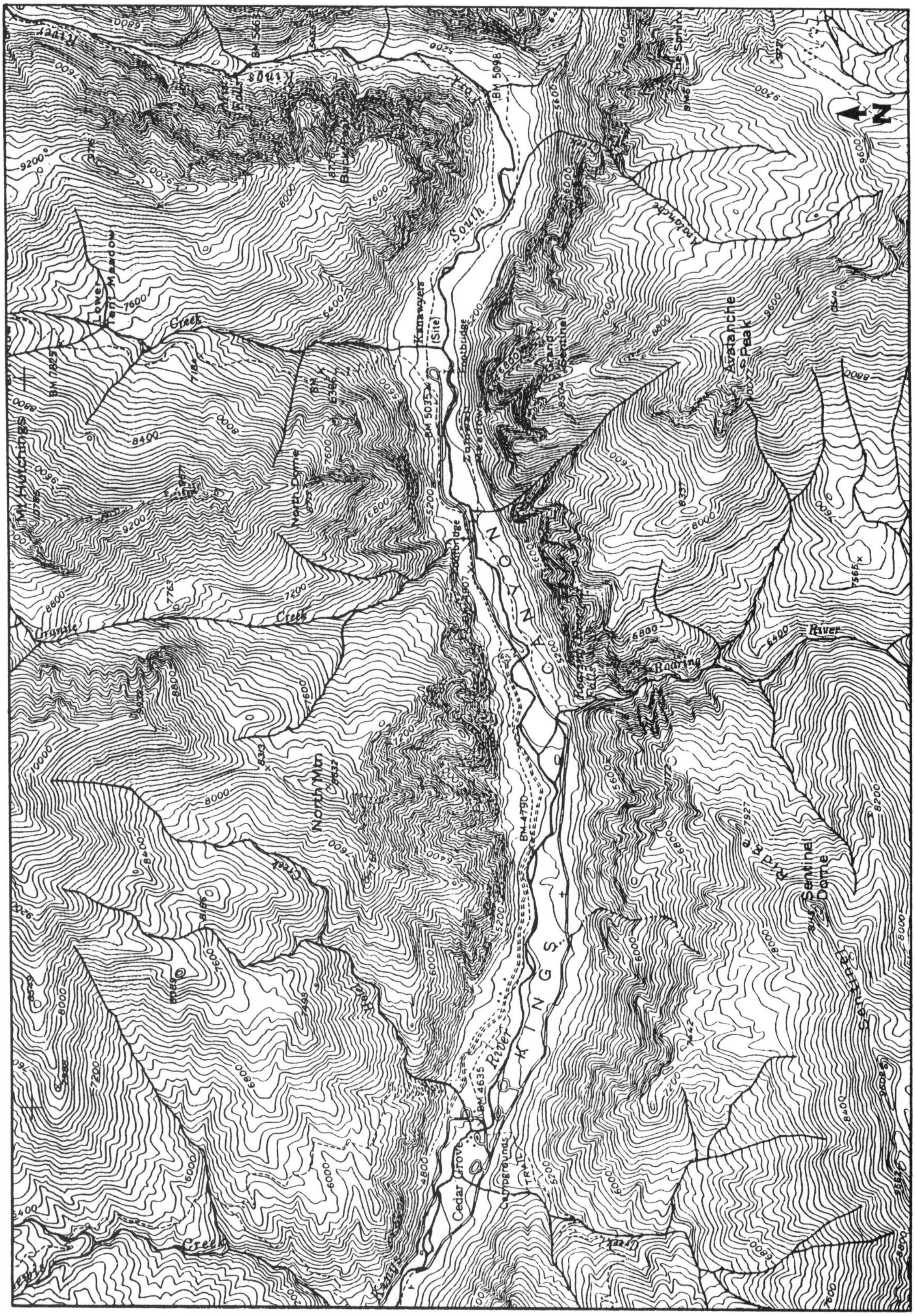


**Map 59. Topographic map of Kings Canyon, which is a longer, deeper, and narrower "Yosemite." Scale 1:62,500.**

what it is today. Instead, incision, rather than being brought to a "grinding halt," would have continued more or less unabated, ultimately creating a narrower, deeper canyon—in short, a Kings Canyon (Map 59). This did not happen, and until cooling and drying began about 33 million years ago, the jointed floor of Yosemite Valley was subjected to tens of millions of years of subsurface chemical weathering. As was mentioned under "Deep glacial lakes," such weathering can extend 2000 feet below a valley's surface, which coincidentally is the maximum depth to bedrock through today's sediments of Yosemite Valley.

Between the time of establishment of the Merced River's local base level at Pulpit Rock and the onset of major glaciation about 2 million years ago, a period of about 63 million years if the river first reached granite at the Cretaceous-Cenozoic boundary, the river would have incised more slowly than it had through the overlying metamorphic rock. And that relatively slow incision rate should have greatly diminished after the climate change at 33 million years ago, when discharge would have diminished, greatly reducing the river's erosive power. Therefore, the average rate of incision through the sill should have been considerably above 25 feet per million years before that 33-million-year date and considerably less after it, the former perhaps an order of magnitude (10 x) greater than the latter. The latter rate may have been about 6 feet per million years, which is in line with the modern rate of the smaller, unglaciated South Fork Kern River, whose rate over the last 2.4 million years has been about 1 foot or less per million years. If this is the case, then in the last 33 million years the Merced River's incision has been about 200 feet, and between 65 and 33 million years ago it would have been about 1750 feet, which amounts to about 55 feet per million years, a rate roughly ten times greater. Under this analysis, then, the Merced River at Pulpit Rock was at about 5550 feet elevation by 65 million years ago, at about 3800 feet by 33 million years ago, and at about 3600 feet today. The first two elevations are by no means exact, just my best estimates based on elevations of the area's metamorphic/granitic contact, on the known rate of the South Fork Kern River's incision, and on observations by geomorphologists that weathering and erosion rates in the tropics are much higher than in our area.

Another issue is how Yosemite Valley, which had achieved nearly modern dimensions by the onset of major glaciation, had widened over time. In the last 2 million years the valley's walls would have retreated mostly through mass wasting, discussed above in Chapter 21's "Glacial sediments." There I estimated that about 460 feet of retreat could have occurred in the last 2 million years, although I offered a number of reasons why retreat must have been much slower before then. For most of the Cenozoic era, mass wasting would have been at a very slow rate, as has been the case in Dome Land Wilderness. Tropical weathering would have been effective in highly jointed rock, so the valley's recesses would have developed early on. In contrast, it would have been negligible in virtually joint-free rock, such as that which comprises both El Capitan and Middle Cathedral Rock. Their faces should have retreated only slowly, suggesting that not much change has occurred since the late Cretaceous uplift. They are separated by about 0.6 mile today, and this width is due not to glaciation or to retreat of massive granite walls. Rather, it is due to subsurface weathering through a *broad* zone of jointed bedrock, which now lies about 1100 feet below the Merced River (review Chapter 21's Figure 32: Yosemite Valley cross profile from El Capitan to Cathedral Rocks). By the climate change at 33 million years ago, most of the subsurface weathering in the valley should have already occurred.

Note that if this analysis is correct, then at 33 million years ago, the valley was already quite wide and it possessed a fairly flat floor of river sediments atop decomposed bedrock. Also back then, when the Merced River may have been at about 3800 feet elevation at Pulpit Rock, it would have been perhaps at about only 3850 to 3900 feet elevation through Yosemite Valley proper. This is because, as in tropical lands, a reach of a stream above a rockbar (i.e., the Pulpit Rock sill) has a very low gradient. Consequently, the valley's broad floor would have been about 100 feet lower than it is today. And because resistant summits such as El Capitan and Middle Cathedral Rock would have been slightly higher than today's, the relief also would have been slightly higher. *In short, Yosemite Valley would have been about as impressive some 33 million years ago as it is today.* And if humans had been around back then, they would have made it a national park, and it would have attracted climbers from all around the world.

Great spirits have always encountered violent opposition from mediochre minds.

—Albert Einstein

# 26

# Reconstructing Yosemite Valley over the last 65 Million Years

We now have all the information necessary to make a fairly accurate reconstruction of Yosemite Valley's geomorphic evolution over time, one that does not contradict any geomorphic evidence, any geologic evidence, any geophysical evidence, any plate-tectonic evidence, any paleontological evidence, or any biogeographical evidence. Such a reconstruction has not been possible before because geoscientists were operating under the wrong set of assumptions. François Matthes made a very detailed reconstruction with precisely drawn, absolutely convincing illustrations, but it is fantasy, based on theory, not on evidence. He did not address any evidence that went contrary to theory—and there was a lot of it. My illustrations are not as precisely drawn, for I am not the artist Matthes was. Additionally, a very detailed reconstruction of a landscape implies that you have very detailed evidence to support it. But the farther back in time you go, the less evidence you have, and so the more speculative your reconstruction becomes.

I've been told to hire a top-rate artist to make illustrations so convincing that no one would doubt them. That is what Matthes did, when in spring 1937, several years after his Professional Paper 160, he engaged Herbert A. Collins, Sr. and Herbert A. Collins, Jr. to paint six Yosemite Valley reconstructions for the Yosemite National Park Museum (these based on his own detailed line drawings). These colorful oil paintings appear in black and white as Plates 2-7 in Matthes' posthumous *The Incomparable Valley*. Four of the line drawings, which were the basis for four of the paintings, were presented in Chapter 9 as Figures 15 through 18. These show how the valley might have looked during the broad-valley stage (mid-to-late Miocene, about 10 million years ago by the latest geologic time scale, not by Matthes'), the mountain-valley stage (mid-to-late Pliocene, about 3 million years ago), the canyon stage (late Pliocene or early Quaternary, about 2 million years ago), and just after the last glacier left it (about 14,000 years ago). In my reconstructions I have used the same viewpoint and perspective that Matthes used so that the reader might be able to compare our different reconstructions.

My first illustration is for Yosemite Valley as it might have appeared around 80 million years ago (Figure 49). Before then, there would have been precious little granitic bedrock exposed, and any reconstruction would be highly speculative. The elevation of the Sierran crest may have been about 5000 to 6500 to feet, although locally it could have been higher. For example, I have drawn an active, waning stratovolcano and flow in the background, centered on today's Johnson Peak of the Cathedral Range, which presumably underlay a volcano at about this time. Around 80 million years ago the last of the plutons in the Yosemite high country should have been solidifying beneath metamorphic bedrock, which in turn had a dying volcanic range atop it. In the Sierra Nevada range, volcanism was ceasing, and the Merced River drainage, if it had not already stopped rearranging in response to massive, repeated volcanic inundation, now did so. As in the Cenozoic era, in the Cretaceous period the volcanic sediments would have buried the preexisting landscape, which would have been largely a metamorphic one. In Yosemite Valley, there might have been some still

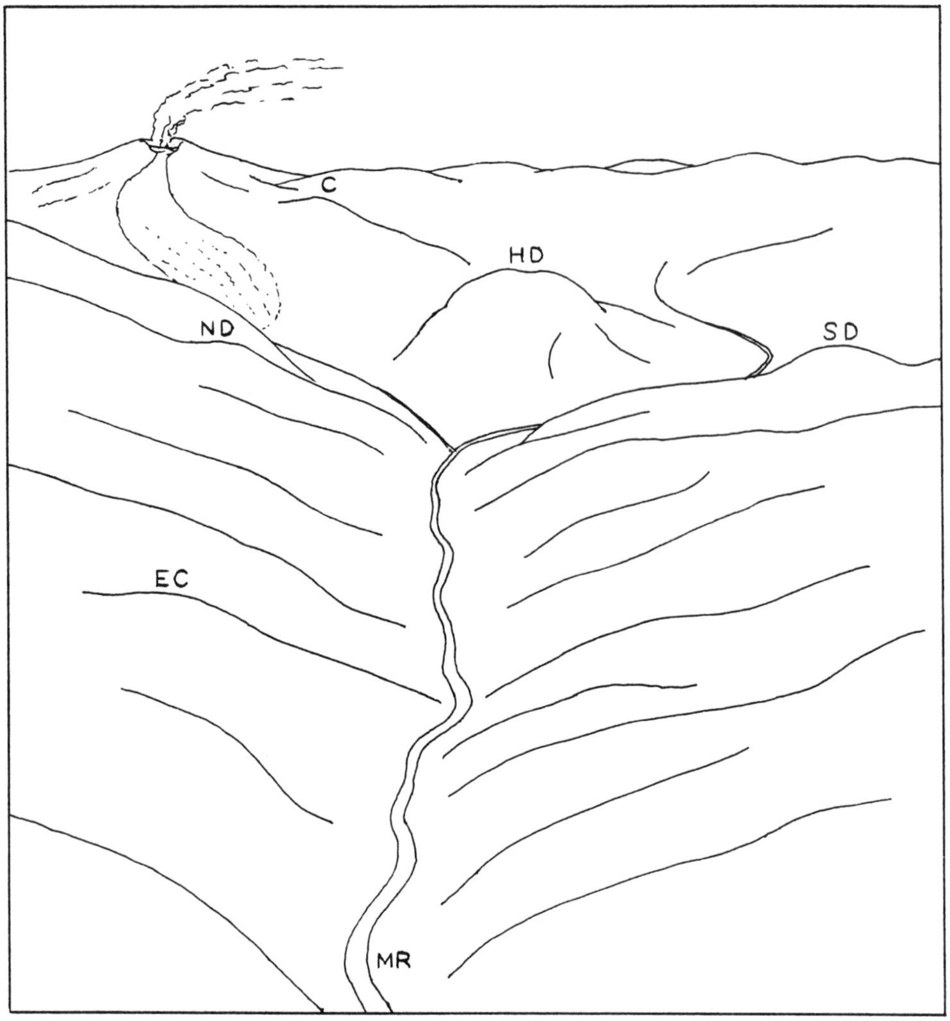


**Figure 49. Yosemite Valley as it might have appeared 80 million years ago. See Chapter 9's Figure 15 caption for the identification of the labeled features that appear on this chapter's figures.**

uneroded volcanic sediments, perhaps a lot like higher up, but then perhaps virtually none at all. Consequently, in Figure 49 I have made the drawing quite sketchy, since evidence for the largely volcanic and metamorphic landscape is essentially absent. While one can say that such a landscape existed, we cannot recreate it with any degree of confidence.

By 80 million years ago, when ceratopsids similar to triceratops may have been preyed upon by tyrannosaurids in the coastal lowlands to the west, the domes around Yosemite Valley, such as Half Dome, Sentinel Dome, and Mt. Starr King, should have been conspicuous granitic features on an otherwise largely metamorphic landscape. I have also shown a granitic Clouds Rest ridge and a granitic El Capitan summit, both somewhat speculative. The lands in the foreground are very speculative, based only on the assumption that wide parts of the valley today began to widen early on. At this time the Merced River may have been incising through lower metamorphic rock, but I have assumed that if it did, the high-density joint pattern in the underlying granitic rock would have continued into the metamorphic rock, which would have caused it to weather and erode quite readily. We can be quite certain that the Merced River had developed a course similar to today's, that is, through Yosemite Valley, but the details are unknown. Still, at this time the river should not have yet encountered the Pulpit Rock sill, so the river should have been actively eroding downward, not meandering behind a rockbar, and through the valley it would have had a moderate gradient.

Late in the Cretaceous between around 75 and 65 million years ago, major uplift occurred in the southern

**Figure 50. Yosemite Valley as it might have appeared 65 million years ago.**

Sierra Nevada due to a collision of a major terrane along the continental edge just south of it. Uplift was greatest in the Mt. Whitney area (or perhaps south of it), and there it was about 8000 to 9500 feet. Amount of uplift decreased farther north, until in the Feather River drainage it was essentially zero. This asymmetrical uplift increased the gradients of north-flowing tributaries, which were oriented more or less parallel to the Sierran crest, and with renewed vigor they expanded their headwaters southward into adjacent drainages. In the vicinity of Yosemite National Park, located roughly 60% of the distance from the Feather River drainage toward Mt. Whitney, the uplift then should have been about 4800 to 5700 feet, which would have raised this part of the range to modern heights.

South of Mt. Whitney the Sierra Nevada subsided in response to extensional faulting either late in the Cretaceous and/or early in the Cenozoic. The northern Sierra Nevada, not significantly uplifted, remained about as low as it is today, and its sluggish rivers accumulated enormous amounts of early and middle Cenozoic gravels. As much as 12 miles of rock was eroded from the Late Cretaceous epoch through the Eocene epoch, most of it being volcanic and metamorphic rocks that overlaid the granitic rocks. During this time of tropical and nearly tropical climates, the Sierra's granitic stepped topography originated and matured. It has changed very little in post-Eocene time. No identifiable, measurable, post-Eocene uplift has occurred, except perhaps in the northernmost part of the range, where I have inferred up to about 500 feet of late Cenozoic uplift.

At the start of the Cenozoic era 65 million years ago, the dinosaurs had met their demise, and small, primitive mammals, including our ancestors, had inherited the giant-free kingdom. At this time the granitic Yosemite Valley, having been uplifted by about 2500 feet to its modern heights, was continuing to evolve under warm, wet climates that generated intense chemical weathering (Figure 50). Perhaps around this time the Merced River had reached the top of the Pulpit Rock sill, west of the valley proper, that spot being at an estimated 5550 feet elevation. It could have been hundreds of feet higher, or lower, given the scarceness of the evidence for this ancient time. The Merced River, having encountered the massive rockbar, greatly slowed its rate of incision. Upvalley from the top of the sill—a rockbar—the fractured granitic bedrock beneath the valley's floor should have already begun significant subsurface weathering. If so, then the river would have had a low gradient and it would have meandered across the valley' river sediments, which lay atop decomposed bedrock. However, it may have been tightly constricted between the bases of El Capitan and Middle Cathedral Rock. Even back then the valley deviated from the classic V-shaped cross profile of late Cenozoic mid-latitude river canyons, developing instead a generally broad U-shaped cross profile characteristic of today's equatorial and tropical, granitic ranges. The valley's floor wouldn't have been as wide as it is today, since mass wasting has obviously occurred over the last 65 million years, but it could have been roughly half as wide.

If the sill's estimated elevation is more or less correct, then at the start of the Cenozoic era the summit of El Capitan would have been about 2000 feet above the west end of the valley, while those of Sentinel Dome, Glacier Point, and Half Dome would have been, respectively, about 2500 feet, 1600 feet, and 3200 feet above the east end. These estimates suggest that the valley had achieved at least half of its present relief. The lower part of Tenaya Canyon also would have achieved a similar relief. All of the valley's major features that today rise above 5600 feet elevation were recognizable, and the slopes of these features were about as steep as those of today's, governed by similar joint patterns. The lower-elevation features, such as Royal Arches, obviously did not exist, if indeed the valley's floor was at about 5600 to 5700 feet elevation. Both Lower Brother and Lower Cathedral Rock barely protruded above the valley's floor.

Since this time, weathering and erosion have removed only about 100 feet of bedrock from the most resistant summits. Broad uplands, where forested, had a higher rate of denudation. Consequently, 65 million years ago they would have been perhaps 200 to 500 feet higher than today's, so the domes would not have projected as much above their surrounding lands as they do today. Just as the Merced River should have encountered the Pulpit Rock sill, around this time it also should have encountered resistant bedrock in Little Yosemite Valley. Then from that valley's west end the river would have plunged down to the eastern end of Yosemite Valley via a series of rapids, cascades, and two small falls—the incipient Nevada and Vernal falls. Other falls existed, especially Upper Yosemite Fall, which could have had about two thirds of its present 1410-foot height if the valley's floor back then was at about 5600 feet elevation. Also today's 1612-foot Ribbon Fall back then may have achieved a similar proportion. These falls, and more-diminutive ones such as Sentinel and Silver Strand falls, leaped from the mouths of high hanging tributary canyons. Stepped topography had already originated. However, the valley floor was high above the brink of today's Bridalveil Fall. That fall would not initiate and then begin to grow in height until millions of years had passed.

Throughout the first half of the Cenozoic era the valley continued to deepen, but only slowly, since the massive bedrock of the Pulpit Rock sill at its western end served as an effective base level for the reach of river through the valley. Behind that base level the jointed bedrock of the valley floor weathered over millions of years, eventually producing deep basins of decomposed end-products of the granitic bedrock. Both immediately above and below the valley the Merced River's gradient was much higher, and its gradient was irregular (stepped). Also, throughout its existence the slopes and cliffs of Yosemite Valley were subject to slow retreat due to mass wasting, which was governed by the spacing and pattern of joint planes. The rate of mass wasting, especially rockfall, generally was too slow to produce talus slopes.

All of the falls in Yosemite Valley proper should have achieved their full heights by the time the climate changed profoundly 33 million years ago, early in the Oligocene epoch, overall becoming cooler and drier (Figure 51). Rates of chemical weathering then decreased substantially. Weathering and erosion would have been more effective in the shallow canyons of the rolling uplands above the valley's rim, but even they would have changed little since the early Oligocene. Today's valley uplands are part of a relict tropical landscape, as are many relatively flat surfaces, such as the floor of Tuolumne Meadows and other broad uplands of the range.

By 33 million years ago, Yosemite Valley may have reached 80% of its width. Resistant summits such as El Capitan, Middle Cathedral Rock, Sentinel Dome, North Dome, Half Dome, and Mt. Starr King stood only about 10 to 30 feet higher than today's summits. The same would apply to lesser features, such as Mt. Broderick and Liberty Cap, which now would have achieved close to their modern size. The prominent joint-controlled gully that separates them today would have been much narrower and not as deep, and their southwest faces would have projected perhaps some 200+ yards beyond their present location. Similar reconstructions apply to other similar faces. The forbidding, exfoliating northwest face of Half Dome, instead of being slightly concave, would

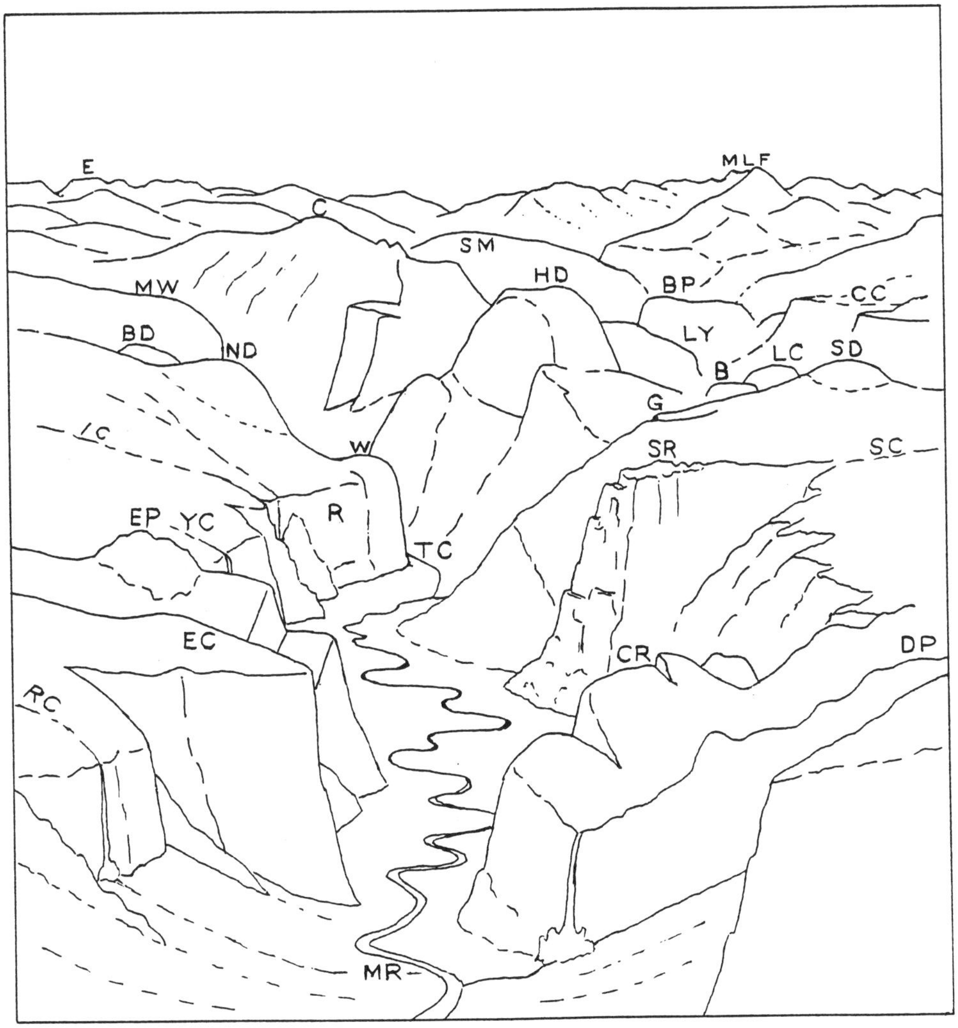


**Figure 51. Yosemite Valley as it might have appeared 33 million years ago.**

have been slightly convex, much like its back side. And I suggest that Royal Arches was an ordinary cliff, somewhat like its western continuation, neither of them having significant arches. Major arches on faces and slight oversteepening of faces would develop much later, in response to repeated depressurization during the two million years of glaciation that much of the range experienced.

Over the 32 million years that had elapsed since the start of the Cenozoic era, the Merced River, incising imperceptibly through the Pulpit Rock sill, had cut down to an elevation of perhaps about 3800 feet. Upstream from this spot, in Yosemite Valley, the river meandered across a flat floor, at perhaps about 3850 to 3900 feet elevation, beneath which lay a very thick layer of decomposed granitic bedrock. Therefore, the valley was slightly deeper than today's, if my analysis is correct, and slightly narrower. Virtually all of the valley's major features existed, including Lower Yosemite and Bridalveil falls. Back then you could have used today's topographic map of the valley to navigate along its forested floors, up its forested recesses and side canyons, which would have lacked talus slopes, and across its forested uplands. The same applies to the rest of the Sierra Nevada from about the northern border of Yosemite National Park southward. The lands have evolved only a relatively minor amount over the last 33 million years, and you could have used today's topographic maps without getting lost. North of the park's border, violent rhyolitic eruptions were beginning to occur, and in time most of the northern Sierran landscape would become buried under various volcanic deposits.

Evidence from temperature-sensitive marine organisms along the California coast and from temperature-indicating plants fossilized in western Nevada indicate that by around 15 million years ago the modern Mediterranean climate had originated, complete with the cold coastal

**Figure 52. Yosemite Valley as it might have appeared 2½ million years ago, just before glaciers entered the valley.**

current and relatively dry summers. In these last 15 million years both chemical and physical weathering have been minimal, and summits such as Yosemite Valley's El Capitan and North Dome have experienced only a few feet of denudation. Unglaciated, granitic Sierran lands at 15 million years ago had achieved topography nearly identical to today's. The main difference appears to have been slight canyon deepening by the major rivers, such as the Tuolumne and the San Joaquin, whose unglaciated lower canyons may have deepened by tens of feet, or perhaps even by about 100 feet. In the lands that would become glaciated, the canyons probably deepened more, but not much more.

Glacial morphology has developed best near the crest of the Sierra Nevada, but this morphology is due more to mass wasting, particularly earthquake-induced rockfall, than to glacial erosion. The principal role of an alpine, or mountain, glacier in any of the world's glaciated ranges is to transport the products of mass wasting, not to erode the bedrock through abrasion and plucking. Since the Sierra Nevada has been high for tens of millions of years, it highest lands could have experienced many minor episodes of cirque glaciation. Such glaciation may have occurred as early as 15 million years ago, when Antarctica first developed a major ice cap. In each glacial episode, rockfall would be removed. By the time major glaciers first appeared about 2 million years ago, the lands of the Sierran crest already should have experienced much glaciation, which should have widened the heads of canyons into cirques. However, the exhumed granitic landscape surrounding the Stanislaus River's Clark Fork Meadow already had a broad, flat-floored cirque when it was buried about 20 million years ago, so perhaps not that much widening has occurred in the High Sierra's cirques. Gla-

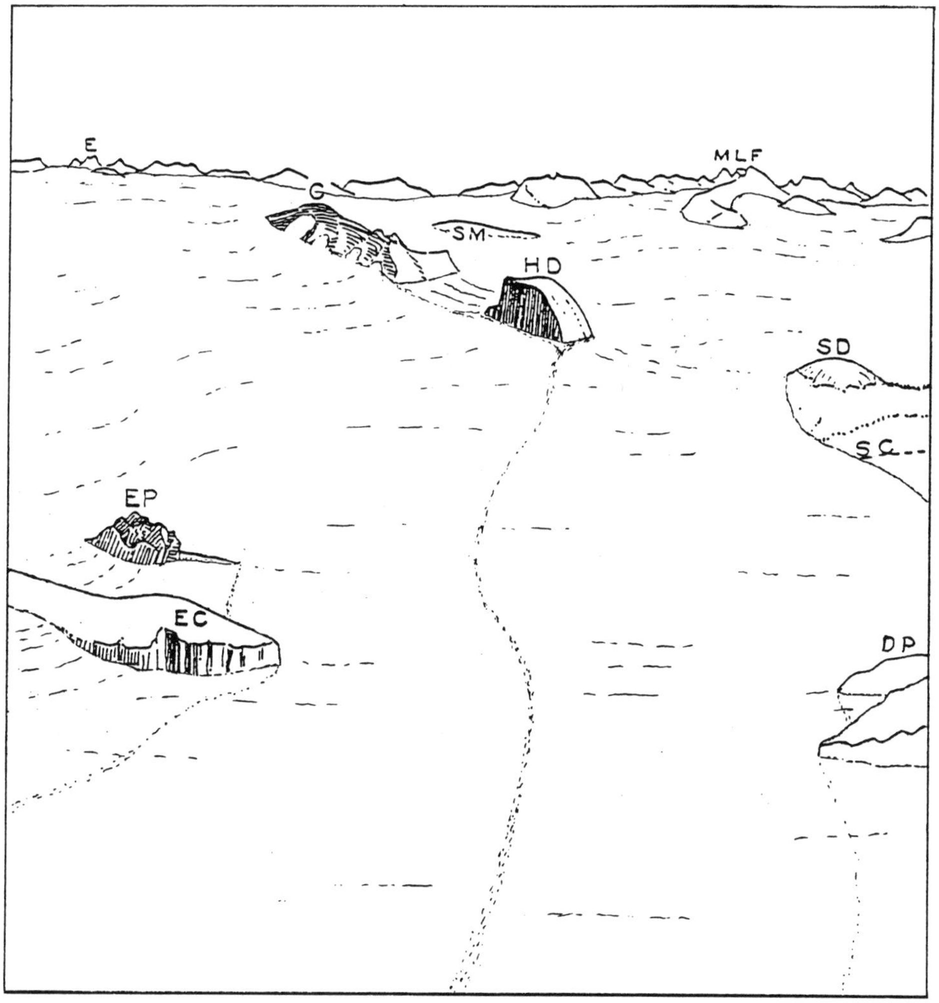


**Figure 53. Yosemite Valley as it might have appeared during the largest pre-Tahoe glaciation.**

cial deepening, also was minimal, for had it occurred, the cirques would have been overdeepened. Instead, the floor of almost every cirque hangs above the floor of the canyon just beyond it. This indicates minimal erosion, despite the cirques having been under glacial ice for far more time than the glaciated canyonlands below them.

The rate of mass wasting greatly increased late in the Cenozoic, judging from the excess amount of post-glacial talus present today, indicating a rate fast enough to fill Yosemite Valley to the rim in 20 million years (at today's volume of space between the rims above the sedimentary floor, not the bedrock floor). This great increase in rate is due to several causes. First, and perhaps least important, was the valley's gradual deepening over time, which led to taller cliffs that had more surface area from which rocks could fall. Second, during glaciations, there would have been some minor glacial abrasion of loose wall rocks. Third, the westward-migrating, post-Laramide-orogeny extensional-fault system generated earthquakes, and as it got closer to the Sierra Nevada, its earthquakes, being closer, were increasingly more effective at dislodging the valley's wall rocks. Fourth, after each glacier left the valley, post-glacial pressure release generated a great amount of mass wasting. Finally, the overabundance of talus in the valley's deep recesses may be due to a misperception. Geoscientists have assumed that all of today's abundant talus formed after the last glacier left the valley, about 14,000 years ago. However, rather than being carried away by glaciers, preexisting talus in these recesses was instead buried by them, and hence much, if not most, of the talus had already accumulated before the last glaciation, not after it. Therefore its average age is much more than we have imagined; hence its rate of production is much less.

**Figure 54. Yosemite Valley as it might have appeared after the Sherwin glaciation.**

Just before the first major glaciers developed in the Sierra Nevada about 2 million years ago, Yosemite Valley had widened, through the mass-wasting retreat of its cliffs, to about 90% of its present width (Figure 52). Since then, its walls, on the average, have retreated an estimated 150 or so yards. The resistant summits were only about 1 to 5 feet higher than today's. In the valley's forested recesses and side canyons, talus accumulation still was minimal, despite earthquakes from faults along the east base of the range, which originated by 3 million years ago, if not considerably earlier. Today, similar unglaciated landscapes, such as those of Dome Land Wilderness, do not develop talus slopes, despite lying somewhat closer the east-side faults than does Yosemite Valley.

When the first major glacier finally entered the valley about 2 million years ago, it would have removed all the loose rock it encountered along the valley's walls. Resistant bedrock would have been virtually unchanged. A main feat the early glaciers accomplished was the removal of the deep, weathered bedrock of the valley floor. Each succeeding early glacier would have removed some of this detritus, excavating a hollow in the sediments that was larger than the one excavated by the previous glacier. As the hollow grew with succeeding glaciations, so too did the post-glacial lake that filled the hollow. But as the base of each succeeding glacier in the valley became progressively lower through excavation of the sediments, the ice surface of each succeeding glacier also would have lowered—assuming that the glaciers were all of about the same size. In actuality, some would have been larger than others. Nevertheless, the overall pattern was one of lowering the early glaciers' ice surfaces over time. The ice surfaces of at least one early glacier overflowed the

**Figure 55. Yosemite Valley as it might have appeared during the Tioga glaciation.**

valley's rim (Figure 53), based on several old erratics left there, but by the time of the maximum known glaciation, tentatively the Sherwin, which lasted from about 900,000 to 800,000 years ago, the ice surface likely was below the valley's rim. That last, large early glacier should have been the final one (or perhaps the only one) to flow along the bedrock floor that today lies deep below the valley's sediments. When the Sherwin glacier finally retreated, it would have left the valley's largest, deepest postglacial Lake Yosemite, dammed in the vicinity of today's Pohono Bridge. This lake would have been about 7 miles long and up to 2000 feet deep (Figure 54).

After the Sherwin glaciation, all of the following glaciers were smaller, perhaps due in part to the significant rain shadow that began to develop in response to the rapid uplift that now began in central California's coastal ranges. These smaller glaciers deposited more sediments than they removed, so after each successive glaciation there was a thicker layer of sediment and a smaller, shallower Lake Yosemite. As the floor filled with sediments, the base of each succeeding glacier was higher, as was its ice surface—the reverse situation of the early glaciers. Two of these post-Sherwin glaciers were larger than any of the others, the Tahoe and Tioga glaciers. These two were larger than previously supposed, advancing into (and possibly just beyond) Merced Gorge. Also both glaciers were thicker, especially in Little Yosemite Valley, where they were twice as thick as both USGS geologists François Matthes and Clyde Wahrhaftig had supposed. As in all other glaciated canyons studied so far, the Tahoe glacier was slightly larger than the Tioga glacier, although in Yosemite Valley the Tioga's ice surface was slightly higher than the Tahoe's simply because some sedimentation had occurred between the two glaciations, giving the Tioga a higher base, and hence a higher ice surface (Figure 55). After the Tioga glacier ultimately left the valley about 14,000 years ago, it might have had a swampy floor, but not a lake. The greatest change between then and now was the significant accumulation of talus, generated mostly through depressurization of the valley's lower slopes when the glacier rapidly left them, after having exerted force against them for some 20,000 years.

Because the landscape evolution of the western slope of the Sierra Nevada from at least the Stanislaus River drainage south to the Kern River drainage has been interpreted mostly in accordance with William Morris Davis' ideas, this landscape needs to be reassessed in accordance with variable climates and mass wasting as well as with stream and glacial erosion. Moraines have been dated incorrectly, and virtually all of them west of the Sierra Nevada crest need to be reexamined. And because there has been no measurable late Cenozoic uplift, all geophysical and plate-tectonic models incorporating such uplift are void. Finally, since many of the world's ranges have had their glacial and/or uplift histories derived from the same erroneous assumptions as those used in Yosemite Valley and the Sierra Nevada, they need to be restudied. Alpine geomorphology, gone astray around the world—not only in the Sierra Nevada—back in the 1860s, is long overdue for a major revision.

# Epilogue

## Science, the Sierra Nevada, and the World

### Riddles in the Rocks

A riddle is a mystifying, misleading, or puzzling question posed as a problem to be solved or guessed; it is a conundrum, an enigma, a mystery. This book began with a series of riddles, which included the following:

- How can three pairs of lateral moraines left by a 20,000-year-old glacier be dated at 20,000, 150,000, and 800,000 years old?
- When are voluminous glacial deposits—moraines—not the result of glacial erosion?
- Where do slabs of rock repeatedly fall from canyon cliffs over millions of years and yet the canyon never widens through cliff retreat?
- How does a steep-walled, flat-floored glacial valley develop in the absence of glaciers?
- How can a block of the earth's crust that was raised above sea level during the last 5 million years possess a mountain landscape that is over 60 million years old?

These are paradoxical questions, that is, they are self-contradictory. For example, in the first riddle, a 20,000-year-old glacier cannot leave moraines of three very distinct ages. In particular it cannot leave moraines that supposedly were deposited long before it ever existed. Riddles in the rocks—they should not exist. We should have questions in science, but not riddles. These exist because alpine geomorphology since its early origins has been formulated on faulty assumptions. Field evidence was available in the late 1800s and early 1900s to refute these assumptions, but it was ignored. Consequently, absurdities developed. Each Sierran canyon had its own climatic history, its own glacial history, and its uplift history—and paradoxical riddles arose. But there are no paradoxes in nature. If we perceive them, we are missing some evidence or we are misinterpreting it. Which brings me to a penultimate riddle:

- How could geologists for more than a century—from the 1860s through today's 1990s—reach conclusions contrary to their very own evidence?

Leslie Ransome, François Matthes, Clyde Wahrhaftig, and N. King Huber—all USGS geologists—ignored their own stratigraphic evidence when they considered Sierran uplift. Others, starting with Josiah D. Whitney in the 1860s, ignored rockfall in Yosemite Valley, until Gerald Wieczorek and I independently began looking at it in the 1990s. Around the world, rockfall was ignored in glaciated mountain ranges. By a consensus that originated in the 1860s, rockfall does not occur where glaciation does. You can quantify stream erosion and glacial erosion—at least in theory—but you cannot quantify rockfall. You cannot say when, or where, or how much of it will break loose. Rockfall came to be ignored. However, a bit of revisionism has appeared in the last few years. It seems that rockfall does occur in glaciated areas, since rocks are required by some hypothetical models to provide the erosive tools to abrade the bedrock of a canyon's floor and walls. However, when these rocks fall onto the glacier, the walls of the canyon do not loose any mass. The walls do not retreat, and the canyon does not widen. This is because by definition widening can happen only through glacial erosion. Likewise, when those glacially transported rocks are deposited atop a moraine, they will be evidence of glacial plucking, for again, by definition they can be derived only through glacial erosion.

If this sounds absurd for 1990's science, it is not the only absurdity. Others exist. For example, take beach erosion along the Atlantic coast. In the May 1996 issue of *GSA Today,* a publication of the Geological Society of America, four perspectives were presented under the title, "Modeling Geology—The Ideal World vs. the Real World." In my opinion, the most enlightening perspective is by Orrin Pilkey of Duke University. He begins with the following paragraph.

> Mathematical models of beach processes fail as predictive tools for practical applications. Currently, the most important coastal management use of such models is the prediction of costs and required sand volumes of prospective replenished beaches. But we are no closer to predicting the 10 year behavior of a beach than we are to the prediction of 10 years of weather.

Predicting 10 years of future weather is an absolute impossibility, so Pilkey implies that with the current modeling, mostly by coastal engineers, prediction of future beach erosion is equally impossible.

Why can't it be done? Like glacial-erosion or uplift models, beach-erosion models are not based on reality. First, models cannot handle unknown dimensions. Waves advance up a sloping beach, carrying sand, then they retreat, also carrying sand. The highest point from which a wave can carry sand away from the beach—thereby eroding it—will be, quite obviously, the highest point the wave reaches. But when sand is carried away from the beach, how far offshore is it transported? To the far edge of the surf zone? To the breaker-free shallow water just beyond it? To the edge of the continental shelf many miles away? The offshore transport of sand is highly variable, so it defies quantification. Therefore, modelers, in their questionable wisdom, have drawn arbitrary boundaries. Some say that no sand is transported beyond the depth of 4 meters (13 feet), while others say none is transported beyond the depth of 10 meters (33 feet). Neither arbitrary barrier has any basis in reality. Then too, there is the problem of storms, especially severe storms such as hurricanes, which occasionally hit the Atlantic coast. One hurricane can do more than ruin your day; it can obliterate your beach. But storms raise the same problems as rockfall: you cannot quantify them. You cannot say when, or where, or how great a storm will hit your beach. Therefore, storms are ignored. For modelers, they simply do not exist. Tell that to coastal residents who may have lost their beachfront property in a major storm.

There is a significant difference between ludicrous beach-erosion modeling and ludicrous alpine-geomophology modeling. The former can cost cities and counties millions of dollars, and result in unexpected environmental damage, which may put in peril not only shoreline real estate, but also the lives of endangered coastal plants and animals. Hopefully, those public agencies footing the bill will ultimately wake up and spend the public's tax dollars more wisely. In contrast, ludicrous modeling of the uplift of a range or of glacial erosion in it has no ill effect. There are no lives lost, no property lost, no lawsuits, no threat to endangered species. Alpine geomorphologists are totally unaccountable, for there is absolutely no threat from anyone to make them address reality. That is why different members of the high-tech Southern Sierra Nevada Continental Dynamics Working Group—perhaps the brainiest team of geoscientists ever assembled for field work—could reach diametrically opposing views on the last 20 million years of the range: one of major uplift, the other of major subsidence. With their soft data, which requires them to make unverifiable assumptions about the nature of the rock deep within the earth over the last 20 million years, they should not even consider whether the range has risen or fallen.

To me, modeling is an addiction. Once you master it, it assumes a reality of its own, while reality becomes a morass of interacting, unfathomable variables that must be avoided at all cost. Furthermore, modeling doesn't have the exertion, pain, and danger that can accompany research in the mountains. Far better to stay in a comfortable office and sip your coffee. Can modelers ever be rehabilitated to address reality? Probably not. You be the judge, based on this passage by Orrin Pilkey.

> Criticism of modeling is usually not well received by modelers; a constructive dialogue between model developers and critics is missing. Resistance to objective oversight is stiff, perhaps in part because it is a highly sensitive bread-and-butter issue. [Their jobs are at stake.] Critical review of models of earth surface processes can be difficult to publish. A common reviewer's response is that "we are already aware of these problems." My coworkers and I have recently published several papers criticizing coastal models, pointing out what are clearly problems fatal to their application. So far we haven't found anyone who disagrees with our criticisms. We also believe that we have had close to zero impact on the use of models.

## Interlude: For lack of rockfall

The pundits of geotruths have decreed:
"There is no great rockfall in our glaciated ranges."
They have turned deaf ears to the thunder of rocks
impaling a glacier's crevassed skin.
They have closed blind eyes to the amassing of debris
defacing a glacier's ermine coat.

The pundits of geotruths have observed:
"Vast moraines are from glaciers in our mountain ranges."
They have shown with their cloistered experiments
how glaciers sand the slickened bedrock.
They have calculated with their erudite equations
how glaciers pluck the loosened blocks.

For lack of rockfall in mountain canyons
the canyons widen through glacial erosion.
For their ability to widen through erosion
glaciers so must deepen through erosion.
For canyons to be enlarged through glaciation
before glaciation they must have been shallow.
For canyons to be initially shallow
their preglacial range would have been low.
So for a lack of rockfall the world's ranges
have suffered great and recent uplift.
But to spite the pundits, the rocks still fall.

## Science gone astray

Whereas beach-erosion modelers at least have been aware of their shortcomings, geoscientists studying alpine geomorphology either have not been or at least have not admitted it. In the world's glaciated mountain ranges, mass wasting (especially rockfall) is the most important geomorphic process, far more so than stream erosion or glacial erosion. Nevertheless it has been belittled to such an extent that geoscientists have essentially acted as if it did not exist, much as beach-erosion modelers disregard hurricanes. There is something seriously wrong with how geoscientists do science if for more than a century many of them have reached conclusions contrary to their very own evidence, and if no one in their field (I am a mountaineering naturalist, not a geologist) has called attention to their errors. Why have geoscientists gone astray? There are a number of reasons.

First, given sufficient time, an unsubstantiated tale can become a genuine truth. I suppose this is how many religions originated; myths, repeated generation after generation, in time become accepted as fact. Prof. William Brewer, a botanist with the Whitney Survey, in 1862 visited the Gold Rush miners working in the Stanislaus River and Tuolumne River drainages, and uncritically accepted their account of a former, large meandering river, based on lava remnants. In this area Table Mountain is composed of a series of remnants of a lengthy lava flow that once had descended through a supposed bedrock-walled canyon and had supposedly buried the river's gravely bed. In the foothills the base of the lava flow is sometimes on bedrock, but gravels, significantly, often are missing. Furthermore, the base often is on volcanic sediments, not bedrock. These two relations were ignored. Josiah Whitney published William Brewer's account, and later on it was embellished, such as in Joseph Le Conte's textbook, *Elements of Geology*. Therefore, by the time Leslie Ransome mapped the lava flow in the 1890s, the myth of a lava flow advancing along the floor of a bedrock-walled canyon had been in existence for about four decades. It did not matter that in the Dardanelles area he himself mapped the base of the flow a full 1000 feet above the base of the oldest, underlying layer. The myth had survived long enough to become an unquestioned fact. Ransome then used his correctly mapped (though incorrectly interpreted) flow as a tool for measuring how much Sierran uplift had occurred since the flow, and the late Cenozoic Sierran uplift myth was on its way to becoming an unquestioned fact.

Second, errors arise because geoscientists, like most humans, prefer simplicity to complexity. And simplicity is what geoscientists proposed for the Sierra Nevada: late Cenozoic uplift caused rivers to cut deep canyons, then glaciers followed, deepening and widening them. All this occurred during a time of essentially modern climates, under which weathering and erosion occurred at essentially modern rates. This simple interpretation has a philosophical and scientific basis known as Occam's razor, formulated back in the 1830s: the simplest of competing hypotheses or theories is to be preferred over more complex ones. This is a good principle, but only if we have sufficient knowledge to objectively judge every competing hypothesis. Geoscientists have known since the late 1870s that wet and warm climates prevailed in the Sierra Nevada, but by assuming that its lands have evolved mostly in the last few million years, they could ignore variable climates, the variable weathering and erosion rates produced by them, and the uncertainty of early Sierran reconstructions due to a paucity of evidence. The more complex hypothesis I present in this book could have been a plausible explanation back in the 1890s, but it was never considered. Simplicity rules.

Third, there has been rampant, uncritical thinking, due in part to geomyths being accepted as fact, as discussed under the first reason. My Figure 41, in Chapter 22, is a cross section through the Dardanelles, which shows a sequence of various volcanic units, the youngest at the top, the oldest at the bottom. If a professor, in giving an exam to elementary geology students, were to ask them to recount the area's history, most would answer correctly: the lower layer was deposited on bedrock, the middle layer was deposited on the lower, and the upper was deposited on the middle. But the professor, being well educated in Sierran uplift history (unlike his undergraduates), would have to believe in the reverse: the upper layer was deposited first, the middle next, and the lower last. This illogical interpretation is what the late Cenozoic uplift hypothesis demands.

Another example of uncritical thinking is the acceptance of a large, deep lake after the last glacier retreated from Yosemite Valley. Early on, visitors and residents observed that the flat floor of Yosemite Valley occasionally floods. During times of high discharge, the water backs up behind the El Capitan moraine. Therefore, the explanation of the flat floor must be glacier related. But if that moraine hadn't existed, the valley still would flood, for the river then would be backed up where it crosses rockfall just downstream from the moraine. The flat floor of Hetch Hetchy valley also flooded, as has been known for a long time, but this observation has gone unnoticed. Where reaches of rivers have a very low gradient, flooding occurs and floodplains develop. This is the simplest explanation for Yosemite Valley's flat floor, and it should have been accepted, if Occam's razor had been applied.

A fourth reason for geoscience going astray is the demise of field work. The best example with regard to this book is the USGS *Geologic Map of Yosemite National Park and Vicinity, California.* For it, Clyde Wahrhaftig mapped most of the area's glacial deposits from evidence perceived on aerial photos, not from evidence seen in the field. This cannot be done with any degree of accuracy. My own personal experience is that about three-fourths of the "glacial deposits" I perceived on aerial photos were either not glacial or were of the wrong glaciation I had assigned to them. In his excellent book, *Tectonics of Suspect Terranes,* David Howell added a short final chapter, "The strategy of a field geologist," which goes to the root of the problem. Field mapping no longer is in vogue. He states that as recently as the early 1960s, a commendable university dissertation might have been to map and interpret the geology of a 7.5' or 15' quadrangle. Do that today and you've just banned yourself from the job market.

The plate-tectonic revolution changed the direction of geology. Overall, this change has been for the good, since we have a far better understanding of geology. However, there still are many major, controversial issues—even in California alone—that will remain unresolved until a lot more basic field work has been done. There are at least two reasons why it is not being done. First, it is very time consuming, and therefore expensive. Second, it is generic, mundane research that is done best by generalists. Today, generalists seem to be going the way of the dinosaurs, while specialists evolve, diversify, and multiply. Like me, a mountaineering naturalist, the general geologist is a jack of all trades, but a master of none. Like me, Howell has noted that being a generalist carries with it a pejorative connotation when it comes to securing a research grant, finding a job, or getting promoted. When in October 1995 about one fourth of the staff of the U.S. Geological Survey were fired, ostensibly to reduce Federal spending, the largest group of those fired was the general geologists.

Recently, my twin brother, who is a senior engineer at a high-tech company in Silicon Valley, commented that earning a Ph.D. in even the country's finest electrical engineering departments is worse than earning an M.S. In the several additional years you spend to secure that prestigious degree, you learn more theory, but you distance yourself from reality. Your designs of complex circuits on microchips are abysmally inferior to those of the competition in the high-tech companies. A similar situation exists in the geosciences. With your Ph.D. in hand, you have become a specialist, perhaps even *the* expert in your very narrow subfield (which you yourself may have defined), but you are sadly deficient in generalities. I wonder just how many Ph.D. graduates being turned out by geology departments today can actually produce a relatively accurate geological field map.

In Cal Berkeley's Geography Department, where I spent too many years as the eldest graduate student, a similar situation exists. The university has a field station on Moorea, just off Tahiti, in the southern Pacific Ocean. There, geography students can study processes of coral reefs. Likewise, the department has expanded its physical-geography labs and has bought increasingly sophisticated equipment, so students can become expert at pollen identification or at soil geochemistry. But from the early 1990s through the present (1996), you could not obtain a basic education in physical geography because the geomorphology professor had retired and had not been replaced.

But then, who needs undergraduates? It is the graduates (getting a good undergraduate education elsewhere) who will do your research and increase your stature among your peers. As a geology professor on my dissertation committee stated: "Seeing [undergraduate] students at the beginning of the semester makes me sick. It means I'm going to have to teach them." Professors are supposed to keep office hours for students; this one rarely does, and once when I actually found him in his office during office hours, he was so incensed that I was taking his precious time that I never returned. Priorities.

In his field-geology chapter, David Howell mentioned another problem, which affects students and professors alike. It is that research monies are most likely to be provided for short-term projects, usually covering only one or two years, and this is hardly enough time to conduct thorough research. The government, especially these days, is not going to give you a grant to do essential long-term research. It wants immediate results. And that is what we taxpayers far too often get for our money—quick results based on flawed research.

Let's follow the initial career of a hypothetical 1990s geoscientist who has just completed a Ph.D. and hopes to become a university professor. Such positions are quite scarce and therefore highly competitive. He or she turns to the "positions open" classified section of a technical journal, say the American Association for the Advancement of Science's weekly journal, *Science,* and in it finds job descriptions from major universities that often say one of the following:

- An extra-murally funded research program is desirable.
- Applicants must have a potential to develop extramural funding.
- It is expected that the candidate will be successful at obtaining extramural grants.
- The successful candidate will be expected to develop an externally funded program.
- The ideal candidate will demonstrate a successful background in fund-raising.

So you greatly improve your chances of becoming a professor if you can make money for the university, which will garner a generous share of whatever grant money you secure. Of course, you will want as much grant money as possible, because you will be notoriously underpaid for the horrendous hours you will be working until you receive tenure.

Tenure—the great impediment to your advancement. You get it by publishing. The old maxim is alive and well, despite reformist efforts over the decades to change it: "publish or perish." Quality does not count, since some people (your friendly associates) think highly of your research, while others (your critics) think otherwise, so how can the deans and department heads objectively assess your work? They can't; therefore, only quantity counts. There is no time to do a carefully thought out, long-term project, which would make a significant contribution to the geosciences. Short-term projects are a must (hopefully each one grant-funded), for they lead you to a higher rate of publication (and hopefully to a higher rate of income). Of course, you already should be quite good at this game, for you never would have gotten your position if you had not first gotten several short papers published while you were a grad student. You did your professor's work and used his or her grant funds. Now you can recruit your own grad students to do your work, using your grant funds.

Your dissertation probably had some merit to it, or else the professors on your committee never would have signed off on it. You probably made some good points, but also you probably made some undetected errors, these perhaps due to logical but faulty assumptions. Those assumptions were not your own, but your professors'. Because they believe them, they of course did not question them nor the extrapolations you have made based on them. (Had you formulated assumptions contradicting theirs, you rapidly would have found yourself *persona non grata,* and you can be academically blacklisted.) Since you need to publish or perish, you lack the luxury to rigorously test such assumptions and to do long-term, carefully reasoned and carefully executed research. Consequently, you may embark on well meaning, but error-prone research. The world's climates seem to be changing, due to an undesirable human attribute—rampant population growth, which has led to major atmospheric pollution and to widescale destruction of habitats. The governments cry out, "We need to know how rapidly the global climate is changing, how this change will impact us. We need to know now, and we will make plenty of grant money available for you to study this problem." Generous grant money—the magic words. If the government is willing to throw dollars at you, why not take them? So you decide to study past climate change, present climate change, and future impact. You'll do it all, get lots of grant money, and likely get tenured.

In the Sierra Nevada there are thousands of lakes and ponds, each with pollen-rich sediments that have been accumulating since their basins became ice-free between around 16,000 and 13,000 years ago. From the pollen in the sediments you'll extract, you will be able to identify dozens of plant species and how each changed in abundance over time. With this knowledge you can reconstruct past Sierran climates and how rapidly and to what degree they changed over time. But which lake should you core? You could spend months, perhaps even years, doing the necessary glacial reconstructions to ensure that you have chosen a suitable lake. So instead you phone the office of the nearest National Forest and ask them what lake in their area can you drive to? You thank them, drive up to the lake, core the next day, and return home. Mission accomplished. You analyze the pollen and you and your grad students publish several papers. *But* if you had expended a great deal of time to carefully examine your area, you would have found that it is absolutely the worst site in the Sierra to study rapid, major climate change.

There is great concern—especially among those who live along beaches or on coral atolls—that the sea level is rising. How to measure it? You need to find the most stable coasts on earth, and then need to accurately measure how sea level has changed over time. There are no such coasts in your earthquake-prone California, and travel to distant shores is costly and time-consuming. Furthermore, odds are you'll only be able to determine sea-level change to the nearest few feet, if you can do it at all. So you devise a simple solution. If the sea level is rising, then in coastal marshes salt water will gradually replace fresh water. So you take an hour and drive to the nearest marsh, extract a core of sediments, which you date and, sure enough, over the last few centuries salt-water intrusion has occurred. You publish another paper. *But* if you had expended a great deal of time to carefully examine your area, you would have found that there are at least four processes operating in the area that could cause salt-water intrusion. The most important of these is the construction of reservoirs to divert water mostly to agricultural fields, from which little water returns back into streams. Consequently, the combined flow of the Sacramento-San Joaquin rivers has been greatly reduced and salt-water from the ocean has become more prevalent in the San Francisco Bay and its adjacent marshes. If this and other processes had been addressed, very likely no sea-level rise could be detected.

Because in your paper you have shown that the sea-level has been rising since the start of the Industrial Revolution, you have established a cause and effect. Serious pollution began with the Industrial Revolution, as did sea-level rise. Therefore, global climate change must be real, and it must be caused by man. So now comes the big question, How will mankind be impacted by climate change? You can't say just how much sea-level rise will occur, so you address a local issue: landsliding. In your earthquake-prone area, substantial slides have occurred in the last decade, especially at times of heavy, prolonged rain, resulting in major losses to hundreds of residences. How will the rain pattern change with climate change? You reason: as the atmosphere warms, tropical climates will expand poleward, displacing the adjacent desert belt poleward. So your part of California will probably become drier. Oaks will wither and will be replaced by grass, so you decide to study soil creep on a grassy slope.

Cosmogenic dating has become a cutting-edge research tool of the 1990s, so you contact another professor who is doing this kind of research in order to learn how it is done. Not only will you be able to predict the effects of climate change on rates of landsliding, but you will do it with impressive, high-tech methods. Tenure is in sight. So what grassy slope should you study? One of your students, who was just out on a field trip, suggests a fine site, less than an hour's drive away and on easily accessible public land. You drive to it, run a line of soil traps up its slope, then over each of the next three storms go back to the site to measure how much soil has accumulated in each trap. Next you must determine how long such soil creep has been occurring at this site, and this is where the cosmogenic dating, which works in open land but not in forests or woodlands, comes in. You carefully extract sediments, have them dated with your grant money, and discover that the grassland has existed, as you had expected, for thousands of years. You publish your results, which are augmented by lines of support. First, statistics. You can demonstrate that within a 90% or better probability your results are accurate. Second, mass balance. You can demonstrate that the amount of soil being eroded from the upper slopes equals that accumulating on the lower slopes. Third, modeling. You use your soil-creep rate for a grassland plus an existing soil-creep rate from someone's oak woodland, and with these two ends in mind, you make a detailed, lifelike series of illustrations that depict the transformation of oak woodland to grassland. Because you have used mathematical equations to produce your model, you greatly impress everyone, even though the equations are rather standard (as those who use them know) and are subject to great misuse (as those who use them also know). You have created a useful model for regional and city planners. You are told that you are a likely recipient for next year's Kirk Bryan Award—the highest honor in American geomorphology. And you are also told that you will be tenured. Your *modus operandi* has worked so well that you will continue to use it as you advance from assistant to associate professor, and then to full professor. By then you will have become an authority, and your grad students will continue your tradition of bountiful research. *But* if you had expended a great deal of time to carefully examine your area, you would have found that the grassland has existed for only a relatively short time.

Today it receives more than enough precipitation to support a dense cover of oaks, so why is there only grass? History has the answer, but geoscientists don't do history. Your area lies near former Mexican land grants, and its owners cleared some oaks in the hills to increase grazing land for their vast herds of cattle. After California became a state, more oaks were felled for construction and for fuel. Then nearby economical coal seams were discovered, and during the mining heyday, the locals razed the remaining oaks. Massive soil erosion ensued, so by the early 1900s only grasses could grow on the denuded slopes. The land was never developed, even though it was lucrative property, because, on account of the mines, it was set aside as a state historic park, which is why you had such easy access to it. So, although you hadn't realized it, your grassland is very young, your soil-creep rates are worthless, and your cosmogenic dating is extremely wrong. Looking for simple answers, you have ignored the effects of major storms, wildfires, prolonged droughts, and especially man, all of which can vastly affect soil-creep rates. In short, there is no way to predict how future climate change will also affect those rates, despite your acclaimed research.

Although this story is fiction, it is based on actual published research by geoscientists. Academia too often encourages consensus, not criticism. Intellectual freedom should reign, but unfortunately defense of name and turf come to play. At Cal Berkeley, home of the Free Speech Movement, censorship ironically is alive and well. There, geologists have put down geomorphologists and paleontologists. When a geologist disagrees with either, he can always resort to: "My evidence is hard and yours is soft." In science's pecking order, some evidence is better than others. To get around your inferiority complex, you imitate your more respected scientists; you give in to peer pressure. They do math; so you will do math. They use statistics; so you will use statistics. They do computer modeling; so you will do computer modeling. In a way, it's like "keeping up with the Jones." Neither math nor statistics nor computer modeling will necessarily make your research better. Indeed, it can make it far worse, because you will have to simplify what you are studying, exchanging real landscapes for imaginary ones. What we naturalists study is too complex to reduce to mathematical formulae, at least in the present and for the near future. You can do this only to a certain extent, such as using math to model the flow of a river or the flow of a glacier. Modeling the evolution of a landscape is much more difficult; the evolution of an ecosystem even more so.

Science, as well as religion, has its share of true believers. Believing is seeing. You search for evidence to support your assumptions—your beliefs. Then, armed perhaps with statistics, mass balance, mathematics, and models, you publish your research. Seeing becomes believing as others are dazzled by your results, particularly your detailed models or reconstructions. François Matthes dazzled everyone with his Professional Paper 160 on Yosemite Valley. Other geoscientists have done likewise with equally contrived research. A litany of false assumptions has arisen. Consequently, alpine geomorphologists and glaciologists, using them in the 1990s, are still trying to solve the same problems that were raised (and solvable) back in the 1890s. Overall, traditional branches of science have made great strides in the twentieth century, although some branches of "soft" science have made questionable advances. Two of them, alpine and desert geomorphology, have made very little. I have not addressed desert geomorphology—a whole book in itself—which has just as many logical, but erroneous, myths.

## Lessons learned from the Sierra Nevada

A very important lesson the evidence in the Sierra Nevada teaches us is that glaciers do not severely erode resistant bedrock. This is a lesson to be applied to all of the world's glaciated ranges. Geoscientists have to measure how much rockfall occurs in a glaciated land. They must accurately determine that volume from talus, from moraines, from till, and from outwash. Only if they succeed in doing all this (which essentially is an impossibility), can they ascribe the remaining volume of sediments, if any, to glacial erosion.

A second, equally important lesson is that determining the timing and uplift of a mountain range can be difficult. Until late in the twentieth century, geoscientists were virtually unanimous in declaring that most of the mountainous lands in western North America experienced significant uplift in late Cenozoic time. These lands include all of California's ranges, the Cascade Range, the mountains of the Basin and Range province (which includes the Great Basin), the Colorado Plateau, and all of the Rocky Mountains. But in 1984 Peter Coney and Tekla Harms questioned the late Cenozoic uplift in the Basin and Range province, suggesting instead that major subsidence had occurred in response to extensional faulting. (Back in 1886, Joseph Le Conte had proposed a similar view.) They were joined several years later by Herbert Meyer, Jack Wolfe, and Howard Schorn, whose study of Cenozoic plant fossils led them to conclude that before the late Cenozoic the lands had been high. And as I have shown in Chapter 23, the Teton Range, in the Rocky Mountains, is not rising, but rather, like the Sierra Nevada, the lands east of the range are sinking. Others recently reevaluating the Rocky Mountains have made similar observations. And in 1995, Roy Dokka has documented the Miocene-epoch collapse of the Mojave Desert's lands. This event affects our interpretation of the Colorado Plateau, a highland that borders the east edge of the desert. Geoscientists have assumed that it rose above the desert's lands in late Cenozoic time, although there never was a plate-tectonic or geophysical mechanism sufficient for this to occur. Now it appears that the plateau did not rise, but that the desert lands sank. This also applies to the east edge of the plateau: the Rio Grande rift valley of northern New Mexico has subsided in response to extensional faulting; the plateau has not risen. In 1996, Richard Young sort of suggested this, stating that most of the uplift must have occurred by late Cretaceous or Paleocene time, not in late Cenozoic time. A new paradigm is emerging: very few ranges are rising; most are stationary or are subsiding.

California's ranges have a mixed late Cenozoic history. The Sierra Nevada range has experienced little or no uplift or subsidence in this time. The Cascade Range, which continues north from it, may have experienced minor uplift, but we can't be sure. West of the Cascade Range lie northwestern California's (and southwestern Oregon's) Klamath Mountains, which supposedly experienced major late Cenozoic uplift, according to Mortimer and Coleman. They inferred this by using tilted strata that lie east of this mountainous province. This is like using tilted strata in the Great Basin to infer uplift of the Sierra Nevada. They also used uplifted marine terraces of the North Coast Ranges, west of the Klamath Mountains. But the North Coast Ranges and Klamath Mountains are two separate geologic provinces separated by a major lineament, the Coast Ranges thrust (which may be a thrust fault, a suture zone, a detachment fault, or possibly something else). One cannot *a priori* assign uplift to a geologic province just because the adjacent one has experienced uplift. Finally, they propose that Condrey Mountain is the center of a domal uplift, yet Mt. Ashland, at the eastern periphery, is higher, as are the much higher southern summits of Mt. Eddy, Thompson Peak, and

Boulder Peak. In short, the Klamath Mountains do not have any evidence of late Cenozoic uplift. This geologic province was displaced from the northern Sierra Nevada during the Sevier (Early Cretaceous) and Laramide (Late Cretaceous-Paleogene) orogenies and a lowland developed between them in which continental deposits accumulated. One could speculate that during this displacement the geologic province actually lost elevation.

In contrast, the North Coast Ranges, which extend south to the San Francisco Bay, and the misnamed South Coast Ranges, which extend south from it along the state's *central* coast, both have demonstrable late Cenozoic uplift. The evidence comes from strata, which record when sedimentation ceased and erosion due to uplift began, as well as from Quaternary-age marine terraces. Likewise, the Transverse Ranges, which are the state's only major assemblage of mountains oriented east-west, have had late Cenozoic uplift, at least in the western quarter of this geologic province, where thick sediments provide documentation. Late Cenozoic uplift probably has occurred in the next two quarters, the San Gabriel Mountains and the San Bernardino Mountains, but it is hard to quantify. The eastern quarter, the Little San Bernardino Mountains, are relatively low, so if uplift has occurred, it wasn't much. What we do know is that the Mojave Desert lands north of them and the Coachella Valley south of them have subsided. Uplift is conjectural. Before the breakup of a major range that included the Klamath Mountains, the Sierra Nevada, parts of the Coast Ranges, the Transverse Ranges, and the Peninsular Ranges, all should have experienced some uplift during the Mesozoic era. That amount should be determined for the San Gabriel Mountains and San Bernardino Mountains before we assign an amount of uplift to late Cenozoic time. The same applies to the Peninsular Ranges, where Late Cretaceous uplift at its northern end has been documented in the San Jacinto Mountains. As I've said in Chapter 23, if anything, the lands of southern California's Peninsular Ranges should be undergoing subsidence, since the lithosphere (crust and upper mantle) of which they are composed is experiencing more extensional forces than compressional forces.

A third lesson from the Sierra Nevada is that if its resistant domes, ridges, benches, and even some valleys are very ancient, so too must these features be in other granitic lands, including in the ranges just mentioned as well as in some desert lands. The fantastic rock-climbing area in central Joshua Tree National Park, in the eastern quarter of the Transverse Ranges, comes to my mind. The pediment, a nearly flat bedrock surface, from which the bedrock masses protrude skyward has evolved only slowly over the last 33 million years. Like the Sierra Nevada's lands, these lands are part of a relict tropical landscape, one that originated in the Cretaceous period. Other examples are less impressive, but nonetheless are very widespread in granitic lands of the Peninsular Ranges. Not all desert lands have ancient pediments; in the Basin and Range Province, which includes the Great Basin, the Mojave Desert, and other lands, gently sloping pediments predominate, which developed in middle and late Cenozoic time. One even exists as a giant, low dome—the Mojave's Cima Dome. These pediments have another origin, detachment faulting. These faults develop deep within the earth's continental (but not oceanic) crust. They began not as faults, but as nearly horizontal, narrow zones of transition between the overlying, brittle crust, in which faulting occurs, and the underlying, ductile crust, in which slow flow occurs. As the post-Laramide-orogeny extensional faulting advanced southwestward, the slow flow of the lower crust caused nearly horizontal faulting within the transition zone above it. The brittle crust was extended and thinned, and in the process it rose. The detachment faults also were raised. Over time they eventually were bowed up to the surface as the brittle crust above them was stripped away, more though faulting than through erosion, creating today's sloping pediments.

A final lesson from the Sierra Nevada is specifically about the distribution of its plants and generally about the distribution of California's and other southwestern states' plants. In the introduction to Part IV's Chapter 23, I mentioned that the presence of species such as giant sequoia, foxtail pine, lodgepole pine, and Shasta red fir in the Sierra Nevada and elsewhere in California is an enigma; these species could not have migrated to their current positions had some of California's ranges—and the adjacent Great Basin—been low until late Cenozoic time. The climates were too warm and the elevations too low to permit establishment of today's mid-to-high-elevation conifers. Additionally, one would have to propose rapid, major evolution of these trees to suit them to their new environments, but conifers are notoriously slow evolvers. In tracing the lineage of conifers, Wilson Stewart observed that essentially modern pines had evolved by *early* in the Cretaceous period, and have changed little since. One also would have to propose even more rapid, major evolution of alpine plants, since their environments appear to have developed, under the current paradigm, in only the last 2½ million years, that is, during the Ice Age. These problems may vanish if the range has been high since Eocene time, if not since the terminal Cretaceous.

Under the current paradigm, one is hard pressed to explain the presence of foxtail pines and Shasta red firs in the Klamath Mountains, a geologic province separated from the Sierra Nevada by the end of the Eocene epoch. Foxtail pines grow in only two restricted, widely separated areas, in the higher lands of the Klamath Mountains and in the higher lands of the southern Sierra Nevada, centered around the upper Kern River basin. Shasta red firs have a greater distribution. In the north they occur on the higher lands of the Klamath Mountains, the higher lands of the North Coast Ranges, the higher lands of southern Oregon, and the upper forested slopes of their namesake, Mt. Shasta. In the south they occur in the upper-forest belt in the upper Kaweah and Kern river drainages. The separation of the two areas for both species is on the order of 300 miles. A high-ridge migration route into the Klamath Mountains never would have existed after the Laramide orogeny. I hypothesize that before the breakup of the Great Basin highlands, foxtail pines and Shasta red firs took two migratory routes from these eastern highlands, one route to the Klamath Mountains and one to the southern Sierra Nevada. Today, the central and northern Sierra Nevada has abundant sites for both species, and they should exist if both species had originally inhabited these lands. One cannot invoke late Cenozoic uplift or major glaciation to explain today's bimodal distributions.

A similar migration problem exists for lodgepole pines of the Peninsular Ranges. These occur only in two small areas, the highlands about San Jacinto Peak and some 200 miles south on the highlands about San Pedro Martir, in Baja California. Both areas are on the order of 10,000 feet high, and their slopes are subalpine refugia separated by a long expanse of semi-desert lands. Under the current paradigm the migration of these pines from the Transverse Ranges to San Jacinto Peak seems improbable; to San Pedro Martir, virtually impossible. While one might imagine a seed-laden grouse, nutcracker, or finch blown from the San Bernardinos southwest to the San Jacintos by strong Santa Ana winds, no such wind could blow a bird some 200 miles south to San Pedro Martir. Furthermore, that peak does not lie directly along a bird migration route.

The greatest plant-migration problem is with giant sequoias. The conventional view is that this species migrated southwestward from interior western North America and reached the Sierra Nevada crest during or after late Miocene time. But back then the prefaulted Owens Valley already had high relief comparable to today's. Consequently the floor should have been too warm and dry for the sequoias, and the high Sierran crest would have presented an insurmountable barrier. Since the time when early Oligocene cooling occurred, roughly 33 million years ago, only a relatively minor amount of denudation has occurred in the Sierra Nevada west of its crest. Consequently, topographic barriers in place today would have been in place by the early Oligocene, if not sooner. For the giant sequoias to migrate across the Sierran crest and over to their current major distribution sites in and about Sequoia and Kings Canyon National Parks, they likely would have had to cross while relief was only moderate, which most likely would have been in the Cretaceous. How early in the Cretaceous they could have migrated into the area is a matter of conjecture. Throughout the Cretaceous period the Sierra Nevada could have had elevations and climates suitable for giant sequoias.

My preferred hypothesis for the southern populations (Kings Canyon, Sequoia, etc.) is that they already were in the range (near its crest) at the time of late Cretaceous uplift of the southern Sierra Nevada, and in response to that uplift they migrated westward away from the crest. Alternatively, the populations could have reached Owens Valley during the Late Cretaceous or the Paleocene epochs, and then ascended to the Sierran crest during the ensuing Eocene warming (and descended from it during the mid-Eocene cooling). In the central Sierra Nevada, where the crest was not elevated as much, one is tempted to believe in a later migration. However, the Calaveras groves of sequoias would have had to surmount the deep East Fork Carson River canyon, which had major relief in pre-volcanic time, that is, at a time contemporaneous with the early Oligocene cooling. So these sequoias (and other central Sierran sequoias) also appear to have crossed the crest at an early time.

The paleofloras tend to confirm the widespread perception that sequoias migrated southwestward during the Oligocene and Miocene epochs, and therefore earlier migrations are refuted. Unfortunately, Nevada is lacking in late Cretaceous and early Cenozoic sediments that could resolve the issue. However, the time-space distribution of paleofloras has another explanation. The Basin and Range province developed its modern topography in response to a southwestward-migrating wave of post-Laramide extensional faulting. This would have disrupted the area's highlands, causing them to sink. The climates would have become too dry for the thirsty sequoias, and consequently the populations would have died out. Therefore, under this interpretation, what the paleofloras demonstrate is not a pattern of southwestern migration, but rather a pattern of southwestern habitat extinction.

The two sequoia-migration hypotheses are *somewhat* testable. Whereas appropriate sediments are lacking in the southern Sierra Nevada and in lands east, they are present in the central and northern part, especially in the Sonora Pass area, where a 3000-foot-thick sequence of volcanic flows and sediments extend back to about 20 million years ago. Paleofloras in the older sediments should reveal if giant sequoias had already existed west of the Sierran crest by about 20 million years ago, instead of much later, as Prof. Daniel Axelrod repeatedly has stated. Axelrod holds that the sequoias migrated along with white firs, which grow at relatively low elevations. However, a review of the fossil evidence shows that sequoias usually migrated along with red firs, which grow at relatively high elevations. Today, however, they grow mostly among white firs. Which brings me to my last riddle, a botanical one:

• Why did giant sequoias switch allegiance from red firs to white firs?

They didn't. There is an obvious explanation for this switch, and it was unwittingly first put forth, in rudimentary form, by John Muir in 1876: glaciers overran most of the trees, eliminating them from their preferred habitat, the red-fir belt. Glaciers removed the belt's deep, preglacial soils, and modern, postglacial ones, while not too thin for red firs, are too thin and water-deficient for sequoias. They do grow in one almost pure, shady, red-fir forest, the Atwell grove, which, extending up to 8800 feet in elevation, is the Sierra's highest grove. They grow there because the slopes never were glaciated.

In addition to resolving the sequoia-migration problem, the 20+ million-year-old Sierran paleofloras should also reveal if the elevations back then were relatively low or high. If high, then late Cenozoic uplift did not occur, and the Sierran crest would have had modern elevations since the late Cretaceous uplift. Under this interpretation, subalpine climates should have existed along the high crest since at least that time, and alpine climates could have come into existence when early Oligocene cooling occurred, roughly 33 million years ago. With the development of modern summer-dry climates about 15 million years ago, alpine climates almost certainly would have existed. But according to Michael Barbour *et al.*, alpine plants evolved quite recently from desert plants. Given that under the old paradigm the Sierra Nevada has had alpine climates only during the last 2½ million years, this suggests anomalously rapid evolution. Furthermore, the desert-to-alpine hypothesis does not explain the presence of water-loving alpine plants, found only in the Sierra Nevada, that obviously lack close affinities to desert plants. An example is the Sierra primrose. From where did it arise? Like the foxtail pine, it occurs only in the highlands of the Sierra Nevada and the Klamath Mountains. Under my interpretation, not only was there a great length of time for plants to adapt to subalpine and alpine climates, but also such plants would have developed long before any desert plants entered Owens Valley a few million years ago.

Consequently, just as alpine and desert geomorphology need to be reevaluated, so too do paleontology and plant (and by implication, animal) biogeography. My research in the Sierra Nevada appears to finally reconcile formerly conflicting evidence in the branches of geomorphology, field geology, plate tectonics, geophysics, glaciology, paleontology, and biogeography. By recognizing that rockfall occurs in mountains, I have turned the field of alpine geomorphology upside down. Mass wasting is major, glacial erosion is minor. Uplift of the world's ranges is not all late Cenozoic; each range has its own uplift history. It is now time for geoscientists to recognize mass wasting and to rewrite the field of alpine geomorphology.

# Appendix

**Table A. Degree of weathering of subsurface boulders in the Tahoe lateral moraine above Pinecrest Lake.**

**A. Dodge Ridge Road cut**

| Maximum diameter, in feet: | 1 | 1½ | 2 | 2½ | 3 | 4 | 5 | 6+ | Total number of boulders |
|---|---|---|---|---|---|---|---|---|---|
| coarse granodiorite, fresh | 4 | 0 | 4 | 0 | 3 | 1 | 2 | 0 | 14 |
| coarse granodiorite, intermediate | 1 | 0 | 2 | 0 | 2 | 0 | 1 | 1 | 7 |
| coarse granodiorite, grussified | 6 | 0 | 6 | 1 | 3 | 1 | 1 | 2 | 20 |
| medium granodiorite, fresh | 0 | 0 | 1 | 0 | 0 | 0 | 0 | 0 | 1 |
| medium granodiorite, intermediate | 1 | 0 | 4 | 0 | 0 | 0 | 1 | 0 | 6 |
| medium granodiorite, grussified | 0 | 0 | 2 | 0 | 0 | 0 | 0 | 0 | 2 |
| fine granodiorite, fresh | 1 | 0 | 1 | 0 | 0 | 0 | 0 | 0 | 2 |
| fine granodiorite, intermediate | 0 | 0 | 0 | 0 | 1 | 0 | 0 | 0 | 1 |
| fine granodiorite, grussified | 0 | 0 | 1 | 0 | 0 | 0 | 0 | 0 | 1 |
| Total number of boulders | 13 | 0 | 21 | 1 | 8 | 4 | 4 | 3 | 54 |

**B. Water Tank cut**

| Maximum diameter, in feet: | 1 | 1½ | 2 | 2½ | 3 | 4 | 5 | 6+ | Total number of boulders |
|---|---|---|---|---|---|---|---|---|---|
| coarse granodiorite, fresh | 0 | 1 | 0 | 0 | 0 | 0 | 0 | 0 | 1 |
| coarse granodiorite, intermediate | 1 | 1 | 3 | 0 | 0 | 1 | 2 | 1 | 9 |
| coarse granodiorite, grussified | 4 | 0 | 2 | 1 | 2 | 0 | 0 | 0 | 9 |
| medium granodiorite, fresh | 0 | 0 | 0 | 0 | 0 | 0 | 0 | 0 | 0 |
| medium granodiorite, intermediate | 0 | 0 | 0 | 0 | 0 | 0 | 0 | 1 | 1 |
| medium granodiorite, grussified | 1 | 0 | 0 | 0 | 0 | 0 | 0 | 0 | 1 |
| fine granodiorite, fresh | 0 | 0 | 0 | 0 | 0 | 0 | 0 | 0 | 0 |
| fine granodiorite, intermediate | 0 | 0 | 0 | 0 | 0 | 0 | 0 | 0 | 0 |
| fine granodiorite, grussified | 2 | 0 | 0 | 0 | 1 | 0 | 0 | 0 | 3 |
| Total number of boulders | 8 | 2 | 5 | 1 | 3 | 1 | 2 | 2 | 24 |

**C. Summary**

| | **Dodge Ridge Road cut** | | **Water Tank cut** | | **Combined cuts** | |
|---|---|---|---|---|---|---|
| | number | % | number | % | number | % |
| fresh | 17 | 31 | 1 | 4 | 18 | 23 |
| intermediate | 14 | 26 | 10 | 42 | 24 | 31 |
| grussified | 23 | 43 | 13 | 54 | 36 | 46 |
| Totals | 54 | 100% | 24 | 100% | 78 | 100% |

**Table B. Degree of weathering of subsurface boulders in the Cascade Creek bridge till.**

| Maximum diameter, in feet: | 1 | 1½ | 2 | 2½ | 3 | 4 | 5 | 6+ | Total number of boulders |
|---|---|---|---|---|---|---|---|---|---|
| coarse granodiorite, fresh | 3 | 2 | 5 | 3 | 2 | 2 | 0 | 1 | 18 |
| coarse granodiorite, intermediate | 0 | 1 | 1 | 0 | 1 | 2 | 0 | 0 | 5 |
| coarse granodiorite, grussified | 0 | 1 | 1 | 0 | 0 | 1 | 0 | 0 | 3 |
| medium granodiorite, fresh | 0 | 0 | 0 | 0 | 1 | 0 | 0 | 0 | 1 |
| medium granodiorite, intermediate | 0 | 2 | 0 | 0 | 0 | 0 | 0 | 0 | 2 |
| medium granodiorite, grussified | 0 | 0 | 0 | 0 | 0 | 0 | 0 | 0 | 0 |
| fine granodiorite, fresh | 0 | 1 | 1 | 0 | 0 | 0 | 0 | 0 | 2 |
| fine granodiorite, intermediate | 0 | 0 | 0 | 0 | 1 | 0 | 0 | 0 | 1 |
| fine granodiorite, grussified | 0 | 0 | 0 | 0 | 0 | 0 | 0 | 0 | 0 |
| Total number of boulders | 3 | 7 | 8 | 3 | 5 | 5 | 0 | 1 | 32 |

**Summary**

| | number | % | | number | % | | number | % |
|---|---|---|---|---|---|---|---|---|
| fresh: | 21 | 66 | intermediate: | 8 | 25 | grussified: | 3 | 9% |

**Table C. Degree of weathering of subsurface boulders in the lateral moraine above Huntington Lake.**

| Maximum diameter, in feet: | 1 | 1½ | 2 | 2½ | 3 | 4 | 5 | 6+ | Total number of boulders |
|---|---|---|---|---|---|---|---|---|---|
| coarse granodiorite, fresh | 3 | 0 | 0 | 0 | 0 | 0 | 0 | 0 | 3 |
| coarse granodiorite, intermediate | 0 | 0 | 0 | 0 | 0 | 1 | 0 | 0 | 1 |
| coarse granodiorite, grussified | 0 | 0 | 0 | 0 | 0 | 0 | 0 | 0 | 0 |
| medium granodiorite, fresh | 11 | 4 | 4 | 0 | 2 | 0 | 2 | 1 | 24 |
| medium granodiorite, intermediate | 17 | 12 | 12 | 1 | 10 | 3 | 1 | 0 | 56 |
| medium granodiorite, grussified | 5 | 3 | 0 | 0 | 0 | 0 | 0 | 0 | 8 |
| fine granodiorite, fresh | 3 | 0 | 2 | 1 | 2 | 0 | 0 | 0 | 8 |
| fine granodiorite, intermediate | 0 | 1 | 2 | 0 | 1 | 0 | 0 | 0 | 4 |
| fine granodiorite, grussified | 0 | 0 | 2 | 0 | 0 | 0 | 0 | 0 | 2 |
| Total number of boulders | 39 | 20 | 22 | 2 | 15 | 4 | 3 | 1 | 106 |

**Summary**

| | number | % | | number | % | | number | % |
|---|---|---|---|---|---|---|---|---|
| fresh: | 35 | 33 | intermediate: | 61 | 58 | grussified: | 10 | 9% |

**Table D. Well logs for two sites along Yosemite Creek.**

West of Yosemite Creek (Map 26, site 9)

| Depth (feet) | Description |
|---|---|
| 0-6 | Soil and decomposed granite |
| 6-43 | Sand, white, fine to coarse |
| 43-48 | Fine white sand |
| 48-65 | Fine to coarse white sand |
| 65-66 | Coarse brown sand |
| 66-90 | Fine white sand |
| 90-423 | Fine blue silt |
| 423-431½ | Boulders and silt |
| 431½-679 | Medium fine grey sand |
| 679-680 | Boulder or layer of rock |
| 680-690 | Medium sand |
| 690-691 | Hard layer |
| 691-780 | Medium sand |
| 780 | 2" hard layer |
| 780-836 | Medium sand |
| 836-836½ | Hard layer |
| 836½-893 | Medium sand |
| 893-893½ | Hard layer |
| 893½-899 | Medium sand |
| 899-905 | Large rocks and sand |
| 905-930 | Medium sand |
| 930 | 2" hard layer |
| 930-934 | Medium sand |
| 934-935 | Hard rock |
| 935-974 | Medium sand |
| 974-980 | Hard rock |
| 980-987 | Loose rock and sand |
| 987-991 | Hard rock |
| 991-995 | Loose rock and sand |
| 995-1002 | Sand |
| 1002-1015 | Loose rock and sand |

East of Yosemite Creek (Map 26, site 10)

| Depth (feet) | Description |
|---|---|
| 0-8 | Soil |
| 8-12 | Gravel and cobbles |
| 12-50 | Coarse sand, becoming finer |
| 50-65 | Fine sand |
| 65-79 | Fine brown sand, some gravel |
| 79-83 | Fine brown sand |
| 83-477 | Glacial flour with some fine sand, cobbles and boulders |
| 477-504 | Uniform sand, approx. 66 mesh |
| 504-545 | Sand and gravel |
| 545-558 | Glacial flour |
| 558-568 | Coarse sand |
| 568-571 | Glacial flour |
| 571-582 | Coarse sand |
| 582-585 | Fine sand |
| 585-600 | Medium sand |
| 600-614 | Fine sand and glacial flour |
| 614-625 | Medium sand |
| 625-636 | Fine sand and glacial flour |
| 636-667 | Medium sand and gravel |
| 667-685 | Layers, glacial flour through coarse sand |
| 685-700 | Very coarse gravel |
| 700-717 | Very coarse gravel with sand |
| 717-960 | Glacial flour, sand and gravel |
| 960-968 | Loose boulders and sand ?* |
| 968-970 | Large boulder or solid granite |

* ? is the driller's query. I underlined "glacial flour" for easy recognition.

**Table E. Strikes of lineaments in the Devils Postpile 15' quadrangle.**

Significant values are **bold-faced**.

| Degrees | | Rock type | |
|---|---|---|---|
| (5°/group) | Metamorphic | Metam./Gran. | Granitic |
| 0-4 | 1 | — | **12** |
| 5-9 | 2 | — | **10** |
| 10-14 | 2 | **4** | **9** |
| 15-19 | 1 | **2** | **14** |
| 20-24 | — | **3** | **12** |
| 25-29 | **4** | — | 2 |
| 30-34 | **4** | 2 | **6** |
| 35-39 | **8** | 1 | 3 |
| 40-44 | **5** | — | 3 |
| 45-49 | **8** | 1 | 2 |
| 50-54 | 3 | — | 3 |
| 55-59 | 1 | 1 | — |
| 60-64 | 3 | — | 2 |
| 65-69 | 1 | 1 | **9** |
| 70-74 | — | 2 | 3 |
| 75-79 | 2 | 1 | **4** |
| 80-84 | 2 | 1 | **4** |
| 85-89 | **4** | — | **5** |
| 90-94 | **7** | 1 | **4** |
| 95-99 | **7** | 2 | **5** |
| 100-104 | 2 | — | 2 |
| 105-109 | 3 | — | 1 |
| 110-114 | 2 | — | — |
| 115-119 | 1 | 1 | 1 |
| 120-124 | 3 | — | 1 |
| 125-129 | 1 | — | — |
| 130-134[1] | **3** | **1** | **—** |
| 135-139 | 1 | — | 2 |
| 140-144[2] | **2** | **—** | **3 =** |
| 145-149 | 1 | — | — |
| 150-154 | 1 | — | 1 |
| 155-159 | 2 | — | — |
| 160-164 | 2 | — | 2 |
| 165-169 | **5** | — | 2 |
| 170-174 | 2 | 1 | 2 |
| 175-179 | 2 | 1 | **12** |
| Total lineaments: | 98 | 26 | 141 |

[1]Bearing of Sierra's crest
[2]Strike of metamorphic strata

# References


Allen, Victor T., 1929, The Ione formation of California: Berkeley, University of California Publications, Bulletin of the Department of Geology, v. 18, p. 347-448.

Alpha, Tau Rho, Wahrhaftig, Clyde, and Huber, N. King, 1987, Oblique map showing maximum extent of 20,000-year-old (Tioga) glaciers, Yosemite National Park, central Sierra Nevada, California: U.S. Geological Survey Miscellaneous Investigations Series Map I-1885, scale variable.

Anderson, R. Scott, 1990, Holocene forest development and paleoclimates within the central Sierra Nevada, California: Journal of Ecology, v. 78, p. 470-489.

Anderson, Robert S., 1990, Evolution of the northern Santa Cruz Mountains by advection of crust past a San Andreas fault bend: Science, v. 249, p. 397-401.

Anderson, S. W., Hunter, T. C., Hoffman, E. B., and Mullen, J. R., 1993, Water resources data—California, water year 1992, volume 3; southern Central Valley basins and the Great Basin from Walker River to Truckee River: U.S. Geological Survey Water-Data Report CA-92-3, 550 p.

Andrews, E. C., 1910, An excursion to the Yosemite (California), or studies in the formation of alpine cirques, "steps," and valley "treads": Royal Society of New South Wales Journal and Proceedings, v. 43, p. 262-315.

Armin, Richard A., John, David A., Moore, William J., and Dohrenwend, John C., 1984, Geologic map of the Markleeville 15-minute quadrangle, Alpine County, California: U.S. Geological Survey Miscellaneous Investigations Series Map I-1474, scale 1:62,500.

Atwater, Brian F., Adam, David P., Bradbury, J. Platt, Forester, Richard M., Mark, Robert K., Lettis, William R., Fisher, G. Reid, Gobalet, Kenneth W., and Robinson, Stephen W., 1986, A fan dam for Tulare Lake, California, and implications for the Wisconsin glacial history of the Sierra Nevada; Geological Society of America Bulletin, v. 97, p. 97-109.

Axelrod, Daniel I., 1957, Late Tertiary floras and the Sierra Nevadan uplift: Geological Society of America Bulletin, v. 68, p. 19-45.

_____, 1959, Late Cenozoic evolution of the Sierran bigtree forest: Evolution, v. 13: 9-23.

_____, 1966, A method for determining the altitudes of Tertiary floras: Paleobotanist, v. 14, p. 144-171.

_____, 1980, The Mt. Reba flora from Alpine County: Berkeley, University of California Publications in Geological Sciences, v. 131, p. 13-75.

_____, 1986, The Sierra redwood (*Sequoiadendron*) forest: end of a dynasty: Geophytology, v. 16, p. 25-36.

Axelrod, Daniel I., and Ting, William S., 1960, Late Pliocene floras east of the Sierra Nevada: Berkeley, University of California Publications in Geological Sciences, v. 39, p. 1-118.

Bach, Andrew J., 1992, Glacial deposits show past dynamics of ice: Mono Lake Newsletter, v. 14, no. 4, p. 16-17.

Bailey, Roy A., 1989, Geologic map of the Long Valley Caldera, Mono-Inyo Craters volcanic chain, and vicinity, eastern California: U.S. Geological Survey Miscellaneous Investigations Series Map I-1933, scale 1:62,500.

Bailey, Roy A., Huber, N. King, and Curry, Robert R., 1990, The diamicton at Deadman Pass, central Sierra Nevada, California: a residual lag and colluvial deposit, not a 3 Ma glacial till: Geological Society of America Bulletin, v. 102, p. 1165-1173.

Barbour, Michael, Pavlik, Bruce, Drysdale, Frank, and Lindstrom, Susan, 1993, California's changing landscapes: diversity and conservation of California vegetation: Sacramento, California Native Plant Society, 246 p.

Bartow, J. Alan, 1984, Geologic map and cross sections of the southeastern margin of the San Joaquin Valley, California: U.S. Geological Survey Miscellaneous Investigations Series Map I-1496, scale 1:125,000.

_____, 1985, Map showing Tertiary stratigraphy and structure of the northern San Joaquin Valley, California: U.S. Geological Survey Miscellaneous Field Studies Map MF-1761, scale 1:250,000.

_____, 1991, The Cenozoic evolution of the San Joaquin Valley, California: U.S. Geological Survey Professional Paper 1501, 40 p.

Bateman, Paul C., 1965, Geologic map of the Blackcap Mountain quadrangle, Fresno County, California: U.S. Geological Survey Geologic Quadrangle Map GQ-428, scale 1:62,500.

_____, 1992, Plutonism in the central part of the Sierra

Nevada batholith, California: U.S. Geological Survey Professional Paper 1483, 186 p.

Bateman, Paul C., and Busacca, A. J., 1982, Geologic map of the Millerton Lake quadrangle, west-central Sierra Nevada, California: U.S. Geological Survey Geologic Quadrangle Map GQ-1555, scale 1:62,500.

Bateman, Paul C., Busacca, A. J., Marchand, D. E., and Sawka, W. N., 1982, Geologic map of the Raymond quadrangle, Madera and Mariposa Counties, California: U.S. Geological Survey Geologic Quadrangle Map GQ-1548, scale 1:62,500.

Bateman, Paul C., Kistler, Ronald W., Peck, Dallas L., and Busacca, Alan, 1983, Geologic map of the Tuolumne Meadows quadrangle, Yosemite National Park, California: U.S. Geological Survey Geologic Quadrangle Map GQ-1570, scale 1:62,500.

Bateman, Paul C., and Krauskopf, Konrad B., 1987, Geologic map of the El Portal quadrangle, west-central Sierra Nevada, California: U.S. Geological Survey Miscellaneous Field Studies Map MF-1998, scale 1:62,500.

Bateman, Paul C., Lockwood, John P., and Lydon, Phillip A., 1971, Geologic map of the Kaiser Peak quadrangle, central Sierra Nevada, California: U.S. Geological Survey Geologic Quadrangle Map GQ-894, scale 1:62,500.

Bateman, Paul C., and Wahrhaftig, Clyde A., 1966, Geology of the Sierra Nevada, *in* Bailey, Edgar H., ed., Geology of Northern California: California Division of Mines and Geology Bulletin 190, p. 107-172.

Becker, George F., 1891, The structure of a portion of the Sierra Nevada of California: Bulletin of the Geological Society of America, v. 2, p. 49-74.

Bennett, John H., Taylor, Gary C., and Toppozada, Tousson R., 1977, Crustal movement in the northern Sierra Nevada: California Geology, v. 30, p. 51-57.

Bergquist, Joel R., Nitkiewicz, A. M., and Tosdal, R. M., 1982, Geologic map of the Domeland Wilderness and contiguous roadless areas, Kern and Tulare Counties, California. U.S. Geological Survey Miscellaneous Field Studies Map MF-1395-A, scale 1:48,000.

Bergquist, Joel R., and Diggles, Michael F., 1986, Potassium-argon ages of volcanic rocks from the Domeland, southern Sierra Nevada, California: Isochron/West, v. 46, p. 6-8.

Bierman, Paul R., 1995, How fast do rocks erode? New answers from atom counting: Geological Society of America Abstracts With Programs, Annual Meeting, v. 27, no. 6, p. A-44.

Bierman, Paul R., Gillespie, Alan R., Whipple, Kelin, and Clark, Douglas, 1991, Quaternary geomorphology and geochronology of Owens Valley, California: Geological Society of America field trip, *in* Walawender, Michael J., and Hanan, Barry B., eds., Geological excursions in southern California and Mexico: Geological Society of America, Guidebook, 1991 Annual Meeting, p. 199-223.

Birkeland, Peter W., 1964, Pleistocene glaciation of the northern Sierra Nevada, north of Lake Tahoe, California: Journal of Geology, v. 72, p. 810-825.

Birkeland, Peter W., Burke, R. M., and Walker, A. L., 1980, Soils and subsurface rock-weathering features of Sherwin and pre-Sherwin glacial deposits, eastern Sierra Nevada, California: Geological Society of American Bulletin, v. 91, p. 238-244.

Birman, Joseph H., 1964, Glacial geology across the crest of the Sierra Nevada, California: Geological Society of America Special Paper 75, 80 p.

Blackwelder, Eliot, 1931, Pleistocene glaciation in the Sierra Nevada and Basin Ranges: Bulletin of the Geological Society of America, v. 42, p. 865-922.

Blake, William P., 1867, Sur l'action des anciens glaciers dans la Sierra Nevada de Californie, et sur l'origine de la vallée de Yo-semite: Compt. Rend., v. 65, p. 179-181.

Bradbury, J. Platt, 1994, A 180,000-year paleolimnological record from Owens Lake, California: Geological Society of America Abstracts With Programs, Annual Meeting, v. 26, no. 7, p. A-237.

Branner, John C., 1903, A topographic feature of the hanging valleys of the Yosemite: Journal of Geology, v. 11, 1903, p. 547-553.

Brewer, William H., 1930 (1966, 3rd edition), Up and Down California in 1860-1864; The Journal of William H. Brewer, Professor of Agriculture in the Sheffield Scientific School from 1864 to 1903: Berkeley, University of California Press, 583 p.

Bryan, Kirk, 1915, Ground water for irrigation in the Sacramento Valley, California: U.S. Geological Survey Water-Supply Paper 375, 49 p.

_____, 1923, Geology and ground-water resources of Sacramento Valley, California: U.S. Geological Survey Water-Supply Paper 495, 285 p.

_____, 1928, Glacial climate in non-glaciated regions: American Journal of Science, fifth series, v. 16, p. 162-164.

_____, 1932, (Review of) Geologic History of the Yosemite Valley (California): Journal of Geology, v. 40, p. 84-87.

_____, 1935, William Morris Davis—leader in geomorphology and geography: Annals of the Association of American Geographers, v. 25, p. 23-31.

Büdel, Julius, 1957, Die "Doppelten Einebnungsflächen" in den feuchten Tropen: Zeitschrift für Geomorphologie; Neue folge, Band 1, Heft 2, p. 201-228.

_____, 1982, Climatic geomorphology: Princeton, Princeton University Press, 443 p.

Bunnell, Lafayette H., 1911 (1990, reprint of 4th edition), Discovery of the Yosemite in 1851: Yosemite National Park: Yosemite Association, 314 p.

Burke, R. M., Birkeland, Peter W., 1979, Reevaluation of multiparameter relative dating techniques and their application to the glacial sequences along the eastern escarpment of the Sierra Nevada, California: Quaternary Research, v. 11, p. 21-51.

Bursick, M. I., and Gillespie, A. R., 1993, Late Pleistocene glaciation of Mono Basin, California: Quaternary Research, v. 39, p. 24-35.

Calkins, Frank C., Huber, N. King, and Roller, Julie A., 1985, Bedrock geologic map of Yosemite Valley, Yosemite National Park, California: U.S. Geological Survey Miscellaneous Investigations Series Map I-1639, scale 1:24,000.

Christensen, Mark N., 1966, Late Cenozoic crustal movements in the Sierra Nevada of California: Geological Society of America Bulletin, v. 77, p. 163-181.

Clark, Douglas H., and Clark, Malcolm M., 1995, New evidence of late-Wisconsin deglaciation in the Sierra Nevada, California refutes the Hilgard glaciation: Geological Society of America Abstracts With Programs, Cordilleran Section, v. 27, no. 5, p. 10.

Clark, Malcolm M., 1976, Evidence for rapid destruction of latest Pleistocene glaciers of the Sierra Nevada, California: Geological Society of America, Cordilleran Section, Abstracts With Programs, p. 361-362.

Clark, William B., Gold districts of California: California Division of Mines and Geology Bulletin 193, 186 p.

Colman, Steven M., and Pierce, Kenneth L., 1981, Weathering rinds on andesitic and basaltic stones as a Quaternary age indicator, Western United States: U.S. Geological Survey Professional Paper 1210, 56 p.

_____, 1992, Varied records of early Wisconsinan alpine glaciation in the western United States derived from weathering-rind thickness: *in* Clark, Peter U., and Lea, Peter D., eds., The last interglacial-glacial transition in North America: Geological Society of America Special Paper 270, p. 269-278.

Coney, Peter J., and Harms, Tekla A., 1984, Cordilleran metamorphic core complexes: Cenozoic extensional relics of Mesozoic compression: Geology, v. 12, p. 550-554.

Cox, Allan, 1973, Plate tectonics and geomagnetic reversals: San Francisco, W. H. Freeman, 702 p.

Crough, S. Thomas, and Thompson, George A., 1977, Upper mantle origin of Sierra Nevada uplift: Geology, v. 5, p. 396-399.

Curtis, Garniss H., 1951, The geology of the Topaz Lake quadrangle and the eastern half of the Ebbetts Pass quadrangle [Ph.D. thesis]: Berkeley, California, University of California, 310 p.

Dalrymple, G. Brent, 1963, Potassium-argon dates of some Cenozoic volcanic rocks of the Sierra Nevada, California: Geological Society of America Bulletin, v. 74, p. 379-390.

_____, 1964, Cenozoic chronology of the Sierra Nevada, California: Berkeley, University of California Publications in Geological Sciences, v. 47, 41 p.

_____, 1964, Potassium-argon dates of three Pleistocene interglacial basalt flows from the Sierra Nevada, California: Geological Society of America Bulletin, v. 75, p. 753-757.

Davis, George H., and Coplen, Tyler B., 1989, Late Cenozoic paleohydrogeology of the western San Joaquin Valley, California, as related to structural movements in the central coast ranges: Geological Society of America Special Paper 234, 40 p.

Davis, William M., 1889, The rivers and valleys of Pennsylvania: National Geographic Magazine, v. 1, p. 183-253.

_____, 1899, The geographical cycle: Geographical Journal, v. 14, p. 481-504

_____, 1904, Complications of the geographical cycle: Eighth International Geographic Congress, p. 150-163.

Dickerson, R. E., 1913, Fauna of the Eocene at Marysville Buttes, California: Berkeley, University of California Publications, Bulletin of the Department of Geology, v. 7, p. 257.

Diggles, Michael F., Dellinger, David A., and Conrad, James E., 1987, Geologic Map of the Owens Peak and Little Lake Canyon wilderness study areas, Inyo and Kern counties, California: U.S. Geological Survey Miscellaneous Field Studies Map MF-1927-A, scale 1:48,000.

Dodge, F. C. W., and Calk, L. C., 1987: Geologic map of the Lake Eleanor quadrangle, central Sierra Nevada, California: U.S. Geological Survey Geologic Quadrangle Map GQ-1639, scale 1:62,500.

Dokka, Roy K., and Ross, Timothy M., 1995, Collapse of southwestern North America and the evolution of early Miocene detachment faults, metamorphic core complexes, the Sierra Nevada orocline, and the San Andreas fault system: Geology, v. 23, p. 1075-1078.

Drewry, David, 1986, Glacial geologic processes: London, Edward Arnold, 276 p.

du Bray, Edward A., and Moore, James G., 1985, Geologic map of the Olancha quadrangle, southern Sierra Nevada, California: U.S. Geological Survey Miscellaneous Field Studies Map MF-1734, scale 1:62,500.

Durrell, Cordell, 1966, Tertiary and Quaternary geology of the northern Sierra Nevada, *in* Bailey, Edgar H., ed., Geology of Northern California: California Division of Mines and Geology Bulletin 190, p. 185-197.

_____, 1987, Geologic history of the Feather River country, California: Berkeley, University of California Press, 337 p.

Elliott-Fiske, Deborah L., 1987, Glacial geomorphology of the White Mountains, California and Nevada: establishment of a glacial chronology: Physical Geography, v. 8, p. 299-323.

Fiske, Richard S., and Tobisch, Othmar T., 1994, Middle

Cretaceous ash-flow tuff and caldera-collapse deposit in the Minarets Caldera, east-central Sierra Nevada, California: Geological Society of America Bulletin, v. 106, p. 582-593.

Fliedner, Moritz M., and Ruppert, Stanley, 1996, Three-dimensional crustal structure of the southern Sierra Nevada from seismic fan profiles and gravity modeling: Geology, v. 24, p. 367-370.

Flint, Richard F., 1971, Glacial and Quaternary geology: New York, John Wiley and Sons, 892 p.

Fullerton, David S., 1986, Chronology and correlation of glacial deposits in the Sierra Nevada, California: *in* Šibrava, V., Bowen, D. Q., and Richmond, G. M., eds., Glaciations in the Northern Hemisphere: Quaternary Science Reviews, v. 5, p. 161-169.

Gannett, Henry, 1898, Lake Chelan: National Geographic Magazine, v. 9, p. 417-428.

_____, 1901, The origin of Yosemite Valley, National Geographic Magazine, v. 12, p. 86-87.

George, Peter G., and Dokka, Roy K., 1994, Major Late Cretaceous cooling events in the eastern Peninsular Ranges, California, and their implications for Cordilleran tectonics: Geological Society of America Bulletin, v. 106, p. 903-914.

Gillespie, Alan R., 1982, Quaternary glaciation and tectonism in the southeastern Sierra Nevada, Inyo County, California [Ph.D. thesis]: Pasadena, California, California Institute of Technology, 695 p.

Giusso, James R., 1981, Preliminary geologic map of the Sonora Pass 15 minute quadrangle, California: U.S. Geological Survey Open-File Report 81-1170, scale 1:62,500.

Gorsline, D. S., and Teng, L. S.-Y., 1989, The California Continental Borderland, *in* Winterer, E. L., Hussong, Donald M., and Decker, Robert W., eds., The eastern Pacific Ocean and Hawaii, The Geology of North America, v. N, p. 471-487.

Gupta, Avijit, 1993, The changing geomorphology of the humid tropics: Geomorphology, v. 7, p. 165-186.

Gutenberg, Beno, and Buwalda, John P., 1938, Geophysical investigation of Yosemite Valley: Proceedings of the Geological Society of America for 1937, p. 240.

Gutenberg, Beno, Buwalda, John P., and Sharp, Robert P., 1956, Seismic explorations on the floor of Yosemite Valley, California: Bulletin of the Geological Society of America, v. 67, p. 1051-1078.

Hake, Benjamin F., 1928, Scarps of the southwestern Sierra Nevada, California: Bulletin of the Geological Society of America, v. 39, p. 1017-1030.

Hall, Clarence A., Jr., 1991, Geology of the Point Sur-Lopez Point region, Coast Ranges, California: a part of the Southern California allochthon: Geological Society of America Special Paper 266, 40 p.

Hamilton, Warren, 1987, Crustal extension in the Basin and Range province, southwestern United States, *in* Coward, M. P., Dewey, J. F., and Hancock, P. L., eds., Continental extensional tectonics (Geological Society Special Publication 28), p. 155-176.

_____, 1988, Tectonic setting and variations with depth of some Cretaceous and Cenozoic structural and magmatic systems of the western United States, *in* Ernst, W. G., ed., Metamorphism and crustal evolution of the western United States (Rubey Volume VII): Englewood Cliffs, Prentice-Hall, p. 1-40.

_____, 1989, Crustal geologic processes of the United States, *in* Pakiser, L. C., and Mooney, Walter D., eds., Geophysical framework of the continental United States: Geological Society of America Memoir 172, p. 743-781.

Haq, B. U., Hardenbol, Jan, and Vail, P. R., 1987, The new chronostratigraphic basis of Cenozoic and Mesozoic sea level cycles, *in* Ross, C. A., and Haman, Drew, eds., Timing and depositional history of eustatic sequences: constraints on seismic stratigraphy: Cushman Foundation for Foraminiferal Research, Special Publication 24, p. 7-13.

Harayama, Satoru, Youngest exposed granitoid pluton on Earth: cooling and rapid uplift of the Pliocene-Quaternary Takidani granodiorite in the Japan Alps, central Japan: Geology, v. 20, p. 657-660.

Harbor, Jonathan M., 1992, Numerical modeling of the development of U-shaped valleys by glacial erosion: Geological Society of America Bulletin, v. 104, p. 1364-1375.

Harland, W. Brian, Armstrong, Richard L., Cox, Allan V., Craig, Lorraine E., Smith, Alan G., and Smith, David G., 1990, A geologic time scale 1989: Cambridge, Cambridge University Press, 263 p.

Harp, Edwin L., Tanaka, Kohei, Sarmiento, John, and Keefer, David K., 1984, Landslides from the May 25-27, 1980, Mammoth Lakes, California, earthquake sequence: U.S. Geological Survey Miscellaneous Investigations Series Map I-1612, scale 1:62,500.

Harwood, David S., and Helley, Edward J., 1987, Late Cenozoic tectonism of the Sacramento Valley, California: U.S. Geological Survey Professional Paper 1359, 45 p.

Hay, Edward A., 1976, Cenozoic uplifting of the Sierra Nevada in isostatic response to North American and Pacific plate interactions: Geology, v. 4, p. 763-766.

Helley, Edward J., and Harwood, David S., 1985, Geologic map of the late Cenozoic deposits of the Sacramento Valley and northern Sierran foothills, California: U.S. Geological Survey Miscellaneous Field Studies Map MF-1790.

Hietanen, Anna, 1973, Geology of the Pulga and Bucks Lake quadrangles, Butte and Plumas Counties, California: U.S. Geological Survey Professional Paper 731, 66 p.

_____, 1976, Metamorphism and plutonism around the

Middle and South forks of the Feather River, California: U.S. Geological Survey Professional Paper 920, 30 p.

_____, 1981, Geology west of the Melones Fault between the Feather and North Yuba Rivers, California: U.S. Geological Survey Professional Paper 1226-A, 35 p.

Hill, E. J., Baldwin, S. L., and Lister, G. S., 1992, Unroofing of active metamorphic core complexes in the D'Entrecasteaux Islands, Papua New Guinea: Geology, v. 20, p. 907-910.

Holmes, Arthur, 1913, The age of the Earth: London, Harper and Brothers, 194 p.

Howell, David G., 1989, Tectonics of suspect terranes: mountain building and continental growth: London, Chapman and Hall, 232 p.

Huber, N. King, 1981, Amount and timing of late Cenozoic uplift and tilt of the central Sierra Nevada, California—evidence from the upper San Joaquin River Basin: U.S. Geological Survey Professional Paper 1197, 28 p.

_____, 1983, Preliminary geologic map of the Dardanelles Cone quadrangle, central Sierra Nevada, California: U.S. Geological Survey Miscellaneous Field Studies Map MF-1436, scale 1:62,500.

_____, 1987, The geologic story of Yosemite National Park: U.S. Geological Survey Bulletin 1595, 64 p.

_____, 1990, The late Cenozoic evolution of the Tuolumne River, central Sierra Nevada, California: Geological Society of America Bulletin, v. 102, p. 102-115.

_____, 1990, Interpreting Yosemite geology—the role of the United States Geological Survey, *in* Elelbrock, Jerry, and Carpenter, Scott, eds., Natural areas and Yosemite: prospects for the future (Yosemite Centennial Symposium Proceedings): Yosemite National Park, p. 523-534.

Huber, N. King, Bateman, Paul C., and Wahrhaftig, Clyde, 1989, Geologic map of Yosemite National Park and vicinity, California: U.S. Geological Survey Miscellaneous Investigations Series Map I-1874, scale 1:125,000.

Huber, N. King, and Rinehart, C. Dean, 1965, Geologic map of the Devils Postpile quadrangle, Sierra Nevada, California: U.S. Geological Survey Geologic Quadrangle Map GQ-437, scale 1:62,500.

_____, 1967, Cenozoic volcanic rocks of the Devils Postpile quadrangle, eastern Sierra Nevada, California: U.S. Geological Survey Professional Paper 554-D, 21 p.

Hudson, Frank S., 1948, Donner Pass zone of deformation, Sierra Nevada, California: Bulletin of the Geological Society of America, v. 59, p. 795-800.

_____, 1951, Mount Lincoln-Castle Peak area, Sierra Nevada, California: Bulletin of the Geological Society of America, v. 62, p. 931-952.

_____, 1955, Measurement of the deformation of the Sierra Nevada, California, since middle Miocene: Bulletin of the Geological Society of America, v. 66, p. 835-869.

_____, 1960, Post-Pliocene uplift of the Sierra Nevada, California: Bulletin of the Geological Society of America, v. 71, p. 1547-1574.

Hutchings, James Mason, 1877, Hutchings' Tourist's Guide to the Yo Semite Valley and the Big Tree Groves for the Spring and Summer of 1877. San Francisco: A. Roman & Co., 102 p.

Ingersoll, Raymond V., 1979, Evolution of the Late Cretaceous forearc basin, northern and central California: Geological Society of America Bulletin, v. 90, p. 813-826.

_____, 1982, Initiation and evolution of the Great Valley forearc basin of northern and central California, U.S.A., *in* Leggett, J. K., ed., Trench-forearc geology: sedimentation and tectonics on modern and ancient active plate margins: Geological Society of London Special Publication 10, p. 459-467.

Jennings, Charles W., 1977, Geologic Map of California: California Division of Mines and Geology Map No. 2, scale 1:750,000.

John, David A., Giusso, James, Moore, William J., and Armin, Richard A., 1981, Reconnaissance geologic map of the Topaz Lake 15 minute quadrangle, California and Nevada: U.S. Geological Survey Open-File Report 81-273, scale 1:62,500.

Johnson, Douglas W., 1911, Hanging valleys of the Yosemite: Bulletin of the American Geographical Society, v. 43, p. 1-25.

Jones, Craig H., 1987, Is extension in Death Valley accommodated by thinning of the mantle lithosphere beneath the Sierra Nevada, California?: Tectonics, v. 6, p. 449-473.

_____, 1991, Isostasy and the southern Sierra Nevada: seismic constraints: Geological Society of America Abstracts With Programs, Cordilleran and Rocky Mountain Sections, v. 23, no. 2, p. 40.

_____, 1993, Isostasy and its implications for the structure of the Sierra Nevada: Geological Society of America Abstracts With Programs, Cordilleran and Rocky Mountain Sections, v. 25, no. 5, p. 59.

Kane, Phillip S., 1975, The glacial geomorphology of the Lassen Volcanic National Park area, California [Ph.D. thesis]: Berkeley, University of California, 224 p.

Keefer, David K., 1994, The importance of earthquake-induced landslides to long-term slope erosion and slope-failure hazards in seismically active regions: Geomorphology, v. 10, p. 265-284.

King, Clarence, 1872 (1970 reprint), Mountaineering in the Sierra Nevada: Lincoln, University of Nebraska Press, 292 p.

King, Lester C., 1953, Canons of landscape evolution:

Bulletin of the Geological Society of America, v. 64, p. 721-752.

Kistler, Ronald W., 1966, Geologic map of the Mono Craters quadrangle, Mono and Tuolumne Counties, California: U.S. Geological Survey Geologic Quadrangle Map GQ-462, scale 1:62,500.

_____, 1973, Geologic map of the Hetch Hetchy Reservoir quadrangle, Yosemite National Park, California: U.S. Geological Survey Geologic Quadrangle Map GQ-1112, scale 1:62,500.

_____, 1993, Mesozoic intrabatholithic faulting, Sierra Nevada, California, in Dunn, George C., and McDougall, Kristin A., eds., Mesozoic paleogeography of the western United States-II: Los Angeles, Society of Economic Paleontologists and Mineralogists, Pacific Section, p. 247-262.

Kistler, Ronald W., and Ross, Donald C., 1990, A strontium isotopic study of plutons and associated rocks of the southern Sierra Nevada and vicinity, California: U.S. Geological Survey Bulletin 1920, 20 p.

Lawson, Andrew C., 1903, The geomorphogeny of the upper Kern Basin: Berkeley, University of California Publications, Bulletin of the Department of Geology, v. 3, p. 291-376.

_____, 1921, Geology of Yosemite National Park, *in* Hall, Ansel F., ed., Handbook of Yosemite National Park, p. 97-122.

Le Conte, Joseph, 1875 (1994 reprint), A journal of ramblings through the High Sierra of California by the University Excursion Party: El Portal, CA, Yosemite Association, 140 p.

_____, 1886, A Post-Tertiary elevation of the Sierra Nevada shown by the river beds: American Journal of Science, v. 32, no. 189, p. 167-181.

_____, 1887, Elements of Geology: a text-book for colleges and for the general reader: New York, D. Appleton and Company, 633 p.

Lesquereux, Leo, 1878, Report on the fossil plants of the auriferous gravel deposits of the Sierra Nevada: Cambridge, Harvard University Museum of Comparative Zoology Memoir 6, no. 2, 68 p.

Lindgren, Waldemar, 1900, Colfax Geologic Folio No. 66: U.S. Geological Survey Geologic Atlas of the United States, 10 p. plus 4 map sheets, scale 1:125,000.

_____, 1911, The Tertiary gravels of the Sierra Nevada of California: U.S. Geological Survey Professional Paper 73, 226 p.

Lliboutry, Louis A., 1994, Monolithologic erosion of hard beds by temperate glaciers: Journal of Glaciology, v. 40, no. 136, p. 433-450.

Loomis, Karen B., and Ingle, James C., Jr., 1994, Subsidence and uplift of the Late Cretaceous-Cenozoic margin of California: new evidence from the Gualala and Point Arena basins: Geological Society of America Bulletin, v. 106, p. 915-931.

Love, J. D., Reed, John C., Jr., and Christiansen, Ann Coe, 1992, Geologic map of Grand Teton National Park, Teton County, Wyoming: U.S. Geological Survey Miscellaneous Field Studies Map MF-2031, scale 1:62,500.

Lucchitta, Ivo, 1990, History of the Grand Canyon and of the Colorado River in Arizona, in Beus, Stanley S., and Morales, Michael, eds., Grand Canyon geology: Oxford: Oxford University Press, p. 311-332.

Luedke, Robert G., and Smith, Robert L., 1981, Map showing distribution, composition, and age of late Cenozoic volcanic centers in California and Nevada: U.S. Geological Survey Miscellaneous Investigations Series Map I-1091-C, scale 1:1,000,000.

Lyell, Charles, 1830 (1990 reprint), Principles of geology, v. I: Chicago: University of Chicago Press, 511 p.

_____, 1833 (1991 reprint), Principles of geology, v. III: Chicago: University of Chicago Press, 398 p. plus 160 p. of additions.

MacDonald, Gordon A., 1941, Geology of the western Sierra Nevada between the Kings and San Joaquin rivers, California. University of California Publications in Geological Sciences, v. 26, no. 2, p. 215-286.

Manger, G. Edward, 1963, Porosity and bulk density of sedimentary rocks: U.S. Geological Survey Bulletin 1144-E, 55 p.

Marchand, Denis E., and Allwardt, Alan, 1981, Late Cenozoic stratigraphic units, northeastern San Joaquin Valley, California: U.S. Geological Survey Bulletin 1470, 70 p.

Mathieson, Scott A., 1981, Pre- and post-Sangamon glacial history of a portion of Sierra and Plumas counties, California [M.A. thesis]: Hawyard, California State University, 258 p.

Matthes, François E., 1900, Glacial sculpture of the Big Horn Mountains, Wyoming: U.S. Geological Survey Annual Report 21, part 2, p. 167-190.

_____, 1904, The Alps of Montana: Appalachia, v. 10, p. 255-276.

_____, 1910, Little studies in the Yosemite Valley. I. The extinct Eagle Peak Falls: Sierra Club Bulletin, v. 7, p. 222-224.

_____, 1911, Little studies in the Yosemite Valley. II. The striped rock floor of the Little Yosemite Valley: Sierra Club Bulletin, v. 8, p. 3-9.

_____, 1911, Little studies in the Yosemite Valley. III. The winds of the Yosemite Valley: Sierra Club Bulletin, v. 8, p. 89-95.

_____, 1912, Sketch of Yosemite National Park and an account of the origin of the Yosemite and Hetch Hetchy valleys: U.S. Department of the Interior, Office of the Secretary, 47 p.

_____, 1913, Little studies in the Yosemite Valley. IV. El Capitan moraine and ancient Lake Yosemite: Sierra Club Bulletin, v. 9, p. 7-15.

_____, 1914, Studying the Yosemite problem: Sierra Club Bulletin, v. 9, p. 136-147.
_____, 1922, The story of the Yosemite Valley, *text on back side of* Map of Yosemite Valley: U.S. Geological Survey, scale 1:24,000.
_____, 1924, Hanging side valleys of the Yosemite and the San Joaquin Canyon: Journal of the Washington Academy of Sciences, v. 14, p. 379.
_____, 1924, The story of the Yosemite Valley: New York, American Museum of Natural History Guide Leaflet No. 60, 21 p.
_____, 1930, Geologic history of the Yosemite Valley: U.S. Geological Survey Professional Paper 160, 137 p.
_____, 1933, Geography and geology of the Sierra Nevada: International Geological Congress, XVI session, Guidebook 16, Excursion C-1, Middle California and western Nevada, p. 26-40.
_____, 1933, Up the western slope of the Sierra Nevada by way of the Yosemite Valley: International Geological Congress, XVI session, Guidebook 16, Excursion C-1, Middle California and western Nevada, p. 67-81.
_____, 1937, The geologic history of Mount Whitney: Sierra Club Bulletin, v. 22, p. 1-18 (also in 1962 entry, below, p. 163-183).
_____, 1938, Avalanche sculpture in the Sierra Nevada: International Association of Hydrology, Bulletin 23, p. 631-637 (also in 1962 entry, below, p. 155-162).
_____, 1947, A geologist's view, *in* Peattie, Roderick, ed., The Sierra Nevada: The Range of Light: New York, Vanguard Press, p. 166-214.
_____, 1950, The Incomparable Valley; a geologic interpretation of the Yosemite: Berkeley, University of California Press, 160 p.
_____, 1950, Sequoia National Park; a geological album: Berkeley, University of California Press, 136 p.
_____, 1960, Reconnaissance of the geomorphology and glacial geology of the San Joaquin basin, Sierra Nevada, California: U.S. Geological Survey Professional Paper 329, 62 p.
_____, 1962, François Matthes and the marks of time; Yosemite and the High Sierra: San Francisco, Sierra Club, 189 p.
_____, 1965, Glacial reconnaissance of Sequoia National Park, California: U.S. Geological Survey Professional Paper 504-A, 58 p.
McIntosh, W. C., Geissman, J. W., Chapin, C. E., Kunk, M. J., and Henry, C. D., 1992, Calibration of the latest Eocene-Oligocene geomagnetic polarity time scale using 40Ar/39Ar dated ignimbrites: Geology, v. 20, p. 459-463.
Meyer, Herbert W., 1986, An evaluation of the methods for estimating paleoaltitudes using Tertiary floras from the Rio Grand rift vicinity, New Mexico and Colorado [Ph.D. thesis]: Berkeley, University of California, 217 p.
_____, 1992, Lapse rates and other variables applied to estimating paleoaltitudes from fossil floras: Palaeogeography, Palaeoclimatology, Palaeoecology, v. 99, p. 71-99.
Mix, Alan C., 1992, The marine oxygen isotope record: constraints on timing and extent of ice-growth events (120-65 ka): *in* Clark, Peter U., and Lea, Peter D., eds., The last interglacial-glacial transition in North America: Geological Society of America Special Paper 270, p. 19-30.
Moore, James G., 1963, Geology of the Mount Pinchot quadrangle, southern Sierra Nevada, California: U.S. Geological Survey Bulletin 1130, 152 p. + 15' geologic quadrangle, scale 1:62,500.
_____, 1978, Geologic Map of the Marion Peak quadrangle, Fresno County, California: U.S. Geological Survey Geologic Quadrangle Map GQ-1399, scale 1:62,500.
_____, 1981, Mount Whitney quadrangle, Inyo and Tulare counties, California—analytic data: U.S. Geological Survey Geologic Quadrangle Map GQ-1545, scale 1:62,500.
Moore, James G., and Nokleberg, Warren J., 1992, Geologic map of the Tehipite Dome quadrangle, Fresno County, California: U.S. Geological Survey Geologic Quadrangle Map GQ-1676, scale 1:62,500.
Moore, James G., and Sisson, Thomas W., 1985, Geologic map of the Kern Peak quadrangle, Tulare County, California: U.S. Geological Survey Geologic Quadrangle Map GQ-1584, scale 1:62,500.
_____, 1987, Geologic map of the Triple Divide Peak quadrangle, Tulare County, California: U.S. Geological Survey Geologic Quadrangle Map GQ-1636, scale 1:62,500.
Morrison, Roger B., ed., 1991, Quaternary nonglacial geology: conterminous U.S.: Geological Society of America, The Geology of North America, v. K-2, 672 p.
Mortimer, N., and Coleman, R. G., 1985, A Neogene structural dome in the Klamath Mountains, California and Oregon: Geology, v. 13, p. 253-256.
Morton, J. L., Silberman, M. L., Bonham, H. F., Jr., Garside, L. J., and Noble, D. C., 1977, K-Ar ages of volcanic rocks, plutonic rocks, and ore deposits in Nevada and eastern California: Isochron/West, v. 20, p. 19-29.
Muir, John, 1874-75, Studies in the Sierra: Overland Monthly. These are:
May 1874, No. I. Mountain sculpture: v. 12, p. 393-403 (also Sierra Club Bulletin, January 1915, v. 9, no. 4, p. 225-239).
June 1874, No. II. Mountain sculpture, origin of Yosemite valleys: v. 12, p. 489-500 (also SCB, January 1916, v. 10, no. 1, p. 62-77).
July 1874, No. III. Ancient glaciers and their pathways: v. 13, p. 67-79 (also SCB, January 1917, v. 10, no. 2, p.

184-201).
August 1874, No. IV. Glacial denudation: v. 13, p. 174-184 (also SCB, January 1918, v. 10, no. 3, p. 304-318).
November 1874, No. V. Post-glacial denudation: v. 13, p. 393-402 (also SCB, January 1919, v. 10, no. 4, p. 414-428).
December 1874, No. VI. Formation of soils: v. 13, p. 530-540 (also SCB, January 1920, v. 11, no. 1, p. 69-85).
January 1875, No. VII. Mountain-building: v. 14, p. 64-73 (also SCB, January 1921, v. 11, no. 2, p. 182-193).

_____, 1876, On the post-glacial history of *Sequoia gigantea*: Proceedings of the American Association for the Advancement of Science, v. 25, p. 242-253.

_____, 1894, The mountains of California: Garden City, Doubleday & Co. (Natural History Library edition with Foreword by Jack McCormick, 1961), 300 p.

_____, 1911, My first summer in the Sierra: Boston, Houghton Mifflin, 272 p.

_____, 1912, The Yosemite: Garden City, Doubleday & Co. (Natural History Library edition with Notes and Introduction by Frederic R. Gunsky, 1962), 225 p.

Nelson, C. A., 1981, Basin and Range province, *in* Ernst, W. G., ed., The geotectonic development of California (Rubey Volume I): Englewood Cliffs, Prentice-Hall, p. 202-216.

Palmer, Allison R., 1983, The Decade of North American Geology 1983 Geologic Time Scale: Geology, v. 11, p. 503-504.

Pierce, Kenneth L., 1979, History and dynamics of glaciation in the northern Yellowstone National Park area: U.S. Geological Survey Professional Paper 729-F, 90 p.

Pilkey, Orrin H., 1996, Mathematical modeling of beach behavior: an impossible task?: GSA Today, v. 6, no. 5, p. 11-12.

Piper, A. M., Gale, H. S., Thomas, H. E., and Robinson, T. W., 1939, Geology and ground-water hydrology of the Mokelumne area, California: U.S. Geological Survey Water-Supply Paper 780, 230 p.

Prothero, Donald R., 1990, Interpreting the stratigraphic record: New York, W. H. Freeman, 410 p.

_____, 1994, The Eocene-Oligocene transition: paradise lost: New York, Columbia University Press, 291 p.

Ransome, F. Leslie, 1896, The Great Valley of California: a criticism of the theory of isostasy: University of California Bulletin of the Department of Geology, v. 1, p. 371-428.

_____, 1898, Some lava flows of the western slope of the Sierra Nevada, California: U.S. Geological Survey Bulletin 89, 74 p.

Rinehart, C. Dean, Ross, Donald C., and Pakiser, L. C., 1964, Geology and mineral deposits of the Mount Morrison quadrangle, Sierra Nevada, California: U.S. Geological Survey Professional Paper 385, 106 p.

Roper, Steve, 1964, A climber's guide to Yosemite Valley: San Francisco, Sierra Club, 190 p.

Ross, Donald C., 1986, Basement-rock correlations across the White Wolf-Breckenridge-Southern Kern Canyon fault zone, southern Sierra Nevada, California: U.S. Geological Survey Bulletin 1651, 25 p.

_____, 1989, The metamorphic and plutonic rocks of the southernmost Sierra Nevada, California, and their tectonic framework: U.S. Geological Survey Professional Paper 1381, 159 p.

Ruddiman, W. F., 1987, Synthesis; the ocean/ice sheet record, *in* Ruddiman, W. F., and Wright, H. E., Jr., eds., North America and adjacent oceans during the last deglaciation: Geological Society of America, The Geology of North America, v. K-3, p. 463-478.

Russell, Israel C., 1889, Quaternary history of Mono Valley, California: U.S. Geological Survey Eighth Annual Report, p. 267-394 (reprinted in 1984, with a new preface, by Artemisia Press, Lee Vining, CA).

Saleeby, Jason B., and Ducea, Mihai N., 1996, A three-dimensional picture of the lower crust and upper mantle structure beneath the southern Sierra Nevada (SSN), California: Geological Society of America Abstracts With Programs, Cordilleran Section, v. 28, no. 5, p. 107.

Saleeby, Jason B., Silver, Leon T., and Wood, David J., 1993, Late Cretaceous tectonics of the southern Sierra Nevada batholith (SNB) viewed from the Tehachapi Mountains (TM), California: Geological Society of America Abstracts With Programs, Cordilleran and Rocky Mountain Sections, v. 25, no. 5, p. 141.

Sargent, Shirley, 1971, John Muir in Yosemite: Yosemite, Flying Spur Press, 48 p.

Saucedo, G. J., and Wagner, D. L., 1992, Geologic map of the Chico quadrangle, California, 1:250,000: California Division of Mines and Geology Regional Geologic Map No. 7A.

Schaffer, Jeffrey P., 1977, Pleistocene Lake Yosemite and the Wisconsin glaciation of Yosemite Valley: California Geology, v. 30, p. 243-248.

_____, 1977, Topographic map of Yosemite Valley: Berkeley, Wilderness Press, scale 1:24,000.

_____, 1977, The ever-changing story of Yosemite Valley, *text on back side of* Topographic map of Yosemite Valley: Berkeley, Wilderness Press, scale 1:24,000.

_____, 1978, Yosemite National Park: a natural-history guide to Yosemite and its trails: Berkeley, Wilderness Press, 296 p.

_____, 1983, Yosemite National Park: a natural-history guide to Yosemite and its trails: Berkeley, Wilderness Press, 272 p.

_____, 1986, A geologic history of Yosemite Valley, *text on back side of* Topographic map of Yosemite Valley: Berkeley, Wilderness Press, scale 1:24,000.

_____, 1987, Past and future uplift: Yosemite, v. 49, no. 2, p. 1-4.

_____, 1989, Past and future uplift (revised): Yosemite, v. 51, no. 2, p. 5.
_____, 1992, Carson-Iceberg Wilderness: Berkeley, Wilderness Press, 182 p.
_____, 1992, Yosemite National Park: a natural-history guide to Yosemite and its trails: Berkeley, Wilderness Press, 274 p.
Schweickert, Richard A., 1981, Tectonic evolution of the Sierra Nevada range, *in* Ernst, W. G., ed., The geotectonic development of California (Rubey Volume I): Englewood Cliffs, Prentice-Hall, p. 87-131.
Segall, Paul, and Pollard, David D., 1983, Nucleation and growth of strike slip faults in granite: Journal of Geophysical Research, v. 88, p. 555-568.
_____, 1983, Joint formation in granitic rock of the Sierra Nevada: Geological Society of America Bulletin, v. 94, p. 563-575.
Seiders, Victor M., Joyce, J. M., Leverett, K. A., and McLean, Hugh, 1983, Geologic map of part of the Ventana Wilderness and the Black Butte, Bear Mountain, and Bear Canyon Roadless Areas, Monterey County, California: U.S. Geological Survey Miscellaneous Field Studies Map MF-1559-B, scale 1:50,000.
Sharp, Robert P., 1968, Sherwin till-Bishop tuff geological relationships, Sierra Nevada, California: Geological Society of America Bulletin, v. 79, p. 351-363.
_____, 1972, Pleistocene glaciation, Bridgeport Basin, California: Geological Society of America Bulletin, v. 83, p. 2233-2260.
_____, 1988, Living Ice: understanding glaciers and glaciation: Cambridge: Cambridge University Press, 225 p.
Sharp, Robert P., and Birman, Joseph H., 1963, Additions to classical sequence of Pleistocene glaciations, Sierra Nevada, California: Geological Society of America Bulletin, v. 74, p. 1079-1086.
Slemmons, David B., 1953, Geology of the Sonora Pass region [Ph.D. thesis]: Berkeley, University of California, 217 p.
_____, 1966, Cenozoic volcanism of the Central Sierra Nevada, California, *in* Bailey, Edgar H., ed., Geology of Northern California: California Division of Mines and Geology Bulletin 190, p. 199-208.
Slemmons, David B., Van Wormer, Douglas, Bell, Elaine J., and Silberman, Miles L., 1979, Recent crustal movements in the Sierra Nevada-Walker Lane region of California-Nevada: Part I, rate and style of deformation: Tectonophysics, v. 52, p. 561-570.
Smith, Susan J., and Anderson, R. Scott, 1992, Late Wisconsin paleoecologic record from Swamp Lake, Yosemite National Park, California: Quaternary Research, v. 38, p. 91-102.
Soles, Stan, Sarna-Wojckcki, A. M., Meyer, C. E., and Wan, Elmira, 1993, Age of the Tahoe moraine at Bloody Canyon, Mono County, California: Geological Society of America Abstracts With Programs, Cordilleran and Rocky Mountain Sections, v. 25, no. 5, p. 148-149.
Spencer, Jon E., and Normark, William R., 1989, Neogene plate-tectonic evolution of the Baja California Sur continental margin and the southern Gulf of California, Mexico, *in* Winterer, E. L., Hussong, Donald M., and Decker, Robert W., eds., The eastern Pacific Ocean and Hawaii, The Geology of North America, v. N, p. 489-497.
Stewart, John H., Carlson, John E., and Johannesen, Dann C., 1982, Geologic map of the Walker Lake 1° by 2° quadrangle, California and Nevada: U.S. Geological Survey Miscellaneous Field Studies Map MF-1382-A, scale 1:250,000.
Stewart, Wilson N., 1983, Paleobotany and the evolution of plants: Cambridge, Cambridge University Press, 405 p.
Stine, Scott, 1994, Extreme and persistent drought in California and Patagonia during mediaeval time: Nature, v. 369, p. 546-549.
Stover, Carl W., and Coffman, Jerry L., 1993, Seismicity of the United States, 1568-1989 (Revised): U.S. Geological Survey Professional Paper 1527, 418 p.
Swisher III, Carl C., Grajales-Nishimura, José, Montanari, Alessandro, Margolis, Stanley V., Claeys, Philippe, Alvarez, Walter, Renne, Paul, Cedillo-Pardo, Esteban, Maurrasse, Florentin J-M. R., Curtis, Garniss H., Smit, Jan, and McWilliams, Michael O., 1992, Coeval $^{40}Ar/^{39}Ar$ ages of 65.0 million years ago from Chicxulub crater melt rock and Cretaceous-Tertiary boundary tektites: Science, v. 257, p. 954-958.
Tarbuck, Edward J., and Lutgens, Frederick K., 1993, The earth: an introduction to physical geology: New York, Macmillan, 654 p.
_____, 1996, The earth: an introduction to physical geology: Upper Saddle River, NJ, Prentice Hall, 605 p.
Tricart, Jean, 1972, The landforms of the humid tropics, forests and savannas: New York, St. Martin's Press, 306 p.
Turner, Henry W., 1902, Post-Tertiary elevation of the Sierra Nevada: Bulletin of the Geological Society of America, v. 13, p. 540-541.
Turner, Henry W., and Ransome, F. Leslie, 1898, Big Trees Geologic Folio No. 51: U.S. Geological Survey Atlas, 8 p., 1 sheet of illus., plus 3 map sheets, scale 1:125,000.
Unruh, Jeffrey R., 1991, The uplift of the Sierra Nevada and implications for late Cenozoic epeirogeny in the western Cordillera: Geological Society of America Bulletin, v. 103, p. 1395-1404.
Wagner, D. L., Jennings, C. W., Bedrossian, T. L., and Bortugno, E. J., 1981, Geologic map of the Sacramento quadrangle, California, 1:250,000: California Division of Mines and Geology Regional Geologic Map No. 1A.
Wahrhaftig, Clyde, 1965, Stepped topography of the

southern Sierra Nevada, California: Geological Society of America Bulletin, v. 76, p. 1165-1189.

_____, 1984, Geomorphology and glacial geology, Wolverton and Crescent Meadow areas and vicinity, Sequoia National Park, California: U.S. Geological Survey Open-File Report 84-400, p. 1-35 [of 52 p.].

Webb, Robert W., 1946, Geomorphology of the middle Kern River Basin, southern Sierra Nevada, California: Bulletin of the Geological Society of America, v. 57, p. 355-382.

Wernicke, Brian, and 18 others, 1996, Origin of high mountains in the continents: the southern Sierra Nevada: Science, v. 271, p. 190-193.

Whitney, Josiah D., 1865, Geology of California, v. 1: Legislature of California, 498 p.

_____, 1868, Yosemite Book: New York: J. Bien, 116 p.

_____, 1869, Yosemite Guide-Book: Cambridge: University Press, Welch, Bigelow & Co., 155 p.

Wieczorek, Gerald F., Snyder, James B., Alger, Christopher S., and Isaacson, Kathleen A., 1992, Rockfalls in Yosemite Valley, California: U.S. Geological Survey Open-File Report 92-387, 186 p.

Wilshire, Howard G., 1956, The history of Tertiary volcanism near Ebbetts Pass, Alpine County, California [Ph.D. thesis]: Berkeley, California, University of California, 123 p.

Wolf, R. A., Farley, K. A., and Silver, L. T., 1997, Assessment of (U-Th)/He thermochronometry: The low-temperature history of the San Jacinto mountains, California: Geology, v. 25, p. 65-68.

Wolfe, Jack A., 1992, An analysis of present-day terrestrial lapse rates in the western conterminous United States and their significance to paleoaltitudinal estimates: U.S. Geological Survey Bulletin 1964, 35 p.

Wolfe, Jack A., and Schorn, Howard E., 1994, Fossil floras indicate high altitude for west-central Nevada at 16 Ma & collapse to ca. present altitudes by 12 Ma: Geological Society of America Abstracts With Programs, Annual Meeting, v. 26, no. 7, p. A521.

Wood, Spencer H., 1975, Holocene stratigraphy and chronology of mountain meadows, Sierra Nevada, California [Ph.D. thesis]: Pasadena, California Institute of Technology, 180 p. [Reproduced as USDA Forest Service, Region 5, Earth Resources Monograph 4].

Yeend, Warren E., 1974, Gold-bearing gravel of the ancestral Yuba River, Sierra Nevada, California: U.S. Geological Survey Professional Paper 772, 44 p.

Young, Richard A., 1996, Cretaceous-Tertiary uplift of the southwest Colorado Plateau: Geological Society of America Abstracts With Programs, Annual Meeting, v. 28, no. 7, p. A508.

Zandt, George, Velasco, Aaron A., and Beck, Susan L., 1994, Composition and thickness of the southern Altiplano crust, Bolivia: Geology, v. 22, p. 1003-1006.

# Index